Problem-Solving Workbook with Solutions

to accompany

Essential Chemistry

Second Edition

Brandon J. Cruickshank
Northern Arizona University

Raymond Chang
Williams College

McGraw Hill

Boston Burr Ridge, IL Dubuque, IA Madison, WI New York San Francisco St. Louis
Bangkok Bogotá Caracas Lisbon London Madrid
Mexico City Milan New Delhi Seoul Singapore Sydney Taipei Toronto

McGraw-Hill Higher Education

A Division of The McGraw-Hill Companies

Problem-Solving Workbook with Solutions to accompany
ESSENTIAL CHEMISTRY, SECOND EDITION

Published by McGraw-Hill Higher Education, an operating unit of The McGraw-Hill Companies, Inc. Printed in the United States of America.

This book is printed on acid-free paper.

1 2 3 4 5 6 7 8 9 0 QPD/QPD 9 0 3 2 1 0 9

ISBN 0-07-290520-4

www.mhhe.com

FOR THE STUDENT

The *Problem-Solving Workbook* is intended for use with the second edition of Raymond Chang's *Essential Chemistry*. For this manual to be of maximum assistance, you should incorporate it into your overall plan for studying general chemistry. The *Problem-Solving Workbook* contains material to help you develop problem-solving skills, material to practice your problem-solving skills, and solutions to the even-numbered problems in the text. Each chapter in the manual corresponds to one in the text. Every chapter of the manual contains the following main features:

Problem-Solving Strategies

You will encounter several different types of problems in each chapter. To make your problem-solving task easier, we will break down each chapter into the most common problem types. We will present step-by-step methods for solving each problem type.

Example Problems

Incorporated into each problem type are example problems with detailed step-by-step tutorial solutions. These example problems and tutorial solutions are intended to reinforce the problem-solving strategy presented with each problem type. The solutions often contain further explanation of important concepts.

Practice Exercises

Throughout all chapters are numerous practice exercises to allow you to test your problem-solving skills and your knowledge of the material. Answers to the practice exercises are found at the end of the problem-solving strategy section of each chapter.

Text Solutions

Solutions to all the even-numbered text problems are presented following the problem-solving strategies for each chapter. Usually one text problem of each problem type is solved in the step-by-step tutorial manner presented in the problem-solving strategy section. The problems with tutorial solutions refer back to the problem type presented in the strategy section. This will allow you to refer back to the appropriate material if you are having difficulty solving the problem.

Similar Problems

Following each problem type is a list of similar even-numbered problems in the text. Problems in bold-face type are solved in the step-by-step tutorial manner presented in the problem-solving strategy section. Similar problems are listed to allow you to practice more problems that follow a comparable strategy.

I hope that the *Problem-Solving Workbook* helps you succeed in general chemistry. I have attempted to present detailed problem-solving strategies to help you become a better problem-solver. Improved problem-solving will not only help you succeed in general chemistry, but also in many other courses throughout your college career. Probably one of the best ways to learn general chemistry is to work through, and sometimes struggle through, the problems assigned by your instructor. It always looks easy when your instructor solves a problem in class, but you really do not learn the material until you prove to yourself that you can solve the problem.

When solving problems, always try to be flexible. Please do not think that all problems that you encounter in general chemistry will fall into one of the problem types presented in each chapter. However, take the knowledge that you gain from the tutorial solutions and learn to apply that knowledge to different and perhaps more difficult problems. Also, try to work a few problems *every* day. (OK, you can take Friday off!) You cannot afford to fall behind in a general chemistry course.

If you have any comments or suggestions regarding the *Problem-Solving Workbook*, I would like to hear from you. My e-mail address is Brandon.Cruickshank@nau.edu or you can write me by snail mail at the following address:

Dr. Brandon Cruickshank
Northern Arizona University
Department of Chemistry
P.O. Box 5698
Flagstaff, AZ 86011-5698

Good luck in all your college endeavors! And try to enjoy chemistry. You might be pleasantly surprised.

ACKNOWLEDGMENTS

My most sincere thanks go to Dr. Raymond Chang who provided solutions for all the new problems in this edition and reviewed the entire manuscript. His insights, helpful suggestions, and corrections greatly improved this workbook.

I also wish to thank Mr. Terrance Stanton, developmental editor at McGraw-Hill, for directing and organizing this project.

Finally, I express my love and thanks to my partner, Liz Hobbs, for her support and encouragement during work on this project.

Brandon J. Cruickshank

CONTENTS

CHAPTER 1
CHEMISTRY: INTRODUCTION

PROBLEM-SOLVING STRATEGIES AND TUTORIAL SOLUTIONS

TYPES OF PROBLEMS

Problem Type 1: Density Calculations.

Problem Type 2: Temperature Conversions.
(a) °C → °F
(b) °F → °C

Problem Type 3: Scientific Notation.
(a) Expressing a number in scientific notation.
(b) Addition and subtraction.
(c) Multiplication and division.

Problem Type 4: Significant Figures.
(a) Addition and subtraction.
(b) Multiplication and division.

Problem Type 5: The Factor-Label Method of Solving Problems.

PROBLEM TYPE 1: DENSITY CALCULATIONS

Density is the mass of an object divided by its volume.

$$\text{density} = \frac{\text{mass}}{\text{volume}}$$

$$d = \frac{m}{V}$$

Densities of solids and liquids are typically expressed in units of grams per cubic centimeter (g/cm^3) or equivalently grams per milliliter (g/mL). Because gases are much less dense than solids and liquids, typical units are grams per liter (g/L).

EXAMPLE 1.1
A lead brick with dimensions of 5.08 cm by 10.2 cm by 20.3 cm has a mass of 11,950 g. What is the density of lead in g/cm^3?

Step 1: You are given the mass of the lead brick in the problem. You need to calculate the volume of the lead brick to solve for the density. The volume of a rectangular object is equal to the length × width × height.

$$\text{Volume} = \text{length} \times \text{width} \times \text{height}$$
$$\text{Volume} = 5.08\text{ cm} \times 10.2\text{ cm} \times 20.3\text{ cm} = \mathbf{1052\ cm^3}$$

Step 2: Calculate the density by substituting the mass and the volume into the equation.

$$\boldsymbol{d} = \frac{m}{V} = \frac{11{,}950\text{ g}}{1052\text{ cm}^3} = \mathbf{11.4\ g/cm^3}$$

PRACTICE EXERCISE

1. Platinum has a density of 21.4 g/cm^3. What is the mass of a small piece of platinum that has a volume of 7.50 cm^3?

Text Problem: 1.18

PROBLEM TYPE 2: TEMPERATURE CONVERSIONS

To convert between the Fahrenheit scale and the Celsius scale, you must account for two differences between the two scales.

(1) The Fahrenheit scale defines the normal freezing point of water to be exactly 32°F, whereas the Celsius scale defines it to be exactly 0°C.

(2) A Fahrenheit degree is 5/9 the size of a Celsius degree.

A. Converting degrees Fahrenheit to degrees Celsius

The equation needed to complete a conversion from degrees Fahrenheit to degrees Celsius is:

$$? ^\circ\text{C} = (^\circ\text{F} - 32^\circ\text{F}) \times \frac{5^\circ\text{C}}{9^\circ\text{F}}$$

32°F is subtracted to compensate for the normal freezing point of water being 32°F, compared to 0° on the Celsius scale. We multiply by (5/9) becauses a Fahrenheit degree is 5/9 the size of a Celsius degree.

EXAMPLE 1.2
Convert 20°F to degrees Celsius.

$$? ^\circ\text{C} = (^\circ\text{F} - 32^\circ\text{F}) \times \frac{5^\circ\text{C}}{9^\circ\text{F}}$$

$$? ^\circ\text{C} = (20^\circ\text{F} - 32^\circ\text{F}) \times \frac{5^\circ\text{C}}{9^\circ\text{F}} = \mathbf{-6.7^\circ C}$$

B. Converting degrees Celsius to degrees Fahrenheit

The equation needed to complete a conversion from degrees Celsius to degrees Fahrenheit is:

$$? ^\circ\text{F} = \left(^\circ\text{C} \times \frac{9^\circ\text{F}}{5^\circ\text{C}}\right) + 32^\circ\text{F}$$

°C is multiplied by (9/5) because a Celsius degree is 9/5 the size of a Fahrenheit degree. 32°F is then added to compensate for the normal freezing point of water being 32°F, compared to 0° on the Celsius scale.

EXAMPLE 1.3
Normal human body temperature on the Celsius scale is 37.0°C. Convert this to the Fahrenheit scale.

$$? ^\circ\text{F} = \left(^\circ\text{C} \times \frac{9^\circ\text{F}}{5^\circ\text{C}}\right) + 32^\circ\text{F}$$

$$? ^\circ\text{F} = \left(37.0^\circ\text{C} \times \frac{9^\circ\text{F}}{5^\circ\text{C}}\right) + 32^\circ\text{F} = \mathbf{98.6^\circ F}$$

PRACTICE EXERCISE

2. Convert –40°F to degrees Celsius.

Text Problems: 1.20, 1.38 d,e

PROBLEM TYPE 3: SCIENTIFIC NOTATION

Scientific notation is typically used when working with small or large numbers. All numbers can be expressed in the form

$$N \times 10^n$$

where N is a number between 1 and 10 and n is an exponent that can be a positive or negative integer, or zero.

A. Expressing a number in scientific notation

Step 1: Determine n by counting the number of places that the decimal point must be moved to give N, a number between 1 and 10.

Step 2: If the decimal point is moved to the left, n is a positive integer, the number you are working with is larger than 10. If the decimal point is moved to the right, n is a negative integer. The number you are working with is smaller than 1.

EXAMPLE 1.4
Express 0.000105 in scientific notation.

Step 1: The decimal point must be moved four places to the right to give N, a number between 1 and 10. In this case,

$$\mathbf{N = 1.05}$$

Step 2: Since 0.000105 is a number less than one, n is a negative integer. In this case, $n = -4$ (The decimal point was moved four places to the right to give $N = 1.05$).

Combining the above two steps:

$$0.000105 = \mathbf{1.05 \times 10^{-4}}$$

Tip: The notation 1.05×10^{-4} means the following: Take 1.05 and multiply by 10^{-4} (0.0001).

$$1.05 \times 0.0001 = 0.000105$$

EXAMPLE 1.5
Express 4224 in scientific notation.

Step 1: The decimal point must be moved three places to the left to give N, a number between 1 and 10. In this case,

$$\mathbf{N = 4.224}$$

Step 2: Since 4,224 is a number greater than one, n is a positive integer. In this case, $n = 3$ (the decimal point was moved three places to the left to give $N = 4.224$).

Combining the above two steps: $4224 = \mathbf{4.224 \times 10^3}$

Tip: The notation 4.224×10^3 means the following: Take 4.224 and multiply by 10^3 (1000).

$$4.224 \times 1000 = 4{,}224$$

PRACTICE EXERCISE

3. Express the following numbers in scientific notation:

(a) 45,781 **(b)** 0.0000430

Text Problems: 1.22, 1.24

B. Addition and subtraction using scientific notation

Step 1: Write each quantity with the same exponent, *n*.

Step 2: Add or subtract the *N* parts of the numbers, keeping the exponent, *n*, the same.

EXAMPLE 1.6

Express the answer to the following calculation in scientific notation.

$$2.43 \times 10^1 + 5.955 \times 10^2 =$$

Step 1: Write each quantity with the same exponent, *n*. Let's write 2.43×10^1 in such a way that $n = 2$.

Tip: We are *increasing* the exponent, *n*, by a factor of 10, so we must *decrease* *N* by a factor of 10. We move the decimal point one place to the left.

$$2.43 \times 10^1 = \mathbf{0.243 \times 10^2}$$

(*n* was increased by 1. Move the decimal point one place to the left.)

Step 2: Add or subtract, as required, the *N* parts of the numbers, keeping the exponent, *n*, the same. In this example, the process is addition.

$$\begin{array}{r} 0.243 \times 10^2 \\ +\ 5.955 \times 10^2 \\ \hline \mathbf{6.198 \times 10^2} \end{array}$$

C. Multiplication and division using scientific notation

Step 1: Multiply or divide the *N* parts of the numbers in the usual way.

Step 2: When multiplying, *add* the exponents. When dividing, *subtract* the exponents.

EXAMPLE 1.7

Divide 4.2×10^{-7} by 5.0×10^{-5}.

Step 1: Divide the *N* parts of the numbers in the usual way.

$$4.2 \div 5.0 = 0.84$$

Step 2: When dividing the 10^n parts, *subtract* the exponents.

$$0.84 \times 10^{-7-(-5)} = 0.84 \times 10^{-7+5} = \mathbf{0.84 \times 10^{-2}}$$

The usual practice is to express *N* as a number between 1 and 10. Therefore, it is more appropriate to move the decimal point of the above number one place to the right, decreasing the exponent by 1.

$$0.84 \times 10^{-2} = \mathbf{8.4 \times 10^{-3}}$$

Tip: In the answer above, we moved the decimal point to the right, *increasing* N by a factor of 10. Therefore, we must change the exponent, n, from –2 to –3.

EXAMPLE 1.8

Multiply 2.2×10^{-3} by 1.4×10^{6}.

Step 1: Multiply the N parts of the numbers in the usual way.

$$2.2 \times 1.4 = 3.1$$

Step 2: When multiplying the 10^n parts, *add* the exponents.

$$3.1 \times 10^{-3+6} = \mathbf{3.1 \times 10^{3}}$$

PRACTICE EXERCISE

4. Express the answer to the following calculations in scientific notation. Try these without using a calculator.

(a) $2.20 \times 10^{3} - 4.54 \times 10^{2} =$

(b) $4.78 \times 10^{5} \div 6.332 \times 10^{-7} =$

Text Problem: 1.26

PROBLEM TYPE 4: SIGNIFICANT FIGURES

See Section 1.6 of your text for guidelines for using significant figures.

A. Addition and subtraction

The number of significant figures to the right of the decimal point in the answer is determined by the lowest number of significant figures to the right of the decimal point in any of the original numbers.

EXAMPLE 1.9

Carry out the following operations and express the answer to the correct number of significant figures.

102.226 + 2.51 + 736.0 =

$$\begin{array}{r} 102.226 \\ 2.51 \\ +\ \underline{736.0} \\ 840.7\mathbf{36} \end{array} \leftarrow \text{fewest digits to the right of the decimal point}$$

The 3 and 6 are nonsignificant digits, since 736.0 only has one digit to the right of the decimal point. The answer should only have one digit to the right of the decimal point.

The correct answer rounded off to the correct number of significant figures is **840.7**

Tip: To round off a number at a certain point, simply drop the digits that follow if the first of them is less than 5. If the first digit following the point of rounding off is equal to or greater than 5, add 1 to the preceding digit.

B. Multiplying and dividing

The number of significant figures in the answer is determined by the original number having the smallest number of significant figures.

EXAMPLE 1.10

Carry out the following operations and express the answer to the correct number of significant figures.

$$\mathbf{12 \times 2143.1 \div 3.11 =}$$

$$12 \times 2143.1 \div 3.11 = 8269.2 = 8.2\mathbf{692} \times 10^3$$

The 6, 9, and 2 (bolded) are nonsignificant digits because the original number 12 only has two significant figures. Therefore, the answer has only two significant figures.

The correct answer rounded off to the correct number of significant figures is $\mathbf{8.3 \times 10^3}$

PRACTICE EXERCISE

5. Carry out the following operations and express the answer to the correct number of significant figures.

(a) $90.25 - 83 + 1.0015 =$

(b) $55.6 \times 3.482 \div 505.34 =$

Text Problem: 1.30

PROBLEM TYPE 5: THE FACTOR-LABEL METHOD OF SOLVING PROBLEMS

In order to convert from one unit to another, you need to be proficient at the factor-label method. See Section 1.7 of your text. Unit conversions can seem daunting, but if you keep track of the units, making sure that the appropriate units cancel, your effort will be rewarded.

Step 1: Map out a strategy to proceed from initial units to final units based on available conversion factors.

Step 2: Use the following method as many times as is necessary to ensure that you obtain the desired unit.

$$\text{Given unit} \times \left(\frac{\text{desired unit}}{\text{given unit}}\right) = \text{desired unit}$$

EXAMPLE 1.11

How long will it take to fly from Denver to New York, a distance of 1631 miles, at a speed of 815 km/h?

Step 1: One conversion factor is given in the problem, 815 km/h. This conversion factor can be used to convert from distance (in km) to time (in hr). If you can convert the distance of 1631 miles to km, then you can use the conversion factor (815 km/h) to convert to time in hours. Another conversion factor that you can look up is

1 mi = 1.61 km

You should come up with the following strategy.

miles → km → hours

Step 2: Carry out the necessary conversions, making sure that units cancel.

$$\textbf{? hours} = 1631\ \text{mi (given)} \times \frac{1.61\ \text{km (desired)}}{1\ \text{mi (given)}} \times \frac{1\ \text{h (desired)}}{815\ \text{km (given)}} = \mathbf{3.22\ h}$$

> **Tip**: In the first conversion factor (km/mi), km is the desired unit. When moving on to the next conversion factor (h/km), km is now given, and the desired unit is h.

EXAMPLE 1.12

The *Voyager II* mission to the outer planets of our solar system transmitted by radio signals many spectacular photographs of Neptune. Radio waves, like light waves, travel at a speed of 3.00×10^8 m/s. If Neptune was 2.75 billion miles from Earth during these transmissions, how many hours were required for radio signals to travel from Neptune to Earth?

Step 1: One conversion factor is given in the problem, 3.00×10^8 m/s. This conversion factor will allow you to convert from distance (in m) to time (in seconds). If you can convert the distance of 2.75 billion miles to meters, then the speed of light (3.00×10^8 m/s) can be used to convert to time in seconds. Other conversion factors that you can look up are:

billion = 10^9
1 mi = 1.61 km
1 km = 1000 m

60 s = 1 min
60 min = 1 h

You should come up with the following strategy.

miles → km → meters → seconds → min → hours

Step 2: Carry out the necessary conversions, making sure that units cancel.

$$? \text{ h} = 2.75 \times 10^9 \cancel{\text{mi}} \times \frac{1.61 \cancel{\text{km}}}{1 \cancel{\text{mi}}} \times \frac{1000 \cancel{\text{m}}}{1 \cancel{\text{km}}} \times \frac{1 \cancel{\text{s}}}{3.00 \times 10^8 \cancel{\text{m}}} \times \frac{1 \cancel{\text{min}}}{60 \cancel{\text{s}}} \times \frac{1 \text{ h}}{60 \cancel{\text{min}}} = \mathbf{4.10 \text{ h}}$$

PRACTICE EXERCISES

6. On a certain day, the concentration of carbon monoxide, CO, in the air over Denver reached 1.8×10^{-5} g/L. Convert this concentration to mg/m^3.

7. Copper (Cu) is a trace element that is essential for nutrition. Newborn infants require 80 μg of Cu per kilogram of body mass per day. The Cu content of a popular baby formula is 0.48 μg of Cu per milliliter. How many milliliters should a 7.0 lb baby consume per day to obtain the minimum daily Cu requirement?

Text Problems: 1.32, 1.34, 1.36, 1.38, 1.40

ANSWERS TO PRACTICE EXERCISES

1. 161 g Pt

2. −40°C

3. **(a)** 4.5781×10^4
(b) 4.30×10^{-5}

4. **(a)** 1.75×10^3
(b) 7.55×10^{11}

5. **(a)** 8
(b) 0.383

6. 18 mg/m^3

7. 530 mL/day

SOLUTIONS TO SELECTED TEXT PROBLEMS

1.8 **(a)** Physical change. The helium isn't changed in any way by leaking out of the balloon.

(b) Chemical change in the battery.

(c) Physical change. The orange juice concentrate can be regenerated by evaporation of the water.

(d) Chemical change. Photosynthesis changes water, carbon dioxide, etc., into complex organic matter.

(e) Physical change. The salt can be recovered unchanged by evaporation.

1.10 **(a)** extensive **(b)** intensive **(c)** intensive

1.12 **(a)** compound **(b)** element **(c)** compound **(d)** element

1.18 Density Calculation, Problem Type 1.

$$\text{density} = \frac{\text{mass}}{\text{volume}}$$

Step 1: You are asked to solve for the mass of mercury. Therefore, solve the above equation algebraically for the mass.

$$\textbf{mass} = \text{density} \times \text{volume}$$

Step 2: Calculate the mass of mercury by substituting the known values into the equation.

$$\textbf{mass of Hg} = 13.6\ \frac{\text{g}}{\cancel{\text{mL}}} \times 95.8\ \cancel{\text{mL}} = \mathbf{1.30 \times 10^3\ g}$$

1.20 Temperature Conversion, Problem Type 2.

(a) Conversion from Fahrenheit to Celsius.

$$?\ ^\circ\text{C} = (^\circ\text{F} - 32^\circ\text{F}) \times \frac{5^\circ\text{C}}{9^\circ\text{F}}$$

$$\textbf{?}\ ^\circ\textbf{C} = (105^\circ\text{F} - 32^\circ\text{F}) \times \frac{5^\circ\text{C}}{9^\circ\text{F}} = \mathbf{41^\circ C}$$

(b) Conversion from Celsius to Fahrenheit.

$$?\ ^\circ\text{F} = \left(^\circ\text{C} \times \frac{9^\circ\text{F}}{5^\circ\text{C}}\right) + 32^\circ\text{F}$$

$$\textbf{?}\ ^\circ\textbf{F} = \left(-11.5\cancel{^\circ\text{C}} \times \frac{9^\circ\text{F}}{5\cancel{^\circ\text{C}}}\right) + 32^\circ\text{F} = \mathbf{11.3^\circ F}$$

(c) Conversion from Celsius to Fahrenheit.

$$?\ ^\circ\text{F} = \left(^\circ\text{C} \times \frac{9^\circ\text{F}}{5^\circ\text{C}}\right) + 32^\circ\text{F}$$

$$\textbf{?}\ ^\circ\textbf{F} = \left(6.3 \times 10^3\ \cancel{^\circ\text{C}} \times \frac{9^\circ\text{F}}{5\cancel{^\circ\text{C}}}\right) + 32\ ^\circ\text{F} = \mathbf{1.1 \times 10^4\,^\circ F}$$

1.22 Expressing a number in scientific notation, Problem Type 3A.

Remember, all numbers can be expressed in the form $N \times 10^n$, where $\underline{N}$ is a number between 1 and 10, and $\underline{n}$ is an exponent that can be a positive or negative integer.

Step 1: Determine $\underline{n}$ by counting the number of places that the decimal point must be moved to give $\underline{N}$, a number between 1 and 10.

Step 2: If the decimal point is moved to the left, $\underline{n}$ is a positive integer. The number you are working with is larger than 10. If the decimal point is moved to the right, $\underline{n}$ is a negative integer. The number you are working with is smaller than 1.

(a) Express 0.749 in scientific notation.

Step 1: The decimal point must be moved one place to give $\underline{N}$, a number between 1 and 10. In this case,

$$N = 7.49$$

Step 2: Since 0.749 is a number less than one, $\underline{n}$ is a negative integer. In this case, $n = -1$.

Combining the above two steps:

$$0.749 = \mathbf{7.49 \times 10^{-1}}$$

(b) Express 802.6 in scientific notation.

Step 1: The decimal point must be moved two places to give $\underline{N}$, a number between 1 and 10. In this case,

$$N = 8.026$$

Step 2: Since 802.6 is a number greater than one, $\underline{n}$ is a positive integer. In this case, $n = 2$.

Combining the above two steps:

$$802.6 = \mathbf{8.026 \times 10^{2}}$$

(c) Express 0.000000621 in scientific notation.

Step 1: The decimal point must be moved seven places to give $\underline{N}$, a number between 1 and 10. In this case,

$$N = 6.21$$

Step 2: Since 0.000000621 is a number less than one, $\underline{n}$ is a negative integer. In this case, $n = -7$.

Combining the above two steps:

$$0.000000621 = \mathbf{6.21 \times 10^{-7}}$$

1.24 **(a)** Express 3.256×10^{-5} in nonscientific notation.

For the above number expressed in scientific notation, $n = -5$. To convert to nonscientific notation, the decimal point must be moved 5 places to the left.

$$3.256 \times 10^{-5} = \mathbf{0.00003256}$$

(b) Express 6.03×10^6 in nonscientific notation.

For the above number expressed in scientific notation, $n = 6$. The decimal place must be moved 6 places to the right to convert to nonscientific notation.

$$6.03 \times 10^6 = \mathbf{6{,}030{,}000}$$

1.26 Scientific Notation, Problem Types 3B and 3C

(a) Addition using scientific notation.

Step 1: Write each quantity with the same exponent, n.

Let's write 0.0095 in such a way that $n = -3$. This means to move the decimal point three places to the right. To decrease n by 10^3, you must increase N by 10^3.

$$0.0095 = \mathbf{9.5 \times 10^{-3}}$$

Step 2: Add the N parts of the numbers, keeping the exponent, n, the same.

$$\begin{array}{r} 9.5 \times 10^{-3} \\ +\ 8.5 \times 10^{-3} \\ \hline \mathbf{18.0 \times 10^{-3}} \end{array}$$

The usual practice is to express N as a number between 1 and 10. Therefore, it is more appropriate to move the decimal point of the above number one place to the left, increasing the exponent by 1.

$$18.0 \times 10^{-3} = \mathbf{1.8 \times 10^{-2}}$$

(b) Division using scientific notation.

Step 1: Make sure that all numbers are expressed in scientific notation.

$$653 = \mathbf{6.53 \times 10^2}$$

Step 2: Divide the N parts of the numbers in the usual way.

$$6.53 \div 5.75 = 1.14$$

Step 3: *Subtract* the exponents, n.

$$1.14 \times 10^{+2-(-8)} = 1.14 \times 10^{+2+8} = \mathbf{1.14 \times 10^{10}}$$

(c) Subtraction using scientific notation.

Step 1: Write each quantity with the same exponent, n.

Let's write 850,000 in such a way that $n = 5$. This means to move the decimal point five places to the left.

$$850{,}000 = \mathbf{8.5 \times 10^5}$$

Step 2: Subtract the N parts of the numbers, keeping the exponent, n, the same.

$$\begin{array}{r} 8.5 \times 10^{5} \\ -\ 9.0 \times 10^{5} \\ \hline \mathbf{-0.5 \times 10^{5}} \end{array}$$

The usual practice is to express N as a number between 1 and 10. Therefore, it is more appropriate to move the decimal point of the above number one place to the right, decreasing the exponent by 1.

$$-0.5 \times 10^5 = \mathbf{-5 \times 10^4}$$

(d) Multiplication using scientific notation.

Step 1: Make sure that all numbers are expressed in scientific notation.

Step 2: Multiply the *N* parts of the numbers in the usual way.

$$3.6 \times 3.6 = 13$$

Step 3: *Add* the exponents, *n*.

$$13 \times 10^{-4 + (+6)} = \mathbf{13 \times 10^2}$$

The usual practice is to express *N* as a number between 1 and 10. Therefore, it is more appropriate to move the decimal point of the above number one place to the left, increasing the exponent by 1.

$$\mathbf{13 \times 10^2 = 1.3 \times 10^3}$$

1.28 **(a)** three **(b)** one **(c)** one or two **(d)** two

1.30 Significant Figures, Problem Types 4B and 4C

(a) Division

The number of significant figures in the answer is determined by the original number having the smallest number of significant figures.

$$\frac{7.310 \text{ km}}{5.70 \text{ km}} = 1.28\mathbf{3}$$

The 3 (bolded) is a nonsignificant digit because the original number 5.70 only has three significant digits. Therefore, the answer has only three significant digits.

The correct answer rounded off to the correct number of significant figures is:

1.28 (Why are there no units?)

(b) Subtraction

The number of significant figures to the right of the decimal point in the answer is determined by the lowest number of digits to the right of the decimal point in any of the original numbers. Writing both numbers with exponents = −3, we have

$$3.26 \times 10^{-3} \text{ mg} - 0.0788 \times 10^{-3} \text{ mg} = \mathbf{3.18 \times 10^{-3} \text{ mg}}$$

Since 3.26×10^{-3} has only two digits to the right of the decimal point, two digits are carried to the right of the decimal point in the final answer.

(c) Addition

The number of significant figures to the right of the decimal point in the answer is determined by the lowest number of digits to the right of the decimal point in any of the original numbers. Writing both numbers with exponents = +7, we have

$$0.402 \times 10^7 \text{ dm} + 7.74 \times 10^7 \text{ dm} = \mathbf{8.14 \times 10^7 \text{ dm}}$$

Since 7.74×10^7 has only two digits to the right of the decimal point, two digits are carried to the right of the decimal point in the final answer.

1.32 Factor-Label Method, Problem Type 5.

(a)

Step 1: You can look up the following conversion factors.

1 lb = 453.6 grams
1 gram = 1000 mg

You should come up with the following strategy.

lb → grams → mg

Step 2: Carry out the necessary conversions, making sure that units cancel.

$$? \text{ mg} = 242 \text{ lb (given)} \times \frac{453.6 \text{ g (desired)}}{1 \text{ lb (given)}} \times \frac{1 \text{ mg (desired)}}{0.001 \text{ g (given)}} = \mathbf{1.10 \times 10^8 \text{ mg}}$$

Tip: In the first conversion factor (g/lb), g is the desired unit. When moving on to the next conversion factor (mg/g), g is now given, and the desired unit is mg.

(b)

Step 1: You can look up or you should know the following conversion factor.

1 m = 100 cm

You should come up with the following strategy.

$cm^3 \rightarrow m^3$

Step 2: Carry out the necessary conversions, making sure that units cancel.

$$? \text{ m}^3 = 68.3 \text{ cm}^3 \text{ (given)} \times \left(\frac{1 \text{ m (desired)}}{100 \text{ cm (given)}}\right)^3 = \mathbf{6.83 \times 10^{-5} \text{ m}^3}$$

Tip: Make sure to cube the unit factor, so that the cm^3 units cancel and the answer has units of m^3.

1.34 Factor-Label Method, Problem Type 5.

Step 1: You should know the following conversion factors.

1 day = 24 hours
1 hr = 60 minutes
1 min = 60 seconds

You should come up with the following strategy.

days → hours → minutes → seconds

Step 2: Carry out the necessary conversions, making sure that units cancel.

$$? \text{ s} = 365.24 \text{ day} \times \frac{24 \text{ h}}{1 \text{ day}} \times \frac{60 \text{ min}}{1 \text{ h}} \times \frac{60 \text{ s}}{1 \text{ min}} = \mathbf{3.1557 \times 10^7 \text{ s}}$$

1.36 **(a)** $$? \text{ in/s} = \frac{1 \text{ mi}}{13 \text{ min}} \times \frac{5280 \text{ ft}}{1 \text{ mi}} \times \frac{12 \text{ in}}{1 \text{ ft}} \times \frac{1 \text{ min}}{60 \text{ s}} = \mathbf{81 \text{ in/s}}$$

(b) $$? \text{ m/min} = \frac{1 \text{ mi}}{13 \text{ min}} \times \frac{1609 \text{ m}}{1 \text{ mi}} = \mathbf{1.2 \times 10^2 \text{ m/min}}$$

(c) $$? \text{ km/h} = \frac{1 \text{ mi}}{13 \text{ min}} \times \frac{1609 \text{ m}}{1 \text{ mi}} \times \frac{1 \text{ km}}{1000 \text{ m}} \times \frac{60 \text{ min}}{1 \text{ h}} = \mathbf{7.4 \text{ km/h}}$$

1.38 (a) $1.42 \text{ yr} \times \frac{365 \text{ day}}{1 \text{ yr}} \times \frac{24 \text{ h}}{1 \text{ day}} \times \frac{3600 \text{ s}}{1 \text{ h}} \times \frac{3.00 \times 10^8 \text{ m}}{1 \text{ s}} \times \frac{1 \text{ mi}}{1609 \text{ m}} = \mathbf{8.35 \times 10^{12} \text{ mi}}$

(b) $32.4 \text{ yd} \times \frac{36 \text{ in}}{1 \text{ yd}} \times \frac{2.54 \text{ cm}}{1 \text{ in}} = \mathbf{2.96 \times 10^3 \text{ cm}}$

(c) $\frac{3.0 \times 10^{10} \text{ cm}}{1 \text{ s}} \times \frac{1 \text{ in}}{2.54 \text{ cm}} \times \frac{1 \text{ ft}}{12 \text{ in}} = \mathbf{9.8 \times 10^8 \text{ ft/s}}$

(d) $?°\text{C} = (47.4 - 32.0)°\text{F}\left(\frac{5°\text{C}}{9°\text{F}}\right) = \mathbf{8.6°C}$

(e) $?°\text{F} = \left(\frac{9°\text{F}}{5°\text{C}}\right)(-273.15°\text{C}) + 32.00°\text{F} = \mathbf{-459.67°F}$

(f) $? \text{ m}^3 = 71.2 \text{ cm}^3 \left(\frac{1 \text{ m}}{10^2 \text{ cm}}\right)^3 = \mathbf{7.12 \times 10^{-5} \text{ m}^3}$

(g) $? \text{ L} = 7.2 \text{ m}^3 \left(\frac{10 \text{ dm}}{1 \text{ m}}\right)^3 \left(\frac{1 \text{ L}}{1 \text{ dm}^3}\right) = \mathbf{7.2 \times 10^3 \text{ L}}$

1.40 $\textbf{density} = \frac{0.625 \text{ g}}{1 \text{ L}} \times \frac{1 \text{ L}}{1000 \text{ mL}} \times \frac{1 \text{ mL}}{1 \text{ cm}^3} = \mathbf{6.25 \times 10^{-4} \text{ g/cm}^3}$

1.42 $(86.6 \times 10^9 \text{ lb of sulfuric acid}) \times \frac{1 \text{ ton}}{2.00 \times 10^3 \text{ lb}} = \mathbf{4.33 \times 10^7 \text{ tons of sulfuric acid}}$

1.44 Volume of rectangular bar = length × width × height

$\textbf{density} = \frac{m}{V} = \frac{52.7064 \text{ g}}{(8.53 \text{ cm})(2.4 \text{ cm})(1.0 \text{ cm})} = \mathbf{2.6 \text{ g/cm}^3}$

1.46 You are asked to solve for the inner diameter of the tube. If you can calculate the volume that the mercury occupies, you can calculate the radius of the cylinder, $V_{\text{cylinder}} = \pi r^2 h$ (r is the inner radius of the cylinder, and h is the height of the cylinder). The cylinder diameter is $2r$.

$$\text{volume of Hg filling cylinder} = \frac{\text{mass of Hg}}{\text{density of Hg}}$$

$$\textbf{volume of Hg filling cylinder} = \frac{105.5 \text{ g}}{13.6 \text{ g/cm}^3} = \mathbf{7.76 \text{ cm}^3}$$

Next, solve for the radius of the cylinder.

$$\text{Volume of cylinder} = \pi r^2 h$$

$$r = \sqrt{\frac{\text{volume}}{\pi \times h}}$$

$$r = \sqrt{\frac{7.76\ \text{cm}^3}{\pi \times 12.7\ \text{cm}}} = \mathbf{0.441\ cm}$$

The cylinder diameter equals $2r$.

$$\textbf{Cylinder diameter} = 2r = 2(0.441\ \text{cm}) = \mathbf{0.882\ cm}$$

1.48 The volume of silver is equal to the volume of water it displaces.

$$\text{Volume of silver} = 260.5\ \text{mL} - 242.0\ \text{mL} = 18.5\ \text{mL} = 18.5\ \text{cm}^3$$

$$\textbf{Density} = \frac{194.3\ \text{g}}{18.5\ \text{cm}^3} = \mathbf{10.5\ g/cm^3}$$

1.50

$$\frac{343\ \cancel{\text{m}}}{1\ \cancel{\text{s}}} \times \frac{1\ \text{mi}}{1609\ \cancel{\text{m}}} \times \frac{3600\ \cancel{\text{s}}}{1\ \text{h}} = \mathbf{767\ mph}$$

1.52 Temperature:

$$?\,^\circ\text{F} = \left(\frac{9^\circ\text{F}}{5^\circ\cancel{\text{C}}}\right)(24.2^\circ\cancel{\text{C}}) + 32^\circ\text{F} = \mathbf{75.6^\circ F}$$

Uncertainty:

$$?\,^\circ\text{F} = \left(\frac{9^\circ\text{F}}{5^\circ\cancel{\text{C}}}\right)(0.1^\circ\cancel{\text{C}}) = \mathbf{0.2^\circ F}$$

$$\mathbf{?\ ^\circ F = 75.6^\circ F \pm 0.2^\circ F}$$

1.54 The key to solving this problem is to realize that all the oxygen needed must come from the 4% difference (20% – 16%) between inhaled and exhaled air.

The 240 mL of pure oxygen/min requirement comes from the 4% of inhaled air that is oxygen.

$$240\ \text{mL of pure oxygen/min} = (0.04)(\text{volume of inhaled air/min})$$

$$\text{Volume of inhaled air/min} = \frac{240\ \text{mL of oxygen / min}}{0.04} = \mathbf{6000\ mL\ of\ inhaled\ air/min}$$

Since there are 12 breaths per min,

$$\textbf{volume of air/breath} = \left(\frac{6000\ \text{mL of inhaled air}}{1\ \cancel{\text{min}}}\right) \times \left(\frac{1\ \cancel{\text{min}}}{12\ \text{breaths}}\right) = \mathbf{5 \times 10^2\ mL/breath}$$

1.56 First, calculate the volume of 1 kg of seawater from the density and the mass. We choose 1 kg of seawater, because the problem gives the amount of Mg in every kg of seawater.

$$\text{volume} = \frac{\text{mass}}{\text{density}}$$

$$\text{volume of 1 kg of seawater} = \frac{1000\ \cancel{\text{g}}}{1.03\ \cancel{\text{g}}/\text{mL}} = \mathbf{971\ mL} = \mathbf{0.971\ L}$$

In other words, there are 1.3 g of Mg in every 0.971 L of seawater.

Next, let's convert tons of Mg to grams of Mg.

$$(8.0 \times 10^4 \text{ tons Mg}) \times \frac{2000 \text{ lb}}{1 \text{ ton}} \times \frac{453.6 \text{ g}}{1 \text{ lb}} = 7.3 \times 10^{10} \text{ g Mg}$$

Volume of seawater needed to extract 8.0 × 10⁴ ton Mg =

$$(7.3 \times 10^{10} \text{ g Mg}) \times \frac{0.971 \text{ L seawater}}{1.3 \text{ g Mg}} = \mathbf{5.5 \times 10^{10} \text{ L of seawater}}$$

1.58 $? \text{ °F} = \left(\text{°C} \times \frac{9\text{°F}}{5\text{°C}}\right) + 32\text{°F}$

Let temperature = t

$$t = \frac{9}{5}t + 32\text{°F}$$

$$t - \frac{9}{5}t = 32\text{°F}$$

$$-\frac{4}{5}t = 32\text{°F}$$

$$\mathbf{t = -40\text{°F} = -40\text{°C}}$$

1.60 **(a)** $\frac{|0.798 \text{ g/mL} - 0.802 \text{ g/mL}|}{0.798 \text{ g/mL}} \times 100\% = \mathbf{0.5\%}$

(b) $\frac{|0.864 \text{ g} - 0.837 \text{ g}|}{0.864 \text{ g}} \times 100\% = \mathbf{3.1\%}$

1.62 $\mathbf{? \text{ g Au}} = \frac{4.0 \times 10^{-12} \text{ g Au}}{1 \text{ mL seawater}} \times \frac{1 \text{ mL}}{0.001 \text{ L}} \times (1.5 \times 10^{21} \text{ L seawater}) = \mathbf{6.0 \times 10^{12} \text{ g Au}}$

$$\textbf{value of gold} = (6.0 \times 10^{12} \text{ g Au}) \times \frac{1 \text{ lb}}{453.6 \text{ g}} \times \frac{16 \text{ oz}}{1 \text{ lb}} \times \frac{\$350}{1 \text{ oz}} = \mathbf{\$7.4 \times 10^{13}}$$

No one has become rich mining gold from the ocean, because the cost of recovering the gold would outweigh the price of the gold.

1.64 10 cm = 0.1 m. We need to find the number of times the 0.1 m wire must be cut in half until the piece left is 1.3×10^{-12} m long. Let n be the number of times we can cut the Cu wire in half. We can write:

$$\left(\frac{1}{2}\right)^n \times 0.1 \text{ m} = 1.3 \times 10^{-12} \text{ m}$$

$$\left(\frac{1}{2}\right)^n = 1.3 \times 10^{-11} \text{ m}$$

Taking the log of both sides of the equation:

$$n\log\left(\frac{1}{2}\right) = \log(1.3 \times 10^{-11})$$

$$\mathbf{n = 36 \text{ times}}$$

1.66 Volume = area × thickness.

From the density, we can calculate the volume of the Al foil.

$$\text{Volume} = \frac{\text{mass}}{\text{density}} = \frac{3.636\ \text{g}}{2.699\ \text{g/cm}^3} = 1.347\ \text{cm}^3$$

Convert the unit of area from ft^2 to cm^2.

$$1.000\ \text{ft}^2 \times \left(\frac{12\ \text{in}}{1\ \text{ft}}\right)^2 \times \left(\frac{2.54\ \text{cm}}{1\ \text{in}}\right)^2 = 929.0\ \text{cm}^2$$

$$\textbf{thickness} = \frac{\text{volume}}{\text{area}} = \frac{1.347\ \text{cm}^3}{929.0\ \text{cm}^2} = 1.450 \times 10^{-3}\ \text{cm} = \mathbf{1.450 \times 10^{-2}\ mm}$$

1.68 The mass of water used by 50,000 people in 1 year is:

$$50{,}000\ \text{people} \times \frac{150\ \text{gal. water}}{1\ \text{person each day}} \times \frac{3.79\ \text{L}}{1\ \text{gal.}} \times \frac{1000\ \text{mL}}{1\ \text{L}} \times \frac{1\ \text{g H}_2\text{O}}{1\ \text{mL H}_2\text{O}} \times \frac{365\ \text{days}}{1\ \text{yr}} = 1.04 \times 10^{13}\ \text{g H}_2\text{O/yr}$$

A concentration of 1 ppm of fluorine is needed. In other words, 1 g of fluorine is needed per million grams of water. NaF is 45.0% fluorine by mass. The amount of NaF needed per year in kg is:

$$1.04 \times 10^{13}\ \text{g H}_2\text{O} \times \frac{1\ \text{g F}}{10^6\ \text{g H}_2\text{O}} \times \frac{100\%\ \text{NaF}}{45\%\ \text{F}} \times \frac{1\ \text{kg}}{1000\ \text{g}} = \mathbf{2.3 \times 10^4\ kg\ NaF}$$

An average person uses 150 gallons of water per day. This is equal to 569 L of water. If only 6 L of water is used for drinking and cooking, 563 L of water is used for purposes in which NaF is not necessary. Therefore the amount of NaF wasted is:

$$\frac{563\ \text{L}}{569\ \text{L}} \times 100\% = \mathbf{99\%}$$

1.70 To calculate the density of the pheromone, you need the mass of the pheromone, and the volume that it occupies. The mass is given in the problem. First, let's calculate the volume of the cylinder. Converting the radius and height to cm gives:

$$0.50\ \text{mi} \times \frac{1609\ \text{m}}{1\ \text{mi}} \times \frac{1\ \text{cm}}{0.01\ \text{m}} = 8.0 \times 10^4\ \text{cm}$$

$$40\ \text{ft} \times \frac{12\ \text{in}}{1\ \text{ft}} \times \frac{2.54\ \text{cm}}{1\ \text{in}} = 1.2 \times 10^3\ \text{cm}$$

$$\text{volume of a cylinder} = \text{area} \times \text{height} = \pi r^2 \times h$$

$$\text{volume} = \pi(8.0 \times 10^4\ \text{cm})^2 \times (1.2 \times 10^3\ \text{cm}) = 2.4 \times 10^{13}\ \text{cm}^3$$

Density of gases are usually expressed in g/L. Let's convert the volume to liters.

$$2.4 \times 10^{13}\ \text{cm}^3 \times \frac{1\ \text{mL}}{1\ \text{cm}^3} \times \frac{1\ \text{L}}{1000\ \text{mL}} = 2.4 \times 10^{10}\ \text{L}$$

$$\textbf{density} = \frac{\text{mass}}{\text{volume}} = \frac{1.0 \times 10^{-8}\ \text{g}}{2.4 \times 10^{10}\ \text{L}} = \mathbf{4.2 \times 10^{-19}\ g/L}$$

1.72 **(a)** Data was collected to indicate a high iridium content in clay deposited above sediments formed during the Cretaceous period. A hypothesis was formulated that the iridium came from a large asteroid. When the asteroid impacted Earth, large amounts of dust and debris blocked the sunlight. Plants eventually died, then many plant eating animals gradually perished, then, in turn, meat eating animals starved.

(b) If the hypothesis is correct, we should find a similarly high iridium content in corresponding rock layers at different locations on Earth. Also, we should expect the simultaneous extinction of other large species in addition to dinosaurs.

(c) Yes. Hypotheses that survive many experimental tests of their validity may evolve into theories. A ***theory*** is a unifying principle that explains a body of facts and/or those laws that are based on them.

(d) Twenty percent of the asteroid's mass turned into dust. First, let's calculate the mass of the dust.

$$? \text{ g dust} = \frac{0.02 \text{ g}}{\text{cm}^2} \times \left(\frac{1 \text{ cm}}{0.01 \text{ m}}\right)^2 \times (5.1 \times 10^{14} \text{ m}^2) = 1.02 \times 10^{17} \text{ g dust}$$

This mass is 20% of the asteroid's mass.

$$\textbf{mass of asteroid} = \frac{1.0 \times 10^{17} \text{ g}}{0.20} = 5.0 \times 10^{17} \text{ g} = \mathbf{5.0 \times 10^{14}\ kg}$$

Converting to tons:

$$5.0 \times 10^{14} \text{ kg} \times \frac{1000 \text{ g}}{1 \text{ kg}} \times \frac{1 \text{ lb}}{453.6 \text{ g}} \times \frac{1 \text{ ton}}{2000 \text{ lb}} = \mathbf{5.5 \times 10^{11}\ tons}$$

From the mass and density of the asteroid, we can calculate its volume. Then, from its volume, we can calculate the radius.

$$\text{Volume} = \frac{\text{mass}}{\text{density}} = \frac{5.0 \times 10^{17} \text{ g}}{2 \text{ g/cm}^3} = 2.5 \times 10^{17} \text{ cm}^3$$

$$\text{Volume}_{\text{sphere}} = \frac{4}{3}\pi r^3$$

$$2.5 \times 10^{17} \text{ cm}^3 = \frac{4}{3}\pi r^3$$

$$r = 4 \times 10^5 \text{ cm} = \mathbf{4 \times 10^3\ m}$$

The diameter of the asteroid is approximately 5 miles!

CHAPTER 2
ATOMS, MOLECULES, AND IONS

PROBLEM-SOLVING STRATEGIES AND TUTORIAL SOLUTIONS

TYPES OF PROBLEMS

Problem Type 1: Atomic number, Mass number, and Isotopes.

Problem Type 2: Empirical and Molecular Formulas.

Problem Type 3: Naming Compounds.
(a) Ionic compounds.
(b) Molecular compounds.
(c) Acids.
(d) Bases.

Problem Type 4: Formulas of Ionic Compounds.

PROBLEM TYPE 1: ATOMIC NUMBER, MASS NUMBER, AND ISOTOPES

The **atomic number (*Z*)** is the number of protons in the nucleus of each atom of an element.

EXAMPLE 2.1
What is the atomic number of an oxygen atom?

The atomic number is listed above each element in the periodic table. For oxygen, the atomic number is 8, meaning that an oxygen atom has eight protons in the nucleus.

The **mass number (*A*)** is the total number of neutrons and protons present in the nucleus of an atom of an element.

mass number = number of protons + number of neutrons
mass number = atomic number + number of neutrons

EXAMPLE 2.2
A particular oxygen atom has nine neutrons in the nucleus. What is the mass number of this atom?

Looking at a periodic table, you should find that every oxygen atom has an atomic number of 8. Thus,

mass number = atomic number + number of neutrons

mass number = 8 + 9 = **17**

Isotopes are atoms that have the same atomic number, but different mass numbers. For example, there are three isotopes of oxygen found in nature, oxygen-16, oxygen-17, and oxygen-18. The accepted way to denote the atomic number and mass number of an element X is as follows:

$$^{A}_{Z}\mathrm{X}$$

where, A = mass number
Z = atomic number

EXAMPLE 2.3

The three isotopes of oxygen found in nature are oxygen-16, -17, and -18. Write their isotopic symbols.

The atomic number of oxygen is 8, so all isotopes of oxygen contain eight protons. The mass numbers are 16, 17, and 18, respectively.

$$^{16}_{8}O \quad ^{17}_{8}O \quad ^{18}_{8}O$$

The number of **electrons** in an *atom* is equal to the number of protons.

number of electrons (atom) = number of protons = atomic number

The number of **electrons** in an *ion* is equal to the number of protons minus the charge on the ion.

number of electrons (ion) = number of protons − charge on the ion

EXAMPLE 2.4

What is the total number of fundamental particles (protons, neutrons, and electrons) in (a) an atom of $^{56}_{26}Fe$ and (b) an $^{56}_{26}Fe^{3+}$ ion?

Both $^{56}_{26}Fe$ and $^{56}_{26}Fe^{3+}$ have the same atomic number and mass number. Thus, for both (a) and (b),

number of protons = atomic number = **26**

and

number of neutrons = mass number − atomic number = 56 − 26 = **30**

However, the number of electrons for the above species differ.

(a) $^{56}_{26}Fe$ is a neutral atom. Therefore,

number of electrons = number of protons = **26**

(b) $^{56}_{26}Fe^{3+}$ is an ion with a +3 charge. Therefore,

number of electrons = number of protons − charge = 26 − (+3) = **23**

PRACTICE EXERCISE

1. How many protons, neutrons, and electrons are contained in each of the following atoms or ions?

   (a) ^{19}F **(b)** $^{79}Se^{2-}$ **(c)** ^{40}Ca **(d)** $^{48}Ti^{4+}$

Text Problems: 2.12, 2.14, 2.16, 2.38

PROBLEM TYPE 2: EMPIRICAL AND MOLECULAR FORMULAS

A *molecular* formula shows the exact number of atoms of each element in the smallest unit of a substance. An *empirical formula* tells us which elements are present and the simplest whole-number ratio of their atoms. Empirical formulas are therefore the simplest chemical formulas; they are always written so that the subscripts in the molecular formulas are converted to the smallest possible whole numbers.

EXAMPLE 2.5

What is the empirical formula of each of the following compounds: (a) H_2O_2, (b) $C_6H_8O_6$, (c) $MgCl_2$, and (d) C_8H_{18}?

(a) The simplest whole number ratio of the atoms in H_2O_2 is **HO**.

(b) The simplest whole number ratio of the atoms in $C_6H_8O_6$ is $\mathbf{C_3H_4O_3}$.

(c) The molecular formula as written contains the simplest whole number ratio of the atoms present. In this case, the molecular formula and the empirical formula are the same.

(d) The simplest whole number ratio of the atoms in C_8H_{18} is $\mathbf{C_4H_9}$.

PRACTICE EXERCISE

2. What is the empirical formula of each of the following compounds?

(a) C_2H_6O (b) $C_6H_{12}O_6$ (c) CH_3COOH (d) $C_{12}H_{22}O_{11}$

Text Problem: 2.34

PROBLEM TYPE 3: NAMING COMPOUNDS

A. Naming ionic compounds

(1) **Metal cation has only one charge**. You should memorize the metal cations that have only one charge when they form ionic compounds. These include the alkali metals (Group 1A), which always have a +1 charge in ionic compounds, the alkaline earth metals (Group 2A), which always have a +2 charge in ionic compounds, and Al^{3+}, Ag^{+}, Cd^{2+}, and Zn^{2+}.

Since the metal cation has only one possible charge, we do not need to specify this charge in the compound. Therefore, the name of this type of ionic compound can be written simply by first naming the metal cation as it appears on the periodic table, followed by the nonmetallic anion. Anions from elements are named by changing the suffix to "-ide".

EXAMPLE 2.6

Name the following ionic compounds: (a) AlF_3, (b) Na_3N, (c) $Ba(NO_3)_2$.

(a) aluminum fluor*ide*
(b) sodium nitr*ide*
(c) barium nitrate

> **Tip:** There are a number of ions that contain more than one atom. These ions are called **polyatomic ions**. Nitrate, NO_3^-, is an example. Ask your instructor which polyatomic ions you should know.

(2) **Metals that form more than one type of cation**. Transition metals typically can form more than one type of cation. For compounds containing these metals, you must specify the charge of the cation. Roman numerals are used for this purpose. The Roman numeral I is used for one positive charge, II for two positive charges, III for three positive charges, and so on.

EXAMPLE 2.7

Name the following compounds: (a) FeO, (b) Fe_2O_3, (c) $HgSO_4$.

(a) In this compound, the iron cation has a +2 charge, since oxide has a −2 charge (see Table 2.2 of your text). Therefore, the compound is named **iron(II) oxide**.

(b) In this compound, the iron cation has a +3 charge. Therefore, the compound is named **iron(III) oxide**.

(c) This compound contains the polyatomic ion sulfate, which has a −2 charge (SO_4^{2-}). Thus, the charge on mercury (Hg) is +2. The compound is named **mercury(II) sulfate**.

B. Naming molecular compounds

Unlike ionic compounds, molecular compounds contain discrete molecular units. They are usually composed of nonmetallic elements. There are two types of molecular compounds to consider.

(1) **Only one compound of the two elements exists.** If this is the case, you simply name the first element in the formula as it appears on the periodic table, followed by naming the second element with an "-ide" suffix.

EXAMPLE 2.8

Name the following compounds: (a) HF, (b) SiC.

(a) hydrogen fluoride
(b) silicon carbide

(2) **More than one compound composed of the two elements exists.** It is quite common for one pair of elements to form several different compounds. Therefore, we must be able to differentiate between the compounds. Greek prefixes are used to denote the number of atoms of each element present. See Table 2.4 in the text for the Greek prefixes used.

EXAMPLE 2.9

Name the following compounds: (a) CO, (b) CO_2, (c) N_2O_5, (d) SF_6.

(a) carbon *mono*xide
(b) carbon *di*oxide
(c) *di*nitrogen *pent*oxide
(d) sulfur *hexa*fluoride

> **Tip:** The prefix "mono-" may be omitted when naming the first element in a molecular compound (see carbon monoxide and carbon dioxide above). The absence of a prefix for the first element usually means that only one atom of that element is present in the molecule.

C. Naming acids

An acid is a substance that yields hydrogen ions (H^+) when dissolved in water. There are two types of acids to consider.

(1) **Acids that do not contain oxygen.** This type of acid contains one or more hydrogen atoms as well as an anionic group. To name these acids, you add the prefix "hydro-" to the anion name, change the "-ide" suffix of the anion to "-ic", and then add the word "acid" at the end.

EXAMPLE 2.10

Name the following binary acids: (a) HF (*aq*), (b) HCN (*aq*).

(a) *hydro*fluor*ic* acid

(b) CN^- is a polyatomic ion called cyanide. In the acid, the "-ide" suffix is changed to "-ic". The correct name is ***hydro*cyan*ic* acid.**

Tip: The (*aq*) above means that the substance is dissolved in water. HF dissolved in water is an acid and is named hydrofluoric acid. However, HF in its pure state is a molecular compound and is named hydrogen fluoride.

(2) **Oxoacids.** This type of acid contains hydrogen, oxygen, and another element. To name oxoacids, you must look carefully at the anion name. If the suffix of the anion is "-ate", change the suffix to "-ic" and add the word "acid" at the end. If the suffix of the anion is "-ite", change the suffix to "-ous" and add the word "acid" at the end.

EXAMPLE 2.11
Name the following oxoacids: (a) HNO_2 (*aq*), (b) $HClO_3$ (*aq*).

(a) The NO_2^- polyatomic ion is called nitr*ite*. Simply change the suffix to "-ous" and add the word "acid". The correct name is **nitr*ous* acid**.

(b) The ClO_3^- polyatomic ion is called chlor*ate*. Simply change the suffix to "-ic" and add the work "acid". The correct name is **chlor*ic* acid**.

Tip: If one O atom is added to the "-ic" acid, the acid is called "per...ic" acid. For example, $HClO_4$ is named perchloric acid. Compare this acid to chloric acid above. If one O atom is removed from the "-ous" acid, the acid is called "hypo...ous" acid. For example, HClO is named hypochlorous acid. Compare this to chlorous acid, $HClO_2$.

D. Naming Bases

A base is a substance that produces the hydroxide ion (OH^-) when dissolved in water. At this point, for naming purposes, we will only consider bases that contain the hydroxide ion. To name this type of base, simply name the metal cation first as it appears on the periodic table, then add "hydroxide".

EXAMPLE 2.12
Name the following bases: (a) KOH, (b) $Sr(OH)_2$.

(a) potassium hydroxide
(b) strontium hydroxide

PRACTICE EXERCISE

3. Name the following compounds:

(a) $MgCl_2$ **(b)** $CuCl_2$ **(c)** HNO_3 **(d)** P_2O_5 **(e)** $Ca(OH)_2$

Text Problem: 2.42

PROBLEM TYPE 4: FORMULAS OF IONIC COMPOUNDS

The formulas of ionic compounds are always the same as their empirical formulas because ionic compounds do not consist of discrete molecular units. See Section 2.6 of your text if you need further information.

Ionic compounds are electrically neutral. In order for ionic compounds to be electrically neutral, the sum of the charges on the cation and anion in each formula unit must add up to zero. There are two possibilities to consider.

(1) If the charges on the cation and anion are numerically equal, no subscripts are necessary in the formula.
(2) If the charges on the cation and anion are numerically different, the subscript of the cation is numerically equal to the charge on the anion, and the subscript of the anion is numerically equal to the charge on the cation.

You should memorize the metal cations that have only one charge when they form ionic compounds. These include the alkali metals (Group 1A), which always have a +1 charge in ionic compounds, the alkaline earth metals (Group 2A), which always have a +2 charge in ionic compounds, and Al^{3+}, Ag^{+}, Cd^{2+}, and Zn^{2+}. All other metals can have more than one possible positive charge when forming ionic compounds. This positive charge will be specified using a Roman numeral in the name of the compound.

You should also memorize the charges of polyatomic ions (see Table 2.3 of your text) and the common charges of monatomic anions based on their positions in the periodic table (see Table 2.2 of your text).

EXAMPLE 2.13

Write the formula for the ionic compound, magnesium oxide.

The magnesium cation is an alkaline earth metal cation which always has a +2 charge in an ionic compound, and the oxide anion is usually −2 in an ionic compound (see Table 2.2 of your text). Mg^{2+} and the oxide anion, O^{2-}, combine to form the ionic compound magnesium oxide. The sum of the charges is +2 + (−2) = 0, so no subscripts are necessary. The formula is **MgO**.

EXAMPLE 2.14

Write the formula for the ionic compound, iron(II) chloride.

An iron cation can either have a +2 or +3 charge in an ionic compound. The Roman numeral, II, specifies that in this compound it is the +2 cation, Fe^{2+}. The chloride anion has a −1 charge in an ionic compound (see Table 2.2 of your text). Fe^{2+} and the chloride anion, Cl^{-}, combine to form the ionic compound iron(II) chloride. The charges on the cation and anion are numerically different, so make the subscript of the cation (Fe^{2+}) numerically equal to the charge of the anion (subscript = 1). Also, make the subscript of the anion (Cl^{-}) numerically equal to the charge of the cation (subscript = 2). The formula is **$FeCl_2$**.

> **Tip:** Check to make sure that the compound is electrically neutral by multiplying the charge of each ion by its subscript and then adding them together. The sum should equal zero.

$$(+2)(1) + (-1)(2) = 0$$

PRACTICE EXERCISE

4. Write the correct formulas for the following ionic compounds:

(a) Sodium oxide **(b)** Copper(II) nitrate **(c)** Aluminum oxide

Text Problems: 2.44 a, b, e, f, g, i, j

ANSWERS TO PRACTICE EXERCISES

1. **(a)** 9p, 10n, 9e
(b) 34p, 45n, 36e
(c) 20p, 20n, 20e
(d) 22p, 26n, 18e

2. **(a)** C_2H_6O
(b) CH_2O
(c) CH_2O
(d) $C_{12}H_{22}O_{11}$

3. **(a)** magnesium chloride
(b) copper(II) chloride
(c) nitric acid
(d) diphosphorus pentoxide
(e) calcium hydroxide

4. **(a)** Na_2O
(b) $Cu(NO_3)_2$
(c) Al_2O_3

SOLUTIONS TO SELECTED TEXT PROBLEMS

2.8 Note that you are given the information to set up the unit factor relating meters and miles.

$$r_{\text{atom}} = 10^4 r_{\text{nucleus}} = 10^4 \times 10\ \text{cm} \times \frac{1\ \text{m}}{100\ \text{cm}} \times \frac{1\ \text{mi}}{1609\ \text{m}} = \mathbf{0.62\ mi}$$

2.12 Problem Type 1, Atomic number, Mass number, and Isotopes.

The 239 in Pu-239 is the mass number. The **mass number (A)** is the total number of neutrons and protons present in the nucleus of an atom of an element. You can look up the atomic number (number of protons) on the periodic table.

mass number = number of protons + number of neutrons

number of neutrons = mass number – number of protons = 239 – 94 = **145**

2.14

Isotope	$^{15}_{7}N$	$^{33}_{16}S$	$^{63}_{29}Cu$	$^{84}_{38}Sr$	$^{130}_{56}Ba$	$^{186}_{74}W$	$^{202}_{80}Hg$
No. Protons	7	16	29	38	56	74	80
No. Neutrons	8	17	34	46	74	112	122
No. Electrons	7	16	29	38	56	74	80

2.16 The accepted way to denote the atomic number and mass number of an element X is as follows:

$$^{A}_{Z}\mathbf{X}$$

where,

A = mass number
Z = atomic number

(a) $^{186}_{74}W$ **(b)** $^{201}_{80}Hg$

2.34 Problem Type 2, Empirical and Molecular Formulas.

An *empirical formula* tells us which elements are present and the simplest whole-number ratio of their atoms.

(a) The simplest whole number ratio of the atoms in Al_2Br_6 is $\mathbf{AlBr_3}$.

(b) The simplest whole number ratio of the atoms in $Na_2S_2O_4$ is $\mathbf{NaSO_2}$.

(c) The molecular formula as written, $\mathbf{N_2O_5}$, contains the simplest whole number ratio of the atoms present. In this case, the molecular formula and the empirical formula are the same.

(d) The molecular formula as written, $\mathbf{K_2Cr_2O_7}$, contains the simplest whole number ratio of the atoms present. In this case, the molecular formula and the empirical formula are the same.

2.38 The **atomic number (*Z*)** is the number of protons in the nucleus of each atom of an element. You can find this on a periodic table. The number of **electrons** in an *ion* is equal to the number of protons minus the charge on the ion.

number of electrons (ion) = number of protons – charge on the ion

Ion	K^+	Mg^{2+}	Fe^{3+}	Br^-	Mn^{2+}	C^{4-}	Cu^{2+}
No. protons	19	12	26	35	25	6	29
No. electrons	18	10	23	36	23	10	27

2.40 Compounds of metals with nonmetals are usually ionic. Nonmetal-nonmetal compounds are usually molecular.

Ionic: $NaBr$, BaF_2, $CsCl$

Molecular: CH_4, CCl_4, ICl, NF_3

2.42 Problem Type 4, Naming Compounds.

(a) This is an ionic compound in which the metal cation (K^+) has only one charge (see Problem Type 4A). The correct name is **potassium hypochlorite**. Hypochlorite is a polyatomic ion with one less O atom than the chlorite ion, ClO_2^-.

(b) **silver carbonate**

(c) This is an ionic compound in which the metal can form more than one cation (see Problem Type 4A). Use a Roman numeral to specify the charge of the Fe ion. Since each chloride ion has a −1 charge, the Fe ion has a +2 charge. The correct name is **iron(II) chloride**.

(d) **potassium permanganate** **(e)** **cesium chlorate** **(f)** **potassium ammonium sulfate**

(g) This is an ionic compound in which the metal can form more than one cation (see Problem Type 4A). Use a Roman numeral to specify the charge of the Fe ion. Since the oxide ion has a −2 charge, the Fe ion has a +2 charge. The correct name is **iron(II) oxide**.

(h) **iron(III) oxide**

(i) This is an ionic compound in which the metal can form more than one cation (see Problem Type 4A). Use a Roman numeral to specify the charge of the Ti ion. Since each of the four chloride ions has a −1 charge (total of −4), the Ti ion has a +4 charge. The correct name is **titanium(IV) chloride**.

(j) **sodium hydride** **(k)** **lithium nitride** **(l)** **sodium oxide**

(m) This is an ionic compound in which the metal cation (Na^+) has only one charge (see Problem Type 4A). The O_2^{2-} ion is called the peroxide ion. Each oxygen has a −1 charge. You can determine that each oxygen only has a −1 charge, because each of the two Na ions has a +1 charge. Compare this to sodium oxide in part (l). The correct name is **sodium peroxide**.

2.44 Problem Types 3 and 4.

(a) The Roman numeral I tells you that the Cu cation has a +1 charge. Cyanide has a −1 charge. Since, the charges are numerically equal, no subscripts are necessary in the formula. The correct formula is **CuCN**.

(b) Strontium is an alkaline earth metal. It only forms a +2 cation. The polyatomic ion chlorite, ClO_2^-, has a −1 charge. Since the charges on the cation and anion are numerically different, the subscript of the cation is numerically equal to the charge on the anion, and the subscript of the anion is numerically equal to the charge on the cation. The correct formula is $\mathbf{Sr(ClO_2)_2}$.

(c) Perbromic tells you that the anion of this oxoacid is perbromate, BrO_4^-. The correct formula is $\mathbf{HBrO_4}$ **(*aq*)**. Remember that (*aq*) means that the substance is dissolved in water.

(d) Hydroiodic tells you that the anion of this binary acid is iodide, I^-. The correct formula is **HI (*aq*)**.

(e) Na is an alkali metal. It only forms a +1 cation. The polyatomic ion ammonium, NH_4^+, has a +1 charge and the polyatomic ion phosphate, PO_4^{3-}, has a −3 charge. To balance the charge, you need 2 Na^+ cations. The correct formula is $\mathbf{Na_2(NH_4)PO_4}$.

(f) The Roman numeral II tells you that the Pb cation has a +2 charge. The polyatomic ion carbonate, CO_3^{2-}, has a −2 charge. Since, the charges are numerically equal, no subscripts are necessary in the formula. The correct formula is $\mathbf{PbCO_3}$.

(g) The Roman numeral II tells you that the Sn cation has a +2 charge. Fluoride has a −1 charge. Since the charges on the cation and anion are numerically different, the subscript of the cation is numerically equal to the charge on the anion, and the subscript of the anion is numerically equal to the charge on the cation. The correct formula is **SnF_2**.

(h) This is a molecular compound. The Greek prefixes tell you the number of each type of atom in the molecule. The correct formula is **P_4S_{10}**.

(i) The Roman numeral II tells you that the Hg cation has a +2 charge. Oxide has a −2 charge. Since, the charges are numerically equal, no subscripts are necessary in the formula. The correct formula is **HgO**.

(j) The Roman numeral I tells you that the Hg cation has a +1 charge. However, this cation exists as Hg_2^{2+}. Iodide has a −1 charge. You need two iodide ion to balance the +2 charge of Hg_2^{2+}. The correct formula is **Hg_2I_2**.

2.46 **(c)** Changing the electrical charge of an atom usually has a major effect on its chemical properties. The two electrically neutral carbon isotopes should have nearly identical chemical properties.

2.48 **(a)** Does this refer to hydrogen atoms or hydrogen molecules? One can't be sure.
(b) NaCl is an ionic compound; it doesn't form molecules.

2.50 The species and their identification are as follows:

(a)	SO_2	molecule and compound	**(g)**	O_3	element and molecule
(b)	S_8	element and molecule	**(h)**	CH_4	molecule and compound
(c)	Cs	element	**(i)**	KBr	compound
(d)	N_2O_5	molecule and compound	**(j)**	S	element
(e)	O	element	**(k)**	P_4	element and molecule
(f)	O_2	element and molecule	**(l)**	LiF	compound

2.52 **(a)** CO_2 (*s*), solid carbon dioxide
(b) NaCl, sodium chloride
(c) N_2O, nitrous oxide
(d) $CaCO_3$, calcium carbonate
(e) CaO, calcium oxide
(f) $Ca(OH)_2$, calcium hydroxide
(g) $NaHCO_3$, sodium bicarbonate
(h) $Mg(OH)_2$, magnesium hydroxide

2.54 **(a)** Ionic compounds are typically formed between metallic and nonmetallic elements.
(b) In general the transition metals, the actinides and lanthanides have variable charges.

2.56 The symbol ^{23}Na provides more information than $_{11}Na$. The mass number plus the chemical symbol identifies a specific isotope of Na (sodium) while combining the atomic number with the chemical symbol tells you nothing new. Can other isotopes of sodium have different atomic numbers?

2.58 Mercury (Hg) and bromine (Br_2)

2.60 H_2, N_2, O_2, O_3, F_2, Cl_2, He, Ne, Ar, Kr, Xe, Rn

2.62 They do not have a strong tendency to form compounds. Helium, neon, and argon are chemically inert.

2.64 All isotopes of radium are radioactive. It is a radioactive decay product of uranium-238. Radium itself does *not* occur naturally on Earth.

2.66 **(a)** NaH, sodium hydride **(b)** B_2O_3, diboron trioxide **(c)** Na_2S, sodium sulfide
(d) AlF_3, aluminum fluoride **(e)** OF_2, oxygen difluoride **(f)** $SrCl_2$, strontium chloride

2.68 Rutherford's experiment is described in detail in Section 2.2 of the text. From the average magnitude of scattering, Rutherford estimated the number of protons (based on electrostatic interactions) in the nucleus.

CHAPTER 3
STOICHIOMETRY

PROBLEM-SOLVING STRATEGIES

TYPES OF PROBLEMS

Problem Type 1: Calculating Average Atomic Mass.

Problem Type 2: Calculations Involving Molar Mass of an Element and Avogadro's Number.
- **(a)** Converting between moles of atoms and mass of atoms.
- **(b)** Calculating the mass of a single atom.
- **(c)** Converting mass in grams to number of atoms.

Problem Type 3: Calculations Involving Molecular Mass.
- **(a)** Calculating molecular mass.
- **(b)** Calculating the number of moles in a given amount of a compound.
- **(c)** Calculating the number of atoms in a given amount of a compound.

Problem Type 4: Calculations Involving Percent Composition
- **(a)** Calculating percent composition of a compound.
- **(b)** Determining empirical formula from percent composition.
- **(c)** Calculating mass from percent composition.

Problem Type 5: Experimental Determination of Empirical Formulas.

Problem Type 6: Determining the Molecular Formula of a Compound.

Problem Type 7: Calculating the Amounts of Reactants and Products.

Problem Type 8: Limiting Reagent Calculations.

Problem Type 9: Calculating the Percent Yield of a Reaction.

PROBLEM TYPE 1: CALCULATING AVERAGE ATOMIC MASS

The atomic mass you look up on a periodic table is an average atomic mass. The reason for this is that most naturally occurring elements have more than one isotope. The average atomic mass can be calculated as follows:

Step 1: Convert the percentage of each isotope to fractions. For example, an isotope that is 69.09 percent abundant becomes 69.09/100 = 0.6909.

Step 2: Multiply the mass of each isotope by its abundance and add them together.

average atomic mass = (fraction of isotope A)(mass of isotope A) + (fraction of isotope B) (mass of isotope B) + . . . + (fraction of isotope Z)(mass of isotope Z).

EXAMPLE 3.1

The element lithium has two isotopes that occur in nature: $^{6}_{3}\mathrm{Li}$ with 7.5 percent abundance and $^{7}_{3}\mathrm{Li}$ with 92.5 percent abundance. The atomic mass of $^{6}_{3}\mathrm{Li}$ is 6.01513 amu and that of $^{7}_{3}\mathrm{Li}$ is 7.01601 amu. Calculate the average atomic mass of lithium.

Step 1: Convert the percentage of each isotope to fractions.

$$^{6}_{3}\mathrm{Li}: 7.5/100 = 0.075$$

$$^{7}_{3}\mathrm{Li}: 92.5/100 = 0.925$$

Step 2: Multiply the mass of each isotope by its abundance and add them together.

$$\textbf{average atomic mass} = (0.075)(6.01513) + (0.925)(7.01601) = \textbf{6.94 amu}$$

PRACTICE EXERCISE

1. The element boron (B) consists of two stable isotopes with atomic masses of 10.0129 amu and 11.0093 amu. The average atomic mass of B is 10.81 amu. Which isotope is more abundant?

Text Problem: 3.10

PROBLEM TYPE 2: CALCULATIONS INVOLVING MOLAR MASS OF AN ELEMENT AND AVOGADRO'S NUMBER

A. Converting between moles of atoms and mass of atoms

In order to convert from one unit to another, you need to be proficient at the factor-label method. See Section 1.7 of your text and Problem Type 5, chapter 1. Unit conversions can seem daunting, but if you keep track of the units, making sure that the appropriate units cancel, your effort will be rewarded.

Step 1: Map out a strategy to proceed from initial units to final units based on available conversion factors.

Step 2: Use the following method to ensure that you obtain the desired unit.

$$\text{Given unit} \times \left(\frac{\text{desired unit}}{\text{given unit}}\right) = \text{desired unit}$$

To convert between moles and mass, you need to use the molar mass of the element as a conversion factor.

$$\text{mol} \times \frac{\text{g}}{\text{mol}} = \text{g}$$

Also, going in the opposite direction

$$\text{g} \times \frac{\text{mol}}{\text{g}} = \text{mol}$$

Tip: Whether you are converting from g → mol or from mol → g, you will need to use the molar mass as the conversion factor. The molar mass of an element can be found directly on the periodic table.

EXAMPLE 3.2

How many grams are there in 0.130 mole of Cu?

Step 1: The conversion factor needed is the molar mass of Cu.

$$\frac{63.55 \text{ g Cu}}{1 \text{ mol Cu}}$$

You should come up with the following strategy.

moles → grams

Step 2:

$$\textbf{? g Cu} = 0.130 \text{ mol Cu} \times \frac{63.55 \text{ g Cu}}{1 \text{ mol Cu}} = \textbf{8.26 g Cu}$$

PRACTICE EXERCISE

2. How many moles of Cu are in 125 g of Cu?

Text Problem: 3.18

B. Calculating the mass of a single atom

To calculate the mass of a single atom, you must use Avogadro's number. The conversion factor is

$$\frac{1 \text{ mol}}{6.022 \times 10^{23} \text{ atoms}}$$

EXAMPLE 3.3

Copper is a minor component of pennies minted since 1981, and it is also used in electrical cables. Calculate the mass (in grams) of a single Cu atom.

Step 1: One conversion factor needed is given above. Another necessary conversion factor is the molar mass of Cu.

$$\frac{63.55 \text{ g Cu}}{1 \text{ mol Cu}}$$

You should come up with the following strategy.

grams/mol of Cu atoms → grams/single Cu atom

Step 2:

$$\textbf{? g/Cu atom} = \frac{63.55 \text{ g Cu}}{1 \text{ mol Cu}} \times \frac{1 \text{ mol Cu}}{6.022 \times 10^{23} \text{ Cu atoms}}$$

$$= \mathbf{1.055 \times 10^{-22}} \textbf{ g/Cu atom}$$

PRACTICE EXERCISE

3. Titanium (Ti) is a transition metal with a very high strength-to-weight ratio. For this reason, titanium is used in the construction of aircraft. What is the mass (in grams) of one Ti atom?

Text Problems: 3.20, 3.24

C. Converting mass in grams to number of atoms

To complete the following conversion, you need to use both molar mass and Avogadro's number as conversion factors.

EXAMPLE 3.4
Zinc is the main component of pennies minted after 1981. How many zinc atoms are present in 20.0 g of Zn?

Step 1: The two conversion factors needed are:

$$\frac{1 \text{ mol Zn}}{65.39 \text{ g Zn}} \qquad \frac{6.022 \times 10^{23} \text{ Zn atoms}}{1 \text{ mol Zn}}$$

You should come up with the following strategy.

grams of Zn → moles of Zn → atoms of Zn

Step 2:

$$\textbf{? atoms of Zn} = 20.0 \text{ g Zn} \times \frac{1 \text{ mol Zn}}{65.39 \text{ g Zn}} \times \frac{6.022 \times 10^{23} \text{ Zn atoms}}{1 \text{ mol Zn}}$$

$$= \mathbf{1.84 \times 10^{23}} \textbf{ Zn atoms}$$

PRACTICE EXERCISE

4. What is the mass (in grams) of 9.09×10^{23} atoms of Zn?

Text Problem: 3.22

PROBLEM TYPE 3: CALCULATIONS INVOLVING MOLECULAR MASS

A. Calculating molecular mass

The molecular mass is simply the sum of the atomic masses (in amu) of all the atoms in the molecule.

EXAMPLE 3.5
Calculate the molecular mass of carbon tetrachloride (CCl_4).

molecular mass CCl_4 = (mass of C) + 4(mass of Cl)

molecular mass CCl_4 = (12.01 amu) + 4(35.45 amu) = **153.8 amu**

PRACTICE EXERCISE

5. Bananas owe their characteristic smell and flavor to the ester, isopentyl acetate [$CH_3COOCH_2CH_2CH(CH_3)_2$]. Calculate the molecular mass of isopentyl acetate.

Text Problem: 3.26

B. Calculating the number of moles in a given amount of a compound

To complete this conversion, the only conversion factor needed is the molar mass in units of g/mol. Remember, the molar mass of a compound (in grams) is numerically equal to its molecular mass (in atomic mass units). For example, the molar mass of CCl_4 is 153.8 g/mol, compared to its molecular mass of 153.8 amu.

EXAMPLE 3.6

How many moles of ethane (C_2H_6) are present in 50.3 g of ethane?

Step 1: First, the molar mass of ethane must be calculated in order to have the appropriate conversion factor for the problem.

$$\text{molar mass of } C_2H_6 = 2(12.01\text{ g}) + 6(1.008\text{ g}) = 30.07\text{ g}$$

Hence, the conversion factor is

$$\frac{1\text{ mol } C_2H_6}{30.07\text{ g } C_2H_6}$$

Step 2: Based on the conversion factor above, you should come up with the following strategy.

$$\text{grams } C_2H_6 \rightarrow \text{moles of } C_2H_6$$

Step 3:

$$\textbf{? moles of } \mathbf{C_2H_6} = 50.3\cancel{\text{g}}\,C_2H_6 \times \frac{1\text{ mol } C_2H_6}{30.07\cancel{\text{g}}\,C_2H_6} = \mathbf{1.67\text{ moles}}$$

PRACTICE EXERCISE

6. What is the mass (in grams) of 0.436 moles of ethane (C_2H_6)?

Text Problem: 3.28

C. Calculating the number of atoms in a given amount of a compound

Again, this is a unit conversion problem. This calculation is more difficult than the conversions above, because you must convert from *grams of compound* to *moles of compound* to *moles of a particular atom* to *number of atoms*. Sound tough? Let's try an example.

EXAMPLE 3.7

How many carbon atoms are present in 50.3 g of ethane (C_2H_6)?

We started this problem in Example 3.6 when we calculated the moles of ethane in 50.3 g ethane. To continue, we need two additional conversion factors. One should represent the mole ratio between moles of C atoms and moles of ethane molecules. The other conversion factor needed is Avogadro's number.

Step 1: The two conversion factors needed are:

$$\frac{2\text{ mol C}}{1\text{ mol } C_2H_6} \qquad \frac{6.022 \times 10^{23}\text{ C atoms}}{1\text{ mol C}}$$

You should come up with the following strategy.

$$\text{grams of } C_2H_6 \rightarrow \text{moles of } C_2H_6 \rightarrow \text{moles of C} \rightarrow \text{atoms of C}$$

Step 2:

$$\textbf{? C atoms} = 50.3\cancel{\text{g}}\,C_2H_6 \times \frac{1\cancel{\text{mol}}\,C_2H_6}{30.07\cancel{\text{g}}\,C_2H_6} \times \frac{2\cancel{\text{mol}}\text{ C}}{1\cancel{\text{mol}}\,C_2H_6} \times \frac{6.022 \times 10^{23}\text{ C atoms}}{1\cancel{\text{mol}}\text{ C}}$$

$$= \mathbf{2.01 \times 10^{24}\text{ C atoms}}$$

PRACTICE EXERCISE

7. Glucose, the sugar used by the cells of our bodies for energy, has the molecular formula, $C_6H_{12}O_6$. How many atoms of *carbon* are present in a 3.50-g sample of glucose?

Text Problem: 3.30

PROBLEM TYPE 4: CALCULATIONS INVOLVING PERCENT COMPOSITION

A. Calculating percent composition of a compound

The *percent composition by mass* is the percent by mass of each element the compound contains. Percent composition is obtained by dividing the mass of each element in 1 mole of the compound by the molar mass of the compound, then multiplying by 100 percent.

$$\textbf{percent by mass of each element} = \frac{\text{mass of element in 1 mol of compound}}{\text{molar mass of compound}} \times 100\%$$

EXAMPLE 3.8
Calculate the percent composition by mass of all the elements in sodium bicarbonate, $NaHCO_3$.

First, calculate the molar mass of sodium bicarbonate.

$$\text{molar mass sodium bicarbonate} = 22.99\text{ g} + 1.008\text{ g} + 12.01\text{ g} + 3(16.00\text{ g}) = 84.01\text{ g}$$

$$\%\text{Na} = \frac{22.99\text{ g}}{84.01\text{ g}} \times 100\% = \mathbf{27.37\%}$$

$$\%\text{H} = \frac{1.008\text{ g}}{84.01\text{ g}} \times 100\% = \mathbf{1.200\%}$$

$$\%\text{C} = \frac{12.01\text{ g}}{84.01\text{ g}} \times 100\% = \mathbf{14.30\%}$$

$$\%\text{O} = \frac{3(16.00\text{ g})}{84.01\text{ g}} \times 100\% = \mathbf{57.14\%}$$

Tip: You can check your work by making sure that the mass percents of all the elements added together equals 100%. Checking above, 27.37% + 1.200% + 14.30% + 57.14% = 100.01% ≈ 100%.

PRACTICE EXERCISE

8. Cinnamic alcohol is used mainly in perfumes, particularly for soaps and cosmetics. Its molecular formula is $C_9H_{10}O$. Calculate the percent composition by mass of *hydrogen* in cinnamic alcohol.

Text Problems: 3.42, 3.46

B. Determining empirical formula from percent composition

The procedure used above to calculate the percent composition of a compound can be reversed. Given the percent composition by mass of a compound, you can determine the empirical formula of the compound.

EXAMPLE 3.9
Dieldrin, like DDT, is an insecticide that contains only C, H, Cl, and O. It is composed of 37.84 percent C, 2.12 percent H, 55.84 percent Cl, and 4.20 percent O. Determine its empirical formula.

Step 1: Assume you have exactly 100 g of substance. 100 g is a convenient amount, because all the percentages sum to 100 percent. In 100 g of dieldrin there will be 37.84 g C, 2.12 g H, 55.84 g Cl, and 4.20 g O.

Step 2: Calculate the number of moles of each element in 100 g of the compound. Remember, an *empirical formula* tells us which elements are present and the simplest whole-number ratio of their atoms. This ratio is also a mole ratio. Let n_C, n_H, n_{Cl}, and n_O be the number of moles of elements present. Use the molar masses of these elements as conversion factors to convert to moles

$$n_C = 37.84\ \text{g C} \times \frac{1 \text{ mol C}}{12.01\ \text{g C}} = \mathbf{3.151\ mol\ C}$$

$$n_H = 2.12\ \text{g H} \times \frac{1 \text{ mol H}}{1.008\ \text{g H}} = \mathbf{2.10\ mol\ H}$$

$$n_{Cl} = 55.84\ \text{g Cl} \times \frac{1 \text{ mol Cl}}{35.45\ \text{g Cl}} = \mathbf{1.575\ mol\ Cl}$$

$$n_O = 4.20\ \text{g O} \times \frac{1 \text{ mol O}}{16.00\ \text{g O}} = \mathbf{0.263\ mol\ O}$$

Thus, we arrive at the formula $C_{3.151}H_{2.10}Cl_{1.575}O_{0.263}$, which gives the identity and the ratios of atoms present. However, chemical formulas are written with whole numbers.

Step 3: Try to convert to whole numbers by dividing all the subscripts by the smallest subscript.

$$\textbf{C: } \frac{3.151}{0.263} = 12.0 \qquad \textbf{H: } \frac{2.10}{0.263} = 7.98 \approx 8 \qquad \textbf{Cl: } \frac{1.575}{0.263} = 5.99 \approx 6 \qquad \textbf{O: } \frac{0.263}{0.263} = 1$$

This gives us the empirical for dieldrin, $\mathbf{C_{12}H_8Cl_6O}$.

> **Tip:** It's not always this easy. Step 3 often does not give all whole numbers. If this is the case, you must multiply all the subscripts by some *integer* to come up with whole number subscripts. Try the practice exercise below.

PRACTICE EXERCISE

9. The substance responsible for the green color on the yolk of a boiled egg is composed of 53.58 percent Fe and 46.42 percent S. Determine its empirical formula.

Text Problems: 3.42, 3.52, **3.54**

C. Calculating mass from percent composition

Step 1: Convert the mass percentage to a fraction. For example, if the mass percent of an element in a compound were 54.73 percent, you would convert this to 54.73/100 = 0.5473.

Step 2: Multiply the fraction by the total mass of the compound. This gives the mass of the particular element in the compound.

EXAMPLE 3.10

Calculate the mass of carbon in exactly 10 g of glucose ($C_6H_{12}O_6$).

Step 1: Calculate the percentage by mass of carbon in glucose, then convert the percentage to a fraction.

$$\text{mass \% C} = \frac{\text{mass of C in 1 mol of glucose}}{\text{molar mass of glucose}} \times 100\%$$

$$\textbf{mass \% C} = \frac{6(12.01)\text{ g}}{180.16\text{ g}} \times 100\% = \mathbf{40.00\%\ C}$$

Converting this percentage to a fraction, we obtain 40.00/100 = **0.4000**.

Step 2: Multiply the fraction by the total mass of the compound.

$$\textbf{? g C in 10 g glucose} = (0.4000)(10\text{ g}) = \textbf{4.000 g C}$$

PRACTICE EXERCISE

10. Calculate the mass of hydrogen in exactly 10 grams of glucose ($C_6H_{12}O_6$).

Text Problem: 3.50

PROBLEM TYPE 5: EXPERIMENTAL DETERMINATION OF EMPIRICAL FORMULAS

See Section 3.6 of your text for a description of the experimental setup. To solve this type of problem, you must recognize that all of the carbon in the sample is converted to CO_2 and all the hydrogen in the sample is converted to H_2O. Then, you can calculate the mass of C in CO_2 and the mass of H in H_2O. Finally, you can calculate the mass of oxygen by difference, if necessary.

EXAMPLE 3.11

When a 0.761-g sample of a compound containing only carbon and hydrogen is burned in an apparatus with CO_2 and H_2O absorbers, 2.23 g CO_2 and 1.37 g H_2O are collected. Determine the empirical formula of the compound.

Step 1: Calculate the mass of C in 2.23 g CO_2, and the mass of H in 1.37 g H_2O. This is a factor-label problem. To calculate the mass of each component, you need the molar masses and the correct mole ratio.

You should come up with the following strategy.

$$\text{g } CO_2 \rightarrow \text{mol } CO_2 \rightarrow \text{mol C} \rightarrow \text{g C}$$

Step 2:

$$?\text{ g C} = 2.23\text{ g } CO_2 \times \frac{1\text{ mol } CO_2}{44.01\text{ g } CO_2} \times \frac{1\text{ mol C}}{1\text{ mol } CO_2} \times \frac{12.01\text{ g C}}{1\text{ mol C}} = 0.609\text{ g C}$$

Similarly,

$$?\text{ g H} = 1.37\text{ g } H_2O \times \frac{1\text{ mol } H_2O}{18.02\text{ g } H_2O} \times \frac{2\text{ mol H}}{1\text{ mol } H_2O} \times \frac{1.008\text{ g H}}{1\text{ mol H}} = 0.153\text{ g H}$$

Step 3: Calculate the number of moles of each element present in the 0.761 g sample. This is also a factor-label problem, using the molar mass as a conversion factor. You should come up with the following strategy. See Problem Type 2A.

$$\text{g C} \rightarrow \text{mol C}$$

$$\textbf{? mol C} = 0.609\text{ g C} \times \frac{1\text{ mol C}}{12.01\text{ g C}} = \textbf{0.0507 mol C}$$

Similarly,

$$\textbf{? mol H} = 0.153\text{ g H} \times \frac{1\text{ mol H}}{1.008\text{ g H}} = \textbf{0.152 mol H}$$

Thus, we arrive at the formula $C_{0.0507}H_{0.152}$, which gives the identity and the ratios of atoms present. However, chemical formulas are written with whole numbers.

Step 4: Try to convert to whole numbers by dividing all the subscripts by the smallest subscript.

$$\textbf{C:}\ \frac{0.0507}{0.0507} = 1.00 \qquad \textbf{H:}\ \frac{0.152}{0.0507} = 3.00$$

This gives us the empirical formula, $\mathbf{CH_3}$.

PRACTICE EXERCISE

11. Diethyl ether, commonly known as "ether", was used as an anesthetic for many years. Diethyl ether contains C, H, and O. When a 1.45-g sample of ether is burned in an apparatus such as that shown in Figure 3.5 of the text, 2.77 g of CO_2 and 1.70 g of H_2O are collected. Determine the empirical formula of diethyl ether.

PROBLEM TYPE 6: DETERMINING THE MOLECULAR FORMULA OF A COMPOUND

To determine the molecular formula of a compound, we must know both the *approximate* molar mass and the empirical formula of the compound. The molecular formula will either be equal to the empirical formula or be some integral multiple of it. Thus, the molar mass divided by the empirical mass will be an integer greater than or equal to one.

$$\frac{\text{molar mass}}{\text{empirical molar mass}} \geq 1 \text{ (integer values)}$$

EXAMPLE 3.12

A mass spectrum obtained on the compound in Example 3.11, shows its molecular mass to be about 31 g/mol. What is its molecular formula?

Step 1: Determine the empirical formula. This was already done in the previous example and was determined to be $\mathbf{CH_3}$.

Step 2: Calculate the empirical molar mass.

$$\text{empirical molar mass} = 12.01\text{ g} + 3(1.008\text{ g}) = 15.03\text{ g/mol}$$

Step 3: Determine the number of (CH_3) units present in the molecular formula. This number is found by taking the ratio

$$\frac{\text{molar mass}}{\text{empirical molar mass}} = \frac{31\text{ g}}{15.03\text{ g}} = 2.1 \approx 2$$

Thus, there are two CH_3 units in each molecule of the compound, so the molecular formula is $(CH_3)_2$, or $\mathbf{C_2H_6}$.

PRACTICE EXERCISE

12. In Example 3.9, the empirical formula of dieldrin was determined to be $C_{12}H_8Cl_6O$. If the molar mass of dieldrin is 381 ± 10 g/mol, what is the molecular formula of dieldrin?

Text Problem: 3.54

PROBLEM TYPE 7: CALCULATING THE AMOUNTS OF REACTANTS AND PRODUCTS

These types of problems are factor-label problems. You must always remember to start this type of problem with a balanced chemical equation. The typical approach is given below. See Section 3.8 of your text for a step-by-step method.

Amount of reactant (expressed in grams, kg, etc.) → Moles of reactant → Moles of product → Amount of product (expressed in grams, kg, etc.)

> **Tip:** Always try to be flexible when solving problems. Most problems of this type will follow an approach similar to the one above, but you may have to modify it sometimes.

EXAMPLE 3.13

Sulfur dioxide can be removed from stack gases by reaction with quicklime (CaO):

$$\mathbf{SO_2\,(g) + CaO\,(s) \longrightarrow CaSO_3\,(s)}$$

If 975 kg of SO_2 are to be removed from stack gases by the above reaction, how many kilograms of CaO are required?

Step 1: Start with a balanced chemical equation. It's given in the problem.

$$SO_2\,(g) + CaO\,(s) \longrightarrow CaSO_3\,(s)$$

Step 2: Map out the following strategy to solve this problem.

$$\text{kg } SO_2 \rightarrow \text{g } SO_2 \rightarrow \text{moles } SO_2 \rightarrow \text{moles CaO} \rightarrow \text{g CaO} \rightarrow \text{kg CaO}$$

Step 3: Calculate the mass of CaO in kg using the strategy above.

$$\textbf{? kg CaO} = 975\text{ kg } SO_2 \times \frac{1000\text{ g } SO_2}{1\text{ kg } SO_2} \times \frac{1\text{ mol } SO_2}{64.07\text{ g } SO_2} \times \frac{1\text{ mol CaO}}{1\text{ mol } SO_2} \times \frac{56.08\text{ g CaO}}{1\text{ mol CaO}} \times \frac{1\text{ kg CaO}}{1000\text{ g CaO}}$$

$$= \mathbf{853\text{ kg CaO}}$$

> **Tip:** Notice that the approach followed was a slight modification of the one given above. We went from mass of one reactant, to moles of that reactant, to moles of a second reactant, and finally to mass of second reactant.

PRACTICE EXERCISE

13. Carbon dioxide in the air of a spacecraft can be removed by its reaction with a lithium hydroxide solution.

$$CO_2\,(g) + 2\,LiOH\,(aq) \longrightarrow Li_2CO_3\,(aq) + H_2O\,(l)$$

On average, a person will exhale about 1 kg of CO_2/day. How many kilograms of LiOH are required to react with 1.0 kg of CO_2?

Text Problems: 3.66, 3.68, 3.70, 3.72, **3.74**, 3.76, 3.86

PROBLEM TYPE 8: LIMITING REAGENT CALCULATIONS

When a chemist carries out a reaction, the reactants are usually not present in exact **stoichiometric amounts**, that is, in the proportions indicated by the balanced equation. The reactant used up first in a reaction is called the **limiting reagent**. When this reactant is used up, no more product can be formed.

Typically, the only difference between this type of problem and Problem Type 7, Calculating the Amounts of Reactants and Products, is that you must first determine which reactant is the limiting reagent.

EXAMPLE 3.14

Phosphine (PH_3) burns in oxygen (O_2) to produce phosphorus pentoxide and water.

$$\mathbf{2\,PH_3\,(\mathit{g}) + 4\,O_2\,(\mathit{g}) \longrightarrow P_2O_5\,(\mathit{s}) + 3\,H_2O\,(\mathit{l})}$$

How many grams of P_2O_5 will be produced when 17.0 g of phosphine are reacted with 16.0 g of O_2?

Step 1: **Determine which reactant is the limiting reagent.** There are two different methods to determine the limiting reagent. One is given in Section 3.9 of your text. A second method involves treating each reactant separately. Calculate the amount of product formed if that reactant reacts completely, assuming the other reactant is in excess. Repeat the procedure for the second reactant. The reactant that produces the *least* amount of product is the *limiting reagent*; it *limits* the amount of product that can be formed.

You should follow the strategy below to calculate the amount of product formed

g reactant → mol reactant → mol product

Calculate moles of P_2O_5 formed if all PH_3 reacts.

$$\textbf{? mol P}_2\textbf{O}_5 = 17.0\,\text{g PH}_3 \times \frac{1\text{ mol PH}_3}{33.99\text{ g PH}_3} \times \frac{1\text{ mol P}_2\text{O}_5}{2\text{ mol PH}_3} = \textbf{0.250 mol P}_2\textbf{O}_5$$

Calculate moles of P_2O_5 formed if all O_2 reacts.

$$\textbf{? mol P}_2\textbf{O}_5 = 16.0\,\text{g O}_2 \times \frac{1\text{ mol O}_2}{32.0\text{ g O}_2} \times \frac{1\text{ mol P}_2\text{O}_5}{4\text{ mol O}_2} = \textbf{0.125 mol P}_2\textbf{O}_5$$

The initial amount of O_2 limits the amount of product that can be formed; therefore, it is the limiting reagent.

Step 2: The problem asks for grams of P_2O_5 produced. We already know the moles of P_2O_5 produced, 0.125 moles. Use the molar mass of P_2O_5 as a conversion factor to convert to grams.

$$\textbf{? g P}_2\textbf{O}_5 = 0.125\text{ mol P}_2\text{O}_5 \times \frac{141.94\text{ g P}_2\text{O}_5}{1\text{ mol P}_2\text{O}_5} = \textbf{17.7 g P}_2\textbf{O}_5$$

PRACTICE EXERCISE

14. Iron can be produced by reacting iron ore with carbon. The iron produced can then be used to make steel. The reaction is

$$2\,Fe_2O_3\,(s) + 3\,C\,(s) \xrightarrow{\text{heat}} 4\,Fe\,(l) + 3\,CO_2\,(g)$$

(a) How many grams of Fe can be produced from a mixture of 200.0 g of Fe_2O_3 and 300.0 g C?
(b) How many grams of excess reagent will remain after the reaction ceases?

Text Problems: 3.80, 3.82

PROBLEM TYPE 9: CALCULATING THE PERCENT YIELD OF A REACTION

The **theoretical yield** is the amount of product that would result if all the limiting reagent reacted. This is the maximum obtainable yield predicted by the balanced equation. However, the amount of product obtained is almost always less than the theoretical yield. The **actual yield** is the quantity of product that actually results from a reaction.

To determine the efficiency of a reaction, chemists often calculate the **percent yield**, which describes the proportion of the actual yield to the theoretical yield. The percent yield is calculated as follows:

$$\%\text{ yield} = \frac{\text{actual yield}}{\text{theoretical yield}} \times 100\%$$

EXAMPLE 3.15

In Example 3.14, the theoretical yield of P_2O_5 was determined to be 17.7 g. If only 12.6 g of P_2O_5 are actually obtained, what is the percent yield of the reaction?

$$\%\text{ yield} = \frac{\text{actual yield}}{\text{theoretical yield}} \times 100\%$$

$$\textbf{\% yield} = \frac{12.6\text{ g}}{17.7\text{ g}} \times 100\% = \textbf{71.2\%}$$

PRACTICE EXERCISE

15. Refer back to Practice Exercise 14 to answer this question. If the actual yield of Fe is 110 g, what is the percent yield of Fe?

Text Problems: 3.86, 3.88

ANSWERS TO PRACTICE EXERCISES

1. ^{11}B

2. 1.97 moles Cu

3. 7.951×10^{-23} g/Ti atom

4. 98.7 g Zn

5. 130.18 amu

6. 13.1 g ethane

7. 7.02×10^{22} C atoms

8. 7.513 percent H by mass

9. Fe_2S_3

10. 0.67 g H

11. C_2H_6O

12. $C_{12}H_8Cl_6O$

13. 1.1 kg LiOH

14. **(a)** 139.9 g Fe
(b) 277 g C

15. 78.6 percent yield

SOLUTIONS TO SELECTED TEXT PROBLEMS

3.10 This is a variation of Problem Type 1, Calculating Average Atomic Mass.

It would seem that there are two unknowns in this problem, the natural abundance of ^{6}Li and the natural abundance of ^{7}Li. However, these two quantities are not independent of each other; they are related by the fact that they must sum to 1. Start by letting x be the natural abundance of ^{6}Li. Since the sum of the two abundance's must be 1, we can write

$$\text{Abundance } ^7\text{Li} = (1 - x)$$

Remember,

average atomic mass = (fraction of isotope A)(mass of isotope A) + (fraction of isotope B)(mass of isotope B) + . . . + (fraction of isotope Z)(mass of isotope Z).

$$\textbf{Average atomic mass of Li} = 6.941\text{ amu} = x(6.0151\text{ amu}) + (1 - x)(7.0160\text{ amu})$$
$$6.941 = 1.0009x + 7.0160$$
$$1.0009x = 0.075$$
$$\boldsymbol{x = 0.075}$$

$x = 0.075$ corresponds to a natural abundance of ^{6}Li of **7.5 percent**. The natural abundance of ^{7}Li is $(1 - x) = 0.925$ or **92.5 percent**.

3.12 The thickness of the book in miles would be:

$$\frac{3.6\times10^{-3}\text{ in}}{1\text{ page}} \times \frac{1\text{ ft}}{12\text{ in}} \times \frac{1\text{ mi}}{5280\text{ ft}} \times 6.022\times10^{23}\text{ pages} = 3.42\times10^{16}\text{ mi}$$

The distance, in miles, traveled by light in one year is:

$$1.00\text{ yr} \times \frac{365\text{ day}}{1\text{ yr}} \times \frac{24\text{ h}}{1\text{ day}} \times \frac{3600\text{ s}}{1\text{ h}} \times \frac{3.00\times10^{8}\text{ m}}{1\text{ s}} \times \frac{1\text{ mi}}{1609\text{ m}} = \mathbf{5.88\times10^{12}\text{ mi}}$$

The thickness of the book in light-years is:

$$3.42\times10^{16}\text{ mi} \times \frac{1\text{ light-yr}}{5.88\times10^{12}\text{ mi}} = \mathbf{5.82\times10^{3}\text{ light-yr}}$$

It will take light 5.82×10^3 years to travel from the first page to the last one!

3.14 The unit factor required is $\left(\dfrac{6.022\times10^{23}\text{ amu}}{1\text{ g}}\right)$

$$\textbf{? amu} = (8.4\text{ g}) \times \left(\frac{6.022\times10^{23}\text{ amu}}{1\text{ g}}\right) = \mathbf{5.1\times10^{24}\text{ amu}}$$

3.16 $$(6.00\times10^{9}\text{ Co atoms}) \times \left(\frac{1\text{ mol Co}}{6.022\times10^{23}\text{ Co atoms}}\right) = \mathbf{9.96\times10^{-15}\text{ mol Co}}$$

3.18 Converting between moles of atoms and mass of atoms, Problem Type 2A.

Step 1: The conversion factor needed is the molar mass of Au.

$$\frac{197.0 \text{ g Au}}{1 \text{ mol Au}}$$

You should come up with the following strategy.

moles → grams

Step 2:

$$\textbf{? g Au} = 15.3 \text{ mol Au} \times \frac{197.0 \text{ g Au}}{1 \text{ mol Au}} = \mathbf{3.01 \times 10^3 \text{ g Au}}$$

3.20 Calculating the mass of a single atom, Problem Type 2B.

(a)

Step 1: One conversion factor needed is Avogadro's number. Another necessary conversion factor is the molar mass of As.

$$\frac{74.92 \text{ g As}}{1 \text{ mol As}}$$

You should come up with the following strategy.

grams/mole of As → grams/single As atom

Step 2:

$$\textbf{?g/As atom} = \frac{74.92 \text{ g As}}{1 \text{ mol As}} \times \frac{1 \text{ mol As}}{6.022 \times 10^{23} \text{ As atoms}} = \mathbf{1.244 \times 10^{-22} \text{ g/As atom}}$$

(b) Follow same method as part (a).

$$\textbf{? g/Ni atom} = \frac{58.69 \text{ g Ni}}{1 \text{ mol Ni}} \times \frac{1 \text{ mol Ni}}{6.022 \times 10^{23} \text{ Ni atoms}} = \mathbf{9.746 \times 10^{-23} \text{ g/Ni atom}}$$

3.22 Converting mass in grams to number of atoms, Problem Type 2a.

Step 1: The two conversion factors needed for this problem are the molar mass of Cu and Avogadro's number.

$$\frac{1 \text{ mol Cu}}{63.55 \text{ g Cu}} \qquad \frac{6.022 \times 10^{23} \text{ Cu atoms}}{1 \text{ mol Cu}}$$

You should come up with the following strategy.

grams of Cu → moles of Cu → atoms of Cu

Step 2:

$$\textbf{? atoms of Cu} = 3.14 \text{ g Cu} \times \frac{1 \text{ mol Cu}}{63.55 \text{ g Cu}} \times \frac{6.022 \times 10^{23} \text{ Cu atoms}}{1 \text{ mol Cu}} = \mathbf{2.98 \times 10^{22} \text{ Cu atoms}}$$

3.24 $$(2 \text{ Pb atoms}) \times \left(\frac{1 \text{ mol Pb}}{6.022\times10^{23} \text{ Pb atoms}}\right) \times \left(\frac{207.2 \text{ g Pb}}{1 \text{ mol Pb}}\right) = \mathbf{6.881\times10^{-22} \text{ g Pb}}$$

$$(5.1\times10^{-23} \text{ mol He})\left(\frac{4.003 \text{ g He}}{1 \text{ mol He}}\right) = \mathbf{2.0\times10^{-22} \text{ g He}}$$

2 atoms of lead has a greater mass than 5.1×10^{-23} mol of helium.

3.26 Calculating molar mass, modification of Problem Type 3A.

(a) **molar mass S_8** = 8(32.07 g) = **256.6 g**

(b) **molar mass CS_2** = 12.01 g + 2(32.07 g) = **76.15 g**

(c) **molar mass $CHCl_3$** = 12.01 g + 1.008 g + 3(35.45 g) = **119.37 g**

(d) **molar mass $C_6H_8O_6$** = 6(12.01 g) + 8(1.008 g) + 6(16.00 g) = **176.12 g**

3.28 Calculating the number of molecules in a given amount of compound, similar to Problem Type 3B.

Step 1: First, the molar mass of ethane must be calculated in order to have the appropriate conversion factor for the problem.

$$\text{molar mass of } C_2H_6 = 2(12.01 \text{ g}) + 6(1.008 \text{ g}) = 30.07 \text{ g}$$

Hence, the conversion factor is:

$$\frac{1 \text{ mol } C_2H_6}{30.07 \text{ g } C_2H_6}$$

Step 2: Based on the conversion factor above, you should come up with the following strategy.

$$\text{grams } C_2H_6 \rightarrow \text{moles of } C_2H_6$$

Step 3:

$$\textbf{? mol of } \mathbf{C_2H_6} = 0.334 \text{ g } C_2H_6 \times \frac{1 \text{ mol } C_2H_6}{30.07 \text{ g } C_2H_6} = \mathbf{0.0111 \text{ mol}}$$

3.30 Calculating the number of atoms in a given amount of a compound, Problem Type 3C.

This is a unit conversion problem. You must convert from *grams of compound* to *moles of compound* to *moles of a particular atom* to *number of atoms*.

Step 1: The conversion factors needed are:

$$\frac{1 \text{ mol urea}}{60.06 \text{ g urea}} \qquad \frac{2 \text{ mol N}}{1 \text{ mol urea}} \qquad \frac{6.022 \times 10^{23} \text{ N atoms}}{1 \text{ mol N}}$$

You should come up with the following strategy.

grams of urea → moles of urea → moles of N → atoms of N

Step 2:

$$\textbf{? atoms of N} = 1.68 \times 10^4 \text{ g urea} \times \frac{1 \text{ mol urea}}{60.06 \text{ g urea}} \times \frac{2 \text{ mol N}}{1 \text{ mol urea}} \times \frac{6.022 \times 10^{23} \text{ N atoms}}{1 \text{ mol N}}$$

$$= \mathbf{3.37 \times 10^{26} \text{ N atoms}}$$

We could calculate the number of atoms of the remaining elements in the same manner, or we can use the atom ratios from the molecular formula. The carbon atom to nitrogen atom ratio in a urea molecule is 1:2, the oxygen atom to nitrogen atom ratio is 1:2, and the hydrogen atom to nitrogen atom ration is 4:2.

$$\textbf{? atoms of C} = 3.37 \times 10^{26}\ \text{N atoms} \times \frac{1\ \text{C atom}}{2\ \text{N atoms}} = \mathbf{1.69 \times 10^{26}\ C\ atoms}$$

$$\textbf{? atoms of O} = 3.37 \times 10^{26}\ \text{N atoms} \times \frac{1\ \text{O atom}}{2\ \text{N atoms}} = \mathbf{1.69 \times 10^{26}\ O\ atoms}$$

$$\textbf{? atoms of H} = 3.37 \times 10^{26}\ \text{N atoms} \times \frac{4\ \text{H atoms}}{2\ \text{N atoms}} = \mathbf{6.74 \times 10^{26}\ H\ atoms}$$

3.32 Mass of water $= (2.56\ \text{mL})\left(\dfrac{1.00\ \text{g}}{1.00\ \text{mL}}\right) = 2.56\ \text{g}$

Molar mass of H_2O = (16.00 g) + 2(1.008 g) = 18.02 g/mol

$$\textbf{? H}_2\textbf{O molecules} = 2.56\ \text{g H}_2\text{O} \times \left(\frac{1\ \text{mol H}_2\text{O}}{18.02\ \text{g H}_2\text{O}}\right) \times \left(\frac{6.022 \times 10^{23}\ \text{molecules H}_2\text{O}}{1\ \text{mol H}_2\text{O}}\right)$$

$$= \mathbf{8.56 \times 10^{22}\ molecules}$$

3.36 Since there are two hydrogen isotopes, they can be paired in three ways: $^1H-^1H$, $^1H-^2H$, and $^2H-^2H$. There will then be three choices for each sulfur isotope. We can make a table showing all the possibilities (masses in amu):

	^{32}S	^{33}S	^{34}S	^{36}S
$^1H_2$	34	35	36	38
$^1H^2H$	35	36	37	39
$^2H_2$	36	37	38	40

There will be seven peaks of the following mass numbers: 34, 35, 36, 37, 38, 39, and 40.

Very accurate (and expensive!) mass spectrometers can detect the mass difference between two ^1H and one ^2H. How many peaks would be detected in such a "high resolution" mass spectrum?

3.42 Calculating the percent composition of a compound, Problem Type 4A.

$$\text{percent by mass of each element} = \frac{\text{mass of element in 1 mol of compound}}{\text{molar mass of compound}} \times 100\%$$

(a) First, calculate the molar mass of NaF.

molar mass of NaF = 22.99 g + 19.00 g = 41.99 g

$$\textbf{\%Na} = \frac{22.99\ \text{g}}{41.99\ \text{g}} \times 100\% = \mathbf{54.75\%}$$

$$\textbf{\%F} = \frac{19.00\ \text{g}}{41.99\ \text{g}} \times 100\% = \mathbf{45.25\%}$$

(b) Molar mass NaCl = 22.99 g + 35.45 g = 58.44 g

$$\textbf{\%Na} = \frac{22.99\text{ g}}{58.44\text{ g}} \times 100\% = \textbf{39.34\%}$$

$$\textbf{\%Cl} = \frac{35.45\text{ g}}{58.44\text{ g}} \times 100\% = \textbf{60.66\%}$$

(c) Molar mass NaBr = 22.99 g + 79.90 g = 102.89 g

$$\textbf{\%Na} = \frac{22.99\text{ g}}{102.89\text{ g}} \times 100\% = \textbf{22.34\%}$$

$$\textbf{\%Br} = \frac{79.90\text{ g}}{102.89\text{ g}} \times 100\% = \textbf{77.66\%}$$

(d) Molar mass NaI = 22.99 g + 126.9 g = 149.9 g

$$\textbf{\%Na} = \frac{22.99\text{ g}}{149.9\text{ g}} \times 100\% = \textbf{15.34\%}$$

$$\textbf{\%I} = \frac{126.9\text{ g}}{149.9\text{ g}} \times 100\% = \textbf{84.66\%}$$

3.44 Determining empirical formula from percent composition, Problem Type 4B.

Step 1: Assume you have exactly 100 g of substance. 100 g is a convenient amount, because all the percentages sum to 100%. In 100 g of allicin there will be 44.4 g C, 6.21 g H, 39.5 g S, and 9.86 g O.

Step 2: Calculate the number of moles of each element in 100 g of the compound. Remember, an <u>empirical formula</u> tells us which elements are present and the simplest whole-number ratio of their atoms. This ratio is also a mole ratio. Let n_C, n_H, n_S, and n_O be the number of moles of elements present in 100 g of compound. Use the molar masses of these elements as conversion factors to convert to moles.

$$n_C = 44.4\text{ g C} \times \frac{1\text{ mol C}}{12.01\text{ g C}} = \textbf{3.70 mol C}$$

$$n_H = 6.21\text{ g H} \times \frac{1\text{ mol H}}{1.008\text{ g H}} = \textbf{6.16 mol H}$$

$$n_S = 39.5\text{ g S} \times \frac{1\text{ mol S}}{32.07\text{ g S}} = \textbf{1.23 mol S}$$

$$n_O = 9.86\text{ g O} \times \frac{1\text{ mol O}}{16.00\text{ g O}} = \textbf{0.616 mol O}$$

Thus, we arrive at the formula $C_{3.70}H_{6.16}S_{1.23}O_{0.616}$, which gives the identity and the ratios of atoms present. However, chemical formulas are written with whole numbers.

Step 3: Try to convert to whole numbers by dividing all the subscripts by the smallest subscript.

C: $\frac{3.70}{0.616} = 6.01 \cong 6$ **H:** $\frac{6.16}{0.616} = 10.0$ **S:** $\frac{1.23}{0.616} = 2.00$ **O:** $\frac{0.616}{0.616} = 1$

This gives us the empirical for allicin, $\mathbf{C_6H_{10}S_2O}$.

To determine the molecular formula, remember that the molar mass/empirical mass will be an integer greater than or equal to one.

$$\frac{\text{molar mass}}{\text{empirical molar mass}} \geq 1 \text{ (integer values)}$$

In this case,

$$\frac{\text{molar mass}}{\text{empirical molar mass}} = \frac{162\text{ g}}{162.3\text{ g}} \cong 1$$

Hence, the molecular formula and the empirical formula are the same, $\mathbf{C_6H_{10}S_2O}$.

3.46

	Compound	Molar mass (g)	N% by mass
(a)	$(NH_2)_2CO$	60.06	$\frac{2(14.01\text{ g})}{60.06\text{ g}} \times 100\% = 46.65\%$
(b)	NH_4NO_3	80.05	$\frac{2(14.01\text{ g})}{80.05\text{ g}} \times 100\% = 35.00\%$
(c)	$HNC(NH_2)_2$	59.08	$\frac{3(14.01\text{ g})}{59.08\text{ g}} \times 100\% = 71.14\%$
(d)	NH_3	17.03	$\frac{14.01\text{ g}}{17.03\text{ g}} \times 100\% = 82.27\%$

Ammonia, NH_3, is the richest source of nitrogen on a mass percentage basis.

3.48 Using unit factors we convert:

g of Hg → mol Hg → mol S → g S

$$?\text{ g S} = 246\text{ g Hg} \times \left(\frac{1\text{ mol Hg}}{200.6\text{ g Hg}}\right) \times \left(\frac{1\text{ mol S}}{1\text{ mol Hg}}\right) \times \left(\frac{32.07\text{ g S}}{1\text{ mol S}}\right) = \mathbf{39.3\text{ g S}}$$

3.50 Calculating mass from percent composition, Problem Type 4C.

Step 1: Calculate the percentage by mass of fluorine in tin(II) fluoride, then convert the percentage to a fraction.

$$\text{mass \% F} = \frac{\text{mass of F in 1 mol SnF}_2}{\text{molar mass of SnF}_2} \times 100\%$$

$$= \frac{2(19.00\text{ g})}{156.7\text{ g}} \times 100\% = 24.25\%\text{ F}$$

Converting this percentage to a fraction, we obtain 24.25/100 = 0.2425.

Step 2: Multiply the fraction by the total mass of the compound.

$$\mathbf{?\text{ g F in 24.6 g SnF}_2} = (0.2425)(24.6\text{ g}) = \mathbf{5.97\text{ g F}}$$

Note: This problem could have been worked in a manner similar to Problem 3.48. You could complete the following conversions:

g of SnF_2 → mol of SnF_2 → mol of F → g of F

3.52 It is convenient to assume that you have 100 g of sample.

The percentage of oxygen is found by difference:

$$100\% - (19.8\% + 2.50\% + 11.6\%) = \mathbf{66.1\%}$$

$$\text{Moles C} = 19.8\,\cancel{g}\,C\left(\frac{1\ \text{mol C}}{12.01\,\cancel{g}\,C}\right) = 1.65\ \text{mol C}$$

$$\text{Moles H} = 2.50\,\cancel{g}\,H\left(\frac{1\ \text{mol H}}{1.008\,\cancel{g}\,H}\right) = 2.48\ \text{mol H}$$

$$\text{Moles N} = 11.6\,\cancel{g}\,N\left(\frac{1\ \text{mol N}}{14.01\,\cancel{g}\,N}\right) = 0.828\ \text{mol N}$$

$$\text{Moles O} = 66.1\,\cancel{g}\,O\left(\frac{1\ \text{mol O}}{16.00\,\cancel{g}\,O}\right) = 4.13\ \text{mol O}$$

The formula is $C_{1.65}H_{2.48}N_{0.828}O_{4.13}$. Dividing the subscripts by 0.828 gives the empirical formula, $\mathbf{C_2H_3NO_5}$. Would you get a different answer if you had assumed an amount of sample different than 100 g?

3.54 Determining empirical formula from percent composition, Problem Type 4B.

Step 1: Assume you have exactly 100 g of substance. 100 g is a convenient amount, because all the percentages sum to 100%. In 100 g of MSG there will be 35.51 g C, 4.77 g H, 37.85 g O, 8.29 g N, and 13.60 g Na.

Step 2: Calculate the number of moles of each element in the compound. Remember, an *empirical formula* tells us which elements are present and the simplest whole-number ratio of their atoms. This ratio is also a mole ratio. Let n_C, n_H, n_O, n_N, and n_{Na} be the number of moles of elements present. Use the molar masses of these elements as conversion factors to convert to moles.

$$n_C = 35.51\,\cancel{g}\,C \times \frac{1\ \text{mol C}}{12.01\,\cancel{g}\,C} = \mathbf{2.957\ mol\ C}$$

$$n_H = 4.77\,\cancel{g}\,H \times \frac{1\ \text{mol H}}{1.008\,\cancel{g}\,H} = \mathbf{4.73\ mol\ H}$$

$$n_O = 37.85\,\cancel{g}\,O \times \frac{1\ \text{mol O}}{16.00\,\cancel{g}\,O} = \mathbf{2.366\ mol\ O}$$

$$n_N = 8.29\,\cancel{g}\,N \times \frac{1\ \text{mol N}}{14.01\,\cancel{g}\,N} = \mathbf{0.592\ mol\ N}$$

$$n_{Na} = 13.60\,\cancel{g}\,Na \times \frac{1\ \text{mol Na}}{22.99\,\cancel{g}\,Na} = \mathbf{0.5916\ mol\ Na}$$

Thus, we arrive at the formula $C_{2.957}H_{4.73}O_{2.366}N_{0.592}Na_{0.5916}$, which gives the identity and the ratios of atoms present. However, chemical formulas are written with whole numbers.

Step 3: Try to convert to whole numbers by dividing all the subscripts by the smallest subscript.

$$\textbf{C:}\ \frac{2.957}{0.5916} = 4.998 \approx 5 \qquad \textbf{H:}\ \frac{4.73}{0.5916} = 8.00 \qquad \textbf{O:}\ \frac{2.366}{0.5916} = 3.999 \approx 4$$

$$\textbf{N:}\ \frac{0.592}{0.5916} = 1.00 \qquad \textbf{Na:}\ \frac{0.5916}{0.5916} = 1$$

This gives us the empirical formula for MSG, $\mathbf{C_5H_8O_4NNa}$.

To determine the molecular formula, remember that the molar mass/empirical mass will be an integer greater than or equal to one.

$$\frac{\text{molar mass}}{\text{empirical molar mass}} \geq 1 \text{ (integer values)}$$

In this case,

$$\frac{\text{molar mass}}{\text{empirical molar mass}} = \frac{169 \text{ g}}{169.11 \text{ g}} \approx 1$$

Hence, the molecular formula and the empirical formula are the same, $\mathbf{C_5H_8O_4NNa}$. It should come as no surprise that the empirical and molecular formulas are the same since MSG stands for *monosodium*glutamate.

3.60 The balanced equations are as follows:

(a) $2KClO_3 \rightarrow 2KCl + 3O_2$

(b) $2KNO_3 \rightarrow 2KNO_2 + O_2$

(c) $NH_4NO_3 \rightarrow N_2O + 2H_2O$

(d) $NH_4NO_2 \rightarrow N_2 + 2H_2O$

(e) $2NaHCO_3 \rightarrow Na_2CO_3 + H_2O + CO_2$

(f) $P_4O_{10} + 6H_2O \rightarrow 4H_3PO_4$

(g) $2HCl + CaCO_3 \rightarrow CaCl_2 + H_2O + CO_2$

(h) $2Al + 3H_2SO_4 \rightarrow Al_2(SO_4)_3 + 3H_2$

(i) $CO_2 + 2KOH \rightarrow K_2CO_3 + H_2O$

(j) $CH_4 + 2O_2 \rightarrow CO_2 + 2H_2O$

3.66 Calculating the Amounts of Reactants and Products, Problem Type 7.

Step 1: Start with a balanced chemical equation. It's given in the problem.

$$Si\ (s) + 2\ Cl_2\ (g) \longrightarrow SiCl_4\ (l)$$

Step 2: Map out the following strategy to solve this problem.

$$\text{moles } SiCl_4 \rightarrow \text{mol } Cl_2$$

Step 3: Calculate the moles of Cl_2 using the stoichiometric ratio from the balanced equation.

$$\mathbf{?\ mol\ Cl_2} = 0.507 \text{ mol } SiCl_4 \times \frac{2 \text{ mol } Cl_2}{1 \text{ mol } SiCl_4} = \mathbf{1.01\ mol\ Cl_2}$$

3.68 Molar mass $NaHCO_3$ = 22.99 g + 1.008 g + 12.01 g + 3(16.00 g) = 84.01 g
Molar mass CO_2 = 12.01 g + 2(16.00 g) = 44.01 g

(a) $2\ NaHCO_3 \longrightarrow Na_2CO_3 + H_2O + CO_2$

(b) The balanced equation shows one mole of CO_2 formed from two moles of $NaHCO_3$.

$$\mathbf{mass\ NaHCO_3} = 20.5 \text{ g } CO_2 \left(\frac{1 \text{ mol } CO_2}{44.01 \text{ g } CO_2}\right)\left(\frac{2 \text{ mol } NaHCO_3}{1 \text{ mol } CO_2}\right)\left(\frac{84.01 \text{ g } NaHCO_3}{1 \text{ mol } NaHCO_3}\right)$$

$$= \mathbf{78.3\ g\ NaHCO_3}$$

3.70 The grams of ethanol that can be obtained are:

$$\mathbf{?\ g\ C_2H_5OH} = 500.4 \text{ g } C_6H_{12}O_6 \times \frac{1 \text{ mol } C_6H_{12}O_6}{180.16 \text{ g } C_6H_{12}O_6} \times \frac{2 \text{ mol } C_2H_5OH}{1 \text{ mol } C_6H_{12}O_6} \times \frac{46.07 \text{ g } C_2H_5OH}{1 \text{ mol } C_2H_5OH}$$

$$= \mathbf{255.9\ g\ C_2H_5OH}$$

The liters of ethanol can be calculated from the density and the mass of ethanol.

$$\text{volume} = \frac{\text{mass}}{\text{density}}$$

$$\textbf{Volume of ethanol obtained} = \frac{255.9\ \text{g}}{0.789\ \text{g/mL}} = 324\ \text{mL} = \mathbf{0.324\ L}$$

3.72 The balanced equation shows that eight moles of KCN are needed to combine with four moles of Au.

$$\textbf{? mol KCN} = 29.0\ \text{g Au} \times \left(\frac{1\ \text{mol Au}}{197.0\ \text{g Au}}\right) \times \left(\frac{8\ \text{mol KCN}}{4\ \text{mol Au}}\right) = \mathbf{0.294\ mol\ KCN}$$

3.74 Calculating the Amounts of Reactants and Products, Problem Type 7.

(a) $NH_4NO_3\,(s) \longrightarrow N_2O\,(g) + 2\,H_2O\,(g)$

(b)
Step 1: Map out the following strategy to solve this problem.

$$\text{mol } NH_4NO_3 \rightarrow \text{mol } N_2O \rightarrow \text{g } N_2O$$

Step 2: The molar mass of nitrous oxide is needed as a conversion factor.

$$\frac{44.02\ \text{g } N_2O}{1\ \text{mol } N_2O}$$

Step 3: Calculate the grams of N_2O using the strategy in step 1.

$$\textbf{? g } \mathbf{N_2O} = 0.46\ \text{mol } NH_4NO_3 \times \frac{1\ \text{mol } N_2O}{1\ \text{mol } NH_4NO_3} \times \frac{44.02\ \text{g } N_2O}{1\ \text{mol } N_2O} = \mathbf{2.0 \times 10^1\ g\ N_2O}$$

3.76 The balanced equation for the decomposition is :

$$2\,KClO_3\,(s) \longrightarrow 2\,KCl\,(s) + 3\,O_2\,(g)$$

$$\textbf{? g } \mathbf{O_2} = 46.0\ \text{g } KClO_3 \times \frac{1\ \text{mol } KClO_3}{122.6\ \text{g } KClO_3} \times \frac{3\ \text{mol } O_2}{2\ \text{mol } KClO_3} \times \frac{32.00\ \text{g } O_2}{1\ \text{mol } O_2} = \mathbf{18.0\ g\ O_2}$$

3.80 Limiting Reagent Calculation, Problem Type 8.

Step 1: Determine which reactant is the limiting reagent.

You should follow the strategy below to calculate the amount of product formed.

$$\text{g reactant} \rightarrow \text{mol reactant} \rightarrow \text{mol product}$$

Calculate moles of NO_2 formed if all O_3 reacts.

$$\text{? mol } NO_2 = 0.740\ \text{g } O_3 \times \frac{1\ \text{mol } O_3}{48.00\ \text{g } O_3} \times \frac{1\ \text{mol } NO_2}{1\ \text{mol } O_3} = 0.0154\ \text{mol } NO_2$$

Calculate moles of NO_2 formed if all NO reacts.

$$\text{? mol } NO_2 = 0.670\ \text{g NO} \times \frac{1\ \text{mol NO}}{30.01\ \text{g NO}} \times \frac{1\ \text{mol } NO_2}{1\ \text{mol NO}} = 0.0223\ \text{mol } NO_2$$

The initial amount of $\mathbf{O_3}$ limits the amount of product that can be formed; therefore, it is the **limiting reagent**.

Step 2: The problem asks for grams of NO_2 produced. We already know the moles of NO_2 produced, 0.0154 moles. Use the molar mass of NO_2 as a conversion factor to convert to grams.

$$\textbf{? mol NO}_2 = 0.0154 \text{ mol NO}_2 \times \frac{46.01 \text{ g NO}_2}{1 \text{ mol NO}_2} = \mathbf{0.709 \text{ g NO}_2}$$

Step 3: The number of moles of NO (excess reagent) remaining at the end of the reaction is

mol NO remaining = mol NO initial − mol NO reacted.

$$\text{mol NO initial} = 0.670 \text{ g NO} \times \frac{1 \text{ mol NO}}{30.01 \text{ g NO}} = 0.0223 \text{ mol NO}$$

$$\text{mol NO reacted} = 0.740 \text{ g } O_3 \times \frac{1 \text{ mol } O_3}{48.00 \text{ g } O_3} \times \frac{1 \text{ mol NO}}{1 \text{ mol } O_3} = 0.0154 \text{ mol NO}$$

or,

$$\text{mol NO reacted} = \text{mol NO}_2 \text{ formed} = 0.0154 \text{ mol NO}$$

$$\textbf{mol NO remaining} = 0.0223 \text{ mol NO} - 0.0154 \text{ mol NO} = \mathbf{0.0069 \text{ mol NO}}$$

3.82 As in all limiting reagent problems, we first must ascertain which reagent is limiting.

(a) We first convert the mass of HCl given to moles

$$48.2 \text{ g HCl}\left(\frac{1 \text{ mol HCl}}{36.46 \text{ g HCl}}\right) = 1.32 \text{ mol HCl}$$

Since one mole of MnO_2 reacts with four moles of HCl, the number of moles of HCl needed to combine with 0.86 mole of MnO_2 is

$$0.86 \text{ mol MnO}_2\left(\frac{4 \text{ mol HCl}}{1 \text{ mol MnO}_2}\right) = 3.4 \text{ mol HCl}$$

Since we do not have that much HCl present, **HCl** must be the **limiting reagent** and MnO_2 the excess reagent.

(b) The mass of Cl_2 formed is then:

$$\textbf{? g Cl}_2 = 1.32 \text{ mol HCl} \times \frac{1 \text{ mol Cl}_2}{4 \text{ mol HCl}} \times \frac{70.90 \text{ g Cl}_2}{1 \text{ mol Cl}_2} = \mathbf{23.4 \text{ g Cl}_2}$$

3.86 This is a combination of Problem Type 7, Calculating the Amounts of Reactants and Products, and Problem Type 9, Calculating the Percent Yield of a Reaction.

(a)
Step 1: Start with a balanced chemical equation. It's given in the problem. We use NG as an abbreviation for nitroglycerin. The molar mass of NG = 227.1 g/mol.

$$4\, C_3H_5N_3O_9 \longrightarrow 6\, N_2 + 12\, CO_2 + 10\, H_2O + O_2$$

Step 2: Map out the following strategy to solve this problem.

$$\text{g NG} \rightarrow \text{mol NG} \rightarrow \text{mol } O_2 \rightarrow \text{g } O_2$$

Step 3: Calculate the grams of O_2 using the strategy above.

$$? \text{ g } O_2 = 2.00 \times 10^2 \text{ g NG} \times \frac{1 \text{ mol NG}}{227.1 \text{ g NG}} \times \frac{1 \text{ mol } O_2}{4 \text{ mol NG}} \times \frac{32.00 \text{ g } O_2}{1 \text{ mol } O_2}$$

$$= \mathbf{7.05 \text{ g } O_2}$$

(b) Calculate the percent yield

$$\% \text{ yield} = \frac{\text{actual yield}}{\text{theoretical yield}} \times 100\%$$

$$\mathbf{\% \text{ yield}} = \frac{6.55 \text{ g } O_2}{7.05 \text{ g } O_2} \times 100\% = \mathbf{92.9\%}$$

3.88 The actual yield of ethylene is 481 g. Let's calculate the yield of ethylene if the reaction is 100 percent efficient. We can calculate this from the definition of percent yield. We can then calculate the mass of hexane that must be reacted.

$$\% \text{ yield} = \frac{\text{actual yield}}{\text{theoretical yield}} \times 100\%$$

$$42.5 \% \text{ yield} = \frac{481 \text{ g } C_2H_4}{\text{theoretical yield}} \times 100\%$$

$$\text{theoretical yield } C_2H_4 = 1.13 \times 10^3 \text{ g } C_2H_4$$

The mass of hexane that must be reacted is:

$$1.13 \times 10^3 \text{ g } C_2H_4 \times \frac{1 \text{ mol } C_2H_4}{28.05 \text{ g } C_2H_4} \times \frac{1 \text{ mol } C_6H_{14}}{1 \text{ mol } C_2H_4} \times \frac{86.15 \text{ g } C_6H_{14}}{1 \text{ mol } C_6H_{14}} = \mathbf{3.47 \times 10^3 \text{ g } C_6H_{14}}$$

3.90 This is a calculation involving percent composition. Remember,

$$\textbf{percent by mass of each element} = \frac{\text{mass of element in 1 mol of compound}}{\text{molar mass of compound}} \times 100\%$$

The molar masses are: Al, 26.98 g/mol; $Al_2(SO_4)_3$, 342.2 g/mol; H_2O, 18.02 g/mol. Thus, using x as the number of H_2O molecules,

$$\text{mass \% Al} = \left(\frac{2(\text{molar mass of Al})}{\text{molar mass of } Al_2(SO_4)_3 + x(\text{molar mass of } H_2O)}\right) \times 100\%$$

$$8.20\% = \left(\frac{2(26.98 \text{ g})}{342.2 \text{ g} + x(18.02 \text{ g})}\right) \times 100\%$$

$$\boldsymbol{x = 17.53}$$

Rounding off to a whole number of water molecules, $\boldsymbol{x = 18}$. Therefore the formula is $\mathbf{Al_2(SO_4)_3 \cdot 18\, H_2O}$.

3.92 The mass of oxygen in MO is 39.46 g – 31.70 g = 7.76 g O. Therefore, for every 31.70 g of M, there is 7.76 g of O in the compound MO. The molecular formula shows a mole ratio of 1 mol M : 1 mol O. First, calculate moles of M that react with 7.76 g O.

$$\text{mol M} = 7.76 \text{ g O} \times \frac{1 \text{ mol O}}{16.00 \text{ g O}} \times \frac{1 \text{ mol M}}{1 \text{ mol O}} = 0.485 \text{ mol M}$$

$$\textbf{molar mass M} = \frac{31.70 \text{ g M}}{0.485 \text{ mol M}} = \textbf{65.4 g/mol}$$

Thus, the atomic mass of M is **65.4 amu**. The metal is most likely **Zn**.

3.94 The wording of the problem suggests that the actual yield is less than the theoretical yield. The percent yield will be equal to the percent purity of the iron(III) oxide. We find the theoretical yield :

$$2.62 \times 10^3 \text{ kg } Fe_2O_3 \left(\frac{1000 \text{ g } Fe_2O_3}{1 \text{ kg } Fe_2O_3}\right)\left(\frac{1 \text{ mol } Fe_2O_3}{159.7 \text{ g } Fe_2O_3}\right)\left(\frac{2 \text{ mol Fe}}{1 \text{ mol } Fe_2O_3}\right)\left(\frac{55.85 \text{ g Fe}}{1 \text{ mol Fe}}\right)\left(\frac{1 \text{ kg Fe}}{1000 \text{ g Fe}}\right)$$

$$= \mathbf{1.83 \times 10^3 \text{ kg Fe}}$$

$$\text{percent yield} = \frac{\text{actual yield}}{\text{theoretical yield}} \times 100\%$$

$$\textbf{percent yield} = \frac{1.64 \times 10^3 \text{ kg Fe}}{1.83 \times 10^3 \text{ kg Fe}} \times 100\% = \mathbf{89.6\%} = \textbf{purity of } \mathbf{Fe_2O_3}$$

3.96 The carbohydrate contains 40 percent carbon; therefore, the remaining 60 percent is hydrogen and oxygen. The problem states that the hydrogen to oxygen ratio is 2:1. We can write this 2:1 ratio as H_2O.

Assume 100 g of compound.

$$40.0 \text{ g C} \times \frac{1 \text{ mol C}}{12.01 \text{ g C}} = 3.33 \text{ mol C}$$

$$60.0 \text{ g } H_2O \times \frac{1 \text{ mol } H_2O}{18.02 \text{ g } H_2O} = 3.33 \text{ mol } H_2O$$

Dividing by 3.33 gives $\mathbf{CH_2O}$ for the empirical formula.

To find the molecular formula, divide the molar mass by the empirical mass.

$$\frac{\text{molar mass}}{\text{empirical mass}} = \frac{178 \text{ g}}{30.03 \text{ g}} \cong 6$$

Thus, there are six CH_2O units in each molecule of the compound, so the molecular formula is $(CH_2O)_6$, or $\mathbf{C_6H_{12}O_6}$.

3.98 **(a)** The following strategy can be used to convert from the volume of the Mg cube to the number of Mg atoms.

$$\text{cm}^3 \rightarrow \text{grams} \rightarrow \text{moles} \rightarrow \text{atoms}$$

$$1.0 \text{ cm}^3 \times \frac{1.74 \text{ g Mg}}{1 \text{ cm}^3} \times \frac{1 \text{ mol Mg}}{24.31 \text{ g Mg}} \times \frac{6.022 \times 10^{23} \text{ Mg atoms}}{1 \text{ mol Mg}} = \mathbf{4.3 \times 10^{22} \textbf{ Mg atoms}}$$

(b) Since 74 percent of the available space is taken up by Mg atoms, 4.3×10^{22} atoms occupy the following volume:

$$0.74 \times 1.0 \text{ cm}^3 = 0.74 \text{ cm}^3$$

We are trying to calculate the radius of a single Mg atom, so we need the volume occupied by a single Mg atom.

$$\text{volume Mg atom} = \frac{0.74 \text{ cm}^3}{4.3 \times 10^{22} \text{ Mg atoms}} = 1.7 \times 10^{-23} \text{ cm}^3/\text{Mg atom}$$

The volume of a sphere is $(4/3)\pi r^3$. Solving for the radius:

$$V = 1.7 \times 10^{-23} \text{ cm}^3 = \frac{4}{3}\pi r^3$$

$$r^3 = 4.1 \times 10^{-24} \text{ cm}^3$$

$$r = 1.6 \times 10^{-8} \text{ cm}$$

Converting to picometers:

$$\textbf{radius Mg atom} = 1.6 \times 10^{-8} \cancel{\text{cm}} \times \frac{0.01 \cancel{\text{m}}}{1 \cancel{\text{cm}}} \times \frac{1 \text{ pm}}{1 \times 10^{-12} \cancel{\text{m}}} = \mathbf{1.6 \times 10^2 \text{ pm}}$$

3.100 The surface area of the water can be calculated assuming that the dish is circular.

$$\text{surface area of water} = \pi r^2 = \pi(10 \text{ cm})^2 = 3.1 \times 10^2 \text{ cm}^2$$

The cross-sectional area of one stearic acid molecule in cm^2 is:

$$0.21 \cancel{\text{nm}^2} \times \left(\frac{1 \times 10^{-9} \cancel{\text{m}}}{1 \cancel{\text{nm}}}\right)^2 \times \left(\frac{1 \text{ cm}}{0.01 \cancel{\text{m}}}\right)^2 = 2.1 \times 10^{-15} \text{ cm}^2/\text{molecule}$$

Assuming that there is no empty space between molecules, we can calculate the number of stearic acid molecules that will fit in an area of $3.1 \times 10^2 \text{ cm}^2$.

$$3.1 \times 10^2 \cancel{\text{cm}^2} \times \frac{1 \text{ molecule}}{2.1 \times 10^{-15} \cancel{\text{cm}^2}} = 1.5 \times 10^{17} \text{ molecules}$$

Next, we can calculate the moles of stearic acid in the 1.4×10^{-4} g sample. Then, we can calculate Avogadro's number – the number of molecules per mole.

$$1.4 \times 10^{-4} \cancel{\text{g}} \text{ stearic acid} \times \frac{1 \text{ mol stearic acid}}{284.5 \cancel{\text{g}} \text{ stearic acid}} = 4.9 \times 10^{-7} \text{ mol stearic acid.}$$

$$\textbf{Avogadro's number } (N_A) = \frac{1.5 \times 10^{17} \text{ molecules}}{4.9 \times 10^{-7} \text{ mol}} = \mathbf{3.1 \times 10^{23} \text{ molecules/mol}}$$

3.102 **(a)** The observations mean either that the amount of the more abundant isotope was increasing or the amount of the less abundant isotope was decreasing. One possible explanation is that the less abundant isotope was undergoing radioactive decay, and thus it's mass would decrease with time.

(b) 16 amu, CH_4 17 amu, NH_3 18 amu, H_2O 64 amu, SO_2

(c) The formula C_3H_8 can also be written as $CH_3CH_2CH_3$. A CH_3 fragment could break off from this molecule giving a peak at 15 amu. No fragment of CO_2 can have a mass of 15 amu. Therefore, the substance responsible for the mass spectrum is most likely C_3H_8.

(d) First, let's calculate the masses of CO_2 and C_3H_8.

molecular mass CO_2 = 12.00000 amu + 2(15.99491 amu) = 43.98982 amu

molecular mass C_3H_8 = 3(12.00000 amu) + 8(1.00797 amu) = 44.06376 amu

These masses differ by only 0.07394 amu. The measurements must be precise to ±0.030 amu.

43.98982 + 0.030 amu = 44.02 amu

44.06376 – 0.030 amu = 44.03 amu

(e) To analyze a gold sample, scientists first heat a tiny spot (about 0.01 cm in diameter and in depth) of the sample with a high-power laser. The vaporized gold and its trace elements are swept by a stream of argon gas into a mass spectrometer. Comparison of the mass spectrum with a library of mass spectra of gold from known origins will identify the source of the gold, much as fingerprints are used to identify individuals. This technique can be used on large objects like bullion and nuggets as well as on small articles of jewelry.

CHAPTER 4
REACTIONS IN AQUEOUS SOLUTIONS

PROBLEM-SOLVING STRATEGIES AND TUTORIAL SOLUTIONS

TYPES OF PROBLEMS

Problem Type 1: Applying Solubility Rules.

Problem Type 2: Writing Molecular, Ionic, and Net Ionic Equations.

Problem Type 3: Acid-Base Reactions.
(a) Identifying Brønsted acids and bases.
(b) Writing acid/base reactions.

Problem Type 4: Oxidation-Reduction Reactions.
(a) Assigning oxidation numbers.
(b) Writing oxidation/reduction half-reactions.
(c) Using an activity series.

Problem Type 5: Concentration of Solutions.

Problem Type 6: Dilution of Solutions.

Problem Type 7: Gravimetric Analysis.

Problem Type 8: Acid-Base Titrations.

PROBLEM TYPE 1: APPLYING SOLUBILITY RULES

Ionic compounds are classified as "soluble", "slightly soluble", or "insoluble". Table 4.2 of your text provides solubility rules that will help you determine how a given compound behaves in aqueous solution.

EXAMPLE 4.1

According to the solubility rules, which of the following compounds are soluble in water?
(a) $MgCO_3$ (b) $AgNO_3$ (c) $MgCl_2$ (d) $Ca_3(PO_4)_2$ (e) KOH

Refer to Table 4.2 of your text to answer this question.

(a) $MgCO_3$ is *insoluble* (Rule 6).
(b) $AgNO_3$ is *soluble* (Rule 3).
(c) $MgCl_2$ is *soluble* (Rule 5).
(d) $Ca_3(PO_4)_2$ is *insoluble* (Rule 6).
(e) KOH is *soluble* (Rule 1).

PRACTICE EXERCISE

1. Predict whether the following ionic compounds are soluble or insoluble in water.

(a) $NaNO_3$ **(b)** AgCl **(c)** $Ba(OH)_2$ **(d)** $CaCO_3$

Text Problems: 4.14, 4.18

PROBLEM TYPE 2: WRITING MOLECULAR, IONIC, AND NET IONIC EQUATIONS

In a *molecular equation*, the formulas are written as though all species existed as molecules or whole units. However, a molecular equation does not accurately describe what actually happens at the microscopic level. To better describe the reaction in solution, the equation should show the dissociation of dissolved ionic compounds into ions. An *ionic equation* shows dissolved ionic compounds in terms of their free ions. A *net ionic equation* shows only the species that actually take part in the reaction.

EXAMPLE 4.2

Write balanced molecular, ionic, and net ionic equations for the reaction that occurs when a $BaCl_2$ solution is mixed with a Na_2SO_4 solution.

Both ionic solids given in the problem are soluble, because the problem states that both ionic solids are in solution. You could also check the solubility rules. The combined solution will contain Na^+, SO_4^{2-}, Ba^{2+}, and Cl^- ions. The two possible ionic products are NaCl and $BaSO_4$. If one (or both) is insoluble, it (they) will precipitate. According to Table 4.2, NaCl is soluble. On the other hand, $BaSO_4$ is insoluble. The *molecular equation* should show $BaSO_4$ (*s*) as a product along with NaCl (*aq*).

$$BaCl_2\,(aq) + Na_2SO_4\,(aq) \longrightarrow BaSO_4\,(s) + 2\,NaCl\,(aq)$$

The *ionic equation* should show dissolved ionic compounds in terms of their free ions.

$$Ba^{2+}\,(aq) + 2\,Cl^-\,(aq) + 2\,Na^+\,(aq) + SO_4^{2-}\,(aq) \longrightarrow BaSO_4\,(s) + 2\,Na^+\,(aq) + 2\,Cl^-\,(aq)$$

As you write out the ionic equation above, you should notice that some ions (Na^+ and Cl^-) are not involved in the overall reaction. These ions are called *spectator ions*. Since the spectator ions appear on both sides of the equation and are unchanged in the chemical reaction, they can be canceled from both sides of the equation. A *net ionic equation* shows only the species that actually take part in the reaction.

Cancel the spectator ions to write the *net ionic equation*.

$$Ba^{2+}\,(aq) + SO_4^{2-}\,(aq) \longrightarrow BaSO_4\,(s)$$

> **Tip:** To help pick out the spectator ions, think about spectators at a sporting event. The spectators are at the stadium, watching the action, but they do *not* participate in the game.

PRACTICE EXERCISE

2. Write the balanced molecular, ionic, and net ionic equations for the following reaction:

$$CaCl_2\,(aq) + Na_2CO_3\,(aq) \longrightarrow$$

Text Problem: 4.16

PROBLEM TYPE 3: ACID-BASE REACTIONS

A. Identifying Brønsted acids and bases

A **Brønsted acid** is a proton donor, and a **Brønsted base** is a proton acceptor. To identify a Brønsted acid, you should look for a substance that contains hydrogen. The formula of inorganic acids will begin with H. For example, HCl (hydrochloric acid), HNO_2 (nitrous acid), H_3PO_4 (phosphoric acid). Carboxylic acids contain the carboxyl group, –COOH. The hydrogen from the carboxyl group can be donated. Examples of carboxylic acids are CH_3COOH (acetic acid) and HCOOH (formic acid).

(b) $HCOOH\ (aq) + NaOH\ (aq) \longrightarrow HCOONa\ (aq) + H_2O\ (l)$

Formic acid, HCOOH (*aq*), is a weak acid and therefore it only ionizes to a small extent in water. In solution, most formic acid molecules remain as HCOOH. NaOH is a strong base and therefore it completely ionizes in solution. The balanced *ionic equation* is:

$$HCOOH\ (aq) + Na^+\ (aq) + OH^-\ (aq) \longrightarrow Na^+\ (aq) + HCOO^-\ (aq) + H_2O\ (l)$$

Cancel the spectator ion (Na^+) to write the *net ionic equation.*

$$HCOOH\ (aq) + OH^-\ (aq) \longrightarrow HCOO^-\ (aq) + H_2O\ (l)$$

PRACTICE EXERCISE

4. Write the balanced molecular, ionic, and net ionic equations for the following acid-base reaction:

$$HNO_2\ (aq) + Ba(OH)_2\ (aq) \longrightarrow$$

Text Problem: 4.28

PROBLEM TYPE 4: OXIDATION-REDUCTION REACTIONS

A. Assigning oxidation numbers

Oxidation numbers are assigned to reactants and products in oxidation-reduction (redox) reactions to keep track of electrons. An oxidation number refers to the number of charges an atom would have in a molecule (or an ionic compound) if electrons were transferred completely.

Rules for assigning oxidation numbers are in Section 4.4 of your text. These rules will be used in the following example.

To assign oxidation numbers you should refer to the following *two* steps:

Step 1: Use the rules in Section 4.4 to assign oxidation numbers to as many atoms as possible.

Step 2: Often times, one atom does not follow any rules outlined in Section 4.4. To assign an oxidation number to this atom, follow rule 6. In a neutral molecule, the sum of the oxidation numbers of all the atoms must be zero. In a polyatomic ion, the sum of the oxidation numbers of all the elements in the ion must be equal to the net charge of the ion.

EXAMPLE 4.5

Assign oxidation numbers to all the atoms in the following compounds and ion:

(a) Na_2SO_4, (b) CuCl, (c) SO_3^{2-}

(a)

Step 1: Na always has an oxidation number of +1 (Rule 2). The oxidation number of oxygen in most compounds is −2 (Rule 3).

Step 2: You can now assign an oxidation number to S based on Na having a +1 oxidation number and O having a −2 oxidation number. This is a neutral ionic compound, so the sum of the oxidation numbers of all the atoms must be zero.

$$2(\text{oxi. no. Na}) + (\text{oxi. no. S}) + 4(\text{oxi. no. O}) = 0$$

$$2(+1) + (\text{oxi. no. S}) + 4(-2) = 0$$

$$\mathbf{(oxi.\ no.\ S)} = 8 - 2 = \mathbf{+6}$$

(b)
Step 1: An oxidation number of −1 can be assigned to Cl (Rule 5).

Step 2: You can now assign an oxidation number to Cu. This is a neutral ionic compound.

$$\text{(oxi. no. Cu)} + \text{(oxi. no. Cl)} = 0$$
$$\text{(oxi. no. Cu)} + (-1) = 0$$
$$\textbf{(oxi. no. Cu)} = \mathbf{+1}$$

(c)
Step 1: An oxidation number of −2 can be assigned to oxygen (Rule 3).

Step 2: You can now assign an oxidation number to S. SO_3^{2-} is a polyatomic ion. The sum of the oxidation numbers of all elements in the ion must be equal to the net charge of the ion, in this case −2.

$$\text{(oxi. no. S)} + 3\text{(oxi. no. O)} = -2$$
$$\text{(oxi. no. S)} + 3(-2) = -2$$
$$\textbf{(oxi. no. S)} = -2 + 6 = \mathbf{+4}$$

PRACTICE EXERCISE

5. Assign oxidation numbers to the underlined atoms in the following molecules or ions:

(a) $\underline{C}O_3^{2-}$ **(b)** $\underline{Cu}Cl_2$ **(c)** $\underline{Ti}O_2$ **(d)** $\underline{N}O_3^-$

Text Problems: 4.36, 4.38, 4.40, 4.42

B. Writing oxidation-reduction half-reactions

In order to break a redox reaction down into an oxidation half-reaction and a reduction half-reaction, you must first assign oxidation numbers to all the atoms in the reaction. In this way, you can determine which element is oxidized (loses electrons) and which element is reduced (gains electrons).

EXAMPLE 4.6
For the following redox reaction, break down the reaction into its half-reactions.

$$\mathbf{2\,Al + Fe_2O_3 \longrightarrow Al_2O_3 + 2\,Fe}$$

Reactants
Step 1: The oxidation number of Al is 0 (Rule 1), and the oxidation number of O in a compound is −2 (Rule 3).

Step 2: Solve for the oxidation number of Fe in Fe_2O_3. This is a neutral ionic compound.

$$2\text{(oxi. no. Fe)} + 3\text{(oxi. no. O)} = 0$$
$$2\text{(oxi. no. Fe)} + 3(-2) = 0$$
$$2\text{(oxi. no. Fe)} = +6$$
$$\textbf{(oxi. no. Fe)} = \mathbf{+3}$$

Products
Step 1: The oxidation number of Fe is 0 (Rule 1), and the oxidation number of O in a compound is −2 (Rule 3).

Step 2: Solve for the oxidation number of Al in Al_2O_3. This is a neutral ionic compound.

$$2\text{(oxi. no. Al)} + 3\text{(oxi. no. O)} = 0$$
$$2\text{(oxi. no. Al)} + 3(-2) = 0$$
$$2\text{(oxi. no. Al)} = +6$$
$$\textbf{(oxi. no. Al)} = \mathbf{+3}$$

$$Al \longrightarrow Al^{3+} + 3\,e^- \quad \text{(oxidation half-reaction, 3 } e^- \text{ lost)}$$

$$Fe^{3+} + 3\,e^- \longrightarrow Fe \quad \text{(reduction half-reaction, 3 } e^- \text{ gained)}$$

Tip: When a species is oxidized, the oxidation number will *increase*. In this example, the oxidation number of Al *increased* from 0 to +3. When a species is reduced, the oxidation number will *decrease*. In this example, the oxidation number of Fe *decreased* from +3 to 0.

PRACTICE EXERCISE

6. The nickel-cadmium (nicad) battery, a popular rechargeable "dry cell" used in battery-operated tools, uses the following redox reaction to generate electricity:

$$Cd\,(s) + NiO_2\,(s) + 2\,H_2O\,(l) \longrightarrow Cd(OH)_2\,(s) + Ni(OH)_2\,(s)$$

Assign oxidation numbers to all the atoms and ions, identify the substances that are oxidized and reduced, and write oxidation and reduction half-reactions.

Text Problem: 4.32

C. Using an activity series

An activity series is used to predict whether a metal or hydrogen displacement reaction will occur (See Figure 4.11). An activity series can be described as a convenient summary of the results of many possible displacement reactions.

(1) Hydrogen displacement. Any metal above hydrogen in the activity series will displace it from water or from an acid. Metals below hydrogen will *not* react with either water or an acid.

(2) Metal displacement. Any metal will react with a compound containing any metal ion listed below it.

EXAMPLE 4.7
Predict the outcome of the reactions represented by the following equations by using the activity series, and balance the equations.

(a) $Mg\,(s) + HCl\,(aq) \longrightarrow$

Since Mg is above hydrogen in the activity series, it will displace hydrogen from the acid.

$$Mg\,(s) + 2\,HCl\,(aq) \longrightarrow MgCl_2\,(aq) + H_2\,(g)$$

(b) $Au\,(s) + H_2O\,(l) \longrightarrow$

Since Au (gold) is below hydrogen in the activity series, it will *not* react with water. You probably already knew this.

$$Au\,(s) + H_2O\,(l) \longrightarrow \text{No reaction}$$

(c) $Cu\,(s) + NiCl_2\,(aq) \longrightarrow$

Any metal will react with a compound containing any metal ion listed *below* it. In this case Ni^{2+} is listed *above* Cu in the activity series. No reaction will occur.

$$Cu\,(s) + NiCl_2\,(aq) \longrightarrow \text{No reaction}$$

PRACTICE EXERCISE

7. Predict the outcome of the following reactions using the activity series. If a reaction occurs, balance the equation.

(a) $Al\,(s) + HCl\,(aq) \longrightarrow$
(b) $Au\,(s) + KCl\,(aq) \longrightarrow$

Text Problem: 4.44

PROBLEM TYPE 5: CONCENTRATION OF SOLUTIONS

Solutions are characterized by their concentration, that is, the amount of solute dissolved in a given quantity of solvent. One of the most common units of concentration in chemistry is **molarity (*M*)**. Molarity is the number of moles of solute in 1 liter of solution:

$$M = \text{molarity} = \frac{\text{moles of solute}}{\text{liters of solution}}$$

Sometimes, it is useful to rearrange the above equation to the following form:

$$\text{moles solute} = \text{molarity} \times \text{liters of solution}$$

EXAMPLE 4.8

What is the molarity of a solution made by dissolving 32.1 g of KNO_3 in enough water to make 500 mL of solution?

Step 1: Based on the definition of molarity, grams of solute must be converted to moles of solute, and mL of solution must be converted to L of solution. For discussion of unit conversions, see Problem Type 5, chapter 1.

$\mathcal{M}(KNO_3) = 101.1$ g/mol

$$?\text{ moles solute} = 32.1\,\cancel{\text{g}}\,KNO_3 \times \frac{1\text{ mol }KNO_3}{101.1\,\cancel{\text{g}}\,KNO_3} = 0.318\text{ mol }KNO_3$$

$$?\text{ liters of solution} = 500\,\cancel{\text{mL}}\text{ solution} \times \frac{1\text{ L}}{1000\,\cancel{\text{mL}}} = 0.500\text{ L solution}$$

Step 2: Substitute the above values into the molarity equation.

$$M = \frac{\text{moles of solute}}{\text{liters of solution}}$$

$$M = \frac{0.318\text{ mol }KNO_3}{0.500\text{ L solution}} = \mathbf{0.636}\ \boldsymbol{M}$$

This is normally written 0.636 M KNO_3.

PRACTICE EXERCISE

8. An aqueous nutrient solution is prepared by adding 50.23 g of KNO_3 (molar mass = 101.1 g/mol) to enough water to fill a 40.0 L container. What is the molarity of the KNO_3 solution?

EXAMPLE 4.9

How many moles of solute are in 2.50×10^2 mL of 0.100 M KCl?

Step 1: Since the problem asks for moles of solute, you must solve the equation algebraically for moles of solute.

$$\text{moles solute} = \text{molarity} \times \text{liters of solution}$$

Step 2: Substitute the molarity and liters of solution into the above equation to solve for moles solute.

$$2.50 \times 10^2 \text{ mL} = 0.250 \text{ L}$$

$$\textbf{? moles KCl solute} = \frac{0.100 \text{ moles solute}}{1 \not{L} \text{ solution}} \times 0.250 \not{L} \text{ solution} = \textbf{0.0250 mol}$$

PRACTICE EXERCISE

9. You need to prepare 1.00 L of a 0.500 *M* NaCl solution. What mass of NaCl (in g) must you weigh out to prepare this solution?

Text Problems: 4.48, 4.50, 4.52, 4.54

PROBLEM TYPE 6: DILUTION OF SOLUTIONS

Dilution refers to the procedure for preparing a less-concentrated solution from a more-concentrated one. The key to solving a dilution problem is to realize that

moles of solute *before* dilution = moles of solute *after* dilution

In Problem Type 5 above, we discussed how to calculate moles of solute from the molarity and the volume of solution.

moles solute = molarity × volume of solution (in L)

Thus,

moles of solute *before* dilution (initial) = moles of solute *after* dilution (final)

$$M_{\text{initial}} V_{\text{initial}} = M_{\text{final}} V_{\text{final}}$$

EXAMPLE 4.10

What volume of a concentrated (12.0 *M*) hydrochloric acid stock solution is needed to prepare 8.00×10^2 mL of 0.120 *M* HCl?

Recognize that the problem asks for the initial volume of stock solution needed to prepare the dilute solution. Solve the above equation algebraically for V_{initial}, then substitute in the appropriate values from the problem.

$$V_{\text{initial}} = \frac{M_{\text{final}} V_{\text{final}}}{M_{\text{initial}}}$$

$$\mathbf{V_{initial}} = \frac{(0.120 \not{M})(8.00 \times 10^2 \text{ mL})}{12.0 \not{M}} = \textbf{8.00 mL}$$

Tip: The units of V_{initial} and V_{final} can be milliliters or liters for a dilution problem as long as they are the same. Be consistent. Also, make sure to check whether your results seem reasonable. Be sure that $M_{\text{initial}} > M_{\text{final}}$ and $V_{\text{final}} > V_{\text{initial}}$.

PRACTICE EXERCISE

10. A 20.0 mL sample of 0.127 *M* $Ca(NO_3)_2$ is diluted to 5.00 L. What is the molarity of the resulting solution?

Text Problems: 4.58, 4.60, 4.62

PROBLEM TYPE 7: GRAVIMETRIC ANALYSIS

Gravimetric analysis is an analytical technique based on the measurement of mass. The type of gravimetric analysis discussed in your text involves the formation, isolation, and mass determination of a precipitate. This procedure is applicable only to reactions that go to completion, or have nearly a 100 percent yield. Thus, the precipitate must be insoluble rather than slightly soluble. See Section 4.6 of your text for further discussion of this technique.

The typical problem involves determining the mass percent of an element in one of the reactants. The element (ion) of interest is completely precipitated from solution. Use the following approach to solve this type of problem.

Step 1: From the measured mass of precipitate, calculate the mass of the element of interest in the precipitate. This is the same amount of the element that was present in the original sample.

Step 2: Calculate the mass percent of the element of interest in the original sample. See Problem Type 4A, Chapter 3.

> **Tip:** Try to be flexible when solving problems. Some gravimetric analysis problems may ask you to calculate the *molar concentration* of the component of interest in the sample, rather than mass percent. For this type of problem, you must modify your approach by converting grams of the component of interest to moles, then dividing by the volume of solution in liters.

EXAMPLE 4.11

A 0.7469 g sample of an ionic compound containing Pb ions is dissolved in water and treated with excess Na_2SO_4. If the mass of $PbSO_4$ that precipitates is 0.6839 g, what is the percent by mass of Pb in the original sample?

Step 1: Calculate the mass of Pb in 0.6839 g of the $PbSO_4$ precipitate.

$$\textbf{? mass of Pb} = 0.6839\ \text{g PbSO}_4 \times \frac{1\ \text{mol PbSO}_4}{303.27\ \text{g PbSO}_4} \times \frac{1\ \text{mol Pb}}{1\ \text{mol PbSO}_4} \times \frac{207.2\ \text{g Pb}}{1\ \text{mol Pb}}$$

$$= \mathbf{0.4673\ g\ Pb}$$

Step 2: Calculate the mass percent of Pb in the unknown sample.

$$\%\text{Pb by mass} = \frac{\text{mass of Pb}}{\text{mass of sample}} \times 100\%$$

$$\textbf{\%Pb by mass} = \frac{0.4673\ \text{g}}{0.7469\ \text{g}} \times 100\% = \mathbf{62.57\%}$$

PRACTICE EXERCISE

11. The concentration of Pb^{2+} ions in tap water could be determined by adding excess sodium sulfate solution to water. Excess sodium sulfate solution is added to 0.250 L of tap water. Write the net ionic equation and calculate the molar concentration of Pb^{2+} in the water sample if 0.01685 g of solid $PbSO_4$ is formed.

Text Problems: 4.66, 4.68

PROBLEM TYPE 8: ACID-BASE TITRATIONS

You must try to convince yourself that a titration problem follows the same thought process as the stoichiometry problems discussed in Chapter 3. The difference is that for acid and base solutions, you will typically be given the molarity rather than grams of substance.

Remember, you cannot directly compare grams of one substance to grams of another. Similarly, you typically cannot compare molarities or volume of one substance to that of another. Therefore, you must convert to *moles* of one substance, then apply the correct mole ratio from the balanced chemical equation to convert to *moles* of the other substance.

A typical approach to a stoichiometry problem is outlined below.

Amount of reactant. Expressed in grams, molarity, etc.	→	moles of reactant	→	moles of product	→	amount of product. Expressed in grams, volume, etc.

EXAMPLE 4.12

What volume of 0.900 *M* HCl is required to completely neutralize 50.0 mL of a 0.500 *M* $Ba(OH)_2$ solution?

Step 1: In order to have the correct mole ratio to solve the problem, you must start with a balanced chemical equation.

$$2\ HCl\ (aq) + Ba(OH)_2\ (aq) \longrightarrow BaCl_2\ (aq) + 2\ H_2O\ (l)$$

Step 2: From the molarity and volume of the $Ba(OH)_2$ solution, you can calculate moles of $Ba(OH)_2$. Then, using the mole ratio from the balanced equation above, you can calculate moles of HCl.

50.0 mL = 0.0500 L

$$\textbf{? mol HCl} = 0.0500\ \cancel{L} \times \frac{0.500\ \cancel{mol}\ Ba(OH)_2}{1\ \cancel{L}\ \text{of solution}} \times \frac{2\ \text{mol HCl}}{1\ \cancel{mol}\ Ba(OH)_2}$$

$$= \textbf{0.0500 mol HCl}$$

Thus, 0.0500 mol of HCl are required to neutralize 50.0 mL of 0.500 *M* $Ba(OH)_2$.

Step 3: Solve the molarity equation algebraically for liters of solution. Then, substitute in the moles of HCl and molarity of HCl to solve for volume of HCl.

$$M = \frac{\text{moles of solute}}{\text{liters of solution}}$$

$$\text{liters of solution} = \frac{\text{moles of solute}}{M}$$

$$\textbf{volume of HCl} = \frac{0.0500\ \cancel{mol}\ HCl}{0.900\ \cancel{mol}/L} = \textbf{0.0556 L}$$

PRACTICE EXERCISE

12. The distinctive odor of vinegar is due to acetic acid, CH_3COOH. Acetic acid reacts with sodium hydroxide in the following fashion:

$$CH_3COOH\ (aq) + NaOH\ (aq) \longrightarrow H_2O\ (l) + CH_3COONa\ (aq)$$

If 2.50 mL of vinegar requires 34.9 mL of 0.0960 *M* NaOH to reach the equivalence point in a titration, how many grams of acetic acid are in the 2.50 mL sample?

Text Problem: 4.72

ANSWERS TO PRACTICE EXERCISES

1. **(a)** soluble **(b)** insoluble **(c)** soluble **(d)** insoluble

2. $CaCl_2\ (aq) + Na_2CO_3\ (aq) \longrightarrow 2\ NaCl\ (aq) + CaCO_3\ (s)$

$Ca^{2+}(aq) + 2Cl^{-}(aq) + 2Na^{+}(aq) + CO_3^{\ 2-}(aq) \longrightarrow 2Na^{+}(aq) + CO_3^{\ 2-}(aq) + CaCO_3(s)$

$Ca^{2+}\ (aq) + CO_3^{\ 2-}\ (aq) \longrightarrow CaCO_3\ (s)$

3. **(a)** acid **(b)** acid **(c)** base **(d)** both

4. $2\ HNO_2\ (aq) + Ba(OH)_2\ (aq) \longrightarrow Ba(NO_2)_2\ (aq) + 2\ H_2O\ (l)$

$2\ HNO_2\ (aq) + Ba^{2+}\ (aq) + 2\ OH^{-}\ (aq) \longrightarrow Ba^{2+}\ (aq) + 2\ NO_2^{\ -}\ (aq) + 2\ H_2O\ (l)$

$2\ HNO_2\ (aq) + 2\ OH^{-}\ (aq) \longrightarrow 2\ NO_2^{\ -}\ (aq) + 2\ H_2O\ (l)$

5. **(a)** +4 **(b)** +2 **(c)** +4 **(d)** +5

6. The oxidation numbers of the atoms and ions are:

$$Cd^{0}\ (s) + Ni^{4+}O_2^{\ 2-}\ (s) + 2\ H_2^{\ +}O^{2-}\ (l) \longrightarrow Cd^{2+}(OH)_2^{\ -}\ (s) + Ni^{2+}(OH)_2^{\ -}\ (s)$$

Cd^{0} (*s*) is oxidized, and Ni^{4+} is reduced. The oxidation and reduction half-reactions are:

$$Cd \longrightarrow Cd^{2+} + 2e^{-}$$

$$Ni^{4+} + 2e^{-} \longrightarrow Ni^{2+}$$

7. **(a)** $2\ Al\ (s) + 6\ HCl\ (aq) \longrightarrow 2\ AlCl_3\ (aq) + 3\ H_2\ (g)$ **(b)** No reaction

8. 0.0124 *M* **9.** 29.22 g NaCl **10.** 5.08×10^{-4} *M*

11. $Pb^{2+}\ (aq) + SO_4^{\ 2-}\ (aq) \longrightarrow PbSO_4\ (s)$

$[Pb^{2+}] = 2.222 \times 10^{-4}\ M$

12. 0.201 g acetic acid **13.** 1.79 g Fe

SOLUTIONS TO SELECTED TEXT PROBLEMS

4.6 Ionic compounds, strong acids, and strong bases (metal hydroxides) are strong electrolytes (completely broken up into ions of the compound). Weak acids and weak bases are weak electrolytes. Molecular substances other than acids or bases are nonelectrolytes.

(a) strong electrolyte (ionic) **(b)** nonelectrolyte

(c) weak electrolyte (weak base) **(d)** strong electrolyte (strong base)

4.8 **(a)** Solid NaCl does not conduct. The ions are locked in a rigid lattice structure.

(b) Molten NaCl conducts. The ions can move around in the liquid state.

(c) Aqueous NaCl conducts. NaCl ionizes completely to Na^+ (*aq*) and Cl^- (*aq*) in water.

4.10 Since HCl dissolved in water conducts electricity, then HCl (*aq*) must actually exists as H^+ (*aq*) cations and Cl^- (*aq*) anions. Since HCl dissolved in benzene solvent does not conduct electricity, then we must assume that the HCl molecules in benzene solvent do not ionize, but rather exist as un-ionized molecules.

4.14 Applying solubility rules, Problem Type 1. Refer to Table 4.2 of your text to solve this problem.

(a) $CaCO_3$ is <u>insoluble</u> (Rule 6).
(b) $ZnSO_4$ is <u>soluble</u> (Rule 7).
(c) $Hg(NO_3)_2$ is <u>soluble</u> (Rule 3).
(d) $HgSO_4$ is <u>insoluble</u> (Rule 7).
(e) NH_4ClO_4 is <u>soluble</u> (Rule 2).

4.16 Writing Molecular, Ionic, and Net Ionic Equations, Problem Type 2.

Recall that an <u>ionic equation</u> shows dissolved ionic compounds in terms of their free ions. A <u>net ionic equation</u> shows only the species that actually take part in the reaction.

(a) <u>Ionic:</u> $2\,Na^+(aq) + S^{2-}(aq) + Zn^{2+}(aq) + 2\,Cl^-(aq) \longrightarrow ZnS(s) + 2\,Na^+(aq) + 2\,Cl^-(aq)$

<u>Net ionic:</u> $Zn^{2+}(aq) + S^{2-}(aq) \longrightarrow ZnS(s)$

Table 4.2 indicates that ZnS is <u>insoluble</u>.

(b) <u>Ionic:</u> $6K^+(aq) + 2PO_4^{3-}(aq) + 3Sr^{2+}(aq) + 6NO_3^-(aq) \longrightarrow Sr_3(PO_4)_2(s) + 6K^+(aq) + 6NO_3^-(aq)$

<u>Net ionic:</u> $3\,Sr^{2+}(aq) + 2\,PO_4^{3-}(aq) \longrightarrow Sr_3(PO_4)_2(s)$

Table 4.2 indicates that $Sr_3(PO_4)_2$ is <u>insoluble</u>.

(c) <u>Ionic:</u> $Mg^{2+}(aq) + 2NO_3^-(aq) + 2Na^+(aq) + 2OH^-(aq) \longrightarrow Mg(OH)_2(s) + 2Na^+(aq) + 2NO_3^-(aq)$

<u>Net ionic:</u> $Mg^{2+}(aq) + 2\,OH^-(aq) \longrightarrow Mg(OH)_2(s)$

Table 4.2 indicates that $Mg(OH)_2$ is <u>insoluble</u>.

To identify a Brønsted base, you should look for soluble hydroxide salts. The hydroxide ion (OH^-) will accept a proton to form H_2O. Also look for weak bases, which are amines. Ammonia (NH_3) is an example. Finally, look for anions from acids. These negatives ions can accept a proton (H^+). Some examples, are $H_2PO_4^{2-}$, NO_2^-, and HCO_3^-.

EXAMPLE 4.3

Identify each of the following species as a Brønsted acid, base, or both: (a) HNO_3, (b) $Ba(OH)_2$, (c) SO_4^{2-}, (d) $CH_3CH_2CH_2COOH$, (e) HPO_4^{2-}.

(a) Brønsted acid. The formula of this compound starts with H; this indicates that it is probably an acid.
(b) Brønsted base. This is a soluble hydroxide salt.
(c) Brønsted base. This negative ion can accept a proton; therefore, it is a base.
(d) Brønsted acid. This is a carboxylic acid. It contains a carboxyl group, –COOH.
(e) Both a Brønsted acid and base. This ion has a proton (H^+) that it can donate. It also has a negative charge and therefore can accept a proton.

PRACTICE EXERCISE

3. Identify each of the following species as a Brønsted acid, base, or both.

(a) CH_3CH_2COOH (b) HF (c) KOH (d) HCO_3^-

Text Problem: 4.26

B. Writing Acid-Base Reactions

An acid-base reaction is called a **neutralization reaction**. The typical products of an acid-base reaction are a salt and water.

$$\text{acid} + \text{base} \longrightarrow \text{salt} + \text{water}$$

Let's consider a generic acid, HA, reacted with a generic base, MOH.

$$HA + MOH \longrightarrow MA\ (\text{salt}) + H_2O$$

The H^+ from the acid combines with OH^- from the base to produce water. The anion from the acid, A^-, combines with the metal cation from the base, M^+, to form the salt, MA.

EXAMPLE 4.4

Complete and balance the following equations and write the corresponding ionic and net ionic equations:

(a) $HBr\ (aq) + Ba(OH)_2\ (aq) \longrightarrow$

(b) $HCOOH\ (aq) + NaOH\ (aq) \longrightarrow$

(a) $2\ HBr\ (aq) + Ba(OH)_2\ (aq) \longrightarrow BaBr_2\ (aq) + 2\ H_2O\ (l)$

HBr (*aq*) is a strong acid and therefore it completely ionizes in solution. $Ba(OH)_2$ (*aq*) is a strong base and therefore it also completely ionizes in solution. The balanced *ionic equation* is:

$$2H^+(aq) + 2Br^-(aq) + Ba^{2+}(aq) + 2OH^-(aq) \longrightarrow Ba^{2+}(aq) + 2Br^-(aq) + 2H_2O(l)$$

Cancel the spectator ions to write the *net ionic equation*.

$$2\ H^+\ (aq) + 2\ OH^-\ (aq) \longrightarrow 2\ H_2O\ (l) \quad \text{or} \quad H^+\ (aq) + OH^-\ (aq) \longrightarrow H_2O\ (l)$$

4.18 **(a)** Add chloride ions. KCl is soluble, but AgCl is not.

(b) Add sulfate ions. Ag_2SO_4 is moderately soluble, but $PbSO_4$ is insoluble.

(c) Add carbonate ions. $(NH_4)_2CO_3$ is soluble, but $CaCO_3$ is insoluble.

(d) Add sulfate ions. $CuSO_4$ is soluble, but $BaSO_4$ is insoluble.

4.26 Identifying Brønsted acids and bases, Problem Type 3A.

(a) PO_4^{3-} in water can accept a proton to become HPO_4^{2-}, and is thus a Brønsted base.

(b) ClO_2^- in water can accept a proton to become $HClO_2$, and is thus a Brønsted base.

(c) NH_4^+ dissolved in water can donate a proton H^+, thus behaving as a Brønsted acid.

(d) HCO_3^- can either accept a proton to become H_2CO_3, thus behaving as a Brønsted base. Or, HCO_3^- can donate a proton to yield H^+ and CO_3^{2-}, thus behaving as a Brønsted acid.

4.28 Writing acid-base reactions, Problem Type 3B.

Recall that strong acids and strong bases are strong electrolytes. They are completely ionized in solution. An <u>ionic equation</u> will show strong acids and strong bases in terms of their free ions. A <u>net ionic equation</u> shows only the species that actually take part in the reaction.

(a) <u>Ionic:</u> $CH_3COOH\ (aq) + K^+\ (aq) + OH^-\ (aq) \longrightarrow K^+\ (aq) + CH_3COO^-\ (aq) + H_2O\ (l)$

<u>Net ionic:</u> $CH_3COOH\ (aq) + OH^-\ (aq) \longrightarrow CH_3COO^-\ (aq) + H_2O\ (l)$

(b) <u>Ionic:</u> $H_2CO_3\ (aq) + 2\ Na^+\ (aq) + 2\ OH^-\ (aq) \longrightarrow 2\ Na^+\ (aq) + CO_3^{2-}\ (aq) + 2\ H_2O\ (l)$

<u>Net ionic:</u> $H_2CO_3\ (aq) + 2\ OH^-\ (aq) \longrightarrow CO_3^{2-}\ (aq) + 2\ H_2O\ (l)$

(c) <u>Ionic:</u> $2H^+\ (aq) + 2\ NO_3^-\ (aq) + Ba^{2+}\ (aq) + 2OH^-\ (aq) \longrightarrow Ba^{2+}\ (aq) + 2NO_3^-\ (aq) + 2H_2O\ (l)$

<u>Net ionic:</u> $2\ H^+\ (aq) + 2\ OH^-\ (aq) \longrightarrow 2\ H_2O\ (l)$

4.32 Writing oxidation/reduction half-reactions, Problem Type 4B.

In each part, the reducing agent is the reactant in the first half-reaction and the oxidizing agent is the reactant in the second half-reaction. The coefficients in each half-reaction have been reduced to smallest whole numbers.

(a) The product is an ionic compound whose ions are Fe^{3+} and O^{2-}.

$$Fe \longrightarrow Fe^{3+} + 3e^-$$
$$O_2 + 4e^- \longrightarrow 2\,O^{2-}$$

O_2 is the oxidizing agent; Fe is the reducing agent.

(b) Na^+ does not change in this reaction. It is a "spectator ion."

$$2\,Br^- \longrightarrow Br_2 + 2e^-$$
$$Cl_2 + 2e^- \longrightarrow 2\,Cl^-$$

Cl_2 is the oxidizing agent; Br^- is the reducing agent.

(c) Assume SiF_4 is made up of Si^{4+} and F^-.

$$Si \longrightarrow Si^{4+} + 4e^-$$

$$F_2 + 2e^- \longrightarrow 2\,F^-$$

F_2 is the oxidizing agent; Si is the reducing agent.

(d) Assume HCl is made up of H^+ and Cl^-.

$$H_2 \longrightarrow 2\,H^+ + 2e^-$$

$$Cl_2 + 2e^- \longrightarrow 2\,Cl^-$$

Cl_2 is the oxidizing agent; H_2 is the reducing agent.

4.36 Assigning oxidation numbers, Problem Type 4A.

Step 1: In each case the oxidation number of oxygen is −2 and that of hydrogen is +1 (rules 3 and 4, Section 4.4 of your text).

Step 2: All the compounds listed are neutral compounds, so the oxidation numbers must sum to zero (Rule 6, Section 4).

Let the oxidation number of P = x.

(a) $x + 1 + (3)(-2) = 0,\ \mathbf{x = +5}$

(b) $x + (3)(+1) + (2)(-2) = 0,\ \mathbf{x = +1}$

(c) $x + (3)(+1) + (3)(-2) = 0,\ \mathbf{x = +3}$

(d) $x + (3)(+1) + (4)(-2) = 0,\ \mathbf{x = +5}$

(e) $2x + (4)(+1) + (7)(-2) = 0,\ 2x = 10,\ \mathbf{x = +5}$

(f) $3x + (5)(+1) + (10)(-2) = 0,\ 3x = 15,\ \mathbf{x = +5}$

The molecules in part (a), (e), and (f) can be made by strongly heating the compound in part (d). Are these oxidation-reduction reactions?

4.38 **(a)** $\underline{Cs}_2O$, +1 **(b)** CaI_2, −1 **(c)** $\underline{Al}_2O_3$, +3 **(d)** $H_3\underline{As}O_3$, +3 **(e)** $\underline{Ti}O_2$, +4 **(f)** $\underline{Mo}O_4^{2-}$, +6 **(g)** $\underline{Pt}Cl_4^{2-}$, +2 **(h)** $\underline{Pt}Cl_6^{2-}$, +4 **(i)** $\underline{Sn}F_2$, +2 **(j)** $\underline{Cl}F_3$, +3 **(k)** $\underline{Sb}F_6^-$, +5

4.40 **(a)** N: −3 **(b)** O: −1/2 **(c)** C: −1 **(d)** C: +4 **(e)** C: +3 **(f)** O: −2 **(g)** B: +3 **(h)** W: +6

4.42 In order to work this problem, you need to assign the oxidation numbers to all the elements in the compounds. In each case oxygen has an oxidation number of −2 (rule 3). These oxidation numbers should then be compared to the range of possible oxidation numbers that each element can have. Molecular oxygen is a powerful oxidizing agent. In SO_3 alone, the oxidation number of the element bound to oxygen (S) is at its maximum value (+6); the sulfur cannot be oxidized further. The other elements bound to oxygen in this problem have less than their maximum oxidation number and can undergo further oxidation.

4.44 Using an activity series, Problem Type 4C.

Hydrogen displacement. Any metal above hydrogen in the activity series will displace it from water or from an acid. Metals below hydrogen will not react with either water or an acid.

Only **(b)** Li and **(d)** Ca are above hydrogen in the activity series, so they are the only metals in this problem that will react with water.

4.48 Concentration of Solutions, Problem Type 5.

Step 1: Based on the definition of molarity, grams of solute must be converted to moles of solute.

$\mathcal{M}$ (NaOH) = 40.00 g/mol.

$$\text{Moles NaOH} = 5.25\ \cancel{\text{g}}\ \text{NaOH} \times \frac{1\ \text{mol NaOH}}{40.00\ \cancel{\text{g}}\ \text{NaOH}} = 0.131\ \text{mol NaOH}$$

Step 2: Substitute the above values into the molarity equation.

$$M = \frac{\text{moles of solute}}{\text{liters of solution}}$$

$$M = \frac{0.131\ \text{mol NaOH}}{1\ \text{L solution}} = \mathbf{0.131\ \mathit{M}}$$

This is normally written 0.131 M NaOH.

4.50 Since the problem asks for grams of solute (KOH), you should be thinking that you can calculate moles of solute from the molarity and volume of solution. Then, you can convert moles of solute to grams of solute.

$$?\ \text{moles KOH solute} = \frac{5.50\ \text{moles solute}}{1\ \cancel{\text{L}}\ \text{solution}} \times 0.0350\ \cancel{\text{L}}\ \text{solution} = 0.193\ \text{moles KOH}$$

The molar mass of KOH is 56.11 g/mol. Use this conversion factor to calculate grams of KOH.

$$\mathbf{?\ grams\ KOH} = 0.193\ \cancel{\text{mol}}\ \text{KOH} \times \frac{56.11\ \text{g KOH}}{1\ \cancel{\text{mol}}\ \text{KOH}} = \mathbf{10.8\ g\ KOH}$$

4.52 **(a)** $?\ \text{mol}\ CH_3OH = 6.57\ \cancel{\text{g}}\ CH_3OH \times \frac{1\ \text{mol}\ CH_3OH}{32.04\ \cancel{\text{g}}\ CH_3OH} = 0.205\ \text{mol}\ CH_3OH$

$$M = \frac{0.205\ \text{mol}\ CH_3OH}{0.150\ \text{L}} = \mathbf{1.37\ \mathit{M}}$$

(b) $?\ \text{mol}\ CaCl_2 = 10.4\ \cancel{\text{g}}\ CaCl_2 \times \frac{1\ \text{mol}\ CaCl_2}{111.0\ \cancel{\text{g}}\ CaCl_2} = 0.0937\ \text{mol}\ CaCl_2$

$$M = \frac{0.0937\ \text{mol}\ CaCl_2}{0.220\ \text{L}} = \mathbf{0.426\ \mathit{M}}$$

(c) $?\ \text{mol}\ C_{10}H_8 = 7.82\ \cancel{\text{g}}\ C_{10}H_8 \times \frac{1\ \text{mol}\ C_{10}H_8}{128.2\ \cancel{\text{g}}\ C_{10}H_8} = 0.0610\ \text{mol}\ C_{10}H_8$

$$M = \frac{0.0610\ \text{mol}\ C_{10}H_8}{0.0852\ \text{L}} = \mathbf{0.716\ \mathit{M}}$$

4.54 A 250 mL sample of 0.100 *M* solution contains 0.0250 mol of solute (mol = $M \times$ L). The computation in each case is the same:

(a) $$0.0250 \text{ mol CsI} \times \frac{259.8 \text{ g CsI}}{1 \text{ mol CsI}} = \mathbf{6.50 \text{ g CsI}}$$

(b) $$0.0250 \text{ mol } H_2SO_4 \times \frac{98.09 \text{ g } H_2SO_4}{1 \text{ mol } H_2SO_4} = \mathbf{2.45 \text{ g } H_2SO_4}$$

(c) $$0.0250 \text{ mol } Na_2CO_3 \times \frac{106.0 \text{ g } Na_2CO_3}{1 \text{ mol } Na_2CO_3} = \mathbf{2.65 \text{ g } Na_2CO_3}$$

(d) $$0.0250 \text{ mol } K_2Cr_2O_7 \times \frac{294.2 \text{ g } K_2Cr_2O_7}{1 \text{ mol } K_2Cr_2O_7} = \mathbf{7.36 \text{ g } K_2Cr_2O_7}$$

(e) $$0.0250 \text{ mol } KMnO_4 \times \frac{158.0 \text{ g } KMnO_4}{1 \text{ mol } KMnO_4} = \mathbf{3.95 \text{ g } KMnO_4}$$

4.58 Dilution of Solutions, Problem Type 6.

You should be able to come up with the following equation.

$$M_{\text{initial}}V_{\text{initial}} = M_{\text{final}}V_{\text{final}}$$

You can solve the equation algebraically for M_{final}. Then substitute in the given quantities to solve the problem.

$$\boldsymbol{M_{\text{final}}} = \frac{M_{\text{initial}} \times V_{\text{initial}}}{V_{\text{final}}} = \frac{0.866 \text{ M} \times 25.0 \text{ mL}}{500 \text{ mL}} = \mathbf{0.0433}\ \boldsymbol{M}$$

4.60 You need to calculate the final volume of the dilute solution. Then, you can subtract 505 mL from this volume to calculate the amount of water that should be added.

$$V_{\text{final}} = \frac{M_{\text{initial}}V_{\text{initial}}}{M_{\text{final}}} = \frac{(0.125\ M)(505 \text{ mL})}{(0.100\ M)} = 631 \text{ mL}$$

Add (631 − 505) mL = **126 mL of water.**

4.62 Moles of calcium nitrate in the first solution:

$$\frac{0.568 \text{ mol}}{1000 \text{ mL soln}} \times 46.2 \text{ mL soln} = 0.0262 \text{ mol } Ca(NO_3)_2$$

Moles of calcium nitrate in the second solution:

$$\frac{1.396 \text{ mol}}{1000 \text{ mL soln}} \times 80.5 \text{ mL soln} = 0.112 \text{ mol } Ca(NO_3)_2$$

The volume of the combined solutions = 46.2 mL + 80.5 mL = 126.7 mL. The concentration of the final solution is:

$$\boldsymbol{M} = \frac{(0.0262 + 0.112) \text{ mol}}{0.1267 \text{ L}} = \mathbf{1.09}\ \boldsymbol{M}$$

4.66 Gravimetric Analysis, Problem Type 7.

We assume the precipitation is quantitative, that is, that all of the barium in the sample has been precipitated as barium sulfate.

Step 1: Calculate the mass of Ba in 0.4105 g of the $BaSO_4$ precipitate.

$$? \text{ mass of Ba} = 0.4105 \text{ g BaSO}_4 \times \frac{1 \text{ mol BaSO}_4}{233.4 \text{ g BaSO}_4} \times \frac{1 \text{ mol Ba}}{1 \text{ mol BaSO}_4} \times \frac{137.33 \text{ g Ba}}{1 \text{ mol Ba}}$$

$$= 0.2415 \text{ g Ba}$$

Step 2: Calculate the mass percent of Ba in the unknown compounds.

$$\% \text{ Ba by mass} = \frac{\text{mass of Ba}}{\text{mass of sample}} \times 100\%$$

$$\textbf{\% Ba by mass} = \frac{0.2415 \text{ g}}{0.6760 \text{ g}} \times 100\% = \textbf{35.72 \%}$$

4.68 The net ionic equation is: $Cu^{2+}(aq) + S^{2-}(aq) \longrightarrow CuS(s)$

The answer sought is the molar concentration of Cu^{2+}, that is, moles of Cu^{2+} ions per liter of solution. The factor-label method is used to convert, in order:

$$\text{g of CuS} \rightarrow \text{moles CuS} \rightarrow \text{moles Cu}^{2+} \rightarrow \text{moles Cu}^{2+} \text{ per liter soln}$$

$$[\mathbf{Cu^{2+}}] = 0.0177 \text{ g CuS} \times \frac{1 \text{ mol CuS}}{95.62 \text{ g CuS}} \times \frac{1 \text{ mol Cu}^{2+}}{1 \text{ mol CuS}} \times \frac{1}{0.800 \text{ L}} = \mathbf{2.31 \times 10^{-4}} \, \boldsymbol{M}$$

4.72 These problems are related to titration problems discussed in Problem Type 8.

(a)

Step 1: In order to have the correct mole ratio to solve the problem, you must start with a balanced chemical equation.

$$HCl(aq) + KOH(aq) \longrightarrow KCl(aq) + H_2O(l)$$

Step 2: From the molarity and volume of the HCl solution, you can calculate moles of HCl. Then, using the mole ratio from the balanced equation above, you can calculate moles of KOH.

10.0 mL = 0.0100 L

$$? \text{ mol KOH} = \frac{0.30 \text{ mol HCl}}{1 \text{ L of solution}} \times (0.0100 \text{ L solution}) \times \frac{1 \text{ mol KOH}}{1 \text{ mol HCl}}$$

$$= 3.0 \times 10^{-3} \text{ mol KOH}$$

Step 3: Solve the molarity equation algebraically for liters of solution. Then, substitute in the moles of KOH and molarity of KOH to solve for volume of the KOH solution.

$$M = \frac{\text{moles of solute}}{\text{liters of solution}}$$

$$\text{liters of solution} = \frac{\text{moles of solute}}{M}$$

$$\textbf{volume of KOH} = \frac{3.0 \times 10^{-3}\ \text{mol KOH}}{0.50\ \text{mol / L}} = \mathbf{6.0 \times 10^{-3}\ L = 6.0\ mL}$$

(b) This problem is similar to part (a). The difference is that the mole ratio between base and acid is 2:1.

Step 1: In order to have the correct mole ratio to solve the problem, you must start with a balanced chemical equation.

$$H_2SO_4\,(aq) + 2\,KOH\,(aq) \longrightarrow K_2SO_4\,(aq) + H_2O\,(l)$$

Step 2: From the molarity and volume of the H_2SO_4 solution, you can calculate moles of H_2SO_4. Then, using the mole ratio from the balanced equation above, you can calculate moles of KOH.

10.0 mL = 0.0100 L

$$?\ \text{mol KOH} = \frac{0.20\ \text{mol } H_2SO_4}{1\ \text{L of solution}} \times (0.0100\ \text{L solution}) \times \frac{2\ \text{mol KOH}}{1\ \text{mol } H_2SO_4}$$

$$= 4.0 \times 10^{-3}\ \text{mol KOH}$$

Step 3: Solve the molarity equation algebraically for liters of solution. Then, substitute in the moles of KOH and molarity of KOH to solve for volume of the KOH solution.

$$M = \frac{\text{moles of solute}}{\text{liters of solution}}$$

$$\text{liters of solution} = \frac{\text{moles of solute}}{M}$$

$$\textbf{volume of KOH} = \frac{4.0 \times 10^{-3}\ \text{mol KOH}}{0.50\ \text{mol / L}} = \mathbf{8.0 \times 10^{-3}\ L = 8.0\ mL}$$

(c) This problem is similar to parts (a) and (b). The difference is that the mole ratio between base and acid is 3:1.

Step 1: In order to have the correct mole ratio to solve the problem, you must start with a balanced chemical equation.

$$H_3PO_4\,(aq) + 3\,KOH\,(aq) \longrightarrow K_3PO_4\,(aq) + 3\,H_2O\,(l)$$

Step 2: From the molarity and volume of the H_3PO_4 solution, you can calculate moles of H_3PO_4. Then, using the mole ratio from the balanced equation above, you can calculate moles of KOH.

15.0 mL = 0.0150 L

$$?\ \text{mol KOH} = \frac{0.25\ \text{mol } H_3PO_4}{1\ \text{L of solution}} \times (0.0150\ \text{L solution}) \times \frac{3\ \text{mol KOH}}{1\ \text{mol } H_3PO_4}$$

$$= 1.1 \times 10^{-2}\ \text{mol KOH}$$

Step 3: Solve the molarity equation algebraically for liters of solution. Then, substitute in the moles of KOH and molarity of KOH to solve for volume of the KOH solution.

$$M = \frac{\text{moles of solute}}{\text{liters of solution}}$$

$$\text{liters of solution} = \frac{\text{moles of solute}}{M}$$

$$\textbf{volume of KOH} = \frac{1.1 \times 10^{-2} \text{ mol KOH}}{0.50 \text{ mol / L}} = \mathbf{2.2 \times 10^{-2}\ L = 22\ mL}$$

4.74 Since both of the original solutions were strong electrolytes, you would expect a mixture of the two solutions to also be a strong electrolyte. However, since the light dims, the mixture must contain fewer ions than the original solution. Indeed, H^+ from the sulfuric acid reacts with the OH^- from the barium hydroxide to form water. The barium cations react with the sulfate anions to form insoluble barium sulfate.

$$2\,H^+\,(aq) + SO_4^{2-}\,(aq) + Ba^{2+}\,(aq) + 2\,OH^-\,(aq) \longrightarrow 2\,H_2O\,(l) + BaSO_4\,(s)$$

Thus, the reaction depletes the solution of ions and the conductivity decreases.

4.76 **Physical Test:** You could test the conductivity of the solutions. Sugar is a nonelectrolyte and an aqueous sugar solution will not conduct electricity; whereas, NaCl is a strong electrolyte when dissolved in water. **Chemical Test:** Silver nitrate could be added to the solutions to see if silver chloride precipitated. Only the NaCl solution would give a AgCl precipitate.

4.78 You would choose a more active metal than Al. Looking at the activity series (Figure 4.11 of your text), metals higher in the activity series than Al are Mg, Na, Ca, Ba, K, Li. All of these metals could be used to reduce Al^{3+} to Al.

4.80 Using the rules of assigning oxidation numbers, we find that the oxidation number for C in CO_2 is +4, which is the maximum oxidation number that carbon can have. Thus it cannot be further oxidized and CO_2 will not burn. In contrast, the oxidation number of C in CO is +2. The carbon in CO can be further oxidized (to what?) and therefore will burn.

4.82 Starting with a balanced chemical equation:

$$Mg\,(s) + 2\,HCl\,(aq) \longrightarrow MgCl_2\,(aq) + H_2\,(g)$$

From the mass of Mg, you can calculate moles of Mg. Then, using the mole ratio from the balanced equation above, you can calculate moles of HCl reacted.

$$4.47\ \text{g Mg} \times \frac{1\ \text{mol Mg}}{24.31\ \text{g Mg}} \times \frac{2\ \text{mol HCl}}{1\ \text{mol Mg}} = 0.368\ \text{mol HCl reacted}$$

Next we can calculate the number of moles of HCl in the original solution.

5.00×10^2 mL = 0.500 L

$$\frac{2.00\ \text{mol HCl}}{1\ \text{L soln}} \times 0.500\ \text{L soln} = 1.00\ \text{mol HCl}$$

$$\text{Moles HCl remaining} = 1.00\ \text{mol} - 0.368\ \text{mol} = 0.63\ \text{mol HCl}$$

$$\textbf{conc. of HCl after reaction} = \frac{\text{mol HCl}}{\text{L soln}} = \frac{0.63\ \text{mol HCl}}{0.500\ \text{L}} = \mathbf{1.3\ mol/L = 1.3}\ \boldsymbol{M}$$

4.84 The balanced equation is:

$$2\ HCl\ (aq) + Na_2CO_3\ (s) \longrightarrow CO_2\ (g) + H_2O\ (l) + 2\ NaCl\ (aq)$$

The mole ratio from the balanced equation is 2 mol HCl : 1 mol Na_2CO_3. The moles of HCl needed to react with 0.256 g of Na_2CO_3 are:

$$0.256\ \text{g}\ Na_2CO_3 \times \frac{1\ \text{mol}\ Na_2CO_3}{106.0\ \text{g}\ Na_2CO_3} \times \frac{2\ \text{mol HCl}}{1\ \text{mol}\ Na_2CO_3} = 4.83 \times 10^{-3}\ \text{mol HCl}$$

$$\textbf{Molarity HCl} = \frac{\text{moles HCl}}{\text{L soln}} = \frac{4.83 \times 10^{-3}\ \text{mol HCl}}{0.0283\ \text{L soln}} = \mathbf{0.171\ mol/L = 0.171}\ \boldsymbol{M}$$

4.86 Starting with a balanced chemical equation:

$$CH_3COOH\ (aq) + NaOH\ (aq) \longrightarrow CH_3COONa\ (aq) + H_2O\ (l)$$

From the molarity and volume of the NaOH solution, you can calculate moles of NaOH. Then, using the mole ratio from the balanced equation above, you can calculate moles of CH_3COOH.

$$5.75 \times 10^{-3}\ \text{L solution} \times \frac{1.00\ \text{mol NaOH}}{1\ \text{L of solution}} \times \frac{1\ \text{mol}\ CH_3COOH}{1\ \text{mol NaOH}} = 5.75 \times 10^{-3}\ \text{mol}\ CH_3COOH$$

$$\textbf{Molarity}\ \mathbf{CH_3COOH} = \frac{5.75 \times 10^{-3}\ \text{mol}\ CH_3COOH}{0.0500\ \text{L}} = \mathbf{0.115}\ \boldsymbol{M}$$

4.88 The balanced equation is: $2\ HCl + Mg(OH)_2 \longrightarrow MgCl_2 + 2\ H_2O$

The moles of $Mg(OH)_2$ in 10 mL of milk of magnesia are:

$$10.0\ \text{mL soln} \times \frac{0.080\ \text{g}\ Mg(OH)_2}{1\ \text{mL soln}} \times \frac{1\ \text{mol}\ Mg(OH)_2}{58.33\ \text{g}\ Mg(OH)_2} = 0.014\ \text{mol}\ Mg(OH)_2$$

$$\text{Moles of HCl reacted} = 0.014\ \text{mol}\ Mg(OH)_2 \times \frac{2\ \text{mol HCl}}{1\ \text{mol}\ Mg(OH)_2} = 0.028\ \text{mol HCl}$$

$$\textbf{Volume of HCl} = \frac{\text{mol solute}}{M} = \frac{0.028\ \text{mol HCl}}{0.035\ \text{mol / L}} = \mathbf{0.80\ L}$$

4.90 First, calculate the number of moles of glucose present.

$$\frac{0.513\ \text{mol glucose}}{1\ \text{L soln}} \times 60.0 \times 10^{-3}\ \text{L} = 0.0308\ \text{mol glucose}$$

$$\frac{2.33\ \text{mol glucose}}{1\ \text{L soln}} \times 120.0 \times 10^{-3}\ \text{L} = 0.280\ \text{mol glucose}$$

Add the moles of glucose, then divide by the total volume of the combined solutions to calculate the molarity.

$$\textbf{Molarity of final solution} = \frac{0.0308\ \text{mol} + 0.280\ \text{mol glucose}}{(60.0 + 120.0)\ \text{mL}} \times \frac{1\ \text{mL}}{0.001\ \text{L}} = 1.73\ \text{mol/L} = \mathbf{1.73}\ \boldsymbol{M}$$

4.92 The solid sodium bicarbonate would be the better choice. The hydrogen carbonate ion, HCO_3^-, behaves as a Brønsted base to accept a proton from the acid.

$$HCO_3^-(aq) + H^+(aq) \longrightarrow H_2CO_3(aq) \longrightarrow H_2O(l) + CO_2(g)$$

The heat generated during the reaction of hydrogen carbonate with the acid causes the carbonic acid, H_2CO_3, that was formed to decompose to water and carbon dioxide.

The reaction of the spilled sulfuric acid with sodium hydroxide would produce sodium sulfate, Na_2SO_4, and water. There is a possibility that the Na_2SO_4 could precipitate. Also, the sulfate ion, SO_4^{2-} is a weak base; therefore, the "neutralized" solution would actually be basic.

$$H_2SO_4(aq) + 2\,NaOH(aq) \longrightarrow Na_2SO_4(aq) + 2\,H_2O(l)$$

Also, NaOH is a caustic substance and therefore is not safe to use in this manner.

4.94 The balanced equation for the reaction is:

$$XCl(aq) + AgNO_3(aq) \longrightarrow AgCl(s) + XNO_3(aq) \qquad \text{where X = Na, or K}$$

From the amount of AgCl produced, we can calculate the moles of XCl reacted (X = Na, or K).

$$1.913\text{ g AgCl} \times \frac{1\text{ mol AgCl}}{143.4\text{ g AgCl}} \times \frac{1\text{ mol XCl}}{1\text{ mol AgCl}} = 0.01334\text{ mol XCl.}$$

Let x = number of moles NaCl. Then, the number of moles of KCl = 0.01334 mol – x. The sum of the NaCl and KCl masses must equal the mass of the mixture, 0.8870 g. We can write:

$$\text{mass NaCl} + \text{mass KCl} = 0.8870\text{ g}$$

$$\left[x\text{ mol NaCl} \times \frac{58.44\text{ g NaCl}}{1\text{ mol NaCl}}\right] + \left[(0.01334 - x)\text{mol KCl} \times \frac{74.55\text{ g KCl}}{1\text{ mol KCl}}\right] = 0.8870\text{ g}$$

$$x = 6.673 \times 10^{-3} = \text{moles NaCl}$$

$$\text{mol KCl} = 0.01334 - x = 0.01334\text{ mol} - 6.673 \times 10^{-3}\text{ mol} = 6.667 \times 10^{-3}\text{ mol KCl}$$

Converting moles to grams:

$$\text{mass NaCl} = 6.673 \times 10^{-3}\text{ mol NaCl} \times \frac{58.44\text{ g NaCl}}{1\text{ mol NaCl}} = 0.3900\text{ g NaCl}$$

$$\text{mass KCl} = 6.667 \times 10^{-3}\text{ mol KCl} \times \frac{74.55\text{ g KCl}}{1\text{ mol KCl}} = 0.4970\text{ g KCl}$$

The percentages by mass for each compound are:

$$\textbf{\% NaCl} = \frac{0.3900\text{ g}}{0.8870\text{ g}} \times 100\% = \textbf{43.97\% NaCl}$$

$$\textbf{\% KCl} = \frac{0.4970\text{ g}}{0.8870\text{ g}} \times 100\% = \textbf{56.03\% KCl}$$

4.96 **(a)** The precipitate $CaSO_4$ formed over Ca prevents the Ca from reacting with the sulfuric acid.

(b) Aluminum is protected by a tenacious oxide layer with the composition Al_2O_3.

(c) These metals react more readily with water.

$$2\ Na\ (s) + 2\ H_2O\ (l) \longrightarrow 2\ NaOH\ (aq) + H_2\ (g)$$

(d) The metal should be placed below Fe and above H.

4.98 **(a)** The precipitation reaction is: $Mg^{2+}\ (aq) + 2\ OH^-\ (aq) \longrightarrow Mg(OH)_2\ (s)$

The acid-base reaction is: $Mg(OH)_2\ (s) + 2\ HCl\ (aq) \longrightarrow MgCl_2\ (aq) + 2\ H_2O\ (l)$

The redox reactions are: $Mg^{2+} + 2\ e^- \longrightarrow Mg\ (s)$

$2\ Cl^- \longrightarrow Cl_2\ (g) + 2\ e^-$

(b) NaOH is much more expensive than CaO.

(c) Dolomite has the advantage of being an additional source of magnesium that can also be recovered.

(d) It is less expensive to "mine" magnesium metal from the ocean than from the Earth's crust. Also, it is less destructive, because you do not have to dig a mine to obtain the mineral.

CHAPTER 5
GASES

PROBLEM-SOLVING STRATEGIES AND TUTORIAL SOLUTIONS

TYPES OF PROBLEMS

Problem Type 1: Pressure Conversions (Also see Problem Type 5, Chapter 1).

Problem Type 2: Calculations Using the Ideal Gas Law.

- **(a)** Given three of the four variable quantities in the equation (P, V, n, and T).
- **(b)** Only one or two variable quantities have fixed values.
- **(c)** Calculations using density or molar mass (Also see Density calculations, Problem Type 1, Chapter 1).
- **(d)** Stoichiometry involving gases (Also see Stoichiometry problems, Problem Type 7, Chapter 3).
- **(e)** Collecting a gas over water (Also see Dalton's Law of Partial Pressures, below).

Problem Type 3: Dalton's Law of Partial Pressures.

Problem Type 4: Root-Mean-Square (RMS) Speed.

Problem Type 5: Deviations from Ideal Behavior.

PROBLEM TYPE 1: PRESSURE CONVERSIONS

(Also see Problem Type 5, Chapter 1)

In order to convert from one unit of pressure to another, you need to be proficient at the factor-label method. See Section 1.7 of your text. Unit conversions can seem daunting, but if you keep track of the units, making sure that the appropriate units cancel, your effort will be rewarded.

EXAMPLE 5.1
Convert 555 mmHg to kPa.

Step 1: Map out a strategy to proceed from initial units to final units based on available conversion factors. Looking in Section 5.2 of your text, you should find the following conversions.

$$1 \text{ atm} = 760 \text{ mmHg}$$
$$1 \text{ atm} = 1.01325 \times 10^2 \text{ kPa}$$

You should come up with the following strategy:

$$\text{mm Hg} \rightarrow \text{atm} \rightarrow \text{kPa}$$

Step 2: Use the following method to ensure that you obtain the desired unit.

$$\text{Given unit} \times \left(\frac{\text{desired unit}}{\text{given unit}}\right) = \text{desired unit}$$

$$\textbf{? kPa} = 555 \text{ mmHg (given)} \times \left(\frac{1 \text{ atm (desired)}}{760 \text{ mmHg (given)}}\right) \times \left(\frac{1.01325 \times 10^2 \text{ kPa (desired)}}{1 \text{ atm (given)}}\right)$$

$$= \textbf{74.0 kPa}$$

Note: In the first conversion factor (atm/mmHg), atm is the desired unit. When moving on to the next conversion factor (kPa/atm), atm is now given, and the desired unit is kPa.

PRACTICE EXERCISE

1. Convert 5.32 atm to mmHg.

Text Problem: 5.14

PROBLEM TYPE 2: CALCULATIONS USING THE IDEAL GAS LAW

You will encounter five different types of problems in this chapter that use the Ideal Gas Law. Each type will be addressed individually below. Remember that R (the gas constant) has the units L·atm/mol·K. Therefore, in problems that contain R, you *must* use the following units: volume in liters (L), pressure in atmospheres (atm), and temperature in Kelvin (K).

A. Given three of the four variable quantities in the equation (*P*, *V*, *n*, and *T*)

If you know the values of any three of the variable quantities, you can solve $PV = nRT$ algebraically for the fourth one. Then, you can calculate its value by substituting in the three known quantities.

EXAMPLE 5.2

Carbon monoxide gas, CO, stored in a 2.00 L container at 25.0°C exerts a pressure of 15.5 atm. How many moles of CO (g) are in the container?

Step 1: You are given *P*, *V*, and *T*. Solve $PV = nRT$ algebraically for *n*.

$$n = \frac{PV}{RT}$$

Step 2: Check that the three known quantities (*P*, *V*, and *T*) have the appropriate units.

P and *V* have the correct units, but *T* must be converted to units of K.

$$T(\text{K}) = °\text{C} + 273°$$
$$T(\text{K}) = 25.0° + 273° = 298\ \text{K}$$

Step 3: Calculate the value of n by substituting the three known quantities into the equation.

$$n = \frac{PV}{RT}$$

$$n = \frac{(15.5\ \cancel{\text{atm}})(2.00\ \cancel{\text{L}})}{298\ \cancel{\text{K}}} \times \frac{\text{mol}\cdot\cancel{\text{K}}}{0.0821\ \cancel{\text{L}}\cdot\cancel{\text{atm}}} = \mathbf{1.27\ mol\ CO}$$

PRACTICE EXERCISE

2. Carbon dioxide gas, CO_2, and water vapor are the two species primarily responsible for the greenhouse effect. Combustion of fossil fuels adds 22 billion tons of carbon dioxide to the atmosphere each year. How many liters of CO_2 is this at 25°C and 1.0 atm pressure?

Text Problems: 5.32, 5.42

B. Only one or two variable quantities have fixed values.

Let's look at the case where n and P are constant (Charles' law).

Step 1: Rearrange the ideal gas equation so that the constants are all on one side.

$$\frac{V}{T} = \frac{nR}{P} \qquad \left(\frac{nR}{P} = \text{constant}\right)$$

$$\text{so,} \quad \mathbf{\frac{V}{T} = constant}$$

Step 2: The individual values of volume and temperature can vary greatly for a given sample of gas. However, as long as the pressure and the amount of gas (n) do *not* change, $\frac{V}{T}$ always equals the same constant. Thus,

$$\frac{V_1}{T_1} = \text{constant} = \frac{V_2}{T_2}$$

$$\text{and,} \quad \mathbf{\frac{V_1}{T_1} = \frac{V_2}{T_2}}$$

This is Charles' law.

Step 3: Solve the equation algebraically for the missing quantity. Then, you can calculate its value by substituting in the known quantities.

EXAMPLE 5.3

Given 10.0 L of neon gas at 5.0°C and 630 mmHg, calculate the new volume at 400°C and 2.5 atm.

Step 1: In this problem, only n is constant. As long as gas is not added or removed, the number of moles of gas (n) is constant. Therefore,

$$\frac{PV}{T} = nR = \text{constant}$$

Step 2: Volume, temperature, and pressure change.

$$\frac{P_1V_1}{T_1} = \text{constant} = \frac{P_2V_2}{T_2}$$

$$\frac{P_1V_1}{T_1} = \frac{P_2V_2}{T_2}$$

Step 3: Solve the equation algebraically for V_2, then substitute in the known quantities to solve for V_2.

$$P_2 = 2.5\ \text{atm} \times \frac{760\ \text{mmHg}}{1\ \text{atm}} = 1.9 \times 10^3\ \text{mmHg}$$

$$V_2 = \frac{P_1V_1T_2}{P_2T_1}$$

$$V_2 = \frac{(630\ \text{mmHg})(10.0\ \text{L})(673\ \text{K})}{(1.9 \times 10^3\ \text{mmHg})(278\ \text{K})} = \mathbf{8.0\ L}$$

Tip: Since R is not in this equation, the units of pressure and volume do *not* have to be atm and liters, respectively. However, temperature must *always* be in units of Kelvin for gas law calculations.

PRACTICE EXERCISE

3. After flying in an airplane, a passenger finds that his sticky hair-gel has opened in his bag making a complete mess. This happens because the cabin is pressurized to about 8000 feet above sea level. During flight, the volume of air inside the bottle increases due to the decreased pressure. Sometimes, this can cause the flip-top cap to open. If the volume of air inside the bottle is 125 mL at sea level (P = 760 mmHg), what volume does the air in the bottle occupy in the airplane (P = 595 mmHg)? Assume that the temperature is kept constant.

Text Problems: 5.18, **5.20**, **5.24**, 5.34, 5.36, 5.38, 5.40

C. Calculations involving density (*d*) or molar mass ($\mathcal{M}$)

(Also see Problem Type 1, Chapter 1)

Step 1: Algebraically solve the ideal gas equation for either density $\left(\frac{m}{V}\right)$ or molar mass ($\mathcal{M}$).

$$PV = nRT$$

$$\frac{n}{V} = \frac{P}{RT} \quad \text{and} \quad n = \frac{m}{\mathcal{M}}$$

Substituting for *n*,

$$\frac{m}{\mathcal{M}V} = \frac{P}{RT}$$

and,

$$\boldsymbol{d = \frac{m}{V} = \frac{P\mathcal{M}}{RT}} \qquad (5.1)$$

Furthermore,

$$\mathcal{M} = \frac{\boldsymbol{dRT}}{\boldsymbol{P}}$$

Step 2: Calculate the value of density or molar mass by substituting in the known quantities.

EXAMPLE 5.4

What is the density of methane gas, CH_4, at STP.

This problem is simplified if you recall that 1 mole of an ideal gas occupies a volume of 22.4 L at STP.

Step 1: Let's assume that we have 1 mole of methane gas. At STP, 1 mole of methane will occupy a volume of 22.4 L. First, we calculate the molar mass of methane.

$$\mathcal{M}(CH_4) = 12.01\text{ g} + 4(1.008\text{ g}) = 16.04\text{ g}$$

Step 2: Knowing the mass of 1 mole of methane and the volume it occupies, we can calculate the density.

$$d = \frac{\text{mass}}{\text{volume}}$$

$$d = \frac{16.04\text{ g}}{22.4\text{ L}} = \mathbf{0.716\text{ g/L}}$$

This problem could also be solved using Equation (5.1) derived above.

Step 1: The necessary equation to derive is:

$$d = \frac{P\mathcal{M}}{RT}$$

Step 2: Calculate the density by substituting in the known quantities.

STP: $P = 1.00$ atm
$T = 273$ K
$\mathcal{M} = 16.04$ g/mol

$$d = \frac{(1.00\ \cancel{\text{atm}})\left(16.04\ \frac{\text{g}}{\cancel{\text{mol}}}\right)}{273\ \cancel{\text{K}}} \times \frac{\cancel{\text{mol}} \cdot \cancel{\text{K}}}{0.0821\ \text{L} \cdot \cancel{\text{atm}}} = \mathbf{0.716\ g/L}$$

PRACTICE EXERCISE

4. An element that exists as a diatomic gas at room temperature has a density of 1.553 g/L at 25°C and 1 atm pressure. Identify the unknown gas.

Text Problems: 5.44, 5.48, 5.50

D. Stoichiometry Involving Gases

(Also see Problem Type 7, Chapter 3)

A typical gas stoichiometry problem involves the following approach:

Amount of reactant. Expressed in grams or volume → moles of reactant → moles of product → amount of product. Expressed in grams, volume, etc.

EXAMPLE 5.5

Oxygen gas was discovered by decomposing mercury (II) oxide.

$$\mathbf{2\ HgO\ (\mathit{s}) \longrightarrow 2\ Hg\ (\mathit{l}) + O_2\ (\mathit{g})}$$

What volume of oxygen gas would be produced by the reaction of 35.2 g of the oxide if the gas is collected at STP?

Step 1: Since stoichiometry uses mole ratios, make sure that you start with a balanced equation. In this problem, you are given the balanced equation.

Step 2: Use the following strategy to solve this problem.

grams HgO → moles HgO → moles O_2 → volume O_2

First, calculate the moles of O_2.

$$?\ \text{mol } O_2 = 35.2\ \cancel{\text{g}}\ \text{HgO} \times \frac{1\ \cancel{\text{mol}}\ \text{HgO}}{216.6\ \cancel{\text{g}}\ \text{HgO}} \times \frac{1\ \text{mol } O_2}{2\ \cancel{\text{mol}}\ \text{HgO}} = 0.0813\ \text{mol } O_2$$

Step 3: Solve the ideal gas equation algebraically for V_{O_2}. Then, calculate the volume by substituting the known quantities into the equation.

STP: $P = 1.00$ atm
$T = 273$ K

$$V_{O_2} = \frac{n_{O_2}RT}{P}$$

$$V_{O_2} = \frac{(0.0813\ \cancel{\text{mol}})(273\ \cancel{\text{K}})}{1.00\ \cancel{\text{atm}}} \times \frac{0.0821\ \text{L} \cdot \cancel{\text{atm}}}{\cancel{\text{mol}} \cdot \cancel{\text{K}}} = \mathbf{1.82\ L}$$

EXAMPLE 5.6

Calculate the volume of methane, CH_4, at STP required to completely consume 3.50 L of oxygen at STP.

Step 1: Write the balanced chemical equation.

$$CH_4\,(g) + 2\,O_2\,(g) \longrightarrow CO_2\,(g) + 2\,H_2O\,(l)$$

Step 2: You could follow the strategy used in the previous example to solve this problem. You could find the moles of oxygen in 3.50 L, apply the correct mole ratio to find moles of methane, and then use the ideal gas equation to find volume of methane. However, there is a short cut if you remember Avogadro's Law. Avogadro's Law states that the volume of a gas is directly proportional to the number of moles of gas at constant temperature and pressure.

Thus, the stoichiometric ratio, $\frac{1 \text{ mol } CH_4}{2 \text{ mol } O_2}$, can be written in terms of volume, $\frac{1 \text{ L } CH_4}{2 \text{ L } O_2}$.

Step 3: Calculate the volume of methane using the above conversion factor.

$$\textbf{? L } \mathbf{CH_4} = 3.50 \cancel{\text{L}}\, O_2 \times \frac{1 \text{ L } CH_4}{2 \cancel{\text{L}}\, O_2} = \mathbf{1.75 \text{ L } CH_4}$$

PRACTICE EXERCISE

5. The decomposition of sodium azide (NaN_3) to sodium and nitrogen gas was one of the first reactions used to inflate air-bag systems in automobiles.

$$2\,NaN_3\,(s) \longrightarrow 2\,Na\,(l) + 3\,N_2\,(g)$$

What mass of NaN_3 in grams would be needed to inflate a 100 L air bag with nitrogen at 25.0°C and 755 mmHg?

Text Problem: 5.80

E. Collecting a gas over water

In this type of problem, you must account for the fact that

$$P_{TOTAL} = P_{SAMPLE} + P_{H_2O(g)}$$

The pressure of the gas sample of interest is:

$$P_{SAMPLE} = P_{TOTAL} - P_{H_2O(g)}$$

Also see Dalton's law of partial pressures, Problem Type 3.

EXAMPLE 5.7

When heated, calcium carbonate decomposes forming solid calcium oxide and carbon dioxide gas. A sample of calcium carbonate is completely decomposed by heating and 50.5 mL of gas is collected at 731.2 torr and 25.0°C. How much did the sample of calcium carbonate weigh?

Step 1: Write the balanced chemical equation.

$$CaCO_3\,(s) \longrightarrow CaO\,(s) + CO_2\,(g)$$

Step 2: Map out the following strategy to solve this problem.

$$\text{volume } CO_2 \rightarrow \text{moles } CO_2 \rightarrow \text{moles } CaCO_3 \rightarrow \text{grams } CaCO_3$$

Step 3: Solve the ideal gas equation algebraically for moles of CO_2. Then, substitute in the known values.

$$P_{CO_2} = 731.2\ \text{torr} \times \frac{1\ \text{atm}}{760\ \text{torr}} = 0.9621\ \text{atm}$$

$$T(\text{K}) = 25.0° + 273° = 298\ \text{K}$$

$$V = 50.5\ \text{mL} \times \frac{1\ \text{L}}{1000\ \text{mL}} = 0.0505\ \text{L}$$

$$n_{CO_2} = \frac{P_{CO_2}V}{RT}$$

$$n_{CO_2} = \frac{(0.9621\ \text{atm})(0.0505\ \text{L})}{298\ \text{K}} \times \frac{\text{mol}\cdot\text{K}}{0.0821\ \text{L}\cdot\text{atm}} = 1.99 \times 10^{-3}\ \text{mol}\ CO_2$$

Step 4: Calculate the mass of $CaCO_3$ using the strategy in Step 2.

$$\textbf{? g CaCO}_3 = 1.99 \times 10^{-3}\ \text{mol}\ CO_2 \times \frac{1\ \text{mol}\ CaCO_3}{1\ \text{mol}\ CO_2} \times \frac{100.09\ \text{g}\ CaCO_3}{1\ \text{mol}\ CaCO_3}$$

$$= \mathbf{0.199\ g\ CaCO_3}$$

PRACTICE EXERCISE

6. Fermentation is one of the oldest studied biochemical processes. The fermentation process is a catabolic pathway, which involves the use of yeast to convert sugars into ethanol and carbon dioxide. In brewing beer, malt sugar is fermented to form ethanol (CH_3CH_2OH) and carbon dioxide (CO_2) according to the following overall reaction:

$$C_6H_{12}O_6\ (aq) \longrightarrow 2\ CH_3CH_2OH\ (aq) + 2\ CO_2\ (g)$$

A micro-brewery is making a batch of beer, and they wish to determine the mass percentage of ethanol in their brew by collecting the CO_2 (g) evolved. They start the process by mixing malt sugar, hops, and yeast in water. The mixture has a total mass of 800 lbs. During fermentation, 10,000 L of CO_2 is collected over water at 25.0°C and 750 mmHg. What is the mass percentage of ethanol in the beer? (The vapor pressure of water at 25°C is 23.76 mm Hg.)

Text Problem: 5.58

PROBLEM TYPE 3: DALTON'S LAW OF PARTIAL PRESSURES

In order to understand how to approach problems with partial pressures, it is important to know (1) the derivation of Dalton's law of partial pressures and (2) the relationship between partial pressure and the total pressure.

(1) Dalton's law for two gases, **A** and **B**, in a container of volume, *V*.

$$P_{Total} = \frac{n_{Total}RT}{V}$$

and, $n_{Total} = n_A + n_B$

$$P_{Total} = \frac{(n_A + n_B)RT}{V}$$

$$P_{Total} = \frac{n_A RT}{V} + \frac{n_B RT}{V}$$

$$\mathbf{P_{Total} = P_A + P_B}$$

In general, $P_{Total} = P_1 + P_2 + P_3 + \ldots + P_n$, where $P_1, P_2, P_3, \ldots$ are the partial pressures of components 1, 2, 3, . . .

(2) The relationship between partial pressure and P_{Total} for two gases, **A** and **B**, in a container of volume, V.

$$\frac{P_A}{P_T} = \frac{\dfrac{n_A RT}{V}}{\dfrac{(n_A + n_B)RT}{V}}$$

$$\frac{P_A}{P_T} = \frac{n_A}{(n_A + n_B)} \quad \text{and} \quad \frac{n_A}{(n_A + n_B)} = X_A$$

where X_A = mole fraction of component A.

$$\frac{P_A}{P_T} = X_A$$

$$P_A = X_A P_T$$

Similarly,

$$P_B = X_B P_T$$

In general, $P_i = X_i P_T$, where X_i is the mole fraction of substance i.

EXAMPLE 5.8

A 2.00 L container at 22.0°C contains a mixture of 1.00 g H_2 (g) and 1.00 g He (g).

(a) What are the partial pressures of H_2 and He? What is the total pressure?

This is a mixture of two gases that obeys Dalton's law of partial pressures.

$$P_T = P_{H_2(g)} + P_{He(g)}$$

Step 1: Calculate the moles of H_2 and He.

$$n_{H_2} = 1.00\,\text{g} \times \frac{1 \text{ mol } H_2}{2.016 \text{ g } H_2} = 0.496 \text{ mol } H_2$$

$$n_{He} = 1.00\,\text{g} \times \frac{1 \text{ mol He}}{4.003 \text{ g He}} = 0.250 \text{ mol He}$$

Step 2: Solve the ideal gas equation algebraically for P, then substitute in the known quantities to solve for P_{H_2} or P_{He}.

$T(\text{K}) = 22.0° + 273° = 295$ K

$$P_{H_2} = \frac{n_{H_2} RT}{V} \qquad P_{He} = \frac{n_{He} RT}{V}$$

$$P_{H_2} = \frac{(0.496 \text{ mol})(295 \text{ K})}{2.00 \text{ L}} \times \frac{0.0821 \text{ L}\cdot\text{atm}}{\text{mol}\cdot\text{K}} \qquad P_{He} = \frac{(0.250 \text{ mol})(295 \text{ K})}{2.00 \text{ L}} \times \frac{0.0821 \text{ L}\cdot\text{atm}}{\text{mol}\cdot\text{K}}$$

$$\mathbf{P_{H_2} = 6.01 \text{ atm}} \qquad \mathbf{P_{He} = 3.03 \text{ atm}}$$

$$P_{Total} = P_{H_2} + P_{He}$$

$$P_{Total} = 6.01 \text{ atm} + 3.03 \text{ atm} = \mathbf{9.04 \text{ atm}}$$

(b) What are the mole fractions of H_2 and He?

Remember that $P_i = X_i P_T$.

$$P_{H_2} = X_{H_2} P_T \qquad P_{He} = X_{He} P_T$$

$$X_{H_2} = \frac{P_{H_2}}{P_T} \qquad X_{He} = \frac{P_{He}}{P_T}$$

$$X_{H_2} = \frac{6.01\ \cancel{atm}}{9.04\ \cancel{atm}} = \mathbf{0.665} \qquad X_{He} = \frac{3.03\ \cancel{atm}}{9.04\ \cancel{atm}} = \mathbf{0.335}$$

PRACTICE EXERCISE

7. About two-thirds of the carbon monoxide (CO) emissions in the United States come from automobiles. CO is extremely toxic to humans because it binds 210 times more strongly to hemoglobin than does O_2. Hemoglobin is the iron-containing protein responsible for oxygen transport in blood. In a typical urban environment, the CO concentration is 10 parts per million (ppm). Assuming an atmospheric pressure of 750 torr, calculate the partial pressure of CO.

Hint: 1 ppm of a gas refers to one part by volume in 1 million volume units of the whole (volume fraction).

Text Problems: 5.54, 5.56, 5.58, 5.60

PROBLEM TYPE 4: ROOT-MEAN-SQUARE (RMS) SPEED

This type of problem typically involves substituting known values into the following equation:

$$u_{rms} = \sqrt{\frac{3RT}{\mathcal{M}}}$$

You must remember to use R in units of $\frac{J}{mol \cdot K}$ and $\mathcal{M}$ in units of kg/mol. 1 joule = $1\frac{kg \cdot m^2}{s^2}$. If these units are used, the units of u_{rms} are m/s.

$$u_{rms}\ (\text{units}) = \sqrt{\frac{\left(\frac{J}{mol \cdot K}\right)(K)}{\left(\frac{kg}{mol}\right)}}$$

$$u_{rms}\ (\text{units}) = \sqrt{\frac{\left(\frac{\cancel{kg} \cdot m^2}{s^2} \times \frac{1}{\cancel{mol} \cdot \cancel{K}}\right)(\cancel{K})}{\left(\frac{\cancel{kg}}{\cancel{mol}}\right)}}$$

$$u_{rms}\ (\text{units}) = \sqrt{\frac{m^2}{s^2}}$$

$$u_{rms}\ \mathbf{(units)} = \frac{\mathbf{m}}{\mathbf{s}}$$

EXAMPLE 5.9

Which has a higher root mean square velocity, H_2 (g) at 150 K or He (g) at 650K?

Step 1: The molar mass ($\mathcal{M}$) must be in units of kg/mol for the units to work out correctly. Convert g/mol to kg/mol.

$$\mathcal{M}_{H_2} = \frac{2.016\ \cancel{g}\ H_2}{1\ \text{mol}\ H_2} \times \frac{1\ \text{kg}}{1000\ \cancel{g}} = 2.016 \times 10^{-3}\ \text{kg/mol}$$

$$\mathcal{M}_{He} = \frac{4.003\ \cancel{g}\ He}{1\ \text{mol He}} \times \frac{1\ \text{kg}}{1000\ \cancel{g}} = 4.003 \times 10^{-3}\ \text{kg/mol}$$

Step 2: Substitute the appropriate values into the equation to solve for u_{rms}.

$$u_{rms}(H_2) = \sqrt{\frac{(3)\left(8.314\frac{J}{mol \cdot K}\right)(150\ K)}{\left(2.016 \times 10^{-3}\frac{kg}{mol}\right)}} \qquad u_{rms}(He) = \sqrt{\frac{(3)\left(8.314\frac{J}{mol \cdot K}\right)(650\ K)}{\left(4.003 \times 10^{-3}\frac{kg}{mol}\right)}}$$

$$\boldsymbol{u_{rms}(H_2) = 1.36 \times 10^3\ m/s} \qquad \boldsymbol{u_{rms}(He) = 2.01 \times 10^3\ m/s}$$

He (g) has the higher root mean square velocity.

PRACTICE EXERCISE

8. What is the root-mean-square velocity of F_2 (*g*) at 298 K?

Text Problem: 5.68

PROBLEM TYPE 5: DEVIATIONS FROM IDEAL BEHAVIOR

Van der Waals' equation is a simple modification of the ideal gas equation. It takes into account intermolecular forces and finite molecular volumes.

Starting with the ideal gas equation:

$$P_{ideal}V_{ideal} = nRT$$

van der Waals made two corrections.

(1) $P_{ideal} = P_{real}$ + correction for attraction between particles

$$P_{ideal} = P_{real} + \frac{an^2}{V_{real}^2} \quad \text{(see Section 5.7 of your text for discussion)}$$

(2) $V_{ideal} = V_{real}$ − correction for finite volume of particles

$$V_{ideal} = V_{real} - nb \quad \text{(see Section 5.7 of your text for discussion)}$$

where,

a and *b* are constants for a particular gas (See Table 5.3)
n = moles of gas

Substituting these corrections into the ideal gas equation leads to the **van der Waals equation.**

$$\left(P_{\text{real}} + \frac{an^2}{V_{\text{real}}^2}\right)(V_{\text{real}} - nb) = nRT$$

EXAMPLE 5.10

The molar volume of isopentane (C_5H_{12}) is 1.00 L at 503 K and 30.0 atm.

(a) Does isopentane behave like an ideal gas?

Calculate the pressure of 1 mol of an ideal gas at 503 K that occupies a volume of 1.00 L.

Step 1: Solve the ideal gas equation algebraically for P, and then substitute in the known values.

$$P = \frac{nRT}{V} = \frac{(1.00\ \text{mol})(503\ \text{K})}{1.00\ \text{L}} \times \frac{0.0821\ \text{L}\cdot\text{atm}}{\text{mol}\cdot\text{K}}$$

$$P = 41.3\ \text{atm}$$

Step 2: Compare the ideal pressure to the actual pressure. Calculating the percentage error would be helpful. Percentage of error is the difference between the two values divided by the actual value.

$$\%\ \text{error} = \frac{41.3\ \text{atm} - 30.0\ \text{atm}}{30.0\ \text{atm}} \times 100 = \mathbf{37.7\ \%}$$

Because of the large percent error, we conclude that under these conditions, C_5H_{12} behaves in a nonideal manner.

(b) Given that $a = 17.0\ \text{L}^2\cdot\text{atm/mol}^2$ and $b = 0.136$ L/mol, calculate the pressure of isopentane as predicted by the van der Waals' equation.

Calculate the pressure of 1 mol of gas at 503 K that occupies 1.00 L using van der Waals' equation.

Step 1: Calculate the correction terms.

$$\frac{an^2}{V_{\text{obs}}^2} = \frac{\left(17.0\ \frac{\text{L}^2\cdot\text{atm}}{\text{mol}^2}\right)(1.00\ \text{mol})^2}{1.00\ \text{L}^2} = 17.0\ \text{atm}$$

$$nb = (1.00\ \text{mol})\left(0.136\ \frac{\text{L}}{\text{mol}}\right) = 0.136\ \text{L}$$

Step 2: Substitute the values into the van der Waals' equation and solve for P_{real}.

$$\left(P_{\text{real}} + \frac{an^2}{V_{\text{real}}^2}\right)(V_{\text{real}} - nb) = nRT$$

$$(P_{\text{real}} + 17.0\ \text{atm})(1.00\ \text{L} - 0.136\ \text{L}) = (1.00\ \text{mol})\left(\frac{0.0821\ \text{L}\cdot\text{atm}}{\text{mol}\cdot\text{K}}\right)(503\ \text{K})$$

$$(P_{\text{real}} + 17.0\ \text{atm})(0.864) = 41.3\ \text{atm}$$

$$P_{\text{real}} + 17.0\ \text{atm} = 47.8\ \text{atm}$$

$$\mathbf{P_{\text{real}}} = 47.8\ \text{atm} - 17.0\ \text{atm} = \mathbf{30.8\ atm}$$

The pressure calculated for 1 mole of this "real" gas at 503 K and 1.00 L using van der Waals' equation is much closer to the actual value of 30.0 atm. The percent error is only 2.7 percent.

PRACTICE EXERCISE

9. Using van der Waals' equation, calculate the pressure exerted by 1.00 mol of water vapor that occupies a volume of 5.55 L at 150°C.

Text Problems: 5.74, 5.76

ANSWERS TO PRACTICE EXERCISES

1. 4.04×10^3 mmHg

2. $V_{CO_2} = 1.1 \times 10^{16}$ L

3. $V_{air} = 160$ mL

4. The gas is fluorine, F_2.

5. 176 g NaN_3

6. mass % ethanol = 5.0 %

7. $P_{CO} = 7.5 \times 10^{-3}$ torr

8. $u_{rms} = 442$ m/s

9. $P_{real} = 6.11$ atm

SOLUTIONS TO SELECTED TEXT PROBLEMS

5.14 Pressure Conversions, Problem Type 1.

The conversion needed to solve this problem is:

$$1 \text{ atm} = 760 \text{ mmHg}$$

$$\textbf{? atm} = 606 \text{ mmHg} \times \frac{1 \text{ atm}}{760 \text{ mmHg}} = \textbf{0.797 atm}$$

5.18 Temperature does not change in this problem. Pressure and volume are the variables; it is a Boyle's law problem.

$V_2 = 0.10\ V_1$

$$P_2 = \frac{P_1V_1}{V_2}$$

$$\boldsymbol{P_2} = \frac{(5.3 \text{ atm})V_1}{0.10V_1} = \textbf{53 atm}$$

5.20 Only one or two variable quantities have fixed values, Problem Type 2B. This is a Boyle's law problem.

(a)

Step 1: n and T are constant. The ideal gas equation is already arranged with all constants on one side.

$$PV = nRT$$

$$PV = \text{constant}$$

Step 2: Pressure and volume change.

$$P_1V_1 = \text{constant} = P_2V_2$$

and,

$$\boldsymbol{P_1V_1 = P_2V_2}$$

(Initial Pressure)(Initial Volume) = (Final Pressure)(Final Volume)

Step 3: Solve the equation algebraically for the missing quantity (V_2). Then, calculate its value by substituting in the known quantities.

$$V_2 = \frac{P_1V_1}{V_2}$$

$$\boldsymbol{V_2} = \frac{(1.2 \text{ atm})(3.8 \text{ L})}{(6.6 \text{ atm})} = \textbf{0.69 L}$$

(b) On this part, you should also come up with $P_1V_1 = P_2V_2$.

Solve the equation algebraically for the missing quantity (P_2). Then, calculate its value by substituting in the known quantities.

$$P_2 = \frac{P_1V_1}{V_2}$$

$$\boldsymbol{P_2} = \frac{(1.2 \text{ atm})(3.8 \text{ L})}{(0.075 \text{ L})} = \textbf{61 atm}$$

5.22 77 K – 273 K = –196°C; the other two are similar: –269°C and 5.7×10^3°C.

5.24 Only one or two variable quantities have fixed values, Problem Type 2B. This is a Charles' law problem.

Step 1: P and n are constant. Remember, as long as gas is not added or removed, the number of moles of gas (n) is constant. Rearrange the ideal gas equation so that the constants are all on one side.

$$\frac{V}{T} = \frac{nR}{P}$$

$$\frac{V}{T} = \text{constant}$$

Step 2: Volume and temperature change.

$$\frac{V_1}{T_1} = \text{constant} = \frac{V_2}{T_2}$$

and,

$$\frac{\boldsymbol{V_1}}{\boldsymbol{T_1}} = \frac{\boldsymbol{V_2}}{\boldsymbol{T_2}}$$

$$\frac{\text{Initial Volume}}{\text{Initial Temperature}} = \frac{\text{Final Volume}}{\text{Final Temperature}}$$

Step 3: Solve the equation algebraically for the missing quantity (T_2). Then, calculate its value by substituting in the known quantities.

$$T_1(\text{K}) = 88° + 273° = 361 \text{ K}$$

$$T_2 = \frac{T_1V_2}{V_1}$$

$$\boldsymbol{T_2} = \frac{(361 \text{ K})(3.4 \cancel{\text{L}})}{(9.6 \cancel{\text{L}})} = \mathbf{1.3 \times 10^2\ K}$$

5.26 This is a gas stoichiometry problem that requires a knowledge of Avogadro's law to solve. Avogadro's law states that the volume of a gas is directly proportional to the number of moles of gas at constant temperature and pressure.

The volume ratio, 1 vol. Cl_2 : 3 vol. F_2 : 2 vol. product, can be written as a mole ratio, 1 mol Cl_2 : 3 mol F_2 : 2 mol product.

Attempt to write a balanced chemical equation. The subscript of F in the product will be three times the Cl subscript, because there are three times as many F atoms reacted as Cl atoms.

$$\mathbf{1}\ Cl_2(g) + \mathbf{3}\ F_2(g) \longrightarrow \mathbf{2}\ Cl_xF_{3x}(g)$$

Balance the equation. The x must equal one so that there are two Cl atoms on each side of the equation. If x = 1, the subscript on F is 3.

$$Cl_2(g) + 3\ F_2(g) \longrightarrow 2\ ClF_3(g)$$

The formula of the product is $\mathbf{ClF_3}$.

5.32 Given three of the four variable quantities in the equation (P, V, n, and T), Problem Type 2A.

Step 1: You are given n, V, and T. Solve $PV = nRT$ algebraically for P.

$$P = \frac{nRT}{V}$$

Step 2: Check that the three known quantities (n, V, and T) have the appropriate units.

T must be converted to units of K

$$T(\text{K}) = {}^\circ\text{C} + 273^\circ$$

$$T(\text{K}) = 62^\circ + 273^\circ = 335\ \text{K}$$

Step 3: Calculate the pressure by substituting the known values into the equation.

$$\mathbf{P} = \frac{(6.9\ \cancel{\text{mol}})\left(0.0821\ \frac{\cancel{\text{L}}\cdot\text{atm}}{\cancel{\text{mol}}\cdot\cancel{\text{K}}}\right)(335\ \cancel{\text{K}})}{30.4\ \cancel{\text{L}}} = \mathbf{6.2\ atm}$$

5.34 In this problem, the moles of gas and the volume the gas occupies are constant. Temperature and pressure change.

$$\frac{P_1}{T_1} = \frac{P_2}{T_2}$$

$$T_2 = \frac{T_1 P_2}{P_1}$$

$$\mathbf{T_2} = \frac{(298\ \text{K})(2.00\ \cancel{\text{atm}})}{(0.800\ \cancel{\text{atm}})} = \mathbf{745\ K = 472^\circ C}$$

5.36 $P_1 = 1.0\ \text{atm}$ $\quad$ $P_2 = ?$

$T_1 = 273\ \text{K}$ $\quad$ $T_2 = (250 + 273)\text{K} = 523\ \text{K}$

$$P_2 = \frac{P_1 T_2}{T_1}$$

$$\mathbf{P_2} = \frac{(1.0\ \text{atm})(523\ \cancel{\text{K}})}{273\ \cancel{\text{K}}} = \mathbf{1.9\ atm}$$

5.38 $$V_2 = \frac{V_1 T_2}{T_1}$$

$$\mathbf{V_2} = \frac{(0.78\ \text{L})(36.5 + 273)\cancel{\text{K}}}{(20.1 + 273)\cancel{\text{K}}} = \mathbf{0.82\ L}$$

5.40 In this problem, the number of moles of gas is constant. Pressure, volume, and temperature change.

$$\frac{P_1 V_1}{T_1} = \frac{P_2 V_2}{T_2}$$

$T_2(\text{K}) = 150^\circ + 273^\circ = 423\ \text{K}$

$$V_2 = \frac{P_1V_1T_2}{P_2T_1}$$

$$V_2 = \frac{(1\ \text{atm})(488\ \text{mL})(423\ \text{K})}{(22.5\ \text{atm})(273\ \text{K})} = \mathbf{33.6\ mL}$$

Note: Since R is not in this equation, the units of pressure and volume do *not* have to be atm and liters, respectively. However, temperature must *always* be in units of Kelvin for gas law calculations.

5.42 The molar mass of CO_2 = 44.01 g/mol. Since $PV = nRT$, we write:

$$P = \frac{nRT}{V}$$

$$\boldsymbol{P} = \frac{\left(0.050\ \text{g} \times \frac{1\ \text{mol}}{44.01\ \text{g}}\right)\left(0.0821\ \frac{\text{L}\cdot\text{atm}}{\text{mol}\cdot\text{K}}\right)(30+273)\ \text{K}}{4.6\ \text{L}} = \mathbf{6.1 \times 10^{-3}\ atm}$$

5.44 Calculations involving molar mass, Problem Type 2C.

Step 1: Algebraically solve the ideal gas equation for molar mass. You should come up with

$$\mathcal{M} = \frac{dRT}{P}$$

Step 2: Calculate the density, and check that the other known quantities (P and T) have the appropriate units.

$$d = \frac{7.10\ \text{g}}{5.40\ \text{L}} = 1.31\ \text{g/L}$$

$$T = 44° + 273° = 317\ \text{K}$$

$$P = 741\ \text{torr} \times \frac{1\ \text{atm}}{760\ \text{torr}} = 0.975\ \text{atm}$$

Step 3: Calculate the molar mass by substituting in the known quantities.

$$\mathcal{M} = \frac{\left(1.31\ \frac{\text{g}}{\text{L}}\right)\left(0.0821\ \frac{\text{L}\cdot\text{atm}}{\text{mol}\cdot\text{K}}\right)(317\ \text{K})}{0.975\ \text{atm}} = \mathbf{35.0\ g/mol}$$

5.46 The number of particles in 1 L of gas at STP is:

$$\text{Number of particles} = 1.0\ \text{L} \times \frac{1\ \text{mol}}{22.414\ \text{L}} \times \frac{6.022 \times 10^{23}\ \text{particles}}{1\ \text{mol}} = 2.7 \times 10^{22}\ \text{particles}$$

$$\textbf{Number of N}_\mathbf{2}\textbf{ molecules} = \left(\frac{78\%}{100\%}\right)(2.7 \times 10^{22}\ \text{particles}) = \mathbf{2.1 \times 10^{22}\ N_2\ molecules}$$

$$\textbf{Number of O}_\mathbf{2}\textbf{ molecules} = \left(\frac{21\%}{100\%}\right)(2.7 \times 10^{22}\ \text{particles}) = \mathbf{5.7 \times 10^{21}\ O_2\ molecules}$$

$$\textbf{Number of Ar atoms} = \left(\frac{1\%}{100\%}\right)(2.7 \times 10^{22}\ \text{particles}) = \mathbf{3 \times 10^{20}\ Ar\ atoms}$$

5.48 The density can be calculated from the ideal gas equation.

$$d = \frac{P\mathcal{M}}{RT}$$

$\mathcal{M} = 1.008\text{ g/mol} + 79.90\text{ g/mol} = 80.91\text{ g/mol}$

$T = 46° + 273° = 319\text{ K}$

$$P = 733\text{ mmHg} \times \frac{1\text{ atm}}{760\text{ mmHg}} = 0.964\text{ atm}$$

$$\boldsymbol{d} = \frac{(0.964\text{ atm})\left(\frac{80.91\text{ g}}{\text{mol}}\right)}{319\text{ K}} \times \frac{\text{mol}\cdot\text{K}}{0.0821\text{ L}\cdot\text{atm}} = \mathbf{2.98\text{ g/L}}$$

5.50 This is an extension of an ideal gas law calculation involving molar mass. If you determine the molar mass of the gas, you will be able to determine the molecular formula from the empirical formula (see Determining the Molecular Formula of a Compound, Problem Type 6, Chapter 3).

$$\mathcal{M} = \frac{dRT}{P}$$

Calculate the density, then substitute its value into the equation above.

$$d = \frac{0.100\text{ g}}{22.1\text{ mL}} \times \frac{1000\text{ mL}}{1\text{ L}} = 4.52\text{ g / L}$$

$T(\text{K}) = 20° + 273° = 293\text{ K}$

$$\mathcal{M} = \frac{\left(4.52\ \frac{\text{g}}{\text{L}}\right)\left(0.0821\ \frac{\text{L}\cdot\text{atm}}{\text{mol}\cdot\text{K}}\right)(293\text{ K})}{1.02\text{ atm}} = 107\text{ g/mol}$$

Compare the empirical mass to the molar mass.

$$\text{empirical mass} = 32.07\text{ g/mol} + 4(19.00\text{ g/mol}) = 108.07\text{ g/mol}$$

Remember, the molar mass will be a whole number multiple of the empirical mass. In this case, the $\frac{\text{molar mass}}{\text{empirical mass}} \cong 1$. Therefore, the molecular formula is the same as the empirical formula, $\mathbf{SF_4}$.

5.54 Dalton's law states that the total pressure of the mixture is the sum of the partial pressures.

(a) $\boldsymbol{P_{total}} = 0.32\text{ atm} + 0.15\text{ atm} + 0.42\text{ atm} = \mathbf{0.89\text{ atm}}$

(b) We know:

$P_1 = 0.15\text{ atm} + 0.42\text{ atm} = 0.57\text{ atm}$	$P_2 = 1.0\text{ atm}$
$T_1 = (15 + 273)\text{ K} = 288\text{ K}$	$T_2 = 273\text{ K}$
$V_1 = 2.5\text{L}$	$V_2 = ?$

$$V_2 = \frac{V_1P_1T_2}{P_2T_1}$$

$$\boldsymbol{V_2} = \frac{(2.5\text{ L})(0.57\text{ atm})(273\text{ K})}{(1.0\text{ atm})(288\text{ K})} = \mathbf{1.4\text{ L at STP}}$$

5.56 $P_{Total} = P_1 + P_2 + P_3 + \ldots + P_n$

In this case,

$$P_{Total} = P_{Ne} + P_{He} + P_{H_2O}$$

$$P_{Ne} = P_{Total} - P_{He} - P_{H_2O}$$

$$P_{Ne} = 745\text{ mm Hg} - 368\text{ mmHg} - 28.3\text{ mmHg} = \mathbf{349\ mmHg}$$

5.58 Collecting a gas over water, Problem Type 2E.

Step 1: Dalton's law of partial pressure states that

$$\boldsymbol{P_{Total} = P_1 + P_2 + P_3 + \ldots + P_n}$$

In this case,

$$P_{Total} = P_{H_2} + P_{H_2O}$$

Step 2: Solve the equation algebraically for P_{H_2}. Then, calculate its value by substituting in the known quantities.

$$P_{H_2} = P_{Total} - P_{H_2O}$$

$$P_{H_2} = 0.980\text{ atm} - (23.8\text{ mmHg})\left(\frac{1\text{ atm}}{760\text{ mmHg}}\right) = 0.949\text{ atm}$$

Step 3: Map out the following strategy to calculate the mass of Zn (s) reacted.

$$\text{volume } H_2 \rightarrow \text{moles } H_2 \rightarrow \text{moles Zn} \rightarrow \text{grams Zn}$$

Step 4: Solve the ideal gas equation algebraically for moles of H_2. Then, calculate its value by substituting in the known quantities.

$$T(\text{K}) = 25° + 273° = 298\text{ K}$$

$$n_{H_2} = \frac{P_{H_2}V}{RT}$$

$$n_{H_2} = \frac{(0.949\text{ atm})(7.80\text{ L})}{298\text{ K}} \times \frac{\text{mol}\cdot\text{K}}{0.0821\text{ L}\cdot\text{atm}} = 0.303\text{ mol } H_2$$

Step 5: Calculate the mass of Zn metal using the strategy in Step 3.

$$\textbf{? g Zn} = 0.303\text{ mol } H_2 \times \frac{1\text{ mol Zn}}{1\text{ mol } H_2} \times \frac{65.38\text{ g Zn}}{1\text{ mol Zn}} = \mathbf{19.8\ g\ Zn}$$

5.60 $P_i = X_iP_T$

We need to determine the mole fractions of each component in order to determine their partial pressures. To calculate mole fraction, write the balanced chemical equation to determine the correct mole ratio.

$$2\,NH_3\,(g) \longrightarrow N_2\,(g) + 3\,H_2\,(g)$$

The mole fractions of H_2 and N_2 are:

$$X_{H_2} = \frac{3 \text{ mol}}{3 \text{ mol} + 1 \text{ mol}} = 0.750$$

$$X_{N_2} = \frac{1 \text{ mol}}{3 \text{ mol} + 1 \text{ mol}} = 0.250$$

The partial pressures of H_2 and N_2 are:

$$P_{H_2} = X_{H_2}P_T = (0.750)(866 \text{ mm Hg}) = \mathbf{650 \text{ mmHg}}$$

$$P_{N_2} = X_{N_2}P_T = (0.250)(866 \text{ mm Hg}) = \mathbf{217 \text{ mm Hg}}$$

5.68 Root-Mean-Square (RMS) Speed, Problem Type 4.

Step 1: Recognize that you need to use the following equation.

$$u_{rms} = \sqrt{\frac{3RT}{\mathcal{M}}}$$

Step 2: Calculate the molar masses ($\mathcal{M}$) of N_2, O_2, and O_3. Remember, $\mathcal{M}$ must be in units of kg/mol.

$$\mathcal{M}_{N_2} = 2(14.01 \text{ g/mol}) = 28.02 \frac{\cancel{g}}{\text{mol}} \times \frac{1 \text{ kg}}{1000 \cancel{g}} = 0.02802 \text{ kg/mol}$$

$$\mathcal{M}_{O_2} = 2(16.00 \text{ g/mol}) = 32.00 \frac{\cancel{g}}{\text{mol}} \times \frac{1 \text{ kg}}{1000 \cancel{g}} = 0.03200 \text{ kg/mol}$$

$$\mathcal{M}_{O_3} = 3(16.00 \text{ g/mol}) = 48.00 \frac{\cancel{g}}{\text{mol}} \times \frac{1 \text{ kg}}{1000 \cancel{g}} = 0.04800 \text{ kg/mol}$$

Step 3: Substitute the known quantities into the equation and solve for u_{rms}.

$T(\text{K}) = -23° + 273° = 250 \text{ K}$

$$u_{rms}(N_2) = \sqrt{\frac{(3)\left(8.314 \frac{\text{J}}{\text{mol} \cdot \text{K}}\right)(250 \text{ K})}{\left(0.02802 \frac{\text{kg}}{\text{mol}}\right)}}$$

$$u_{rms}(N_2) = \mathbf{472 \text{ m/s}}$$

Similarly,

$$u_{rms}(O_2) = \mathbf{441 \text{ m/s}} \qquad u_{rms}(O_3) = \mathbf{360 \text{ m/s}}$$

5.74 Deviations from Ideal Behavior, Problem Type 5. In this problem we wish to determine if the gas deviates from ideal behavior, by comparing the ideal pressure with the actual pressure.

Step 1: Calculate the pressure of 10.0 moles of an ideal gas at 27°C in a 1.50 L container. Solve the ideal gas equation algebraically for P, then substitute in the known values.

$T(\text{K}) = 27°\text{C} + 273° = 300 \text{ K}$

$$P = \frac{nRT}{V} = \frac{(10.0 \cancel{\text{mol}})\left(0.0821 \frac{\cancel{\text{L}} \cdot \text{atm}}{\cancel{\text{mol}} \cdot \cancel{\text{K}}}\right)(300 \cancel{\text{K}})}{1.50 \cancel{\text{L}}} = \mathbf{164 \text{ atm}}$$

Step 2: Compare the ideal pressure to the actual pressure by calculating the percent error.

$$\% \text{ error} = \frac{164 \text{ atm} - 130 \text{ atm}}{130 \text{ atm}} \times 100 = \mathbf{26.2\ \%}$$

Based on the large percent error, we conclude that under this condition of high pressure, the gas behaves in a non-ideal manner.

5.76 The PV products are:

P (atm)	V (L)	PV (atm·L)
0.0500	448.2	22.41
0.100	223.8	22.38
0.151	148.8	22.47
0.202	110.8	22.38
0.252	89.0	22.43

In the range of pressures given, the $P \times V$ product of carbon dioxide is constant to three significant figure accuracy. In this range, carbon dioxide behaves like an ideal gas.

5.78 You can map out the following strategy to solve for the total volume of gas.

grams nitroglycerin → moles nitroglycerin → moles products → volume of products

$$? \text{ mol products} = 2.6 \times 10^2 \cancel{\text{g}} \text{ nitroglycerin} \times \frac{1 \cancel{\text{mol}} \text{ nitroglycerin}}{227.09 \cancel{\text{g}} \text{ nitroglycerin}} \times \frac{29 \text{ mol product}}{4 \cancel{\text{mol}} \text{ nitroglycerin}} = 8.3 \text{ mol}$$

Calculating the volume of products:

$$V_{\text{product}} = \frac{n_{\text{product}}RT}{P} = \frac{(8.3 \cancel{\text{mol}})\left(0.0821 \frac{\text{L} \cdot \cancel{\text{atm}}}{\cancel{\text{mol}} \cdot \cancel{\text{K}}}\right)(298 \cancel{\text{K}})}{(1.2 \cancel{\text{atm}})} = \mathbf{1.7 \times 10^2\ L}$$

The relationship between partial pressure and P_{total} is:

$$P_i = X_i P_T$$

Calculate the mole fraction of each gaseous product, then calculate its partial pressure using the equation above.

$$X_{\text{component}} = \frac{\text{moles component}}{\text{total moles all components}}$$

$$X_{CO_2} = \frac{12 \text{ mol } CO_2}{29 \text{ mol product}} = \mathbf{0.41}$$

Similarly, $X_{H_2O} = 0.34$, $X_{N_2} = 0.21$, and $X_{O_2} = 0.034$

$$P_{CO_2} = X_{CO_2} P_T$$

$$P_{CO_2} = (0.41)(1.2 \text{ atm}) = \mathbf{0.49\ atm}$$

Similarly, $P_{H_2O} = 0.41$ atm, $P_{N_2} = 0.25$ atm, and $P_{O_2} = 0.041$ atm.

5.80 Stoichiometry involving gases, Problem Type 2D.

Step 1: Write the balanced chemical equation.

(a) $NH_4NO_2\ (s) \longrightarrow N_2\ (g) + 2\ H_2O\ (l)$

(b)

Step2: Map out the following strategy to solve the problem.

$$\text{volume } N_2 \rightarrow \text{moles } N_2 \rightarrow \text{moles } NH_4NO_2 \rightarrow \text{grams } NH_4NO_2$$

First, calculate the moles of N_2 using the ideal gas equation.

$T(\text{K}) = 22° + 273° = 295 \text{ K}$

$$V = 86.2 \text{ mL} \times \frac{1 \text{ L}}{1000 \text{ mL}} = 0.0862 \text{ L}$$

$$n_{N_2} = \frac{P_{N_2}V}{RT}$$

$$n_{N_2} = \frac{(1.20 \text{ atm})(0.0862 \text{ L})}{295 \text{ K}} \times \frac{\text{mol}\cdot\text{K}}{0.0821 \text{ L}\cdot\text{atm}} = \mathbf{4.27 \times 10^{-3}\ mol}$$

Step 3: Use the mole ratio from the balanced equation to convert from moles of N_2 to moles of NH_4NO_2. Then, use the molar mass of NH_4NO_2 to convert to grams of NH_4NO_2.

$$\textbf{? g } \mathbf{NH_4NO_2} = 4.27 \times 10^{-3} \text{ mol } N_2 \times \frac{1 \text{ mol } NH_4NO_2}{1 \text{ mol } N_2} \times \frac{64.05 \text{ g } NH_4NO_2}{1 \text{ mol } NH_4NO_2} = \mathbf{0.273\ g}$$

5.82 No, because an ideal gas cannot be liquefied, since the assumption is that there are no intermolecular forces in an ideal gas.

5.84 The partial pressure of carbon dioxide is higher in the winter because carbon dioxide is utilized less by photosynthesis in plants.

5.86 **(a)** Write a balanced chemical equation.

$$2\,NaHCO_3\,(s) \longrightarrow Na_2CO_3\,(s) + CO_2\,(g) + H_2O\,(g)$$

First, calculate the moles of CO_2 produced.

$$\text{? mol } CO_2 = 5.0 \text{ g } NaHCO_3 \times \frac{1 \text{ mol } NaHCO_3}{84.01 \text{ g } NaHCO_3} \times \frac{1 \text{ mol } CO_2}{2 \text{ mol } NaHCO_3} = 0.030 \text{ mol}$$

Next, calculate the volume of CO_2 produced using the ideal gas equation.

$T(\text{K}) = 180° + 273° = 453 \text{ K}$

$$V_{CO_2} = \frac{n_{CO_2}RT}{P}$$

$$V_{CO_2} = \frac{(0.030 \text{ mol})\left(0.0821 \frac{\text{L}\cdot\text{atm}}{\text{mol}\cdot\text{K}}\right)(453 \text{ K})}{(1.3 \text{ atm})} = \mathbf{0.86\ L}$$

(b) The balanced chemical equation for the decomposition of NH_4CO_3 is

$$NH_4CO_3\,(s) \longrightarrow NH_3\,(g) + CO_2\,(g) + H_2O\,(g)$$

The advantage in using the ammonium salt is that more gas is produced per gram of reactant. The disadvantage is that one of the gases is ammonia. The strong odor of ammonia would <u>not</u> make the ammonium salt a good choice for baking.

5.88 First, calculate the moles of H_2 formed.

$$? \text{ mol } H_2 = 3.12\,\cancel{g}\text{ Al} \times \frac{1\,\cancel{mol}\text{ Al}}{26.98\,\cancel{g}} \times \frac{3\text{ mol } H_2}{2\,\cancel{mol}\text{ Al}} = 0.173\text{ mol}$$

Next, calculate the volume of H_2 produced using the ideal gas equation.

$$V_{H_2} = \frac{n_{H_2}RT}{P} = \frac{(0.173\,\cancel{mol})\left(0.0821\,\frac{L\cdot \cancel{atm}}{\cancel{mol}\cdot \cancel{K}}\right)(273\,\cancel{K})}{(1..00\,\cancel{atm})} = \mathbf{3.88\ L}$$

5.90 **(a)** $C_3H_8\,(g) + 5\,O_2\,(g) \longrightarrow 3\,CO_2\,(g) + 4\,H_2O\,(g)$

(b) From the balanced equation, we see that there is a 1:3 mole ratio between C_3H_8 and CO_2.

$$\mathbf{?\ L\ CO_2} = 7.45\,\cancel{g}\,C_3H_8 \times \frac{1\,\cancel{mol}\,C_3H_8}{44.09\,\cancel{g}\,C_3H_8} \times \frac{3\,\cancel{mol}\,CO_2}{1\,\cancel{mol}\,C_3H_8} \times \frac{22.414\text{ L } CO_2}{1\,\cancel{mol}\,CO_2} = \mathbf{11.4\ L\ CO_2}$$

5.92 Calculate the initial number of moles of NO and O_2 using the ideal gas equation.

$$n_{NO} = \frac{P_{NO}V}{RT} = \frac{(0.500\,\cancel{atm})(4.00\,\cancel{L})}{\left(0.0821\,\frac{\cancel{L}\cdot \cancel{atm}}{mol\cdot \cancel{K}}\right)(298\,\cancel{K})} = 0.0817\text{ mol NO}$$

$$n_{O_2} = \frac{P_{O_2}V}{RT} = \frac{(1.00\,\cancel{atm})(2.00\,\cancel{L})}{\left(0.0821\,\frac{\cancel{L}\cdot \cancel{atm}}{mol\cdot \cancel{K}}\right)(298\,\cancel{K})} = 0.0817\text{ mol } O_2$$

Determine which reactant is the limiting reagent.

Calculate moles of NO_2 formed if all NO reacts.

$$? \text{ mol } NO_2 = 0.0817\,\cancel{mol}\text{ NO} \times \frac{2\text{ mol } NO_2}{2\,\cancel{mol}\text{ NO}} = 0.0817\text{ mol } NO_2$$

Calculate moles of NO_2 formed if all O_2 reacts.

$$? \text{ mol } NO_2 = 0.0817\,\cancel{mol}\,O_2 \times \frac{2\text{ mol } NO_2}{1\,\cancel{mol}\,O_2} = 0.163\text{ mol } NO_2$$

The initial amount of NO limits the amount of product that can be formed; therefore, it is the limiting reagent.

Determine the molar amounts of NO, O_2, and NO_2 after complete reaction.

mol NO = 0 moles (All NO is consumed during reaction)

mol NO_2 = 0.0817 moles (Calculated above)

$$\text{mol } O_2 \text{ consumed} = 0.0817 \text{ mol NO} \times \frac{1 \text{ mol } O_2}{2 \text{ mol NO}} = 0.0409 \text{ mol } O_2 \text{ consumed}$$

$$\textbf{mol } \mathbf{O_2} \textbf{ remaining} = 0.0817 \text{ mol } O_2 \text{ initial} - 0.0409 \text{ mol } O_2 \text{ consumed} = \mathbf{0.0408 \text{ mol } O_2}$$

Calculate the partial pressures of O_2 and NO_2 using the ideal gas equation.

Volume of entire apparatus = 2.00 L + 4.00 L = 6.00 L
T(K) = 25° + 273° = 298 K

$$P_{O_2} = \frac{n_{O_2}RT}{V} = \frac{(0.0408 \text{ mol})\left(0.0821 \frac{\text{L}\cdot\text{atm}}{\text{mol}\cdot\text{K}}\right)(298 \text{ K})}{(6.00 \text{ L})} = \mathbf{0.166 \text{ atm}}$$

$$P_{NO_2} = \frac{n_{NO_2}RT}{V} = \frac{(0.0817 \text{ mol})\left(0.0821 \frac{\text{L}\cdot\text{atm}}{\text{mol}\cdot\text{K}}\right)(298 \text{ K})}{(6.00 \text{ L})} = \mathbf{0.333 \text{ atm}}$$

5.94 **(a)** $CaO\ (s) + CO_2\ (g) \longrightarrow CaCO_3\ (s)$

$BaO\ (s) + CO_2\ (g) \longrightarrow BaCO_3\ (s)$

(b) First, we need to find the number of moles of CO_2 consumed in the reaction. We can do this by calculating the initial moles of CO_2 in the flask and then comparing it to the CO_2 remaining after the reaction.

Initially:
$$n_{CO_2} = \frac{PV}{RT} = \frac{\left(746 \text{ mmHg} \times \frac{1 \text{ atm}}{760 \text{ mmHg}}\right)(1.46 \text{ L})}{\left(0.0821 \frac{\text{L}\cdot\text{atm}}{\text{mol}\cdot\text{K}}\right)(35 + 273)\text{K}} = 0.0567 \text{ mol } CO_2$$

Remaining:
$$n_{CO_2} = \frac{PV}{RT} = \frac{\left(252 \text{ mmHg} \times \frac{1 \text{ atm}}{760 \text{ mmHg}}\right)(1.46 \text{ L})}{\left(0.0821 \frac{\text{L}\cdot\text{atm}}{\text{mol}\cdot\text{K}}\right)(35 + 273)\text{K}} = 0.0191 \text{ mol } CO_2$$

Thus, the amount of CO_2 consumed in the reaction is: (0.0567 mol − 0.0191 mol) = 0.0376 mol CO_2.

Since the mole ratio between CO_2 and both reactants (CaO and BaO) is 1:1, 0.0376 mole of the mixture must have reacted. We can write:

mol CaO + mol BaO = 0.0376 mol

Let x = mass of CaO in the mixture, then $(4.88 - x)$ = mass of BaO in the mixture. We can write:

$$\left[x \text{ g CaO} \times \frac{1 \text{ mol CaO}}{56.08 \text{ g CaO}}\right] + \left[(4.88 - x)\text{ g BaO} \times \frac{1 \text{ mol BaO}}{153.3 \text{ g BaO}}\right] = 0.0376 \text{ mol}$$

$$0.01783x - 0.006523x + 0.0318 = 0.0376$$

$$x = 0.513\ \text{g} = \text{mass of CaO in the mixture}$$

$$\text{mass of BaO in the mixture} = 4.88 - x = 4.37\ \text{g}$$

The percent compositions by mass in the mixture are:

$$\text{CaO: } \frac{0.513\ \text{g}}{4.88\ \text{g}} \times 100\% = \mathbf{10.5\%} \qquad \text{BaO: } \frac{4.37\ \text{g}}{4.88\ \text{g}} \times 100\% = \mathbf{89.5\%}$$

5.96 The volume of one alveoli is:

$$V = \frac{4}{3}\pi r^3 = \frac{4}{3}\pi(0.0050\ \text{cm})^3 = 5.2 \times 10^{-7}\ \text{cm}^3 \times \frac{1\ \text{mL}}{1\ \text{cm}^3} \times \frac{0.001\ \text{L}}{1\ \text{mL}} = 5.2 \times 10^{-10}\ \text{L}$$

The number of moles of air in one alveoli can be calculated using the ideal gas equation.

$$n = \frac{PV}{RT} = \frac{(1.0\ \text{atm})(5.2 \times 10^{-10}\ \text{L})}{\left(0.0821 \frac{\text{L}\cdot\text{atm}}{\text{mol}\cdot\text{K}}\right)(37 + 273)\text{K}} = 2.0 \times 10^{-11}\ \text{mol of air}$$

Since the air inside the alveoli is 14 percent oxygen, the moles of oxygen in one alveoli equals:

$$2.0 \times 10^{-11}\ \text{mol of air} \times \frac{14\%\ \text{oxygen}}{100\%\ \text{air}} = 2.8 \times 10^{-12}\ \text{mol } O_2$$

Converting to O_2 molecules:

$$2.8 \times 10^{-12}\ \text{mol } O_2 \times \frac{6.022 \times 10^{23}\ O_2\ \text{molecules}}{1\ \text{mol } O_2} = \mathbf{1.7 \times 10^{12}\ O_2\ molecules}$$

5.98 **(a)** No. Temperature is a statistical concept.

(b) (i) Since the two He samples are at the same temperature, the rms speeds and the average kinetic energies are the same.

(ii) The He atoms in V_1 (smaller volume) collide with the walls more frequently. Since the average kinetic energies are the same, the force exerted in the collisions is the same in both flasks.

(c) (i) The rms speed is greater at the higher temperature, T_2.

(ii) The He atoms at the higher temperature, T_2, collide with the walls with greater frequency and with greater force.

(d) (i) False. The rms speed is greater for the lighter gas, He.

(ii) True. The gases are at the same temperature.

(iii) True. $$u_{rms} = \sqrt{\frac{(3)\left(8.314 \frac{\text{J}}{\text{mol}\cdot\text{K}}\right)(74 + 273)\text{K}}{4.003 \times 10^{-3}\ \text{kg/mol}}} = 1.47 \times 10^3\ \text{m/s}$$

CHAPTER 6
ENERGY RELATIONSHIPS IN CHEMICAL REACTIONS

PROBLEM-SOLVING STRATEGIES AND TUTORIAL SOLUTIONS

TYPES OF PROBLEMS

Problem Type 1: Thermochemical Equations.

Problem Type 2: Calculating Heat Absorbed or Released Using Specific Heat Data.

Problem Type 3: Calorimetry.
- **(a)** Constant-volume calorimetry.
- **(b)** Constant-pressure calorimetry.

Problem Type 4: Standard Enthalpy of Formation and Reaction.
- **(a)** Calculating the standard enthalpy of reaction.
- **(b)** Direct method of calculating the standard enthalpy of formation.
- **(c)** Indirect method of calculating the standard enthalpy of formation, Hess's law.

Problem Type 5: The First Law of Thermodynamics.
- **(a)** Applying the First Law of Thermodynamics.
- **(b)** Calculating the work done in gas expansion.
- **(c)** Enthalpy and the First Law of Thermodynamics. Calculating the internal energy change of a gaseous reaction.

PROBLEM TYPE 1: THERMOCHEMICAL EQUATIONS

Equations showing both the mass and enthalpy relations are called **thermochemical equations**. The following guidelines are helpful in writing and interpreting thermochemical equations:

1. The stoichiometric coefficients always refer to the number of moles of each substance.
2. When an equation is reversed, the roles of reactants and products change. Consequently, the magnitude of ΔH for the equation remains the same, but its sign changes.
3. If both sides of a thermochemical equation are multiplied by a factor n, then ΔH must also be multiplied by the same factor.
4. When writing thermochemical equations, the physical states of all reactants and products must be specified, because they help determine the actual enthalpy changes.

See Section 6.3 of your text for further discussion.

EXAMPLE 6.1
Given the thermochemical equation

$$\mathbf{SO_2\,(\mathit{g}) + 1/2\;O_2\,(\mathit{g}) \longrightarrow SO_3\,(\mathit{g})\qquad \Delta \mathit{H} = -99\;kJ}$$

how much heat is evolved when (a) 1/2 mol of SO_2 reacts and (b) 3 mol of SO_2 reacts?

(a) Guideline 3 above states that if both sides of a thermochemical equation are multiplied by a factor n, then ΔH must also be multiplied by the same factor. If 1/2 mole of SO_2 reacts, that means that we are multiplying the equation by 1/2. Therefore, we must multiply ΔH by 1/2.

$$\textbf{heat evolved} = 1/2(99\text{ kJ}) = \mathbf{5.0 \times 10^1\ kJ}$$

(b) Following the same argument as in part a:

$$\textbf{heat evolved} = 3(99\text{ kJ}) = \mathbf{3.0 \times 10^2\ kJ}$$

Why isn't there a negative sign in our answer? The sign convention for an exothermic reaction (energy, as heat, is released by the system) is negative ($-\Delta H$). However, in the above example, we state that heat is evolved, so a negative sign is unnecessary.

Tip: Remember that the heat evolved or absorbed (q) by a reaction carried out under constant-pressure conditions is equal to the enthalpy change of the system, ΔH.

PRACTICE EXERCISE

1. Given the thermochemical equation

$$SO_2\,(g) + 1/2\,O_2\,(g) \longrightarrow SO_3\,(g) \qquad \Delta H = -99\text{ kJ}$$

how much heat is evolved when 75 g of SO_2 is combusted?

Text Problem: 6.50

PROBLEM TYPE 2: CALCULATING HEAT ABSORBED OR RELEASED USING SPECIFIC HEAT DATA

If the specific heat (s) and the amount of substance is known, then the change in the sample's temperature (Δt) will tell us the amount of heat (q) that has been absorbed or released in a particular process. The equation for calculating the heat change is given by:

$$q = ms\Delta t \qquad (6.1)$$

or

$$q = C\Delta t$$

where m is the mass of the sample and Δt is the temperature change.

$$\Delta t = t_{\text{final}} - t_{\text{initial}}$$

EXAMPLE 6.2

How much heat is absorbed by 80.0 g of iron (Fe) when its temperature is raised from 25°C to 500°C? The specific heat of iron is 0.444 J/g·°C.

Notice that the specific heat of iron is given in the problem. Therefore, we should use Equation (6.1) to solve this problem.

Step 1: Calculate Δt.

$$\Delta t = t_{\text{final}} - t_{\text{initial}}$$
$$\Delta t = 500°\text{C} - 25°\text{C} = \mathbf{475°C}$$

Step 2: Substitute the known values into Equation (6.1).

$$q = m_{\text{Fe}}s_{\text{Fe}}\Delta t$$

$$q = (80.0\,\cancel{\text{g}})(0.444\text{ J}/\cancel{\text{g}}\cdot\cancel{°\text{C}})(475\cancel{°\text{C}}) = \mathbf{1.69 \times 10^4\ J}$$

PRACTICE EXERCISE

2. A piece of iron initially at a temperature of 25°C absorbs 10.0 kJ of heat. If its mass is 50.0 g, calculate the final temperature of the piece of iron. The specific heat of iron is 0.444 J/g·°C.

Text Problems: 6.20, 6.22

PROBLEM TYPE 3: CALORIMETRY

A. Constant-volume calorimetry

For a discussion of constant-volume calorimetry, see Section 6.4 of your text. Heat of combustion is usually measured in a constant-volume calorimeter. The heat released during combustion is absorbed both by the water in the calorimeter and the calorimeter itself. Because no heat enters or leaves the system throughout the process, we can write:

$$q_{\text{system}} = 0 = q_{\text{water}} + q_{\text{bomb}} + q_{\text{rxn}}$$

or,

$$q_{\text{rxn}} = -(q_{\text{water}} + q_{\text{bomb}})$$

The heat absorbed by the water can be calculated using the equation

$$q_{\text{water}} = m_{\text{water}} s_{\text{water}} \Delta t$$

The heat absorbed by the bomb calorimeter can be calculated using the heat capacity of the bomb calorimeter.

$$q_{\text{bomb}} = C_{\text{bomb}} \Delta t$$

Note: The negative sign for q_{rxn} indicates that heat was released during the combustion. You should expect this, because all combustion processes are exothermic. Thermal energy is transferred from the system to the surroundings.

EXAMPLE 6.3

0.500 g of ethanol [CH_3CH_2OH (*l*)] was burned in a bomb calorimeter containing 2.000×10^3 g of water. The heat capacity of the bomb calorimeter was 950 J/°C, and the temperature rise was found to be 1.6°C.

(a) Write a balanced equation for the combustion of ethanol.

(b) Calculate the molar heat of combustion of ethanol.

(a) Recall that a **combustion reaction** is typically a vigorous and exothermic reaction that takes place between certain substances and oxygen. If the reactant contains only C, H, and/or O, then the products are CO_2 and H_2O. Therefore, the balanced equation for the combustion of ethanol is:

$$CH_3CH_2OH\,(l) + 3\,O_2\,(g) \longrightarrow 2\,CO_2\,(g) + 3\,H_2O\,(g)$$

(b)

Step 1: First, we must calculate the heat absorbed by both the water and the bomb calorimeter.

$$q_{\text{water}} = m_{\text{water}} s_{\text{water}} \Delta t$$

$$q_{\text{water}} = (2.000 \times 10^3\ \text{g})(4.184\ \text{J/g}\cdot{}^\circ\text{C})(1.6^\circ\text{C}) = \mathbf{1.3 \times 10^4\ J}$$

$$q_{\text{bomb}} = C_{\text{bomb}} \Delta t$$

$$q_{\text{bomb}} = (950\ \text{J/}^\circ\text{C})(1.6^\circ\text{C}) = \mathbf{1.5 \times 10^3\ J}$$

Step 2: Calculate the heat of the reaction.

$$q_{\text{rxn}} = -(q_{\text{water}} + q_{\text{bomb}})$$
$$q_{\text{rxn}} = -(1.3 \times 10^4\text{ J} + 1.5 \times 10^3\text{ J}) = \mathbf{-1.5 \times 10^4\text{ J}}$$

Step 3: The problem asks for the **molar heat of combustion** of ethanol. The heat of reaction calculated in Step 2 is for the combustion of 0.500 g of ethanol. We can use the molar mass of ethanol as a conversion factor to convert from J/0.500 g ethanol to J/mol ethanol.

$$\mathcal{M}_{\text{ethanol}} = 2(12.01\text{ g}) + 6(1.008\text{ g}) + 16.00\text{ g} = 46.07\text{ g}$$

$$\textbf{Molar heat of combustion} = \frac{-1.5 \times 10^4\text{ J}}{0.500\text{ g ethanol}} \times \frac{46.07\text{ g ethanol}}{1\text{ mol ethanol}} = \mathbf{-1.4 \times 10^6\text{ J/mol ethanol}}$$

PRACTICE EXERCISE

3. The combustion of benzoic acid is often used as a standard source of heat for calibrating combustion bomb calorimeters. The heat of combustion of benzoic acid has been accurately determined to be 26.42 kJ/g. When 0.8000 g of benzoic acid was burned in a calorimeter containing 9.50×10^2 g of water, a temperature rise of 4.08°C was observed. What is the heat capacity of the bomb calorimeter?

B. Constant-pressure calorimetry

A simpler device than the constant-volume calorimeter is the constant-pressure calorimeter that is used to determine the heat changes for noncombustion reactions. The reactions usually occur in solution. Because the measurements are carried out under constant atmospheric pressure conditions, the heat change for the process (q_{rxn}) is equal to the enthalpy change (ΔH).

The heat released during reaction is absorbed both by the solution in the calorimeter and the calorimeter itself. Because no heat enters or leaves the system throughout the process, we can write:

$$q_{\text{system}} = 0 = q_{\text{soln}} + q_{\text{calorimeter}} + q_{\text{rxn}}$$

or,

$$q_{\text{rxn}} = -(q_{\text{soln}} + q_{\text{calorimeter}})$$

The heat absorbed by the solution can be calculated using the equation

$$q_{\text{soln}} = m_{\text{soln}} s_{\text{soln}} \Delta t$$

The heat absorbed by the calorimeter can be calculated using the heat capacity of the calorimeter

$$q_{\text{calorimeter}} = C_{\text{calorimeter}} \Delta t$$

EXAMPLE 6.4
The heat of neutralization for the following reaction is –56.2 kJ/mol.

$$\mathbf{NaOH}\,(aq) + \mathbf{HCl}\,(aq) \longrightarrow \mathbf{NaCl}\,(aq) + \mathbf{H_2O}\,(l)$$

1.00×10^2 mL of 1.50 *M* HCl is mixed with 1.00×10^2 mL of 1.50 *M* NaOH in a constant-pressure calorimeter having a heat capacity of 15.2 J/°C. The initial temperature of the HCl and NaOH solutions is the same, 23.2°C. Calculate the final temperature of the mixed solution. Assume that the density and specific heat of the mixed solution is the same as for water (1.00 g/mL and 4.184 J/g·°C, respectively).

Step 1: We know that:

$$q_{rxn} = -(q_{soln} + q_{calorimeter})$$

and

$$q_{rxn} = -(m_{soln}s_{soln}\Delta t + C_{calorimeter}\Delta t)$$

We can calculate the mass of the solution; s_{soln} is given; $C_{calorimeter}$ is given. If we can calculate q_{rxn}, then we can solve for Δt. Once we solve for Δt, we can calculate t_{final}, since we know $t_{initial}$.

Step 2: Calculate q_{rxn}. We are given the heat of neutralization in units of kJ/mol. If we can calculate the moles of HCl and NaOH that are reacted, we can calculate the heat of neutralization (q_{rxn}) for this particular reaction.

$$1.00 \times 10^2 \text{ mL} = 1.00 \times 10^{-1} \text{ L} = 0.100 \text{ L}$$

$$\text{mol HCl} = \text{mol NaOH} = M_{soln} \times L_{soln}$$

$$\textbf{mol HCl} = \left(1.50 \frac{\text{mol}}{\cancel{\text{L}}}\right) \times 0.100 \cancel{\text{L}} = \textbf{0.150 mol HCl}$$

$$q_{\mathbf{rxn}} = \left(-56.2 \frac{\text{kJ}}{\cancel{\text{mol}}}\right) \times 0.150 \cancel{\text{mol}} = \mathbf{-8.43\ kJ} = \mathbf{-8.43 \times 10^3\ J}$$

Step 3: Calculate the mass of the solution. You are told to assume that the density of the solution is the same as the density of water, 1.00 g/mL.

$$V_{soln} = V_{HCl} + V_{NaOH} = 1.00 \times 10^2 \text{ mL} + 1.00 \times 10^2 \text{ mL} = 2.00 \times 10^2 \text{ mL}$$

$$\textbf{Mass}_{\mathbf{soln}} = (2.00 \times 10^2 \cancel{\text{mL}}) \times \frac{1.00 \text{ g}}{1 \cancel{\text{mL}}} = \mathbf{2.00 \times 10^2\ g}$$

Step 4: Substitute the known values into the equation derived in Step 1, then solve for t_{final}.

$$q_{rxn} = -(m_{soln}s_{soln}\Delta t + C_{calorimeter}\Delta t)$$

$$-8.43 \times 10^3 \text{ J} = -[(2.00 \times 10^2 \text{ g})(4.184 \text{ J/g}\cdot°\text{C})\Delta t + (15.2 \text{ J/°C})\Delta t]$$

$$-8.43 \times 10^3 \text{ J} = -[837\ \Delta t + 15.2\ \Delta t]\text{J/°C}$$

$$-8.43 \times 10^3 \text{ J} = -852\ \Delta t\ (\text{J/°C})$$

$$\Delta t = 9.89°\text{C}$$

$$\Delta t = t_{final} - t_{initial}$$

$$t_{\mathbf{final}} = \Delta t + t_{initial} = (9.89 + 23.2)°\text{C} = \mathbf{33.1°C}$$

PRACTICE EXERCISE

4. A 10.4 g sample of an unknown metal at 99.0°C was placed in a constant-pressure calorimeter containing 75.0 g of water at 23.5°C. The final temperature of the system was found to be 25.6°C. Calculate the specific heat of the metal, then use Table 6.1 of your text to predict the identity of the metal. (The heat capacity of the calorimeter is 12.6 J/°C.)

Text Problem: 6.26

PROBLEM TYPE 4: STANDARD ENTHALPY OF FORMATION AND REACTION

The **standard enthalpy of formation** (ΔH_f°) is defined as the heat change that results when one mole of a compound is formed from its elements at a pressure of 1 atm. The standard enthalpy of formation of any element in its most stable form is zero.

A. Calculating the standard enthalpy of reaction

From standard enthalpies of formation, we can calculate the **standard enthalpy of reaction, ΔH_{rxn}°.**

Consider the hypothetical reaction

$$aA + bB \longrightarrow cC + dD$$

where a, b, c, and d are stoichiometric coefficients.

The standard enthalpy of reaction is given by

$$\Delta H_{rxn}^\circ = [c\,\Delta H_f^\circ(C) + d\,\Delta H_f^\circ(D)] - [a\,\Delta H_f^\circ(A) + b\,\Delta H_f^\circ(B)]$$

where a, b, c, and d all have the unit mol.

The equation can be written in the general form:

$$\Delta H_{rxn}^\circ = \Sigma n\,\Delta H_f^\circ(\text{products}) - \Sigma m\,\Delta H_f^\circ(\text{reactants})$$

where m and n denote the stoichiometric coefficients for the reactants and products, and Σ (sigma) means "the sum of".

EXAMPLE 6.5
A reaction used for rocket engines is

$$N_2H_4\,(l) + 2\,H_2O_2\,(l) \longrightarrow N_2\,(g) + 4\,H_2O\,(l)$$

What is the standard enthalpy of reaction in kilojoules? The standard enthalpies of formation are $\Delta H_f^\circ[N_2H_4\,(l)] = 95.1$ kJ, $\Delta H_f^\circ[H_2O_2\,(l)] = -187.8$ kJ, and $\Delta H_f^\circ[H_2O\,(l)] = -285.8$ kJ.

$$\Delta H_{rxn}^\circ = \Sigma n\,\Delta H_f^\circ(\text{products}) - \Sigma m\,\Delta H_f^\circ(\text{reactants})$$

$$\Delta H_{rxn}^\circ = \Delta H_f^\circ[N_2\,(g)] + 4\,\Delta H_f^\circ[H_2O\,(l)] - \{\Delta H_f^\circ[N_2H_4\,(l)] + 2\,\Delta H_f^\circ[H_2O_2\,(l)]\}$$

Remember, the standard enthalpy of formation of any element in its most stable form is zero. Therefore, $\Delta H_f^\circ[N_2(g)] = 0$.

$$\Delta H_{rxn}^\circ = [0 + 4(-285.8\text{ kJ})] - [95.1\text{ kJ} + 2(-187.8\text{ kJ})] = \mathbf{-862.7\text{ kJ}}$$

PRACTICE EXERCISE

5. The combustion of methane, the main component of natural gas, occurs according to the equation

$$CH_4\,(g) + 2\,O_2\,(g) \longrightarrow CO_2\,(g) + 2\,H_2O\,(l) \qquad \Delta H_{rxn}^\circ = -890\text{ kJ}$$

Use standard enthalpies of formation for CO_2 and H_2O to determine the standard enthalpy of formation of methane.

Text Problems: 6.44, 6.46, 6.48, 6.52

B. Direct method of calculating the standard enthalpy of formation

This method of measuring ΔH_f° applies to compounds that can be readily synthesized from their elements. The best way to describe this direct method is to look at an example.

EXAMPLE 6.6

The combustion of sulfur occurs according to the following thermochemical equation:

$$\textbf{S (rhombic)} + \mathbf{O_2}\ (g) \longrightarrow \mathbf{SO_2}\ (g) \qquad \Delta H^\circ_{\text{rxn}} = -296\ \text{kJ}$$

What is the enthalpy of formation of SO_2?

Step 1: Recall that the enthalpy of reaction carried out under standard-state conditions is given by

$$\Delta H^\circ_{\text{rxn}} = \Sigma n\, \Delta H_f^\circ\ (\text{products}) - \Sigma m\, \Delta H_f^\circ\ (\text{reactants})$$

$$\Delta H^\circ_{\text{rxn}} = [\Delta H_f^\circ\ (SO_2)] - [\Delta H_f^\circ\ (S) + \Delta H_f^\circ\ (O_2)]$$

Step 2: Recall that the standard enthalpy of formation of any element in its most stable form is zero. Therefore, ΔH_f° [S(rhombic)] = 0 and $\Delta H_f^\circ\ [O_2(g)] = 0$.

$$\Delta H^\circ_{\text{rxn}} = [\Delta H_f^\circ\ (SO_2)] - [\Delta H_f^\circ\ (S) + \Delta H_f^\circ\ (O_2)]$$

$$-296\ \text{kJ} = [\Delta H_f^\circ\ (SO_2)] - [0 + 0]$$

$$\mathbf{\Delta H_f^\circ\ (SO_2) = -296\ kJ/mol\ SO_2}$$

> **Note:** You should recognize that this chemical equation as written meets the definition of a *formation* reaction. Thus, ΔH_{rxn} *is* ΔH_f° of SO_2 (*g*).

PRACTICE EXERCISE

6. Hydrogen bromide (HBr) can be produced according to the following equation:

$$H_2\ (g) + Br_2\ (l) \longrightarrow 2\ HBr\ (g) \qquad \Delta H^\circ_{\text{rxn}} = -72.8\ \text{kJ/mol}$$

What is the enthalpy of formation (ΔH_f°) of HBr?

Text Problem: 6.34

C. Indirect method of calculating the standard enthalpy of formation, Hess's law

Many compounds cannot be directly synthesized from their elements. In these cases, ΔH_f° can be determined by an indirect approach using **Hess's law**. Hess's law states that when reactants are converted to products, the change in enthalpy is the same whether the reaction takes place in one step or in a series of steps. This means that if we can break down the reaction of interest into a series of reactions for which $\Delta H^\circ_{\text{rxn}}$ can be measured, we can calculate $\Delta H^\circ_{\text{rxn}}$ for the overall reaction. Let's look at an example.

EXAMPLE 6.7

From the following heats of combustion with fluorine, calculate the enthalpy of formation of methane, CH_4.

(a) $CH_4\ (g) + 4\ F_2\ (g) \longrightarrow CF_4\ (g) + 4\ HF\ (g)$ $\qquad \Delta H^\circ_{\text{rxn}} = -1942\ \text{kJ}$

(b) $\text{C (graphite)} + 2\ F_2\ (g) \longrightarrow CF_4\ (g)$ $\qquad \Delta H^\circ_{\text{rxn}} = -933\ \text{kJ}$

(c) $H_2\ (g) + F_2\ (g) \longrightarrow 2\ HF\ (g)$ $\qquad \Delta H^\circ_{\text{rxn}} = -542\ \text{kJ}$

Step 1: The enthalpy of formation of methane can be determined from the following equation.

$$\text{C (graphite)} + 2\,\text{H}_2\,(g) \longrightarrow \text{CH}_4\,(g) \qquad \Delta H^\circ_{\text{rxn}} = ?$$

If we can calculate $\Delta H^\circ_{\text{rxn}}$ for this reaction, we can calculate $\Delta H^\circ_f\,(\text{CH}_4)$ because the enthalpies of formation of C and H_2 are zero. Why? Remember that the standard enthalpy of formation of any element in its most stable form is zero. See Example 6.6.

Step 2: Since we want to obtain one equation containing only C, H_2, and CH_4, we need to eliminate F_2, CF_4, and HF from the first three equations (a–c). First, we note that equation (a) contains methane, CH_4, on the reactant side. Let's reverse (a) to get CH_4 on the product side.

$$\text{CF}_4\,(g) + 4\,\text{HF}\,(g) \longrightarrow \text{CH}_4\,(g) + 4\,\text{F}_2\,(g) \qquad \Delta H^\circ_{\text{rxn}} = +1942\text{ kJ}$$

Note: $\Delta H^\circ_{\text{rxn}}$ changed sign when reversing the direction of the reaction. See Problem Type 1 for discussion.

Next, we notice that equation (b) has C (graphite) on the reactant side so let's keep that equation as written. Also, notice that equation (c) has H_2 on the reactant side. We need two moles of H_2 in the balanced equation for the formation of methane, so let's multiply this equation by two. If we add these three equations together, we end up with the following:

$$\cancel{\text{CF}_4}\,(g) + 4\,\cancel{\text{HF}}\,(g) \longrightarrow \text{CH}_4\,(g) + 4\,\cancel{\text{F}_2}\,(g) \qquad \Delta H^\circ_{\text{rxn}} = +1942\text{ kJ}$$
$$\text{C (graphite)} + 2\,\cancel{\text{F}_2}\,(g) \longrightarrow \cancel{\text{CF}_4}\,(g) \qquad \Delta H^\circ_{\text{rxn}} = -933\text{ kJ}$$
$$2\,\text{H}_2\,(g) + 2\,\cancel{\text{F}_2}\,(g) \longrightarrow 4\,\cancel{\text{HF}}\,(g) \qquad \Delta H^\circ_{\text{rxn}} = -1084\text{ kJ}$$

$$\text{C (graphite)} + 2\,\text{H}_2\,(g) \longrightarrow \text{CH}_4\,(g) \qquad \Delta H^\circ_{\text{rxn}} = -75\text{ kJ}$$

Step 3: Since the above equation represents the synthesis of CH_4 from its elements, we have

$$\Delta H^\circ_{\text{rxn}} = [\Delta H^\circ_f\,(\text{CH}_4)] - [\Delta H^\circ_f\,(\text{C}) + 2\,\Delta H^\circ_f\,(\text{H}_2)]$$

$$-75\text{ kJ} = [\Delta H^\circ_f\,(\text{CH}_4)] - [0 + 0]$$

$$\boldsymbol{\Delta H^\circ_f\,(\text{CH}_4) = -75\text{ kJ/mol CH}_4}$$

PRACTICE EXERCISE

7. From the following enthalpies of reaction, calculate the enthalpy of combustion of methane (CH_4) with F_2:

$$\text{CH}_4\,(g) + 4\,\text{F}_2\,(g) \longrightarrow \text{CF}_4\,(g) + 4\,\text{HF}(g) \qquad \boldsymbol{\Delta H^\circ_{\text{rxn}} = ?}$$

$$\text{C (graphite)} + 2\,\text{H}_2\,(g) \longrightarrow \text{CH}_4\,(g) \qquad \Delta H^\circ_{\text{rxn}} = -75\text{ kJ}$$
$$\text{C (graphite)} + \text{F}_2\,(g) \longrightarrow \text{CF}_4\,(g) \qquad \Delta H^\circ_{\text{rxn}} = -933\text{ kJ}$$
$$\text{H}_2\,(g) + \text{F}_2\,(g) \longrightarrow 2\,\text{HF}\,(g) \qquad \Delta H^\circ_{\text{rxn}} = -542\text{ kJ}$$

Text Problems: 6.38, 6.40

PROBLEM TYPE 5: THE FIRST LAW OF THERMODYNAMICS

A. Applying the First Law of Thermodynamics

The **First Law of Thermodynamics** states that energy can be converted from one form to another, but cannot be created or destroyed. Another way of stating the first law is that the energy of the universe is constant. The universe is composed of both the system and the surroundings.

$$\Delta E_{sys} + \Delta E_{surr} = 0$$

where,

the subscripts "sys" and "surr" denote system and surroundings, respectively.

However, in chemistry, we are normally interested in the changes associated with the *system* (which may be a flask containing reactants and products), not with its surroundings. Therefore, a more useful form of the first law is

$$\Delta E = q + w$$

where,

ΔE is the change in the internal energy of the system
q is the heat exchange between the system and surroundings
w is the work done on (or by) the system

Using the sign convention for thermochemical processes (see Section 6.2 of your text for discussion), q is positive for an endothermic process and negative for an exothermic process. For work, w is positive for work done *on* the system *by* the surroundings and negative for work done *by* the system *on* the surroundings. Try to understand the sign convention in this manner. If a *system* loses heat to the surroundings or does work on the surroundings, we expect its internal energy to decrease since both processes are energy depleting. Conversely, if heat is added to the *system* or if work is done on the *system*, then the internal energy of the system would increase.

EXAMPLE 6.8

A system does 975 kJ of work on its surroundings while at the same time it absorbs 625 kJ of heat. What is the change in energy, ΔE, for the system?

To solve this problem, you must make sure to get the sign convention correct. The system does work on the surroundings; this is an energy-depleting process.

$$w = \mathbf{-975\ kJ}$$

The system absorbs 625 kJ of heat. Therefore, the internal energy of the system would increase.

$$q = \mathbf{+625\ kJ}$$

Finally,

$$\Delta E = q + w = 625\text{ kJ} + (-975\text{ kJ}) = \mathbf{-350\ kJ}$$

PRACTICE EXERCISE

8. The surroundings do 455 kJ of work on the system while at the same time the system releases 253 kJ of heat. What is the change in energy, ΔE, for the system?

Text Problem: 6.58

B. Calculating the work done in gas expansion

A useful example of mechanical work is the expansion of a gas. Picture a gas-filled cylinder that is fitted with a weightless, frictionless, movable piston, at a certain temperature, pressure, and volume. As the gas expands, it pushes the piston upward against a constant, opposing, external atmospheric pressure, *P*. The gas (system) is doing work on the surroundings. The work can be calculated as follows:

$$w = -P\Delta V \qquad (6.2)$$

where,

P is the external pressure
ΔV is the change in volume ($V_f - V_i$)

> **Note:** The minus sign in the equation takes care of the sign convention for w. For gas expansion, $\Delta V > 0$, so $-P\Delta V$ is a negative quantity. When a gas expands, it's doing work on the surroundings; the internal energy of the system decreases. For gas compression, $\Delta V < 0$, so $-P\Delta V$ is a positive quantity. When a gas is compressed, the surroundings are doing work on the system, increasing the internal energy.

EXAMPLE 6.9

A gas initially at a pressure of 10.0 atm and occupying a volume of 5.0 L is allowed to expand at constant temperature against a constant external pressure of 4.0 atm. After expansion, the gas occupies a volume of 12.5 L. Calculate the work done by the gas on the surroundings.

Step 1: We are given the external pressure in the problem, but we must calculate ΔV.

$$\Delta V = V_f - V_i = 12.5\text{ L} - 5.0\text{ L} = \mathbf{7.5\text{ L}}$$

Step 2: Substitute P and ΔV into Equation (6.2) and solve for w.

$$w = -P\Delta V = -(4.0\text{ atm})(7.5\text{ L}) = \mathbf{-3.0 \times 10^1\text{ L·atm}}$$

It would be more convenient to express w in units of joules. The following conversion factor can be obtained from Appendix 1:

$$1\text{ L·atm} = 101.3\text{ J}$$

Thus, we can write:

$$w = -3.0 \times 10^1 \cancel{\text{L·atm}} \times \frac{101.3\text{ J}}{1 \cancel{\text{L·atm}}} = \mathbf{-3.0 \times 10^3\text{ J}}$$

PRACTICE EXERCISE

9. Calculate the work done on the system when 6.0 L of a gas is compressed to 1.0 L by a constant external pressure of 2.0 atm.

Text Problem: 6.60

C. Enthalpy and the First Law of Thermodynamics—Calculating the internal energy change of a gaseous reaction

Let's return to the following form of the first law of thermodynamics.

$$\Delta E_{sys} = q + w$$

Under constant-pressure conditions we can write:

$$\Delta E = q_p + w$$

Recall that the heat evolved or absorbed (q) by a reaction carried out under constant-pressure conditions is equal to the enthalpy change of the system, ΔH.

Thus,

$$\Delta E = \Delta H + w \tag{6.3}$$

Also, we know that for gas expansion or compression under a constant external pressure, $w = -P\Delta V$. Substituting into Equation (6.3), we have:

$$\Delta E = \Delta H - P\Delta V \tag{6.4}$$

Also, for an ideal gas at constant pressure,

$$P\Delta V = \Delta(nRT)$$

Substituting into Equation (6.4),

$$\Delta E = \Delta H - \Delta(nRT)$$

Finally, at constant temperature,

$$\mathbf{\Delta E = \Delta H - RT\Delta n} \tag{6.5}$$

where Δn is defined as

Δn = number of moles of product gases – number of moles of reactant gases

EXAMPLE 6.10

Calculate the change in internal energy when 1 mole of H_2 and 1/2 mole of O_2 are converted to 1 mole of H_2O at 1 atm and 25°C.

$$\mathbf{H_2(g) + 1/2\ O_2(g) \longrightarrow H_2O(l) \quad \Delta H° = -286\ kJ}$$

Step 1: Calculate the total change in the number of moles of gas.

$$\Delta n = 0 \text{ mol} - (1 \text{ mol} + 1/2 \text{ mol}) = -1.5 \text{ mol}$$

Step 2: Substitute the values for $\Delta H°$ and Δn into Equation (6.5).

$T = 25° + 273° = 298$ K

$$\Delta E° = \Delta H° - RT\Delta n$$

$$\Delta E° = -286 \text{ kJ} - \left(8.314 \frac{\text{J}}{\text{mol}\cdot\text{K}}\right)(298 \text{ K})(-1.5 \text{ mol})\left(\frac{1 \text{ kJ}}{1000 \text{ J}}\right) = \mathbf{-282\ kJ}$$

PRACTICE EXERCISE

10. Calculate the change in the internal energy when 1.0 mole of water vaporizes at 1.0 atm and 100°C. Assume that water vapor is an ideal gas and that the volume of liquid water is negligible compared with that of steam at 100°C. [$\Delta H_{vap}(H_2O)$ = 40.67 kJ/mol at 100°C].

Text Problem: 6.68

ANSWERS TO PRACTICE EXERCISES

1. heat evolved $= 1.2 \times 10^2$ kJ

2. 475°C

3. $C_{\text{bomb}} = \dfrac{4.9 \text{ kJ}}{4.08 \text{ °C}} = 1.2 \text{ kJ/°C}$

4. $s_{\text{metal}} = 0.898$ J/g°C. The metal is probably aluminum.

5. $\Delta H_f^\circ(CH_4) = -75.1$ kJ

6. $\Delta H_f^\circ(HBr) = -36.4$ kJ/mol

7. $\Delta H_{\text{rxn}}^\circ = -1942$ kJ

8. $\Delta E = 202$ kJ

9. $w = 1.0 \times 10^3$ J

10. $\Delta E = 37.57$ kJ

SOLUTIONS TO SELECTED TEXT PROBLEMS

6.20 $q = m_{Cu}s_{Cu}\Delta t = (6.22 \times 10^3 \text{ g})(0.385 \text{ J/g}\cdot°\text{C})(324.3°\text{C} - 20.5°\text{C}) =$ **728 kJ**

6.22 Calculating Heat Absorbed Using Specific Heat Data, Problem Type 2.

The hint in this problem gives you some important information. The heat gained by the gold must be equal to the heat lost by the iron.

$$q_{Au} = -q_{Fe}$$

If we can calculate the heat gained by the gold and the heat lost by the iron, we can calculate the final temperature of the combined metals.

Step 1: Calculate the heat gained by the gold and the heat lost by the iron.

Heat gained by the gold $= q_{Au} = m_{Au}s_{Au}\Delta t = (10.0 \text{ g})(0.129 \text{ J/g}\cdot°\text{C})(t_f - 18.0)°\text{C}$

Heat lost by the iron $= m_{Fe}s_{Fe}\Delta t = (20.0 \text{ g})(0.444 \text{ J/g}\cdot°\text{C})(t_f - 55.6)°\text{C}$

Step 2: The heat gained by the gold is the opposite sign of the heat lost by the iron.

$$q_{Au} = -q_{Fe}$$

$$(10.0 \text{ g})(0.129 \text{ J/g}\cdot°\text{C})(t_f - 18.0)°\text{C} = -(20.0 \text{ g})(0.444 \text{ J/g}\cdot°\text{C})(t_f - 55.6)°\text{C}$$

$$1.29\, t_f\, (\text{J/°C}) - 23.2 \text{ J} = -8.88\, t_f\, (\text{J/°C}) + 494 \text{ J}$$

$$10.2\, t_f\, (\text{J/°C}) = 517 \text{ J}$$

$$\mathbf{t_f = 50.7°C}$$

Must the final temperature be between the two starting values?

6.26 Constant-pressure calorimetry, Problem Type 3B.

Step 1: We know that

$$q_{rxn} = -(q_{soln} + q_{calorimeter})$$

and

$$q_{rxn} = -(m_{soln}s_{soln}\Delta t + C_{calorimeter}\Delta t)$$

$C_{calorimeter}$ is given. If we assume that the solution has the same density as water, we can calculate the mass of the solution. From the moles of HCl and the heat of neutralization, we can calculate q_{rxn}. Also, we need to assume that the solution has the same specific heat as water. Once we have all this information, we can solve for Δt. Knowing $t_{initial}$, we can solve for t_{final}.

Step 2: Calculate the mass of the solution. If we assume that the density of the solution is the same as the density of water, 1.00 g/mL, we have:

$$V_{soln} = V_{HCl} + V_{Ba(OH)_2} = 2.00 \times 10^2 \text{ mL} + 2.00 \times 10^2 \text{ mL} = 4.00 \times 10^2 \text{ mL}$$

$$\text{Mass}_{soln} = (4.00 \times 10^2 \text{ mL})\left(\frac{1.00 \text{ g}}{1 \text{ mL}}\right) = 4.00 \times 10^2 \text{ g}$$

Step 3: The equation for the neutralization is

$$2\,HCl\,(aq) + Ba(OH)_2\,(aq) \longrightarrow 2\,H_2O\,(l) + BaCl_2\,(aq)$$

One mole of water is formed from each mole of HCl. There is exactly enough $Ba(OH)_2$ to neutralize all the HCl. The number of moles of HCl is:

$$200.\ \text{mL} \times \frac{1\ \text{L}}{1000\ \text{mL}} \times \frac{0.862\ \text{mol HCl}}{1\ \text{L}} = 0.172\ \text{mol HCl}$$

The amount of heat liberated during the reaction is

$$q_{rxn} = 0.0172\ \text{mol} \times \frac{-56.2 \times 10^3\ \text{J}}{1\ \text{mol}} = -9.67 \times 10^3\ \text{J}$$

Step 4: Substitute the known values into the equation derived in Step 1, then solve for t_{final}.

$$q_{rxn} = -(m_{soln}s_{soln}\Delta t + C_{calorimeter}\Delta t)$$

$$-9.67 \times 10^3\ \text{J} = -[(4.00 \times 10^2\ \text{g})(4.184\ \text{J/g}\cdot{}^\circ\text{C})\Delta t + (453\ \text{J/}^\circ\text{C})\Delta t]$$

$$-9.67 \times 10^3\ \text{J} = -(1.67 \times 10^3 + 453)\Delta t\ (\text{J/}^\circ\text{C})$$

$$\Delta t = 4.55^\circ\text{C}$$

$$\Delta t = t_{final} - t_{initial}$$

$$t_{final} = \Delta t + t_{initial} = (4.55 + 20.48)^\circ\text{C} = \mathbf{25.03^\circ C}$$

6.30 The standard enthalpy of formation of any element in its most stable form is zero. Therefore, since $\Delta H_f^\circ(O_2) = 0$, O_2 is the more stable form of the element oxygen at this temperature.

6.32 **(a)** $Cl_2\,(g)$ is the most stable form of chlorine at 25°C; therefore, $\Delta H_f^\circ\,[Cl_2\,(g)] = 0$. Since $Cl_2\,(l)$ is less stable than $Cl_2\,(g)$, $\Delta H_f^\circ\,[Cl_2\,(l)] > 0$.

(b) $Br_2\,(l)$ is the most stable form of bromine at 25°C; therefore, $\Delta H_f^\circ\,[Br_2\,(l)] = 0$. Since $Br_2\,(g)$ is less stable than $Br_2\,(l)$, $\Delta H_f^\circ\,[Br_2\,(g)] > 0$.

(c) $I_2\,(s)$ is the most stable form of iodine at 25°C; therefore, $\Delta H_f^\circ\,[I_2\,(s)] = 0$. Since $I_2\,(g)$ is less stable than $I_2\,(s)$, $\Delta H_f^\circ\,[I_2\,(g)] > 0$.

6.34 Direct method of calculating the standard enthalpy of formation, Problem Type 4B.

This method of measuring ΔH_f° applies to compounds that can be readily synthesized from their elements.

Step 1: Let's write a balanced equation that shows the formation of $Ag_2O\,(s)$ from its elements.

$$2\,Ag\,(s) + (1/2)\,O_2\,(g) \longrightarrow Ag_2O\,(s)$$

Step 2: Recall that the enthalpy of reaction carried out under standard-state conditions is given by

$$\Delta H^\circ_{rxn} = \Sigma n \Delta H^\circ_f \text{(products)} - \Sigma m \Delta H^\circ_f \text{(reactants)}$$

thus,

$$\Delta H^\circ_{rxn} = [\Delta H^\circ_f (Ag_2O)] - [2\Delta H^\circ_f (Ag) + (1/2)\Delta H^\circ_f (O_2)]$$

Step 3: Recall that the standard enthalpy of formation of any element in its most stable form is zero. Therefore, $\Delta H^\circ_f [Ag(s)] = 0$ and $\Delta H^\circ_f [O_2(g)] = 0$.

$$\Delta H^\circ_{rxn} = [\Delta H^\circ_f (Ag_2O)] - [2\Delta H^\circ_f (Ag) + (1/2)\,\Delta H^\circ_f (O_2)]$$

$$\Delta H^\circ_{rxn} = [\Delta H^\circ_f (Ag_2O)] - [0 + 0]$$

$$\boldsymbol{\Delta H^\circ_f (Ag_2O) = \Delta H^\circ_{rxn}}$$

In a similar manner, you should be able to show that $\boldsymbol{\Delta H^\circ_f (CaCl_2) = \Delta H^\circ_{rxn}}$ for the reaction

$$Ca\,(s) + Cl_2\,(g) \longrightarrow CaCl_2\,(s)$$

6.38 Indirect method of calculating the standard enthalpy of formation, Hess's law. Problem Type 4C.

$$2\,C\,(\text{graphite}) + 3\,H_2\,(g) \longrightarrow C_2H_6\,(g) \qquad \Delta H^\circ_{rxn} = ? \qquad (6.1)$$

We want to obtain one equation containing only C and H_2 on the reactant side and C_2H_6 on the product side. First, we note that the first equation contains 1 mole of C (graphite) on the reactant side. Two moles of C are on the reactant side of equation (6.1), so let's multiply the first equation by 2.

$$2\,C\,(\text{graphite}) + 2\,O_2\,(g) \longrightarrow 2\,CO_2\,(g) \qquad \Delta H^\circ_{rxn} = 2(-393.5\text{ kJ}) = -787.0\text{ kJ}$$

Next, notice that the second equation contains 1 mole of H_2 on the reactant side. Three moles of H_2 are on the reactant side of Equation (6.1), so let's multiply the second equation by 3.

$$3\,H_2\,(g) + \frac{3}{2}\,O_2\,(g) \longrightarrow 3\,H_2O\,(l) \qquad \Delta H^\circ_{rxn} = 3(-285.8\text{ kJ}) = -857.4\text{ kJ}$$

Next, notice that the third equation contains 2 moles of ethane, C_2H_6, on the reactant side. One mole of ethane is on the product side of equation (6.1), so let's reverse the third equation and multiply by 1/2.

$$2\,CO_2\,(g) + 3\,H_2O\,(l) \longrightarrow C_2H_6\,(g) + \frac{7}{2}\,O_2\,(g) \qquad \Delta H^\circ_{rxn} = 1/2(3119.6\text{ kJ}) = 1559.8\text{ kJ}$$

Putting these three equations together, we have:

Reaction	ΔH° (kJ)
$2\,C\,(\text{graphite}) + 2\,\cancel{O_2}\,(g) \longrightarrow 2\,\cancel{CO_2}\,(g)$	−787.0
$3\,H_2\,(g) + \frac{3}{2}\,\cancel{O_2}\,(g) \longrightarrow 3\,\cancel{H_2O}\,(l)$	−857.4
$2\,\cancel{CO_2}\,(g) + 3\,\cancel{H_2O}\,(l) \longrightarrow C_2H_6\,(g) + \frac{7}{2}\,\cancel{O_2}\,(g)$	1559.8
$\mathbf{2\,C\,(graphite) + 3\,H_2\,(g) \longrightarrow C_2H_6\,(g)}$	$\boldsymbol{\Delta H^\circ = -84.6}$ **kJ**

6.40 The second and third equations can be combined to give the first equation.

$$2\text{Al}(s) + \frac{3}{2}\cancel{O_2}(g) \longrightarrow \text{Al}_2\text{O}_3(s) \qquad \Delta H° = -1601 \text{ kJ}$$

$$\text{Fe}_2\text{O}_3(s) \longrightarrow 2\text{ Fe}(s) + \frac{3}{2}\cancel{O_2}(g) \qquad \Delta H° = 821 \text{ kJ}$$

$$2\text{ Al}(s) + \text{Fe}_2\text{O}_3(s) \longrightarrow 2\text{ Fe}(s) + \text{Al}_2\text{O}_3(s) \qquad \mathbf{\Delta H° = -7.80 \times 10^2\ kJ}$$

6.44 Calculating the standard enthalpy of reaction, Problem Type 4A.

$$\Delta H°_{\text{rxn}} = \Sigma n\,\Delta H°_{\text{f}}(\text{products}) - \Sigma m\,\Delta H°_{\text{f}}(\text{reactants})$$

The neutralization reaction is:

$$\text{H}^+(aq) + \text{OH}^-(aq) \rightarrow \text{H}_2\text{O}(l)$$

and,

$$\Delta H°_{\text{rxn}} = \Delta H°_{\text{f}}[\text{H}_2\text{O}(l)] - [\Delta H°_{\text{f}}(\text{H}^+) + \Delta H°_{\text{f}}(\text{OH}^-)]$$

$$\Delta H°_{\text{f}}[\text{H}_2\text{O}(l)] = -285.8 \text{ kJ/mol} \quad \text{(see Table 6.3 of your text)}$$

$$\mathbf{\Delta H°_{rxn}} = (1\text{ mol})(-285.8\text{ kJ/mol}) - [(1\text{ mol})(0\text{ kJ/mol}) + (1\text{ mol})(-229.6\text{ kJ/mol})] = \mathbf{-56.2\ kJ}$$

6.46 **(a)** $\Delta H° = [2\,\Delta H°_{\text{f}}(\text{CO}_2) + 2\,\Delta H°_{\text{f}}(\text{H}_2\text{O})] - [\Delta H°_{\text{f}}(\text{C}_2\text{H}_4) + 3\,\Delta H°_{\text{f}}(\text{O}_2)]$

$\Delta H° = [(2\text{ mol})(-393.5\text{ kJ/mol}) + (2\text{ mol})(-285.8\text{ kJ/mol})] - [(1\text{ mol})(52.3\text{ kJ/mol}) + (3\text{ mol})(0)]$

$\mathbf{\Delta H° = -1411\ kJ}$

(b) $\Delta H° = [2\,\Delta H°_{\text{f}}(\text{H}_2\text{O}) + 2\,\Delta H°_{\text{f}}(\text{SO}_2)] - [2\,\Delta H°_{\text{f}}(\text{H}_2\text{S}) + 3\,\Delta H°_{\text{f}}(\text{O}_2)]$

$\Delta H° = [(2\text{ mol})(-285.8\text{ kJ/mol}) + (2\text{ mol})(-296.1\text{ kJ/mol})] - [(2\text{ mol})(-20.15\text{ kJ/mol}) + (3\text{ mol})(0)]$

$\mathbf{\Delta H° = -1124\ kJ}$

6.48 $\Delta H°_{\text{rxn}} = \Sigma n\,\Delta H°_{\text{f}}(\text{products}) - \Sigma m\,\Delta H°_{\text{f}}(\text{reactants})$

The reaction is:

$$\text{H}_2(g) \longrightarrow \text{H}(g) + \text{H}(g)$$

and,

$$\Delta H°_{\text{rxn}} = [\Delta H°_{\text{f}}(\text{H}) + \Delta H°_{\text{f}}(\text{H})] - \Delta H°_{\text{f}}(\text{H}_2)$$

$$\Delta H°_{\text{f}}(\text{H}_2) = 0$$

$$\Delta H°_{\text{rxn}} = 436.4\text{ kJ} = 2\,\Delta H°_{\text{f}}(\text{H}) - (1\text{ mol})(0\text{ kJ/mol})$$

$$\mathbf{\Delta H°_f(H)} = \frac{436.4\text{ kJ}}{2\text{ mol}} = \mathbf{218.2\ kJ/mol}$$

6.50 Thermochemical Equations, Problem Type 1.

The equation as written shows that 879 kJ of heat is released when two moles of ZnS react. We want to calculate the amount of heat released when 1 g of ZnS reacts.

Let $\Delta H°$ be the heat change per gram of ZnS roasted. We write:

$$\boldsymbol{\Delta H°} = \frac{-879\text{ kJ}}{2\text{ mol ZnS}} \times \frac{1\text{ mol ZnS}}{97.46\text{ g ZnS}} = \mathbf{-4.51\ kJ/g\ ZnS}$$

6.52 $\Delta H°_{rxn} = \Sigma n\, \Delta H°_f\text{(products)} - \Sigma m\, \Delta H°_f\text{(reactants)}$

The balanced equation for the reaction is:

$$CaCO_3(s) \longrightarrow CaO(s) + CO_2(g)$$

$$\Delta H°_{rxn} = [\Delta H°_f(CaO) + \Delta H°_f(CO_2)] - \Delta H°_f(CaCO_3)$$

$$\Delta H°_{rxn} = [-635.6\text{ kJ/mol} + -393.5\text{ kJ/mol}] - (-1206.9\text{ kJ/mol})$$

$$\Delta H°_{rxn} = 177.8\text{ kJ/mol}$$

The enthalpy change calculated above is the enthalpy change if 1 mole of CO_2 is produced. The problem asks for the enthalpy change if 66.8 g of CO_2 are produced. We need to use the molar mass of CO_2 as a conversion factor.

$$\boldsymbol{\Delta H°} = 66.8\text{ g } CO_2 \times \frac{1\text{ mol } CO_2}{44.01\text{ g } CO_2} \times \frac{177.8\text{ kJ}}{1\text{ mol } CO_2} = \mathbf{2.70 \times 10^2\ kJ}$$

6.58 Applying the First Law of Thermodynamics, Problem Type 5A.

To solve this problem, you must make sure to get the sign convention correct. The surroundings do work on the system; this will increase the internal energy of the system.

$$w = +74\text{ kJ}$$

The system gives off 26 kJ of heat. This is an energy-depleting process.

$$q = -26\text{ kJ}$$

Finally,

$$\boldsymbol{\Delta E} = q + w = -26\text{ kJ} + 74\text{ kJ} = \mathbf{48\ kJ}$$

6.60 Calculating the work done in gas expansion, Problem Type 5B.

The work can be calculated as follows:

$$w = -P\Delta V \qquad (6.2)$$

Step 1: First, we need to calculate the volume that the water vapor will occupy (V_f).

Using the ideal gas equation:

$$V_{H_2O} = \frac{n_{H_2O}RT}{P} = \frac{(1\text{ mol})\left(0.0821\dfrac{\text{L}\cdot\text{atm}}{\text{mol}\cdot\text{K}}\right)(373\text{ K})}{(1.0\text{ atm})} = 31\text{ L}$$

Step 2: It is given that the volume occupied by liquid water is negligible. Therefore,

$$\Delta V = V_f - V_i = 31\text{ L} - 0\text{ L} = 31\text{ L}$$

Step 3: Substitute P and ΔV into Equation (6.2) and solve for w.

$$w = -P\Delta V = -(1.0\text{ atm})(31\text{ L}) = \mathbf{-31\ L{\cdot}atm}$$

It would be more convenient to express w in units of joules. The following conversion factor can be obtained from Appendix 1:

$$1\text{ L}\cdot\text{atm} = 101.3\text{ J}$$

Thus, we can write:

$$w = -31\cancel{\text{L}\cdot\text{atm}} \times \frac{101.3\text{ J}}{1\cancel{\text{L}\cdot\text{atm}}} = \mathbf{-3.1 \times 10^3\ J}$$

Why is the sign negative?

6.62 Rearrange the equations as necessary so they can be added to yield the desired equation.

$$\begin{array}{lll} 2\text{B} \longrightarrow \cancel{\text{A}} & & -\Delta H_1 \\ \cancel{\text{A}} \longrightarrow \text{C} & & \Delta H_2 \\ \hline 2\text{B} \longrightarrow \text{C} & & \mathbf{\Delta H^\circ = \Delta H_2 - \Delta H_1} \end{array}$$

6.64 **(a)** $\Delta H^\circ_{rxn} = \Sigma n\,\Delta H^\circ_f\text{(products)} - \Sigma m\,\Delta H^\circ_f\text{(reactants)}$

$$\Delta H^\circ_{rxn} = [4\,\Delta H^\circ_f(NH_3) + \Delta H^\circ_f(N_2)] - 3\,\Delta H^\circ_f(N_2H_4)$$

$$\mathbf{\Delta H^\circ_{rxn}} = [(4\text{ mol})(-46.3\text{ kJ/mol}) + (0)] - (3\text{ mol})(50.42\text{ kJ/mol}) = \mathbf{-336.5\ kJ}$$

(b) The balanced equations are:

(1) $N_2H_4\,(l) + O_2\,(g) \longrightarrow N_2\,(g) + 2\,H_2O\,(l)$

(2) $4\,NH_3\,(g) + 3\,O_2\,(g) \longrightarrow 2\,N_2\,(g) + 6\,H_2O\,(l)$

The standard enthalpy change for equation (1) is:

$$\Delta H^\circ_{rxn} = \Delta H^\circ_f(N_2) + 2\,\Delta H^\circ_f[H_2O\,(l)] - \{\Delta H^\circ_f[N_2H_4\,(l)] + \Delta H^\circ_f(O_2)\}$$

$$\mathbf{\Delta H^\circ_{rxn}} = [(1\text{ mol})(0) + (2\text{ mol})(-285.8\text{ kJ/mol})] - [(1\text{ mol})(50.442\text{ kJ/mol}) + (1\text{ mol})(0)] = \mathbf{-622.0\ kJ}$$

The standard enthalpy change for equation (2) is:

$$\Delta H^\circ_{rxn} = [2\,\Delta H^\circ_f(N_2) + 6\,\Delta H^\circ_f(H_2O)] - [4\,\Delta H^\circ_f(NH_3) + 3\,\Delta H^\circ_f(O_2)]$$

$$\mathbf{\Delta H^\circ_{rxn}} = [(2\text{ mol})(0) + (6\text{ mol})(-285.8\text{ kJ/mol})] - [(4\text{ mol})(-46.3\text{ kJ/mol}) - (3\text{ mol})(0)] = \mathbf{-1529.6\ kJ}$$

We can now calculate the enthalpy change per kilogram of each substance. ΔH°_{rxn} above is in units of kJ/mol. We need to convert to kJ/kg.

$N_2H_4\ (l)$: $\quad \Delta H^\circ_{\mathbf{rxn}} = \dfrac{-622.0\text{ kJ}}{1\text{ mol }N_2H_4} \times \dfrac{1\text{ mol }N_2H_4}{32.05\text{ g }N_2H_4} \times \dfrac{1000\text{ g}}{1\text{ kg}} = \mathbf{-1.941 \times 10^4\ kJ/kg\ N_2H_4}$

$NH_3(g)$: $\quad \Delta H^\circ_{\mathbf{rxn}} = \dfrac{-1529.6\text{ kJ}}{4\text{ mol }NH_3} \times \dfrac{1\text{ mol }NH_3}{17.03\text{ g }NH_3} \times \dfrac{1000\text{ g}}{1\text{ kg}} = \mathbf{-2.245 \times 10^4\ kJ/kg\ NH_3}$

Since ammonia releases more energy per kilogram of substance, it would be a better fuel.

6.66 **(a)** Although we cannot measure $\Delta H^\circ_{\text{rxn}}$ for this reaction, the reverse process, is the combustion of glucose. We could easily measure $\Delta H^\circ_{\text{rxn}}$ for this combustion in a bomb calorimeter.

$$C_6H_{12}O_6\,(s) + 6\,O_2\,(g) \longrightarrow 6\,CO_2\,(g) + 6\,H_2O\,(l)$$

(b) We can calculate $\Delta H^\circ_{\text{rxn}}$ using standard enthalpies of formation.

$$\Delta H^\circ_{\text{rxn}} = \Delta H^\circ_f[C_6H_{12}O_6\,(s)] + 6\,\Delta H^\circ_f[O_2\,(g)] - \{6\,\Delta H^\circ_f[CO_2\,(g)] + 6\,\Delta H^\circ_f[H_2O\,(l)]\}$$

$$\Delta H^\circ_{\mathbf{rxn}} = [(1\text{ mol})(-1274.5\text{ kJ/mol}) + 0] - [(6\text{ mol})(-393.5\text{ kJ/mol}) + (6\text{ mol})(-285.8\text{ kJ/mol})]$$

$$= \mathbf{2801.3\ kJ}$$

$\Delta H^\circ_{\text{rxn}}$ has units of kJ/1 mol glucose. We want the ΔH° change for 7.0×10^{14} kg glucose. We need to calculate how many moles of glucose are in 7.0×10^{14} kg glucose. You should come up with the following strategy to solve the problem.

kg glucose → g glucose → mol glucose → kJ (ΔH°)

$$\Delta H^\circ = (7.0 \times 10^{14}\text{ kg}) \times \frac{1000\text{ g}}{1\text{ kg}} \times \frac{1\text{ mol }C_6H_{12}O_6}{180.2\text{ g }C_6H_{12}O_6} \times \frac{2801.3\text{ kJ}}{1\text{ mol }C_6H_{12}O_6} = \mathbf{1.1 \times 10^{19}\ kJ}$$

6.68 Calculating the internal energy change of a gaseous reaction, Problem Type 5C.

We initially have 6 moles of gas (3 moles of chlorine and 3 moles of hydrogen). Since our product is 6 moles of hydrogen chloride, there is no change in the number of moles of gas. Therefore there is no volume change; $\Delta V = 0$.

$$w = -P\Delta V = [-1\text{ atm} \times (0\text{ L})] = \mathbf{0}$$

$$\Delta E^\circ = \Delta H^\circ - P\Delta V$$

$-P\Delta V = 0$, so

$$\mathbf{\Delta E = \Delta H}$$

$$\Delta H = 3\,\Delta H^\circ_{\text{rxn}} = 3(-184.6\text{ kJ}) = -553.8\text{ kJ}$$

We need to multiply $\Delta H^\circ_{\text{rxn}}$ by three, because the question involves the formation of 6 moles of HCl; whereas, the equation as written only produces 2 moles of HCl.

$$\Delta E^\circ = \Delta H^\circ = \mathbf{-553.8\ kJ = -5.538 \times 10^5\ J}$$

6.70 $H(g) + Br(g) \longrightarrow HBr(g) \qquad \Delta H^\circ_{rxn} = ?$

Rearrange the equations as necessary so they can be added to yield the desired equation.

$H(g) \longrightarrow 1/2\,H_2(g)$	$\Delta H^\circ_{rxn} = 1/2(-436.4\text{ kJ}) = -218.2\text{ kJ}$
$Br(g) \longrightarrow 1/2\,Br_2(g)$	$\Delta H^\circ_{rxn} = 1/2(-192.5\text{ kJ}) = -96.25\text{ kJ}$
$1/2\,H_2(g) + 1/2\,Br_2(g) \longrightarrow HBr(g)$	$\Delta H^\circ_{rxn} = 1/2(-104\text{ kJ}) = -52.0\text{ kJ}$
$H(g) + Br(g) \longrightarrow 2\,HBr(g)$	$\boldsymbol{\Delta H^\circ = -366.5\text{ kJ}}$

6.72 $q_{system} = 0 = q_{metal} + q_{water} + q_{calorimeter}$

$q_{metal} + q_{water} + q_{calorimeter} = 0$

$m_{metal}s_{metal}(t_{final} - t_{initial}) + m_{water}s_{water}(t_{final} - t_{initial}) + C_{calorimeter}(t_{final} - t_{initial}) = 0$

All the needed values are given in the problem. All you need to do is plug in the values and solve for s_{metal}.

$$(44.0\text{ g})(s_{metal})(28.4 - 99.0)^\circ\text{C} + (80.0\text{ g})(4.184\text{ J/g}\cdot^\circ\text{C})(28.4 - 24.0)^\circ\text{C} + (12.4\text{ J/}^\circ\text{C})(28.4 - 24.0)^\circ\text{C} = 0$$

$$(-3.11 \times 10^3)s_{metal}\,(\text{g}\cdot^\circ\text{C}) = -1.53 \times 10^3\text{ J}$$

$$\boldsymbol{s_{metal} = 0.492\text{ J/g}\cdot^\circ\text{C}}$$

6.74 A good starting point would be to calculate the standard enthalpy change for both reactions.

Calculate the standard enthalpy for the reaction: $C(s) + \frac{1}{2}O_2(g) \longrightarrow CO(g)$

This reaction corresponds to the standard enthalpy of formation of CO, so we use the value of −110.5 kJ.

Calculate the standard enthalpy change for the reaction: $C(s) + H_2O(g) \longrightarrow CO(g) + H_2(g)$

$$\Delta H^\circ_{rxn} = [\Delta H^\circ_f(CO) + \Delta H^\circ_f(H_2)] - [\Delta H^\circ_f(C) + \Delta H^\circ_f(H_2O)]$$

$$\Delta H^\circ_{rxn} = [(1\text{ mol})(-110.5\text{ kJ/mol}) + (1\text{ mol})(0)] - [(1\text{ mol})(0) + (1\text{ mol})(-241.8\text{ kJ/mol})] = 131.3\text{ kJ}$$

The first reaction, which is exothermic, can be used to promote the second reaction, which is endothermic. Thus, the two gases are produced alternately.

6.76 First, calculate the energy produced by 1 mole of octane, C_8H_{18}.

$$C_8H_{18}(l) + \frac{25}{2}O_2(g) \longrightarrow 8CO_2(g) + 9H_2O(l)$$

$$\Delta H^\circ_{rxn} = 8\,\Delta H^\circ_f(CO_2) + 9\,\Delta H^\circ_f[H_2O(l)] - [\Delta H^\circ_f(C_8H_{18}) + \frac{25}{2}\,\Delta H^\circ_f(O_2)]$$

$$\Delta H^\circ_{rxn} = [(8\text{ mol})(-393.5\text{ kJ/mol}) + (9\text{ mol})(-285.8\text{ kJ/mol})] - [(1\text{ mol})(-249.9\text{ kJ/mol}) + (\frac{25}{2}\text{ mol})(0)$$

$$= -5470\text{ kJ}$$

The problem asks for the energy produced by the combustion of 1 gallon of octane. ΔH°_{rxn} above has units of kJ/mol octane. We need to convert from kJ/mol octane to kJ/gallon octane. The heat of combustion for 1 gallon of octane is:

$$\Delta H° = \frac{-5470\text{ kJ}}{1\text{ mol octane}} \times \frac{1\text{ mol octane}}{114.2\text{ g octane}} \times \frac{2660\text{ g}}{1\text{ gal}} = \mathbf{-1.274 \times 10^5\ kJ/gal}$$

The combustion of hydrogen corresponds to the standard heat of formation of water:

$$H_2(g) + \frac{1}{2}O_2(g) \longrightarrow H_2O(l)$$

Thus, $\Delta H°_{rxn}$ is the same as $\Delta H°_f$ for H_2O (*l*), which has a value of −285.8 kJ. The number of moles of hydrogen required to produce 1.274×10^5 kJ of heat is:

$$n_{H_2} = (1.274 \times 10^5\text{ kJ}) \times \frac{1\text{ mol }H_2}{285.8\text{ kJ}} = 445.8\text{ mol }H_2$$

Finally, use the ideal gas law to calculate the volume of gas corresponding to 445.8 moles of H_2 at 25°C and 1 atm.

$$V_{H_2} = \frac{n_{H_2}RT}{P} = \frac{(445.8\text{ mol})\left(0.0821\frac{\text{L}\cdot\text{atm}}{\text{mol}\cdot\text{K}}\right)(298\text{ K})}{(1\text{ atm})} = \mathbf{1.09 \times 10^4\ L}$$

That is, the volume of hydrogen that is energy-equivalent to 1 gallon of gasoline is over **10,000 liters** at 1 atm and 25°C!

6.78 The combustion reaction is: $C_2H_6(l) + \frac{7}{2}O_2(g) \longrightarrow 2CO_2(g) + 3H_2O(l)$

The heat released during the combustion of 1 mole of ethane is:

$$\Delta H°_{rxn} = [2\Delta H°_f(CO_2) + 3\Delta H°_f(H_2O)] - [\Delta H°_f(C_2H_6) + \frac{7}{2}\Delta H°_f(O_2)]$$

$$\Delta H°_{rxn} = [(2\text{ mol})(-393.5\text{ kJ/mol}) + (3\text{ mol})(-285.8\text{ kJ/mol})] - [(1\text{ mol})(-84.7\text{ kJ/mol} + (\frac{7}{2}\text{ mol})(0)]$$

$$= -1560\text{ kJ}$$

The heat required to raise the temperature of the water to 98°C is:

$$q = m_{H_2O}s_{H_2O}\Delta t = (855\text{ g})(4.184\text{ J/g}\cdot°\text{C})(98.0 - 25.0)°\text{C} = 2.61 \times 10^5\text{ J} = 261\text{ kJ}$$

The combustion of 1 mole of ethane produces 1560 kJ; the number of moles required to produce 261 kJ is:

$$261\text{ kJ} \times \frac{1\text{ mol ethane}}{1560\text{ kJ}} = 0.167\text{ mol ethane}$$

The volume of ethane is:

$$V_{\text{ethane}} = \frac{nRT}{P} = \frac{(0.167\text{ mol})\left(0.0821\frac{\text{L}\cdot\text{atm}}{\text{mol}\cdot\text{K}}\right)(296\text{ K})}{(752/760)(\text{atm})} = \mathbf{4.10\ L}$$

6.80 The heat gained by the liquid nitrogen must be equal to the heat lost by the water.

$$q_{N_2} = -q_{H_2O}$$

If we can calculate the heat lost by the water, we can calculate the heat gained by 60.0 g of the nitrogen.

$$\text{Heat lost by the water} = q_{H_2O} = m_{H_2O}s_{H_2O}\Delta t$$

$$q_{H_2O} = (2.00 \times 10^2\ \cancel{g})(4.184\ \text{J}/\cancel{g\cdot°C})(41.0 - 55.3)\cancel{°C} = \mathbf{-1.20 \times 10^4\ J}$$

The heat gained by 60.0 g nitrogen is the opposite sign of the heat lost by the water.

$$q_{N_2} = -q_{H_2O}$$

$$q_{N_2} = 1.20 \times 10^4\ \text{J}$$

The problem asks for the molar heat of vaporization of liquid nitrogen. Above, we calculated the amount of heat necessary to vaporize 60.0 g of liquid nitrogen. We need to convert from J/60.0 g N_2 to J/mol N_2.

$$\Delta H_{vap} = \frac{1.20 \times 10^4\ \text{J}}{60.0\ \cancel{g\ N_2}} \times \frac{28.02\ \cancel{g\ N_2}}{1\ \text{mol}\ N_2} = \mathbf{5.60 \times 10^3\ J/mol = 5.60\ kJ/mol}$$

6.82 $w = -P\Delta V = -(1.0\ \text{atm})(0.0196 - 0.0180)\text{L} = -1.6 \times 10^{-3}\ \text{L}\cdot\text{atm}$

Using the conversion factor 1 L·atm = 101.3 J:

$$w = (-1.6 \times 10^{-3}\ \cancel{L\cdot atm})\left(\frac{101.3\ \text{J}}{\cancel{L\cdot atm}}\right) = \mathbf{-0.16\ J}$$

0.16 J of work are done by water as it expands on freezing.

6.84 The equation corresponding to the standard enthalpy of formation of diamond is:

$$\text{C (graphite)} \longrightarrow \text{C (diamond)}$$

Adding the equations:

C (graphite) + $\cancel{O_2}$ (g) ⟶ $\cancel{CO_2}$ (g)	$\Delta H° = -393.5$ kJ
$\cancel{CO_2}$ (g) ⟶ C (diamond) + $\cancel{O_2}$ (g)	$\Delta H° = 395.4$ kJ
C (graphite) ⟶ C (diamond)	**$\Delta H° = 1.9$ kJ**

Since the reverse reaction of changing diamond to graphite is exothermic, need you worry about any diamonds that you might have changing to graphite?

6.86 The balanced equation is:

$$C_6H_{12}O_6\ (s) \longrightarrow 2\ C_2H_5OH\ (l) + 2\ CO_2\ (g)$$

$$\Delta H°_{rxn} = [2\ \Delta H°_f(C_2H_5OH) + 2\ \Delta H°_f(CO_2)] - \Delta H°_f(C_6H_{12}O_6)$$

$$\mathbf{\Delta H°_{rxn}} = [(2\ \text{mol})(-276.98\ \text{kJ/mol}) + (2\ \text{mol})(-393.5\ \text{kJ/mol})] - (1\ \text{mol})(-1274.5\ \text{kJ/mol})$$

$$= \mathbf{-66.5\ kJ}$$

6.88 One conversion factor needed to solve this problem is the molar mass of water. The other conversion factor is given in the problem. It takes 44.0 kJ of energy to vaporize 1 mole of water.

$$\frac{1\text{ mol }H_2O}{44.0\text{ kJ}}$$

You should come up with the following strategy to solve the problem.

$$4000\text{ kJ} \rightarrow \text{mol }H_2O \rightarrow \text{g }H_2O$$

$$\textbf{? g H}_2\textbf{O} = 4000\text{ kJ} \times \frac{1\text{ mol }H_2O}{44.0\text{ kJ}} \times \frac{18.02\text{ g }H_2O}{1\text{ mol }H_2O} = \mathbf{1.64 \times 10^3\ g\ H_2O}$$

6.90 First calculate the final volume of CO_2 gas:

$$V = \frac{nRT}{P} = \frac{[19.2\text{ g}/(44.01\text{ g/mol})](0.0821\text{ L}\cdot\text{atm/K}\cdot\text{mol})(295\text{ K})}{0.995\text{ atm}} = 10.6\text{ L}$$

$$w = -P\Delta V = -(0.995\text{ atm})(10.6\text{ L}) = -10.5\text{ L}\cdot\text{atm}$$

$$w = -10.5\text{ L}\cdot\text{atm} \times \frac{101.3\text{ J}}{1\text{ L}\cdot\text{atm}} = \mathbf{-1.06\ kJ}$$

The expansion work done is 1.06 kJ.

6.92 The reaction we are interested in is the formation of ethanol from its elements.

$$2\text{ C (graphite)} + 1/2\ O_2(g) + 3\ H_2(g) \longrightarrow C_2H_5OH(l)$$

Along with the reaction for the combustion of ethanol, we can add other reactions together to end up with the above reaction.

Reversing the reaction representing the combustion of ethanol gives:

$$2\ CO_2(g) + 3\ H_2O(l) \longrightarrow C_2H_5OH(l) + 3\ O_2(g) \qquad \Delta H° = +1367.4\text{ kJ}$$

We need to add equations that add C (graphite) and remove H_2O from the reactants side of the equation. We write:

$$2\ CO_2(g) + 3\ H_2O(l) \longrightarrow C_2H_5OH(l) + 3\ O_2(g) \qquad \Delta H° = +1367.4\text{ kJ}$$

$$2\text{ C (graphite)} + 2\ O_2(g) \longrightarrow 2\ CO_2(g) \qquad \Delta H° = 2(-393.5\text{ kJ})$$

$$3\ H_2(g) + 3/2\ O_2(g) \longrightarrow 3\ H_2O(l) \qquad \Delta H° = 3(-285.8\text{ kJ})$$

$$\mathbf{2\ C\ (graphite) + 1/2\ O_2(g) + 3\ H_2(g) \longrightarrow C_2H_5OH(l)} \qquad \mathbf{\Delta H_f^\circ = -277.0\ kJ/mol}$$

6.94 From Chapter 5, we saw that the kinetic energy (or internal energy) of 1 mole of a gas is $(3/2)RT$. For 1 mole of an ideal gas, $PV = RT$. We can write:

$$\text{internal energy} = \frac{3}{2}RT = \frac{3}{2}PV$$

$$= \frac{3}{2}(1.2 \times 10^5\text{ Pa})(5.5 \times 10^3\text{ m}^3)$$

$$= 9.9 \times 10^8\text{ Pa}\cdot\text{m}^3$$

$$1 \text{ Pa·m}^3 = 1 \frac{\text{N}}{\text{m}^2}\text{m}^3 = 1 \text{ N·m} = 1 \text{ J}$$

Therefore, the internal energy is **9.9 × 10⁸ J**.

The final temperature of the copper metal can be calculated.

$$q = m_{\text{Cu}}s_{\text{Cu}}\Delta t$$

$$9.9 \times 10^8 \text{ J} = (9.07 \times 10^6 \text{ g})(0.385 \text{ J/g°C})(t_\text{f} - 21°\text{C})$$

$$3.49 \times 10^6\, t_\text{f} = 1.06 \times 10^9$$

$$\mathbf{t_f = 304°C}$$

6.96 Water has a larger specific heat than air. Thus cold, damp air can extract more heat from the body. By the same token, hot, humid air can deliver more heat to the body.

CHAPTER 7
THE ELECTRONIC STRUCTURE OF ATOMS

PROBLEM-SOLVING STRATEGIES AND TUTORIAL SOLUTIONS

TYPES OF PROBLEMS

Problem Type 1: Calculating the Frequency and Wavelength of an Electromagnetic Wave.

Problem Type 2: Calculating the Energy of a Photon.

Problem Type 3: Calculating the Energy, Wavelength, or Frequency in the Emission Spectrum of a Hydrogen Atom.

Problem Type 4: The De Broglie Equation: Calculating the Wavelengths of Particles.

Problem Type 5: Quantum Numbers.
(a) Labeling an atomic orbital.
(b) Counting the number of orbitals associated with a principal quantum number.
(c) Assigning quantum numbers to an electron.
(d) Counting the number of electrons in a principal level.

Problem Type 6: Writing Electron Configurations and Orbital Diagrams.

PROBLEM TYPE 1: CALCULATING THE FREQUENCY AND WAVELENGTH OF AN ELECTROMAGNETIC WAVE

All types of *electromagnetic radiation* move through a vacuum at a speed of about 3.00×10^8 m/s, which is called the speed of light (c). Speed is an important property of a wave traveling through space and is equal to the product of the wavelength and the frequency of the wave. For electromagnetic waves

$$c = \lambda\nu \qquad (7.1)$$

Equation (7.1) can be rearranged as necessary to solve for either the wavelength (λ) or the frequency (ν).

EXAMPLE 7.1

A certain AM radio station broadcasts at a frequency of 6.00×10^2 kHz. What is the wavelength of these radio waves in meters?

Step 1: Solve Equation (7.1) algebraically for the wavelength (λ).

$$c = \lambda\nu$$

$$\lambda = \frac{c}{\nu} \qquad (7.2)$$

Step 2: Since the speed of light has units of m/s, we must convert the frequency from units of kHz to Hz (s^{-1})

$$6.00 \times 10^2\ \cancel{\text{kHz}} \times \frac{1000\ \text{Hz}}{1\ \cancel{\text{kHz}}} = 6.00 \times 10^5\ \text{Hz} = 6.00 \times 10^5\ \text{s}^{-1}$$

Step 3: Calculate the value of λ by substituting the known quantities into Equation (7.2).

$$\lambda = \frac{3.00 \times 10^8 \ \frac{\text{m}}{\cancel{\text{s}}}}{6.00 \times 10^5 \ \frac{1}{\cancel{\text{s}}}} = \mathbf{5.00 \times 10^2\ m}$$

EXAMPLE 7.2
What is the frequency of light that has a wavelength of 665 nm?

Step 1: Solve Equation (7.1) algebraically for the frequency (ν).

$$c = \lambda\nu$$

$$\nu = \frac{c}{\lambda} \qquad (7.3)$$

Step 2: Since the speed of light has units of m/s, we must convert the wavelength from units of nm to m.

$$665\ \cancel{\text{nm}} \times \frac{10^{-9}\ \text{m}}{1\ \cancel{\text{nm}}} = 6.65 \times 10^{-7}\ \text{m}$$

Step 3: Calculate the value of ν by substituting the known quantities into Equation (7.3).

$$\nu = \frac{3.00 \times 10^8\ \frac{\cancel{\text{m}}}{\text{s}}}{6.65 \times 10^{-7}\ \cancel{\text{m}}} = \mathbf{4.51 \times 10^{14}\ s^{-1} = 4.51 \times 10^{14}\ Hz}$$

PRACTICE EXERCISE

1. Domestic microwave ovens generate microwaves with a frequency of 2.450 GHz. What is the wavelength of this microwave radiation?

Text Problems: 7.8, 7.12, **7.16**, 7.20

PROBLEM TYPE 2: CALCULATING THE ENERGY OF A PHOTON

Max Planck said that atoms and molecules could emit (or absorb) energy only in discrete quantities. Planck gave the name *quantum* to the smallest quantity of energy that can be emitted (or absorbed) in the form of electromagnetic radiation. The energy E of a single quantum of energy is given by

$$E = h\nu \qquad (7.4)$$

where,

h is Planck's constant $= 6.63 \times 10^{-34}$ J·s
ν is the frequency of radiation

EXAMPLE 7.3
The yellow light given off by a sodium vapor lamp has a wavelength of 589 nm. What is the energy of a single photon of this radiation?

Step 1: You are given wavelength in this problem. Equation (7.4) shows the relationship between energy and frequency. However, there is a relationship between frequency and wavelength [see Equation (7.3)].

$$\nu = \frac{c}{\lambda}$$

Substituting for the frequency in Equation (7.4), we have

$$E = \frac{hc}{\lambda} \tag{7.5}$$

Step 2: Since the speed of light is in units of m/s, we must convert the wavelength from units of nm to m.

$$589 \text{ nm} \times \frac{10^{-9} \text{ m}}{1 \text{ nm}} = 5.89 \times 10^{-7} \text{ m}$$

Step 3: Calculate the value of E by substituting the known quantities into Equation (7.5).

$$E = \frac{(6.63 \times 10^{-34} \text{ J}\cdot\text{s})\left(3.00 \times 10^{8} \frac{\text{m}}{\text{s}}\right)}{5.89 \times 10^{-7} \text{ m}} = \mathbf{3.38 \times 10^{-19} \text{ J}}$$

PRACTICE EXERCISE

2. The red line in the spectrum of lithium occurs at 670.8 nm. What is the energy of a photon of this light? What is the energy of 1 mol of these photons?

3. The light-sensitive compound in most photographic films is silver bromide (AgBr). When the film is exposed, assume that the light energy absorbed dissociates the molecule into atoms. (The actual process is more complex.) If the energy of dissociation of AgBr is 1.00×10^2 kJ/mol, find the wavelength of light that is just able to dissociate AgBr.

Text Problems: 7.16, 7.18, 7.20, 7.30

PROBLEM TYPE 3: CALCULATING THE ENERGY, WAVELENGTH, OR FREQUENCY IN THE EMISSION SPECTRUM OF A HYDROGEN ATOM

Using arguments based on electrostatic interaction and Newton's laws of motion, Neils Bohr showed that the energies that the electron in the hydrogen atom can possess are given by:

$$E_n = -R_H\left(\frac{1}{n^2}\right) \tag{7.5}$$

where,

R_H is the Rydberg constant $= 2.18 \times 10^{-18}$ J
n is the principal quantum number that has integer values

During the emission process in a hydrogen atom, an electron initially in an excited state characterized by the principal quantum number n_i drops to a lower energy state characterized by the principal quantum number n_f. This lower energy state may be either another excited state or the ground state. The difference between the energies of the initial and final states is:

$$\Delta E = E_f - E_i$$

Substituting Equation (7.5) into the above equation gives:

$$\Delta E = \left(\frac{-R_H}{n_f^2}\right) - \left(\frac{-R_H}{n_i^2}\right)$$

$$\Delta E = R_H\left(\frac{1}{n_i^2} - \frac{1}{n_f^2}\right)$$

Furthermore, since this transition results in the emission of a photon of frequency ν and energy $h\nu$ (See Problem Type 2), we can write:

$$\Delta E = h\nu = R_H\left(\frac{1}{n_i^2} - \frac{1}{n_f^2}\right) \tag{7.6}$$

EXAMPLE 7.4

What wavelength of radiation will be emitted during an electron transition from the $n = 5$ state to the $n = 1$ state in the hydrogen atom? What region of the electromagnetic spectrum does this wavelength correspond to?

Step 1: Using Equation (7.6), we can calculate the energy change for the transition.

$$\Delta E = R_H\left(\frac{1}{n_i^2} - \frac{1}{n_f^2}\right)$$

$$\Delta E = (2.18 \times 10^{-18}\ \text{J})\left(\frac{1}{5^2} - \frac{1}{1^2}\right)$$

$$\Delta E = -2.09 \times 10^{-18}\ \text{J}$$

Step 2: The negative sign for ΔE indicates that energy is released to the surroundings during the emission process. To calculate the wavelength, we omit the minus sign for ΔE because the wavelength of the photon must be positive. We know that

$$\Delta E = h\nu$$

We also know that $\nu = \frac{c}{\lambda}$. Substituting into the above equation gives:

$$\Delta E = \frac{hc}{\lambda}$$

Solving the equation algebraically for the wavelength, then substituting in the known values gives:

$$\lambda = \frac{hc}{\Delta E} = \frac{(6.63 \times 10^{-34}\ \cancel{\text{J}}\cdot\cancel{\text{s}})\left(3.00 \times 10^8\ \frac{\text{m}}{\cancel{\text{s}}}\right)}{2.09 \times 10^{-18}\ \cancel{\text{J}}} = \mathbf{9.52 \times 10^{-8}\ m}$$

Step 3: To determine the region of the electromagnetic spectrum that this wavelength corresponds to, we should convert the wavelength from units of meters to nanometers, and then compare the value to Figure 7.2 of your text.

$$9.52 \times 10^{-8}\ \cancel{\text{m}} \times \frac{1\ \text{nm}}{10^{-9}\ \cancel{\text{m}}} = 95.2\ \text{nm}$$

Checking Figure 7.2, we see that the ultraviolet region of the spectrum is centered at a wavelength of 10 nm. Therefore, this emission is in the ultraviolet (UV) region of the electromagnetic spectrum.

PRACTICE EXERCISE

4. A hydrogen emission line in the ultraviolet region of the spectrum at 95.2 nm corresponds to a transition from a higher energy level n_i to the $n = 1$ level. What is the value of n_i for the higher energy level?

Text Problems: 7.30, **7.32,** 7.34

PROBLEM TYPE 4: THE DE BROGLIE EQUATION: CALCULATING THE WAVELENGTHS OF PARTICLES

Albert Einstein showed that light (electromagnetic radiation) can possess particle like properties. De Broglie reasoned that if waves can behave like particles, then particles can exhibit wave properties. De Broglie deduced that the particle and wave properties are related by the expression:

$$\lambda = \frac{h}{mu} \tag{7.7}$$

where,

λ is the wavelength associated with the moving particle
m is the mass of the particle
u is the velocity of the particle

Equation (7.7) implies that a particle in motion can be treated as a wave, and a wave can exhibit the properties of a particle.

EXAMPLE 7.5
When an atom of Th-232 undergoes radioactive decay, an alpha particle, which has a mass of 4.0 amu, is ejected from the Th nucleus with a velocity of 1.4 × 10^7 m/s. What is the De Broglie wavelength of the alpha particle?

Step 1: Because Planck's constant has units of J·s, and 1 J = 1 kg·m^2·s^{-2}, the mass of the alpha particle must be expressed in kilograms. Since one particle has a mass of 4.0 amu, a mole of alpha particles will have a mass of 4.0 g. A reasonable strategy to complete the conversion is:

g/mol → kg/mol → kg/particle

$$\frac{4.0\ \cancel{g}}{1\ \cancel{mol}} \times \frac{1\ kg}{1000\ \cancel{g}} \times \frac{1\ \cancel{mol}}{6.022 \times 10^{23}\ particles} = 6.6 \times 10^{-27}\ kg/particle$$

Step 2: Substitute the known quantities into Equation (7.7).

$$\lambda = \frac{h}{mu} = \frac{(6.63 \times 10^{-34}\ \cancel{J \cdot s}) \times \left(\dfrac{1\ \cancel{kg} \cdot m^{\cancel{2}} / \cancel{s^2}}{1\ \cancel{J}}\right)}{(6.6 \times 10^{-27}\ \cancel{kg})\left(1.4 \times 10^7\ \dfrac{\cancel{m}}{\cancel{s}}\right)} = \mathbf{7.2 \times 10^{-15}\ m}$$

The wavelength is smaller than the diameter of the thorium nucleus, which is about 2 × 10^{-14} m. Is this what you would expect?

PRACTICE EXERCISE

5. The average kinetic energy of a neutron at 25°C is 6.2 × 10^{-21} J. What is the de Broglie wavelength of an average neutron? The mass of a neutron is 1.008 amu. (**Hint:** kinetic energy $= \frac{1}{2} mu^2$)

Text Problems: 7.40, 7.42

PROBLEM TYPE 5: QUANTUM NUMBERS

See Section 7.7 of your text for a complete discussion. In quantum mechanics, three quantum numbers are required to describe the distribution of electrons in atoms.

(1) **The principal quantum number (n).** In a hydrogen atom, the value of n determines the energy of an orbital. It can have integral values 1, 2, 3, and so forth.

(2) **The angular momentum quantum number (l)** tells us the "shape" of the orbitals. l has possible integral values from 0 to $(n - 1)$. The value of l is generally designated by the letters s, p, d, ..., as follows:

l	0	1	2	3	4	5
Name of orbital	s	p	d	f	g	h

(3) **The magnetic quantum number (m_l)** describes the orientation of the orbital in space. For a certain value of l, there are $(2l + 1)$ integral values of m_l, as follows:

$$-l, (-l + 1), \ldots 0, \ldots (+l - 1), +l$$

The number of m_l values indicates the number of orbitals in a subshell with a particular l value.

Finally, there is a fourth quantum number that tells us the spin of the electron. The **electron spin quantum number (m_s)** has values of +1/2 or −1/2, which correspond to the two spinning motions of the electron.

A. Labeling an atomic orbital

To "label" an atomic orbital, you need to specify the three quantum numbers (n, l, m_l) that give information about the distribution of electrons in orbitals. Remember, m_s tells us the spin of the electron, which tells us nothing about the orbital.

EXAMPLE 7.6

List the values of n, l, and m_l for orbitals in the 2p subshell.

The number given in the designation of the subshell is the principal quantum number, so in this case $n = 2$. For p orbitals, $l = 1$. m_l can have integer values from $-l$ to $+l$. Therefore, m_l can be −1, 0, and +1. (The three values for m_l correspond to the three p orbitals.)

PRACTICE EXERCISE

6. List the values of n, l, and m_l for orbitals in the 4f subshell.

Text Problems: 7.54, 7.56, 7.60

B. Counting the number of orbitals associated with a principal quantum number

To work this type of problem, you must take into account the energy level (n), the types of orbitals in that energy level (l), and the number of orbitals in a subshell with a particular l value (m_l).

EXAMPLE 7.7
What is the total number of orbitals associated with the principal quantum number $n = 2$?

For $n = 2$, there are only two possible values of l, 0 and 1. Thus, there is one 2*s* orbital, and there are three 2*p* orbitals. (For $l = 1$, there are three possible m_l values, −1, 0, and +1.)

Therefore, the total number of orbitals in the $n = 2$ energy level is 1 + 3 = **4**.

> **Tip:** The total number of orbitals with a given n value is n^2. For Example 7.7, the total number of orbitals in the $n = 2$ level equals $2^2 = \mathbf{4}$.

PRACTICE EXERCISE

7. What is the total number of orbitals associated with the principal quantum number $n = 4$?

Text Problems: 7.56, 7.60

C. Assigning quantum numbers to an electron

In assigning quantum numbers to an electron, you need to specify all four quantum numbers. In most cases, there will be more than one possible set of quantum numbers that can designate an electron.

EXAMPLE 7.8
List the different ways to write the four quantum numbers that designate an electron in a 4*s* orbital.

To begin with, we know that the principal quantum number n is 4, and the angular momentum quantum number l is 0 (*s* orbital). For $l = 0$, there is only one possible value for m_l, also 0. Since the electron spin quantum number m_s can be either +1/2 or −1/2, we conclude that there are two possible ways to designate the electron:

(4, 0, 0, +1/2) (4, 0, 0, −1/2)

PRACTICE EXERCISE

8. List the different ways to write the four quantum numbers that designate an electron in a 3*d* orbital.

Text Problem: 7.54

D. Counting the number of electrons in a principal level

To work this type of problem, you need to know that the number of orbitals with a particular l value is $(2l + 1)$. Also, each orbital can accommodate two electrons.

EXAMPLE 7.9
What is the maximum number of electrons that can be present in the principal level for which $n = 4$?

When $n = 4$, $l = 0$, 1, 2, and 3. The number of orbitals for each l value is given by

Value of l	Number of orbitals $(2l + 1)$
0	1
1	3
2	5
3	7

The total number of orbitals in the principal level $n = 4$ is sixteen. Since each orbital can accommodate two electrons, the maximum number of electrons that can reside in the orbitals is 2 × 16 = **32**.

> **Tip:** The above result can be generalized by the formula $2n^2$. For Example 7.9, we have $n = 4$, so $2(4)^2 = 32$.

PRACTICE EXERCISE

9. What is the maximum number of electrons that can be present in the principal level for which $n = 2$?

Text Problem: 7.62

PROBLEM TYPE 6: WRITING ELECTRON CONFIGURATIONS AND ORBITAL DIAGRAMS

The electron configuration of an atom tells us how the electrons are distributed among the various atomic orbitals. To write electron configurations, you should follow the four rules or guidelines given below.

(1) The electron configurations of all elements except hydrogen and helium are represented by a *noble gas core*, which shows (in brackets) the noble gas element that most nearly precedes the element being considered. The noble gas core is followed by the electron configurations of filled or partially filled subshells in the outermost shells.

(2) *The Aufbau or "building up" principle* states that electrons are added to atomic orbitals starting with the lowest energy orbital and "building up" to higher energy orbitals.

(3) In many-electron atoms, the subshells are filled in the order shown in Figure 7.21.

(4) Each orbital can hold only *two* electrons.

The electron configuration can be represented in a more detailed manner called an ***orbital diagram*** that shows the spin of the electron. For orbital diagrams, you need to follow two additional rules given below.

(5) The *Pauli exclusion principle* states that no two electrons in an atom can have the same four quantum numbers. This means that electrons occupying the same orbital *cannot* have the same spin.

(6) *Hund's rule* states that the most stable arrangement of electrons in subshells is the one with the greatest number of parallel spins, without violating the Pauli exclusion principle.

EXAMPLE 7.10
Write the ground-state electron configuration and the orbital diagram for selenium.

Step 1: Selenium has 34 electrons. These electrons need to be placed in atomic orbitals.

Step 2: The noble gas element that most nearly proceeds Se is Ar. Therefore, the *noble gas core* is [Ar]. This core accounts for 18 electrons.

Step 3: See Figure 7.21 to check the order of filling subshells past the Ar noble gas core. You should find that the order of filling is 4*s*, 3*d*, then 4*p*. There are 16 remaining electrons to distribute among these orbitals.

The 4*s* orbital can hold *two* electrons. Each of the five 3*d* orbitals can hold *two* electrons for a total of *ten* electrons. This leaves *four* electrons to fill the 4*p* orbitals.

The electron configuration for Se is:

$$[\mathrm{Ar}]\ 4s^2 3d^{10} 4p^4$$

Step 4: To write an *orbital diagram*, we must also specify the spin of the electrons. The 4*s* and 3*d* orbitals are filled, so according to the Pauli exclusion principle, the paired electrons in the 4*s* orbital and in each of the 3*d* orbitals *must* have opposite spins.

$$[\mathrm{Ar}]\ \underset{4s^2}{\underline{\uparrow\downarrow}} \quad \underset{3d^{10}}{\underline{\uparrow\downarrow}\ \underline{\uparrow\downarrow}\ \underline{\uparrow\downarrow}\ \underline{\uparrow\downarrow}\ \underline{\uparrow\downarrow}}$$

Step 5: Now, let's deal with the 4*p* electrons. Hund's rule states that the most stable arrangement of electrons in subshells is the one with the greatest number of parallel spins. In other words, we want to keep electrons unpaired if possible with parallel spins. Since there are three *p* orbitals, three of the *p* electrons can be placed individually in each of the *p* subshells with parallel spins.

↑ ↑ ↑
$4p^3$

Step 6: Finally, the fourth *p* electron must be paired up in one of the 4*p* orbitals. The complete orbital diagram is:

[Ar] ↑↓ ↑↓ ↑↓ ↑↓ ↑↓ ↑↓ ↑↓ ↑ ↑
$4s^2$ $3d^{10}$ $4p^4$

PRACTICE EXERCISE

10. Write the electron configuration and the orbital diagram for iron (Fe).

Text Problems: 7.78, 7.80, **7.82**, 7.84

ANSWERS TO PRACTICE EXERCISES

1. $\lambda = 0.1224$ m

2. $E = 2.97 \times 10^{-19}$ J
$E = 1.79 \times 10^5$ J/mol

3. $\lambda = 1.20 \times 10^{-6}$ m $= 1.20 \times 10^3$ nm

4. $n_i = 5$

5. $\lambda = 1.5 \times 10^{-10}$ m

6. $n = 4, l = 3, m_l = -3, -2, -1, 0, 1, 2, 3$

7. Total number of orbitals $= n^2 = 4^2 = 16$

8.
(3, 2, −2, +1/2) (3, 2, −2, −1/2)
(3, 2, −1, +1/2) (3, 2, −1, −1/2)
(3, 2, 0, +1/2) (3, 2, 0, −1/2)
(3, 2, +1, +1/2) (3, 2, +1, −1/2)
(3, 2, +2, +1/2) (3, 2, +2, −1/2)

9. 8 electrons

10. $[Ar]4s^23d^6$

[Ar] ↑↓ ↑↓ ↑ ↑ ↑ ↑
$4s^2$ $3d^6$

SOLUTIONS TO SELECTED TEXT PROBLEMS

7.8 Calculating the Frequency and Wavelength of an Electromagnetic Wave, Problem Type 1.

(a)

Step 1: $c = \lambda\nu$. Solve this equation algebraically for the frequency (ν).

$$\nu = \frac{c}{\lambda} \qquad (7.1)$$

Step 2: Since the speed of light has units of m/s, we must convert the wavelength from units of nm to m.

$$456 \text{ nm} \times \frac{1 \times 10^{-9} \text{ m}}{1 \text{ nm}} = 4.56 \times 10^{-7} \text{ m}$$

Step 3: Calculate the value of ν by substituting the known quantities into Equation (7.1).

$$\nu = \frac{3.00 \times 10^{8} \frac{\text{m}}{\text{s}}}{4.56 \times 10^{-7} \text{ m}} = \mathbf{6.58 \times 10^{14} \ s^{-1}} = \mathbf{6.58 \times 10^{14} \ Hz}$$

(b)

Step 1: $c = \lambda\nu$. Solve this equation algebraically for the wavelength (λ).

$$\lambda = \frac{c}{\nu} \qquad (7.2)$$

Step 2: Calculate the value of λ by substituting in the known quantities into Equation (7.2).

$$\lambda = \frac{3.00 \times 10^{8} \frac{\text{m}}{\text{s}}}{2.20 \times 10^{9} \frac{1}{\text{s}}} = 0.136 \text{ m}$$

Step 3: Convert the wavelength in units of meters to nanometers.

$$\lambda = 0.136 \text{ m} \times \frac{1 \text{ nm}}{1 \times 10^{-9} \text{ m}} = \mathbf{1.36 \times 10^{8} \ nm}$$

7.10 $\text{time} = \dfrac{\text{distance}}{\text{speed}}$

A radio wave is an electromagnetic wave, which travels at the speed of light. The speed of light is in units of m/s, so let's convert distance from units of miles to meters.

$$? \text{ distance (m)} = (2.8 \times 10^{7} \text{ mi}) \times \frac{1.61 \text{ km}}{1 \text{ mi}} \times \frac{1000 \text{ m}}{1 \text{ km}} = 4.5 \times 10^{10} \text{ m}$$

$$\textbf{time} = \frac{\text{distance}}{\text{speed}} = \frac{4.5 \times 10^{10} \text{ m}}{3.00 \times 10^{8} \text{ m/s}} = \mathbf{1.5 \times 10^{2} \ s}$$

7.12 The wavelength is:

$$\lambda = \frac{1\text{ m}}{1{,}650{,}763.73\text{ wavelengths}} = 6.057802106 \times 10^{-7}\text{ m}$$

$$\nu = \frac{c}{\lambda} = \frac{3.00 \times 10^{8}\ \cancel{\text{m}}/\text{s}}{6.057802106 \times 10^{-7}\ \cancel{\text{m}}} = \mathbf{4.95 \times 10^{14}\ s^{-1}}$$

7.16 Calculating the wavelength of electromagnetic radiation and calculating the energy of a photon, Problem Types 1 and 2.

(a)
Step 1: $c = \lambda\nu$. Solve this equation algebraically for the wavelength (λ).

$$\lambda = \frac{c}{\nu} \tag{7.2}$$

Step 2: Calculate the value of λ by substituting the known quantities into Equation (7.2).

$$\lambda = \frac{3.00 \times 10^{8}\ \frac{\text{m}}{\cancel{\text{s}}}}{7.5 \times 10^{14}\ \frac{1}{\cancel{\text{s}}}} = \mathbf{4.0 \times 10^{-7}\ m} = \mathbf{4.0 \times 10^{2}\ nm}$$

(b) $E = h\nu$. Substitute the frequency (ν) into this equation to solve for the energy of a single photon associated with this frequency.

$$E = h\nu = (6.63 \times 10^{-34}\ \text{J}\cdot\cancel{\text{s}})(7.5 \times 10^{14}\ \frac{1}{\cancel{\text{s}}}) = \mathbf{5.0 \times 10^{-19}\ J}$$

7.18 The energy given in this problem is for *1 mole* of photons. To apply $E = h\nu$, we must divide by Avogadro's number. The energy of one photon is:

$$E = \frac{1.0 \times 10^{3}\ \cancel{\text{kJ}}}{1\ \cancel{\text{mol}}} \times \frac{1\ \cancel{\text{mol}}}{6.022 \times 10^{23}\text{ photons}} \times \frac{1000\text{ J}}{1\ \cancel{\text{kJ}}} = 1.7 \times 10^{-18}\text{ J/photon}$$

The wavelength of this photon can be found using the relationship $E = hc/\lambda$.

$$\lambda = \frac{hc}{E} = \frac{(6.63 \times 10^{-34}\ \cancel{\text{J}}\cdot\cancel{\text{s}})\left(3.00 \times 10^{8}\ \frac{\cancel{\text{m}}}{\cancel{\text{s}}}\right)}{1.7 \times 10^{-18}\ \cancel{\text{J}}} \times \left(\frac{1\text{ nm}}{1 \times 10^{-9}\ \cancel{\text{m}}}\right) = \mathbf{1.2 \times 10^{2}\ nm}$$

The radiation is in the ultraviolet region (see Figure 7.2 of the text).

7.20 **(a)** $\lambda = \dfrac{c}{\nu}$

$$\lambda = \frac{3.00 \times 10^{8}\ \frac{\text{m}}{\cancel{\text{s}}}}{8.11 \times 10^{14}\ \frac{1}{\cancel{\text{s}}}} = \mathbf{3.70 \times 10^{-7}\ m} = \mathbf{3.70 \times 10^{2}\ nm}$$

(b) Checking Figure 7.2 of your text, you should find that the visible region of the spectrum runs from 400 to 700 nm. 370 nm is in the **ultraviolet** region of the spectrum.

(c) $E = h\nu$. Substitute the frequency (ν) into this equation to solve for the energy of one quantum associated with this frequency.

$$E = h\nu = (6.63 \times 10^{-34}\ \text{J}\cdot\cancel{\text{s}})(8.11 \times 10^{14}\ \frac{1}{\cancel{\text{s}}}) = \mathbf{5.38 \times 10^{-19}\ J}$$

7.26 The emitted light could be analyzed by passing it through a prism.

7.28 Excited atoms of the chemical elements emit the same characteristic frequencies or lines in a terrestrial laboratory, in the sun, or in a star many light-years distant from earth.

7.30 We use more accurate values of h and c for this problem.

$$E = h\nu = \frac{hc}{\lambda} = \frac{(6.6256 \times 10^{-34}\ \text{J}\cdot\cancel{\text{s}})(2.998 \times 10^{8}\ \cancel{\text{m}}/\cancel{\text{s}})}{656.3 \times 10^{-9}\ \cancel{\text{m}}} = \mathbf{3.027 \times 10^{-19}\ J}$$

7.32 Calculating the Energy, Wavelength, or Frequency in the Emission Spectrum of a Hydrogen Atom, Problem Type 3.

Step 1: Using the following equation, we can calculate the energy change for the transition.

$$\Delta E = R_H\left(\frac{1}{n_i^2} - \frac{1}{n_f^2}\right)$$

$$\Delta E = (2.18 \times 10^{-18}\ \text{J})\left(\frac{1}{4^2} - \frac{1}{2^2}\right)$$

$$\Delta E = -4.09 \times 10^{-19}\ \text{J}$$

Step 2: The negative sign for ΔE indicates that energy is released to the surroundings during the emission process. To calculate the wavelength, we omit the minus sign for ΔE because the frequency of the photon must be positive. We know that

$$\Delta E = h\nu$$

Rearranging the equation and substituting in the known values,

$$\nu = \frac{\Delta E}{h} = \frac{(4.09 \times 10^{-19}\ \cancel{\text{J}})}{(6.63 \times 10^{-34}\ \cancel{\text{J}}\cdot\text{s})} = \mathbf{6.17 \times 10^{14}\ s^{-1}\ or\ Hz}$$

Step 3: We also know that $\lambda = \frac{c}{\nu}$. Substituting the frequency calculated above into this equation gives:

$$\lambda = \frac{\left(3.00 \times 10^{8}\ \frac{\text{m}}{\cancel{\text{s}}}\right)}{\left(6.17 \times 10^{14}\ \frac{1}{\cancel{\text{s}}}\right)} = \mathbf{4.86 \times 10^{-7}\ m = 486\ nm}$$

7.34 $\Delta E = R_H\left(\frac{1}{n_i^2} - \frac{1}{n_f^2}\right)$

n_f is given in the problem and R_H is a constant, but we need to calculate ΔE. The photon energy is:

$$E = \frac{hc}{\lambda} = \frac{(6.63 \times 10^{-34}\ \text{J}\cdot\text{s})(3.00 \times 10^{8}\ \text{m/s})}{434 \times 10^{-9}\ \text{m}} = 4.58 \times 10^{-19}\ \text{J}$$

Since this is an emission process, the energy change ΔE must be negative, or -4.58×10^{-19} J.

Substitute ΔE into the following equation, and solve for n_i.

$$\Delta E = R_H\left(\frac{1}{n_i^2} - \frac{1}{n_f^2}\right)$$

$$-4.58 \times 10^{-19}\ \text{J} = (2.18 \times 10^{-18}\ \text{J})\left(\frac{1}{n_i^2} - \frac{1}{2^2}\right)$$

$$\frac{1}{n_i^2} = \left(\frac{-4.58 \times 10^{-19}\ \text{J}}{2.18 \times 10^{-18}\ \text{J}}\right) + \frac{1}{2^2} = -0.210 + 0.250 = 0.040$$

$$\boldsymbol{n_i} = \frac{1}{\sqrt{0.040}} = \mathbf{5}$$

7.40 The De Broglie Equation: Calculating the Wavelengths of Particles, Problem Type 4.

Step 1: The equation needed to solve this problem is the De Broglie equation.

$$\lambda = \frac{h}{mu}$$

Substitute the known values into the equation to solve for the wavelength of the proton.

$$\lambda = \frac{h}{mu} = \frac{(6.63 \times 10^{-34}\ \text{J}\cdot\text{s}) \times \left(\frac{1\ \text{kg}\cdot\text{m}^2/\text{s}^2}{1\ \text{J}}\right)}{(1.673 \times 10^{-27}\ \text{kg})(2.90 \times 10^{8}\ \text{m/s})} = 1.37 \times 10^{-15}\ \text{m}$$

Step 2: Convert wavelength in units of meters to nanometers.

$$\lambda = 1.37 \times 10^{-15}\ \text{m} \times \frac{1\ \text{nm}}{1 \times 10^{-9}\ \text{m}} = \mathbf{1.37 \times 10^{-6}\ nm}$$

7.42 First, we convert mph to m/s.

$$\frac{35\ \text{mi}}{1\ \text{h}} \times \frac{1.61\ \text{km}}{1\ \text{mi}} \times \frac{1000\ \text{m}}{1\ \text{km}} \times \frac{1\ \text{h}}{3600\ \text{s}} = 16\ \text{m/s}$$

$$\lambda = \frac{h}{mu} = \frac{(6.63 \times 10^{-34}\ \text{J}\cdot\text{s}) \times \left(\frac{1\ \text{kg}\cdot\text{m}^2/\text{s}^2}{1\ \text{J}}\right)}{(2.5 \times 10^{-3}\ \text{kg})(16\ \text{m/s})} = \mathbf{1.7 \times 10^{-32}\ m} = \mathbf{1.7 \times 10^{-23}\ nm}$$

7.54 Quantum Numbers, Problem Type 5.

When $n = 3$, there are three possible values of l: 0, 1, and 2, corresponding to the s, p, and d orbitals, respectively. The values of m_l for each l value are:

$l = 0$: $m_l = 0$ $l = 1$: $m_l = -1, 0, 1$ $l = 2$: $m_l = -2, -1, 0, 1, 2$

7.56 Labeling an atomic orbital and counting the number of orbitals associated with a principal quantum number, Problem Types 5A and 5B.

(a) The number given in the designation of the subshell is the principal quantum number, so in this case $n = 4$. For p orbitals, $l = 1$. m_l can have integer values from $-l$ to $+l$. Therefore, m_l can be -1, 0, and $+1$.

The three values for m_l correspond to the *three p* orbitals.

Following the same reasoning as part **(a)**

(b) $3d$: $n = 3, l = 2, m_l = 2, 1, 0, -1$, or -2 (5 orbitals)

(c) $3s$: $n = 3, l = 0, m_l = 0$ (1 orbital)

(d) $5f$: $n = 5, l = 3, m_l = 3, 2, 1, 0, -1, -2$, or -3 (7 orbitals)

7.58 The two orbitals are identical in size, shape, and energy. They differ only in their orientation with respect to each other.

Can you assign a specific value of the magnetic quantum number to these orbitals? What are the allowed values of the magnetic quantum number for the $2p$ subshell?

7.60 For $n = 6$, the allowed values of l are 0, 1, 2, 3, 4, and 5 [$l = 0, ..., (n - 1)$, integer values]. These l values correspond to the $6s$, $6p$, $6d$, $6f$, $6g$, and $6h$ subshells. These subshells each have 1, 3, 5, 7, 9, and 11 orbitals, respectively (number of orbitals = $2l + 1$).

7.62

n value	orbital sum	total number of electrons
1	1	2
2	1 + 3 = 4	8
3	1 + 3 + 5 = 9	18
4	1 + 3 + 5 + 7 = 16	32
5	1 + 3 + 5 + 7 + 9 = 25	50
6	1 + 3 + 5 + 7 + 9 + 11 = 36	72

In each case the total number of orbitals is just the square of the n value (n^2). The total number of electrons is $2n^2$.

7.64 The electron configurations for the elements are

(a) N: $1s^22s^22p^3$ There are three p-type electrons.

(b) Si: $1s^22s^22p^63s^23p^2$ There are six s-type electrons.

(c) S: $1s^22s^22p^63s^23p^4$ There are no d-type electrons.

7.66 In the many-electron atom, the $3p$ orbital electrons are more effectively shielded by the inner electrons of the atom (that is, the $1s$, $2s$, and $2p$ electrons) than the $3s$ electrons. The $3s$ orbital is said to be more "penetrating" than the $3p$ and $3d$ orbitals. In the hydrogen atom there is only one electron, so the $3s$, $3p$, and $3d$ orbitals have the same energy.

7.68 **(a)** $2s < 2p$ **(b)** $3p < 3d$ **(c)** $3s < 4s$ **(d)** $4d < 5f$

7.78 Writing Electron Configurations, Problem Type 6.

For aluminum, there are not enough electrons in the $2p$ subshell. (The $2p$ subshell holds six electrons.) The number of electrons (13) is correct. The electron configuration should be $1s^22s^22p^63s^23p^1$. The configuration shown might be an excited state of an aluminum atom.

For boron, there are too many electrons. (Boron only has five electrons.) The electron configuration should be $1s^22s^22p^1$. What would be the electric charge of a boron ion with the electron arrangement given in the problem?

For fluorine, there are also too many electrons. (Fluorine only has nine electrons.) The configuration shown is that of the F^- ion. The correct electron configuration is $1s^22s^22p^5$.

7.80 You should write the electron configurations for each of these elements to answer this question. In some cases, an orbital diagram may be helpful.

B: $[He]2s^22p^1$ (1 unpaired electron)
P: $[Ne]3s^23p^3$ (3 unpaired electrons)
Mn: $[Ar]4s^23d^5$ (5 unpaired electrons)
Zr: $[Kr]5s^24d^2$ (2 unpaired electrons)
Cd: $[Kr]5s^24d^{10}$ (0 unpaired electrons)
W: $[Xe]6s^24f^{14}5d^4$ (4 unpaired electrons)
Ce: $[Xe]6s^25d^14f^1$ (2 unpaired electrons)

Ne: (0 unpaired electrons, Why?)
Sc: $[Ar]4s^23d^1$ (1 unpaired electron)
Se: $[Ar]4s^23d^{10}4p^4$ (2 unpaired electrons)
Ru: $[Kr]5s^14d^7$ (4 unpaired electrons)
I: $[Kr]5s^24d^{10}5p^5$ (1 unpaired electron)
Pb: $[Xe]6s^24f^{14}5d^{10}6p^2$ (2 unpaired electrons)
Ho: $[Xe]6s^24f^{11}$ (3 unpaired electrons)

7.82 Writing Electron Configurations, Problem Type 6.

Step 1: Germanium (Ge) has 32 electrons. These electrons need to be placed in atomic orbitals.

Step 2: The noble gas element that most nearly proceeds Se is Ar. Therefore, the *noble gas core* is [Ar]. This core accounts for 18 electrons.

Step 3: See Figure 7.22 of your text to check the order of filling subshells past the Ar noble gas core. You should find that the order of filling is $4s$, $3d$, then $4p$. There are 14 remaining electrons to distribute among these orbitals.

The $4s$ orbital can hold *2* electrons. Each of the five $3d$ orbitals can hold *2* electrons for a total of *10* electrons. This leaves *2* electrons to fill the $4p$ orbitals.

The electrons configuration for Ge is:

$$[Ar]\ 4s^23d^{10}4p^2$$

You should follow the same reasoning for the remaining atoms.

Fe: $[Ar]4s^23d^6$
Zn: $[Ar]4s^23d^{10}$
Ru: $[Kr]5s^14d^7$
W: $[Xe]6s^24f^{14}5d^4$
Tl: $[Xe]6s^24f^{14}5d^{10}6p^1$

7.84

↑↓ $3s^2$	↑ ↑ ↑ $3p^3$	↑↓ $3s^2$	↑↓ ↑ ↑ $3p^4$	↑↓ $3s^2$	↑↓ ↑↓ ↑ $3p^5$

S^+ (5 valence electrons)
3 unpaired electrons

S (6 valence electrons)
2 unpaired electrons

S^- (7 valence electrons)
1 unpaired electron

S^+ has the most unpaired electrons

7.86 Part **(b)** is correct in the view of contemporary quantum theory. Bohr's explanation of emission and absorption line spectra appears to have universal validity. Parts **(a)** and **(c)** are artifacts of Bohr's early planetary model of the hydrogen atom and are *not* considered to be valid today.

7.88 **(a)** With $n = 2$, there are n^2 orbitals $= 2^2 = 4$. $m_s = +1/2$, specifies **1** electron per orbital, for a total of **4 electrons**.

(b) $n = 4$ and $m_l = +1$, specifies one orbital in each subshell with $l = 1$, 2, or 3 (i.e., a 4*p*, 4*d*, and 4*f* orbital). Each of the three orbitals holds 2 electrons for a total of **6 electrons**.

(c) If $n = 3$ and $l = 2$, m_l has the values 2, 1, 0, −1, or −2. Each of the five orbitals can hold 2 electrons for a total of **10 electrons** (2 e^- in each of the five 3*d* orbitals).

(d) If $n = 2$ and $l = 0$, then m_l can only be zero. $m_s = -1/2$ specifies 1 electron in this orbital for a total of **1 electron** (one e^- in the 2*s* orbital).

(e) $n = 4$, $l = 3$ and $m_l = -2$, specifies one 4*f* orbital. This orbital can hold **2 electrons**.

7.90 The wave properties of electrons are used in the operation of an electron microscope.

7.92 **(a)** First convert 100 mph to units of m/s.

$$\frac{100\ \text{mi}}{1\ \text{h}} \times \frac{1\ \text{h}}{3600\ \text{s}} \times \frac{1.61\ \text{km}}{1\ \text{mi}} \times \frac{1000\ \text{m}}{1\ \text{km}} = 44.7\ \text{m/s}$$

Using the de Broglie equation:

$$\lambda = \frac{h}{mu} = \frac{(6.63 \times 10^{-34}\ \text{J}\cdot\text{s}) \times \left(\frac{1\ \text{kg}\cdot\text{m}^2/\text{s}^2}{1\ \text{J}}\right)}{(0.141\ \text{kg})(44.7\ \text{m/s})} = 1.05 \times 10^{-34}\ \text{m} = \mathbf{1.05 \times 10^{-25}\ nm}$$

(b) The average mass of a hydrogen atom is:

$$1.008\ \frac{\text{g}}{\text{mol}} \times \frac{1\ \text{mol}}{6.022 \times 10^{23}\ \text{atoms}} = 1.674 \times 10^{-24}\ \text{g/H atom} = 1.674 \times 10^{-27}\ \text{kg}$$

$$\lambda = \frac{h}{mu} = \frac{(6.63 \times 10^{-34}\ \text{J}\cdot\text{s}) \times \left(\frac{1\ \text{kg}\cdot\text{m}^2/\text{s}^2}{1\ \text{J}}\right)}{(1.674 \times 10^{-27}\ \text{kg})(44.7\ \text{m/s})} = 8.86 \times 10^{-9}\ \text{m} = \mathbf{8.86\ nm}$$

7.94 First, let's find the energy needed to photodissociate one water molecule.

$$\frac{285.8\ \text{kJ}}{1\ \text{mol}} \times \frac{1\ \text{mol}}{6.022 \times 10^{23}\ \text{molecules}} = 4.746 \times 10^{-22}\ \text{kJ/molecule} = 4.746 \times 10^{-19}\ \text{J/molecule}$$

The maximum wavelength of a photon that would provide the above energy is:

$$\lambda = \frac{hc}{E} = \frac{(6.63 \times 10^{-34}\ \text{J}\cdot\text{s})(3.00 \times 10^{8}\ \text{m/s})}{4.746 \times 10^{-19}\ \text{J}} = 4.19 \times 10^{-7}\ \text{m} = \mathbf{419\ nm}$$

This wavelength is in the visible region of the electromagnetic spectrum. Since water is continuously being struck by visible radiation *without* decomposition, it seems unlikely that photodissociation of water by this method is feasible.

7.96 The Balmer series corresponds to transitions to the $n = 2$ level.

For He$^+$:

$$\Delta E = R_{\text{He}^+}\left(\frac{1}{n_i^2} - \frac{1}{n_f^2}\right) \qquad \lambda = \frac{hc}{\Delta E} = \frac{(6.63 \times 10^{-34}\ \text{J}\cdot\text{s})(3.00 \times 10^{8}\ \text{m/s})}{\Delta E}$$

For the transition, $n = 3 \rightarrow 2$

$$\Delta E = (8.72 \times 10^{-18}\ \text{J})\left(\frac{1}{3^2} - \frac{1}{2^2}\right) = -1.21 \times 10^{-18}\ \text{J} \qquad \lambda = \frac{1.99 \times 10^{-25}\ \text{J}\cdot\text{m}}{1.21 \times 10^{-18}\ \text{J}} = 1.64 \times 10^{-7}\ \text{m} = \mathbf{164\ nm}$$

For the transition, $n = 4 \rightarrow 2$, $\Delta E = -1.64 \times 10^{-18}$ J $\lambda =$ **121 nm**

For the transition, $n = 5 \rightarrow 2$, $\Delta E = -1.83 \times 10^{-18}$ J $\lambda =$ **109 nm**

For the transition, $n = 6 \rightarrow 2$, $\Delta E = -1.94 \times 10^{-18}$ J $\lambda =$ **103 nm**

For H, the calculations are identical to those above, except the Rydberg constant for H is 2.18×10^{-18} J.

For the transition, $n = 3 \rightarrow 2$, $\Delta E = -3.03 \times 10^{-19}$ J $\lambda =$ **657 nm**

For the transition, $n = 4 \rightarrow 2$, $\Delta E = -4.09 \times 10^{-19}$ J $\lambda =$ **487 nm**

For the transition, $n = 5 \rightarrow 2$, $\Delta E = -4.58 \times 10^{-19}$ J $\lambda =$ **434 nm**

For the transition, $n = 6 \rightarrow 2$, $\Delta E = -4.84 \times 10^{-19}$ J $\lambda =$ **411 nm**

All the Balmer transitions for He$^+$ are in the ultraviolet region; whereas, the transitions for H are all in the visible region. Note the negative sign for energy indicating that a photon has been emitted.

7.98 Since the energy corresponding to a photon of wavelength λ_1 equals the energy of photon of wavelength λ_2 plus the energy of photon of wavelength λ_3, then the equation must relate the wavelength to energy.

energy of photon 1 = (energy of photon 2 + energy of photon 3)

Since $E = \dfrac{hc}{\lambda}$, then:

$$\frac{hc}{\lambda_1} = \frac{hc}{\lambda_2} + \frac{hc}{\lambda_3}$$

Dividing by hc:

$$\frac{1}{\lambda_1} = \frac{1}{\lambda_2} + \frac{1}{\lambda_3}$$

7.100 We use Heisenberg's uncertainty principle with the equality sign to calculate the minimum uncertainty.

$$\Delta x \Delta p = \frac{h}{4\pi}$$

The momentum (p) is equal to the mass times the velocity.

$$p = mu \quad \text{or} \quad \Delta p = m\Delta u$$

We can write:

$$\Delta p = m\Delta u = \frac{h}{4\pi\Delta x}$$

Finally, the uncertainty in the velocity of the oxygen molecule is:

$$\boldsymbol{\Delta u} = \frac{h}{4\pi m\Delta x} = \frac{(6.63 \times 10^{-34}\ \text{J}\cdot\text{s})}{4\pi(5.3 \times 10^{-26}\ \text{kg})(5.0 \times 10^{-5}\ \text{m})} = \mathbf{2.0 \times 10^{-5}\ m/s}$$

7.102 As an estimate, we can equate the energy for ionization ($Fe^{13+} \rightarrow Fe^{14+}$) to the average kinetic energy ($3/2RT$) of the ions.

$$\frac{3.5 \times 10^4\ \text{kJ}}{1\ \text{mol}} \times \frac{1000\ \text{J}}{1\ \text{kJ}} = 3.5 \times 10^7\ \text{J}$$

$$IE = \frac{3}{2}RT$$

$$3.5 \times 10^7\ \text{J/mol} = \frac{3}{2}(8.314\ \text{J/mol}\cdot\text{K})T$$

$$\boldsymbol{T = 2.8 \times 10^6\ \text{K}}$$

The actual temperature can be, and most probably is, higher than this.

7.104 (a) **False.** $n = 2$ is the first excited state.

(b) **False.** In the $n = 4$ state, the electron is (on average) further from the nucleus and hence easier to remove.

(c) **True.**

(d) **False.** The $n = 4$ to $n = 1$ transition is a higher energy transition, which corresponds to a *shorter* wavelength.

(e) **True.**

CHAPTER 8
THE PERIODIC TABLE

PROBLEM-SOLVING STRATEGIES AND TUTORIAL SOLUTIONS

TYPES OF PROBLEMS

Problem Type 1: Writing an Electron Configuration and Identifying an Element (see Problem Type 7.6).

Problem Type 2: Electron Configurations of Cations and Anions.

Problem Type 3: Comparing the Sizes of Atoms.

Problem Type 4: Comparing the Sizes of Ions.

Problem Type 5: Comparing the Ionization Energies of Elements.

Problem Type 6: Electron Affinity.

PROBLEM TYPE 1: WRITING AN ELECTRON CONFIGURATION AND IDENTIFYING AN ELEMENT

See Problem Type 7.6 for information on writing electron configurations. When examining the electron configurations of the elements in a particular group, a clear pattern emerges. Elements in the same group have the same electron configuration of their *outer* electrons. The outer electrons of an atom, which are those involved in chemical bonding, are called **valence electrons**. Having the same number of valence electrons accounts for the similarities in chemical behavior among the elements within each of these groups.

In regards to identifying an element, your text considers a number of items. First, according to the type of subshell being filled, the elements can be divided into categories--the representative elements, the noble gases, the transition metals, the lanthanides, and the actinides.

> The **representative elements** are the elements in Groups 1A through 7A, all of which have *incompletely* filled *s* or *p* subshells of the highest principal quantum number.
>
> The **noble gases**, with the exception of helium, all have a *completely* filled *p* subshell.
>
> The **transition metals** are the elements in Groups 1B and 3B through 8B that have *incompletely* filled *d* subshells, or readily produce cations with *incompletely* filled *d* subshells.
>
> The **lanthanides** and **actinides** are sometimes called *f*-block transition elements because they have *incompletely* filled *f* subshells.

Second, you might be able to classify the element as a metal, nonmetal, or metalloid. However, sometimes you need to be careful in making this classification. For example in Group 4A, carbon is a nonmetal, silicon and germanium are metalloids, and tin and lead are metals.

Third, the problem might ask whether the element is paramagnetic or diamagnetic. In Chapter 7, you learned that an element that contains unpaired electrons is *paramagnetic*, and an element in which all electrons are paired is *diamagnetic*.

EXAMPLE 8.1

A neutral atom of a certain element has 19 electrons. Without consulting a periodic table, answer the following questions: (a) What is the electron configuration of the element? (b) How should the element be classified? (c) Are the atoms of this element diamagnetic or paramagnetic?

(a) The electron configuration is $1s^22s^22p^63s^23p^64s^1$ or $[Ar]4s^1$. See Problem Type 7.6 if you are having difficulty writing electron configurations.

(b) Since the 4*s* subshell is not completely filled, this is a *representative element*. Without consulting a periodic table, you might know that the alkali metal family has one valence electron in the *s* subshell. You could then further classify this element as an *alkali metal*.

(c) There is *one* unpaired electron in the *s* subshell. Therefore, the atoms of this element are paramagnetic.

Tip: Any atom that contains an odd number of electrons *must* be paramagnetic. Why?

PRACTICE EXERCISE

1. A neutral atom of a certain element has 36 electrons. Without consulting a periodic table, answer the following questions: (a) What is the electron configuration of the element? (b) How should the element be classified? (c) Are the atoms of this element diamagnetic or paramagnetic?

Text Problems: 8.18, 8.20, 8.22

PROBLEM TYPE 2: ELECTRON CONFIGURATIONS OF CATIONS AND ANIONS

Writing an electron configuration for a cation or anion requires only a slight extension of the method used for neutral atoms. Let's group the ions in two categories.

(1) Ions Derived from Representative Elements

(a) In the formation of a **cation** from the neutral atom of a representative element, one or more electrons are removed from the highest occupied *n* shell. The number of electrons removed is equal to the charge of the ion.

Examples: Following are the electron configurations of some neutral atoms and their corresponding cations.

Mg: $[Ne]3s^2$	Mg^{2+}: [Ne]
K: $[Ar]4s^1$	K^+: [Ar]

Note: Each ion has a stable noble gas electron configuration.

(b) In the formation of an **anion** from the neutral atom of a representative element, one or more electrons are added to the highest partially filled *n* shell. The number of electrons added is equal to the charge of the ion.

Examples: Following are the electron configurations of some neutral atoms and their corresponding anions.

O: $[He]2s^22p^4$	O^{2-}: [Ne]
Cl: $[Ne]3s^23p^5$	Cl^-: [Ar]

Note: Each ion has a stable noble gas electron configuration.

(2) Cations Derived from Transition Metals

In a neutral transition metal atom, the ns orbital is filled before the $(n-1)d$ orbitals. See Figure 7.21 of your text. However, in a transition metal ion, the $(n-1)d$ orbitals are more stable than the ns orbital. Hence, when a cation is formed from an atom of a transition metal, electrons are *always* removed first from the *ns* orbital and then from the $(n-1)d$ orbitals if necessary. See Section 8.2 of your text for a more complete discussion.

Examples: Following are the electron configurations of some neutral transition metals and their corresponding cations.

Fe: $[Ar]4s^2 3d^6$	Fe^{2+}: $[Ar]3d^6$
Fe: $[Ar]4s^2 3d^6$	Fe^{3+}: $[Ar]3d^5$
Mn: $[Ar]4s^2 3d^5$	Mn^{7+}: [Ar]

PRACTICE EXERCISE

2. Write electron configurations for the following ions: N^{3-}, Ba^{2+}, Zn^{2+}, and, V^{5+}.

Text Problems: 8.28, 8.30, 8.32

PROBLEM TYPE 3: COMPARING THE SIZES OF ATOMS

The general periodic trends in atomic size are:

(1) Moving from left to right across a row (period) of the periodic table, the atomic radius ***decreases*** due to an increase in effective nuclear charge.

(2) Moving down a column (group) of the periodic table, the atomic radius ***increases*** since the orbital size increases with increasing principal quantum number.

For a more detailed discussion, see Section 8.3 of your text.

EXAMPLE 8.2
Which one of the following has the smallest atomic radius? (a) Li, (b) Na, (c) Be, (d) Mg

Atomic radii *increase* going down a group of the periodic table; therefore, Li atoms have a smaller radius than Na atoms, and Be atoms have a smaller radius than Mg atoms. We have narrowed our choices to Li and Be. Atomic radii decrease when moving from left to right across a row of the periodic table. Thus, **Be atoms** are smaller than Li atoms as well as the other choices given.

PRACTICE EXERCISE

3. Which atom should have the largest atomic radius? (a) Br, (b) Cl, (c) Se, (d) Ge, (e) C.

Text Problems: 8.38, 8.40, 8.42

PROBLEM TYPE 4: COMPARING THE SIZES OF IONS

When a neutral atom is converted to an ion, we expect a change in size. When forming an *anion* from an atom, its size (or radius) *increases*, because the nuclear charge remains the same but the repulsion resulting from the additional electron(s) enlarges the domain of the electron cloud. Conversely, when forming a *cation* from an atom, its size *decreases*, because removing one or more electrons reduces electron-electron repulsion but the nuclear charge remains the same. Thus, the electron cloud shrinks.

The general periodic trends in ionic size are:

(1) Similar to atomic size, ionic radii increase when moving down a column (group) of the periodic table.

(2) The next trend only applies to **isoelectronic ions**. Isoelectronic ions have the same number of electrons and the same electron configuration.

(a) Cations are smaller than anions. For example, Mg^{2+} is smaller than O^{2-}. Both ions have the same number of electrons (10), but Mg (Z = 12) has more protons then O (Z = 8). The larger effective nuclear charge of Mg^{2+} results in a smaller radius.

(b) Considering only *isoelectronic cations*, the radius of a tripositive ion is smaller than the radius of a dipositive ion, which is smaller than the radius of a unipositive ion. For example, Al^{3+} is smaller than Mg^{2+}, which is smaller than Na^{+}. Each of the cations has 10 electrons, but Al^{3+} has 13 protons, Mg^{2+} has 12 protons, and Na^{+} has 11 protons. As the effective nuclear charge increases, the ionic radius decreases.

(c) Considering only *isoelectronic anions*, the radius increases as we go from an ion with a uninegative charge, to one with a dinegative charge, and so on. For example, N^{3-} is larger than O^{2-}, which is larger than F^{-}. Each of the anions has 10 electrons, but N^{3-} has 7 protons, O^{2-} has 8 protons, and F^{-} has 9 protons. As the effective nuclear charge increases, ionic radius decreases.

EXAMPLE 8.3

In each of the following pairs, choose the ion with the *larger* ionic radius: (a) K^{+} and Na^{+}; (b) K^{+} and Ca^{2+}; (c) K^{+} and Cl^{-}.

(a) K^{+} and Na^{+} are in the same group of the periodic table (Group 1A, alkali metals). As you proceed down a group of the periodic table, ionic radii increase. This occurs because orbital size increases with increasing principal quantum number. Therefore, **K^{+}** has the larger ionic radius.

(b) K^{+} and Ca^{2+} are isoelectronic cations. Both ions have 18 electrons, but Ca^{2+} has one more proton than K^{+}. The greater effective nuclear charge of Ca^{2+}, pulls the 18 electrons more strongly toward the nucleus. Therefore, **K^{+}** has the larger ionic radius. For isoelectronic cations, the radius of a dipositive ion is smaller than the radius of a unipositive ion.

(c) K^{+} and Cl^{-} are isoelectronic species. Both ions have 18 electrons, but K^{+} has the greater effective nuclear charge, with 19 protons compared to 17 protons for Cl^{-}. Therefore, **Cl^{-}** has the larger ionic radius.

PRACTICE EXERCISE

4. Which of the following has the largest radius: Na^{+}, Mg^{2+}, Al^{3+}, S^{2-}, or Cl^{-}?

Text Problems: 8.44, 8.46

PROBLEM TYPE 5: COMPARING THE IONIZATION ENERGIES OF ELEMENTS

Ionization energy is the minimum energy required to remove an electron from a gaseous atom in its ground state. An equation that represents this process is

$$\text{energy} + X(g) \longrightarrow X^{+}(g) + e^{-}$$

where,

X represents a gaseous atom of any element

e^{-} is an electron

The more "tightly" the electron is held in the atom, the more difficult it will be to remove the electron. Hence, the more "tightly" the electron is held, the higher the ionization energy.

The general periodic trends for ionization energy are:

(1) Moving from left to right across a row of the periodic table, the ionization energy ***increases*** due to an increase in effective nuclear charge. As the effective nuclear charge increases, the electrons will be held more "tightly" by the nucleus, making them more difficult to remove.

(2) Moving down a column (group) of the periodic table, the ionization energy ***decreases***. As the principal quantum number n increases, so does the average distance of a valence electron from the nucleus. A greater separation between the electron and the nucleus results in a weaker attraction; hence, a valence electron becomes increasingly easier to remove as we proceed down a group of the periodic table.

As with most periodic trends, there are exceptions. For a more detailed discussion, see Section 8.4 of your text.

EXAMPLE 8.4
Which of the following has the highest ionization energy: K, Br, Cl, or S?

The ionization energy increases from left to right across a row of the periodic table. Thus, the ionization energy of Cl is greater than for S, and the ionization energy of Br is greater than for K. We have narrowed our choices to Cl and Br. Moving down a group of the periodic table, the ionization decreases. Therefore, **Cl** has a higher ionization energy than Br as well as the other choices given.

PRACTICE EXERCISE

5. Based on periodic trends, which of the following elements has the greatest ionization energy: Cl, K, S, Se, or Br?

Text Problems: 8.52, **8.54,** 8.56

PROBLEM TYPE 6: ELECTRON AFFINITY

Electron affinity is the energy change when an electron is accepted by an atom in the gaseous state. An equation that represents this process is:

$$X(g) + e^- \longrightarrow X^-(g)$$

where,

X represents a gaseous atom of any element

e^- is an electron

A positive electron affinity signifies that energy is liberated when an electron is added to an atom. The more positive the electron affinity, the greater the tendency of the atom to accept an electron. Electron affinity is positive if the reaction is exothermic and negative if the reaction is endothermic.

The general periodic trends for electron affinity are:

(1) The tendency to accept electrons increases (that is, electron affinity values become more positive) as we move from left to right across a period.

(2) The electron affinities of metals are generally less than those of nonmetals.

(3) Electron affinity values vary little within a given group.

EXAMPLE 8.5
Explain why the electron affinities of the halogens are all positive.

The outer-shell electron configuration of the halogens is ns^2np^5. For the process

$$X(g) + e^- \longrightarrow X^-(g)$$

where X denotes a member of the halogen family, the accepted electron would fill the outer shell, giving the halogen ion a stable noble gas electron configuration. This is a favorable process; consequently, halogens have a strong tendency to accept an extra electron (i.e., a highly positive electron affinity).

Text Problems: **8.60**, 8.62

ANSWERS TO PRACTICE EXERCISES

1. **(a)** $1s^22s^22p^63s^23p^64s^23d^{10}4p^6$

   (b) Since the 4*p* subshell is completely filled, this is a *noble gas*. All noble gases are *nonmetals*.

   (c) Since the outer shell is completely filled, there are *no* unpaired electrons. Therefore, the atoms of this element are diamagnetic.

2. N^{3-}: [Ne] Ba^{2+}: [Xe] Zn^{2+}: $[Ar]3d^{10}$ V^{5+}: [Ar]

3. **Ge** has the largest atomic radius of the group.

4. The ions are isoelectronic. S^{2-} has the largest radius.

5. **Cl**

SOLUTIONS TO SELECTED TEXT PROBLEMS

8.18 Elements that have the same number of valence electrons will have similarities in chemical behavior. Looking at the periodic table, elements with the same number of valence electrons are in the same group. Therefore, the pairs that would represent similar chemical properties of their atoms are:

(a) and **(d)** **(b)** and **(e)** **(c)** and **(f)**.

8.20 **(a)** Group 1A **(b)** Group 5A **(c)** Group 8A **(d)** Group 8B
Identify the elements.

8.22 You should realize that the metal ion in question is a transition metal ion because it has five electrons in the 3*d* subshell. Remember that in a transition metal ion, the (*n*–1)*d* orbitals are more stable than the *ns* orbital. Hence, when a cation is formed from an atom of a transition metal, electrons are *always* removed first from the *ns* orbital and then from the (*n*–1)*d* orbitals if necessary. Since the metal ion has a +3 charge, three electrons have been removed. Since the 4*s* subshell is less stable than the 3*d*, two electrons would have been lost from the 4*s* and one electron from the 3*d*. Therefore, the electron configuration of the neutral atom is $[Ar]4s^23d^6$. This is the electron configuration of iron. Thus, the metal is **iron**.

8.28 Electron Configurations of Cations and Anions, Problem Type 2.

In the formation of a **cation** from the neutral atom of a representative element, one or more electrons are *removed* from the highest occupied *n* shell. In the formation of a **anion** from the neutral atom of a representative element, one or more electrons are *added* to the highest partially filled *n* shell. Representative elements typically gain or lose electrons to achieve a stable noble gas electron configuration. When a cation is formed from an atom of a transition metal, electrons are *always* removed first from the *ns* orbital and then from the (*n*–1)*d* orbitals if necessary.

(a) [Ne]

(b) same as (a). Do you see why?

(c) [Ar]

(d) Same as (c). Do you see why?

(e) Same as (c)

(f) $[Ar]3d^6$. Why isn't it $[Ar]4s^23d^4$?

(g) $[Ar]3d^9$. Why not $[Ar]4s^23d^7$?

(h) $[Ar]3d^{10}$. Why not $[Ar]4s^23d^8$?

8.30 **(a)** Cr^{3+} **(b)** Sc^{3+} **(c)** Rh^{3+} **(d)** Ir^{3+}

8.32 Isoelectronic means that the species have the same number of electrons and the same electron configuration.

Be^{2+} and He (2 e^-) F^- and N^{3-} (10 e^-) Fe^{2+} and Co^{3+} (24 e^-) S^{2-} and Ar (18 e^-)

8.38 Comparing the Sizes of Atoms, Problem Type 3.

Recall that the general periodic trends in atomic size are:

(1) Moving from left to right across a row (period) of the periodic table, the atomic radius ***decreases*** due to an increase in effective nuclear charge.

(2) Moving down a column (group) of the periodic table, the atomic radius ***increases*** since the orbital size increases with increasing principal quantum number.

The atoms that we are considering are all in the same period of the periodic table. Hence, the atom furthest to the left in the row will have the largest atomic radius, and the atom furthest to the right in the row will have the smallest atomic radius. Arranged in order of decreasing atomic radius, we have,

$$Na > Mg > Al > P > Cl$$

8.40 Fluorine is the smallest atom in Group 7A. Atomic radius increases moving down a group since the orbital size increases with increasing principal quantum number, *n*.

8.42 The atomic radius is largely determined by how strongly the outer-shell electrons are held by the nucleus. The larger the effective nuclear charge, the more strongly the electrons are held and the smaller the atomic radius. For the second period, the atomic radius of Li is largest because the 2*s* electron is well shielded by the filled 1*s* shell. The effective nuclear charge that the outermost electrons feel increases across the period as a result of incomplete shielding by electrons in the same shell. Consequently, the orbital containing the electrons is compressed and the atomic radius decreases.

8.44 Comparing the Sizes of Ions, Problem Type 4.

The ions listed are all isoelectronic. They each have ten electrons. The ion with the fewest protons will have the largest ionic radius, and the ion with the most protons will have the smallest ionic radius. The effective nuclear charge increases with increasing number of protons. The electrons are attracted more strongly by the nucleus, decreasing the ionic radius. The order of increasing atomic radius is:

$$Mg^{2+} < Na^{+} < F^{-} < O^{2-} < N^{3-}$$

8.46 Both selenium and tellurium are Group 6A elements. Since atomic radius increases going down a column in the periodic table, it follows that Te^{2-} must be larger than Se^{2-}.

8.48 We assume the approximate boiling point of argon is the mean of the boiling points of neon and krypton, based on its position in the periodic table being between Ne and Kr in Group 8A.

$$\textbf{b.p.} = \frac{-245.9°\text{C} + (-152.9°\text{C})}{2} = \mathbf{-199.4°C}$$

The actual boiling point of argon is −185.7°C.

8.52 The Group 3A elements (such as Al) all have a single electron in the outermost *p* subshell, which is well shielded from the nuclear charge by the inner electrons and the ns^2 electrons. Therefore, less energy is needed to remove a single *p* electron than to remove a paired *s* electron from the same principal energy level (such as for Mg).

8.54 Comparing the Ionization Energies of Elements, Problem Type 5.

A noble gas electron configuration, such as $1s^22s^22p^6$, is a very stable configuration, making it extremely difficult to remove an electron. The high ionization energy of 2080 kJ/mol would be associated with the element having this noble gas electron configuration.

The lone electron in the 3*s* orbital will be much easier to remove. Therefore, the ionization energy of 496 kJ/mol is paired with the electron configuration $1s^22s^22p^63s^1$.

8.56 The atomic number of mercury is 80. We write:

$$\Delta E = (2.18 \times 10^{-18}\ \text{J})(80^2)\left(\frac{1}{1^2} - \frac{1}{\infty^2}\right) = 1.40 \times 10^{-14}\ \text{J/ion}$$

$$\boldsymbol{\Delta E} = \left(1.40 \times 10^{-14}\ \frac{\text{J}}{\text{ion}}\right)\left(\frac{6.022 \times 10^{23}\ \text{ions}}{1\ \text{mol}}\right)\left(\frac{1\ \text{kJ}}{1000\ \text{J}}\right) = \mathbf{8.43 \times 10^6\ kJ/mol}$$

8.60 Electron Affinity, Problem Type 6.

One of the general periodic trends for electron affinity is that the tendency to accept electrons increases (that is, electron affinity values become more positive) as we move from left to right across a period. However, this trend does not include the noble gases. We know that noble gases are extremely stable, and they do not want to gain or lose electrons.

Based on the above periodic trend, Cl would be expected to have the highest electron affinity. Addition of an electron to Cl forms Cl^-, which has a stable noble gas electron configuration.

8.62 Alkali metals have a valence electron configuration of ns^1 so they can accept another electron in the ns orbital. On the other hand, alkaline earth metals have a valence electron configuration of ns^2. Alkaline earth metals have little tendency to accept another electron, as it would have to go into a higher energy *p* orbital.

8.66 Since ionization energies decrease going down a column in the periodic table, francium should have the lowest first ionization energy of all the alkali metals. As a result, Fr should be the most reactive of all the Group 1A elements toward water and oxygen. The reaction with oxygen would probably be similar to that of K, Rb, or Cs.

What would you expect the formula of the oxide to be? The chloride?

8.68 The Group 1B elements are much less reactive than the Group 1A elements. The 1B elements are more stable because they have much higher ionization energies resulting from incomplete shielding of the nuclear charge by the inner *d* electrons. The ns^1 electron of a Group 1A element is shielded from the nucleus more effectively by the completely filled noble gas core. Consequently, the outer *s* electrons of 1B elements are more strongly attracted by the nucleus.

8.70 **(a)** Lithium oxide is a basic oxide. It reacts with water to form the metal hydroxide:

$$Li_2O\,(s) + H_2O\,(l) \longrightarrow 2\,LiOH\,(aq)$$

(b) Calcium oxide is a basic oxide. It reacts with water to form the metal hydroxide:

$$CaO\,(s) + H_2O\,(l) \longrightarrow Ca(OH)_2\,(aq)$$

(c) Carbon dioxide is an acidic oxide. It reacts with water to form carbonic acid:

$$CO_2\,(g) + H_2O\,(l) \longrightarrow H_2CO_3\,(aq)$$

8.72 As we move down a column, the metallic character of the elements increases. Since magnesium and barium are both Group 2A elements, we expect barium to be more metallic than magnesium and BaO to be more basic than MgO.

8.74 **(a)** bromine **(b)** nitrogen **(c)** rubidium **(d)** magnesium

8.76 This is an isoelectronic series with ten electrons in each species. The nuclear charge interacting with these ten electrons ranges from +8 for oxygen to +12 for magnesium. Therefore the +12 charge in Mg^{2+} will draw in the ten electrons more tightly than the +11 charge in Na^+, than the +9 charge in F^-, than the +8 charge in O^{2-}. Recall that the largest species will be the *easiest* to ionize.

(a) increasing ionic radius: $Mg^{2+} < Na^+ < F^- < O^{2-}$

(b) increasing ionization energy: $O^{2-} < F^- < Na^+ < Mg^{2+}$

8.78 According to the *Handbook of Chemistry and Physics* (1966-67 edition), potassium metal has a melting point of 63.6°C, bromine is a reddish brown liquid with a melting point of −7.2°C, and potassium bromide (KBr) is a colorless solid with a melting point of 730°C. M is potassium (K) and X is bromine (Br).

8.80 O^+ and N Ar and S^{2-} Ne and N^{3-} Zn and As^{3+} Cs^+ and Xe

8.82 **(a)** and **(d)**

8.84 Fluorine is a yellow-green gas that attacks glass; chlorine is a pale yellow gas; bromine is a fuming red liquid; and iodine is a dark, metallic-looking solid.

8.86 Fluorine

8.88 H^- and He are isoelectronic species with two electrons. Since H^- has only one proton compared to two protons for He, the nucleus of H^- will attract the two electrons less strongly compared to He. Therefore, H^- is larger.

8.90

Oxide	Name	Property
Li_2O	lithium oxide	basic
BeO	beryllium oxide	amphoteric
B_2O_3	boron oxide	acidic
CO_2	carbon dioxide	acidic
N_2O_5	dinitrogen pentoxide	acidic

Note that only the highest oxidation states are considered.

8.92 In its chemistry, hydrogen can behave like an alkali metal (H^+) and like a halogen (H^-).

8.94 Rearrange the given equation to solve for ionization energy.

$$IE = h\nu - \frac{1}{2}mu^2$$

or,

$$IE = \frac{hc}{\lambda} - KE$$

The kinetic energy of the ejected electron is given in the problem. Substitute *h*, *c*, and λ into the above equation to solve for the ionization energy.

$$IE = \frac{(6.63 \times 10^{-34}\ \text{J}\cdot\cancel{\text{s}})(3.00 \times 10^{8}\ \cancel{\text{m}}/\cancel{\text{s}})}{162 \times 10^{-9}\ \cancel{\text{m}}} - (5.34 \times 10^{-19}\ \text{J})$$

$$\boldsymbol{IE = 6.94 \times 10^{-19}\ \text{J}}$$

We might also want to express the ionization energy in kJ/mol.

$$\frac{6.94 \times 10^{-19}\ \cancel{\text{J}}}{1\ \cancel{\text{photon}}} \times \frac{6.022 \times 10^{23}\ \cancel{\text{photons}}}{1\ \text{mol}} \times \frac{1\ \text{kJ}}{1000\ \cancel{\text{J}}} = \mathbf{418\ kJ/mol}$$

To ensure that the ejected electron is the valence electron, UV light of the *longest* wavelength (lowest energy) should be used that can still eject electrons.

8.96 Plotting the melting point versus the atomic number and extrapolating the curve to francium, the estimated melting point is 23°C.

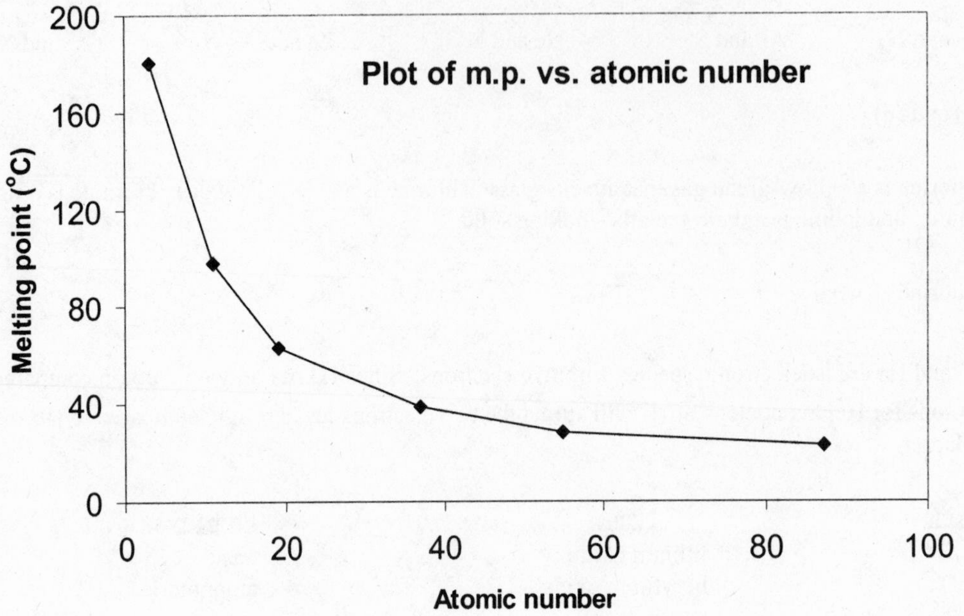

8.98 **(a)** Mg in $Mg(OH)_2$ **(b)** Na, liquid **(c)** Mg in $MgSO_4 \cdot 7H_2O$ **(d)** Na in $NaHCO_3$ **(e)** K in KNO_3 **(f)** Mg **(g)** Ca in CaO **(h)** Ca **(i)** Na in NaCl; Ca in $CaCl_2$

8.100 Physical characteristics: Solid; metallic appearance like iodine; melting point greater than 114°C.

Reaction with sulfuric acid:

$$2NaAt(s) + H_2SO_4(aq) \rightarrow Na_2SO_4(aq) + 2HAt(g)$$

8.102 $Na(g) \rightarrow Na^+(g) + e^-$ $\qquad I_1 = 495.9\ kJ/mol$

Energy needed to ionize one Na atom:

$$\frac{495.9 \times 10^3\ J}{1\ \cancel{mol}} \times \frac{1\ \cancel{mol}}{6.022 \times 10^{23}\ atoms} = 8.235 \times 10^{-19}\ J/atom$$

The corresponding wavelength is:

$$\lambda = \frac{hc}{I_1} = \frac{(6.63 \times 10^{-34}\ \cancel{J} \cdot \cancel{s})(3.00 \times 10^8\ m/\cancel{s})}{8.235 \times 10^{-19}\ \cancel{J}} = 2.42 \times 10^{-7}\ m = \mathbf{242\ nm}$$

8.104 **(a)** F_2 **(b)** Na **(c)** B **(d)** N_2 **(e)** Al

CHAPTER 9
CHEMICAL BONDING I:
THE COVALENT BOND

PROBLEM-SOLVING STRATEGIES AND TUTORIAL SOLUTIONS

TYPES OF PROBLEMS

Problem Type 1: Classifying Chemical Bonds.

Problem Type 2: Writing Lewis Structures.

Problem Type 3: Formal Charges.
(a) Assigning formal charges.
(b) Choosing the most plausible Lewis structure based on formal charges.

Problem Type 4: Drawing Resonance Structures.

Problem Type 5: Exceptions to the Octet Rule.
(a) The incomplete octet.
(b) Odd-electron molecules.
(c) The expanded octet.

Problem Type 6: Using Bond Energies to Estimate the Enthalpy of a Reaction.

PROBLEM TYPE 1: CLASSIFYING CHEMICAL BONDS

We can classify bonds as three different types: ionic, polar covalent, or covalent.

In a **covalent bond**, the electron pair of the bond is shared *equally* by the two atoms.

In a **polar covalent bond**, the electron pair of the bond is shared *unequally* by the two atoms. The electrons spend more time in the vicinity of one atom than the other.

Covalent bonds (polar or nonpolar) are typically formed between two nonmetals.

In an **ionic bond**, the electron or electrons are nearly completely transferred from one atom to another. Ionic bonds are typically formed between a metal cation and a nonmetal anion or a metal cation and a polyatomic anion.

A property that helps us distinguish a nonpolar covalent bond from a polar covalent bond is **electronegativity**, the ability of an atom to attract toward itself the electrons in a chemical bond. Elements with high electronegativities have a greater tendency to attract electrons than elements with low electronegativities. Linus Pauling devised a method for calculating *relative* electronegativities of most elements. These values are shown in Figure 9.4 of the text.

Only atoms of the same element, which have the same electronegativity, can be joined by a pure *covalent bond.* Atoms of elements with similar electronegativities tend to form *polar covalent bonds* with each other because the shift in electron density is usually small. There is no sharp distinction between a polar bond and an ionic bond, but the following rule is helpful in distinguishing them. An *ionic bond* forms when the electronegativity difference between the two bonding atoms is 2.0 or more. This rule applies to most but not all ionic compounds.

EXAMPLE 9.1

Classify the following bonds as ionic, polar covalent, or covalent: (a) the bond in KCl, (b) the OH bond in H_2O, and (c) the OO bond in oxygen gas (O_2).

(a) In Figure 9.4 of the text, we see that the electronegativity difference between K and Cl is 2.2, above the 2.0 guideline. Therefore, the bond between K and Cl is ionic. Remember that an ionic bond is typically formed between a metal cation and a nonmetal anion or a metal cation and a polyatomic anion.

(b) The electronegativity difference between O and H is 1.4, which is appreciable but not large enough to qualify H_2O as an ionic compound. Therefore, the bond between O and H is polar covalent. Recall that polar covalent compounds are typically formed between two different nonmetals.

(c) The two O atoms are identical in every respect. Therefore, the bond between them is purely covalent.

PRACTICE EXERCISE

1. Classify the following bonds as ionic, polar covalent, or covalent: (a) the NH bond in NH_3, (b) the OO bond in hydrogen peroxide (H_2O_2), and (c) the bond in NaF.

Text Problems: 14, 16, 18

PROBLEM TYPE 2: WRITING LEWIS STRUCTURES

The general rules for writing Lewis structures are given below. For more detailed rules, see Section 9.4 of the text.

1. Write the skeletal structure of the compound, using chemical symbols and placing bonded atoms next to one another. In general, the least electronegative atom occupies the central position. Hydrogen and fluorine usually occupy the terminal (end) positions in the Lewis structure.
2. Count the number of valence electrons present. For polyatomic anions, add the number of negative charges to that total. For polyatomic cations, subtract the number of positive charges from the number of valence electrons.
3. Draw a single covalent bond between the central atom and each of the surrounding atoms. Complete the octets of the atoms bonded to the central atom with lone pairs. The total number of electrons to be used is that determined in step 2.
4. Sometimes there will not be enough electrons to satisfy the octet rule of the central atom by placing lone pairs. Try adding double or triple bonds between the surrounding atoms and the central atom, using the lone pairs from the surrounding atoms.

EXAMPLE 9.2

Write the Lewis structure for SO_2.

Step 1: Sulfur is less electronegative than oxygen, so it occupies the central position. The skeletal structure is:

O S O

Step 2: The outer-shell electron configurations of O and S are $2s^22p^4$ and $3s^23p^4$, respectively. Thus, there are

$$6 + (2 \times 6) = 18 \text{ valence electrons}$$

Tip: For the representative elements (Group 1A – 7A), the number of valence electrons is equal to the group number. For example, both O and S are in Group 6A; thus, they both have 6 valence electrons.

Step 3: We draw a single covalent bond between S and each O, and then complete the octets for the O atoms. We place the remaining two electrons on S.

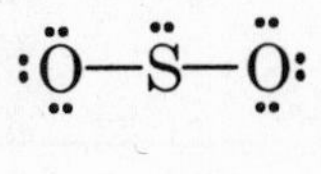

Step 4: The octet rule is satisfied for the oxygen atoms; however, the S atom does *not* satisfy the octet rule. Let's try making a sulfur-to-oxygen double bond by moving a lone pair from one of the O atoms to form another bond with S.

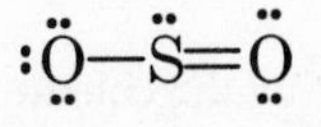

Now, the octet rule is also satisfied for the S atom. As a final check, we verify that there are 18 valence electrons in the Lewis structure of SO_2.

EXAMPLE 9.3

Write the Lewis structure for NO_3^-.

Step 1: Nitrogen is less electronegative than oxygen, so it occupies the central position. The skeletal structure is:

O

O N O

Step 2: Nitrate is a polyatomic anion, so we must add the negative charge to the number of valence electrons. The outer-shell electron configurations of O (Group 4A) and N (Group 5A) are $2s^22p^4$ and $2s^22p^3$, respectively. Thus, there are

$$5 + (3 \times 6) + 1 = 24 \text{ valence electrons}$$

Step 3: We draw a single covalent bond between N and each O, and then complete the octets for the O atoms.

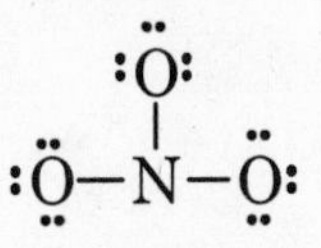

Step 4: The octet rule is satisfied for the oxygen atoms; however, the N atom does *not* satisfy the octet rule. Let's try making a nitrogen-to-oxygen double bond by moving a lone pair from one of the O atoms to form another bond with N.

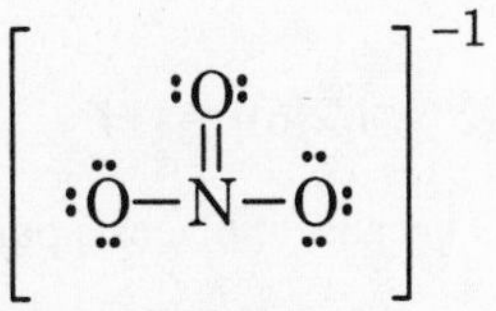

Now, the octet rule is also satisfied for the N atom. As a final check, we verify that there are 24 valence electrons in the Lewis structure of NO_3^-.

Also notice that we draw a bracket with the charge of the polyatomic ion around the Lewis structure. This is to distinguish an ion from a neutral molecule.

PRACTICE EXERCISES

2. Write the Lewis structure for $AsCl_3$.

3. Write the Lewis structure for cyanide ion, CN^-.

Text Problems: 22, 24

PROBLEM TYPE 3: FORMAL CHARGES

By comparing the number of electrons in an isolated atom (valence electrons) with the number of electrons that are associated with the same atom in a Lewis structure, we can determine the distribution of electrons in the molecule and draw the most plausible Lewis structure. This difference between the valence electrons in an isolated atom and the number of electrons assigned to that atom in a Lewis structure is called the atom's **formal charge**.

We can calculate the formal charge on an atom in a molecule using the equation:

$$\text{formal charge} = (\text{no. of valence } e^- \text{ in free atom}) - [(\text{no. of nonbonding } e^-) + 1/2(\text{no. of bonding } e^-)]$$

When you write formal charges, the following rules are helpful.

- For neutral molecules, the sum of the formal charges must add up to zero.
- For cations, the sum of the formal charges must equal the positive charge.
- For anions, the sum of the formal charges must equal the negative charge.

A. Assigning formal charges

EXAMPLE 9.4
Assign formal charges to the atoms in the following Lewis structures.

(a) :C≡O:

(b) 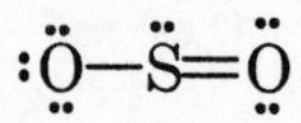

The formula used to calculate the formal charge of an atom is:

$$\text{formal charge} = (\text{no. of valence } e^- \text{ in free atom}) - [(\text{no. of nonbonding } e^-) + 1/2(\text{no. of bonding } e^-)]$$

(a) For the C atom,

$$\text{formal charge} = 4 - [2 + 1/2(6)] = -1$$

For the O atom,

$$\text{formal charge} = 6 - [2 + 1/2(6)] = +1$$

Some chemists do not approve of this structure for CO because it places a positive formal charge on the more electronegative oxygen atom.

(b) For the S atom,

$$\text{formal charge} = 6 - [2 + 1/2(6)] = +1$$

For the O atom on the left,

$$\text{formal charge} = 6 - [6 + 1/2(2)] = -1$$

For the O atom on the right,

$$\text{formal charge} = 6 - [4 + 1/2(4)] = 0$$

PRACTICE EXERCISE

4. Assign formal charges to the atoms in the following Lewis structures.

(a) 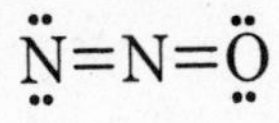

(b) 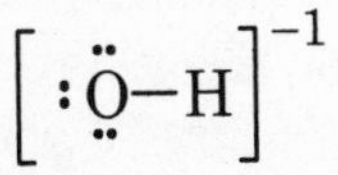

Text Problems: 22, 30, 32, 34

B. Choosing the most plausible Lewis structure based on formal charges

The following guidelines show how to use formal charges to select a plausible Lewis structure for a given compound.

- For neutral molecules, a Lewis structure in which there are no formal charges is preferable to one in which formal charges are present.
- Lewis structures with large formal charges are less plausible than those with small formal charges.
- When comparing two structures with similar magnitudes of formal charges, the most plausible structure is the one in which negative formal charges are placed on the more electronegative atoms.

EXAMPLE 9.5

Two possible Lewis structures for BF_3 are shown below. Which is the more reasonable structure in terms of formal charges?

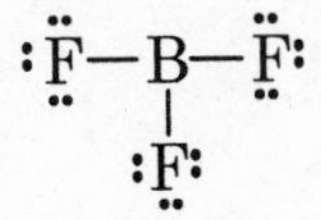

(a)

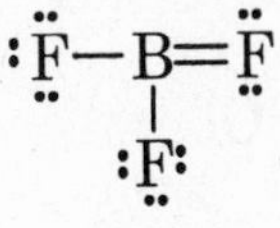

(b)

The formal charges in (a) are:

B atom:	formal charge = 3 − [0 + 1/2(6)] = 0
F atoms:	formal charge = 7 − [6 + 1/2(2)] = 0

The formal charges in (b) are

B atom:	formal charge = 3 − [0 + 1/2(8)] = −1
F atom (double bond):	formal charge = 7 − [4 + 1/2(4)] = +1
F atoms (single bond):	formal charge = 7 − [6 + 1/2(2)] = 0

The rule used to establish the most plausible structure is: For neutral molecules, a Lewis structure in which there are no formal charges is preferable to one in which formal charges are present. Thus, structure (a) is preferred over structure (b). We could also rule out structure (b) as the most plausible structure because there is a positive formal charge on the very electronegative F atom.

PRACTICE EXERCISE

5. Consider the following Lewis structures for the sulfate ion. Which is the more reasonable structure in terms of formal charges?

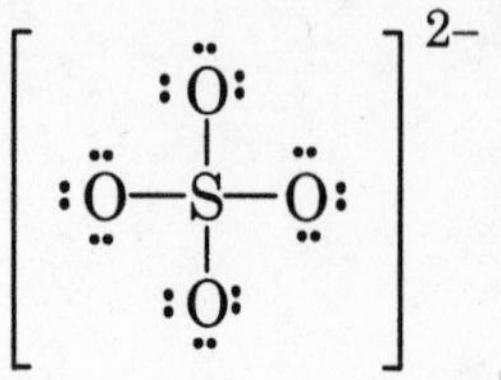

(a)

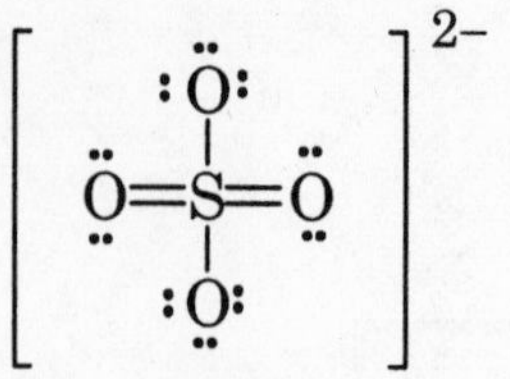

(b)

Text Problem: 40

PROBLEM TYPE 4: DRAWING RESONANCE STRUCTURES

The Lewis structure for SO_3 is shown below.

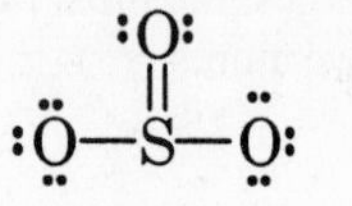

We can draw two more equivalent Lewis structures with the double bond between S and a different oxygen atom.

Which is the correct structure? Let's consider experimental data. We would expect the S–O bond to be longer than the S=O bond because double bonds are known to be shorter than single bonds. Yet experimental evidence shows that all three sulfur-to-oxygen bonds are equal in length. Therefore, none of the three structures shown accurately represents the molecule. We resolve this conflict by using all *three* Lewis structures to represent SO_3.

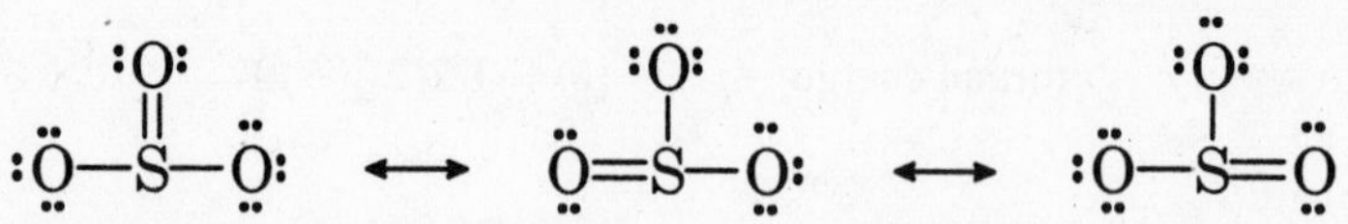

Each of the three structures is called a **resonance structure.** A resonance structure is one of two or more Lewis structures for a single molecule that cannot be described fully with only one Lewis structure. The symbol ⟷ indicates that the structures shown are resonance structures.

EXAMPLE 9.6

We drew the Lewis structure for nitrate ion, NO_3^-, in Example 9.3. However, experimental evidence shows that all N–O bonds are equivalent. Draw resonance structures to indicate the equivalence of the N–O bonds.

The Lewis structure drawn in Example 9.3 was:

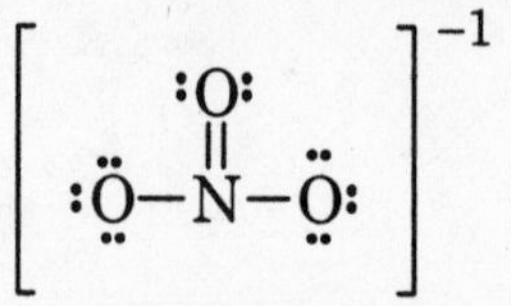

This structure, while a correct Lewis structure, does not show the equivalence of all three N–O bonds. Three contributing resonance structures can be drawn.

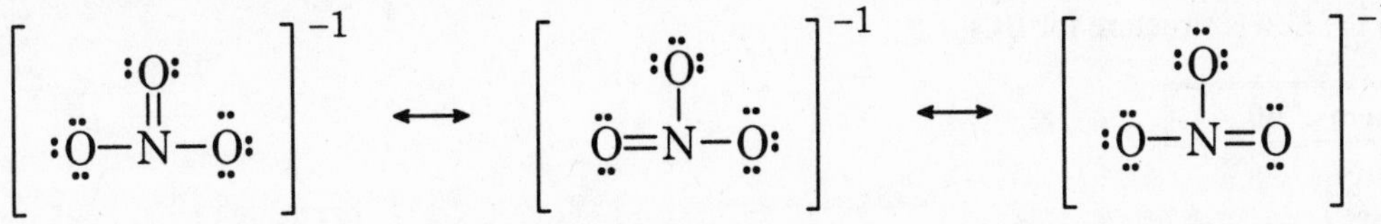

Resonance does not mean that the nitrate ion shifts quickly back and forth from one resonance structure to the other. Keep in mind that *none* of the resonance structures adequately represents the actual molecule, which has its own unique, stable structure. The actual structure is an average or hybrid of the above three structures. Resonance is a human invention, designed to address the limitations in these simple bonding models.

PRACTICE EXERCISE

6. Draw all the resonance structures for N_2O. The skeletal structure is N–N–O.

Text Problems: 30, 32, 34

PROBLEM TYPE 5: EXCEPTIONS TO THE OCTET RULE

A. The incomplete octet

In some compounds the number of electrons surrounding the central atom in a stable molecule is fewer than eight. **Be**, **B**, and **Al** tend to form compounds in which they are surrounded by fewer than eight electrons.

EXAMPLE 9.7

Draw the Lewis structure for GaI_3.

Step 1: Gallium is less electronegative than iodine, so it occupies the central position. The skeletal structure is:

I

I Ga I

Step 2: The outer-shell electron configurations of Ga (Group IIIA) and I (Group VIIA) are $4s^24p^1$ and $5s^25p^5$, respectively. Thus, there are

$$3 + (3 \times 7) = 24 \text{ valence electrons}$$

Step 3: We draw a single covalent bond between Ga and each I, and then complete the octets for the I atoms.

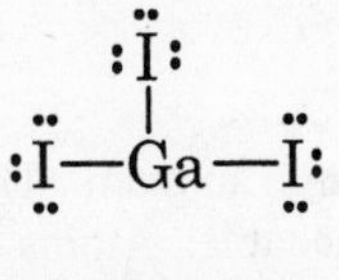

The octet rule is satisfied for the iodine atoms; however, the Ga atom does *not* satisfy the octet rule. A resonance structure with a double bond between Ga and I can be drawn that satisfies the octet rule for Ga. However, the properties of GaI_3 are more consistent with a Lewis structure in which there are single bonds between Ga and each I, as shown above.

Based on formal charges, is the structure shown above more plausible than the structure that contains one double bond between Ga and an I atom?

PRACTICE EXERCISE

7. Write the Lewis structure for BCl_3.

Text Problems: 40, 44

B. Odd-electron molecules

Some molecules contain an **odd** number of electrons. To satisfy the octet rule, we need an even number of electrons. Therefore, the octet rule cannot be satisfied in a molecule that has an odd number of electrons. When drawing a Lewis structure, if an atom has fewer than eight electrons, make sure that it is the least electronegative atom in the compound.

EXAMPLE 9.8

Draw the Lewis structure for NO_2 (all bonds are equivalent).

Step 1: Nitrogen is less electronegative than oxygen, so it occupies the central position. The skeletal structure is:

O N O

Step 2: The outer-shell electron configurations of N (Group 5A) and O (Group 6A) are $2s^22p^3$ and $2s^22p^4$, respectively. Thus, there are

$$5 + (2 \times 6) = 17 \text{ valence electrons}$$

Step 3: We have an odd number of valence electrons, so the octet rule cannot be satisfied for at least one of the atoms in the molecule. Either nitrogen or oxygen in the structure will have fewer than eight electrons, because a second-row element cannot exceed an octet of electrons. Since N is less electronegative than O, nitrogen should be electron deficient. We draw a single covalent bond between N and each O, and then complete the octets for the O atoms.

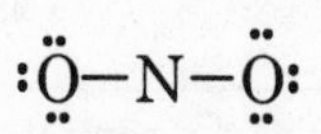

Step 4: We have one electron left to place on the molecule. Placing the electron on N only gives five electrons around N. However, if we make a nitrogen-to-oxygen double bond by moving a lone pair from one of the O atoms to form another bond with N, nitrogen will be surrounded by seven electrons.

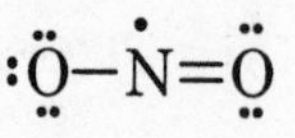

We could also draw resonance structures in which oxygen is electron deficient with seven electrons. However, the above structure is the most plausible since the least electronegative element is electron deficient.

C. The expanded octet

A number of compounds contain more than eight electrons around an atom. These **expanded octets** only occur for atoms of elements in and beyond the third period of the periodic table. Atoms from the third period on can accommodate more than eight electrons because they also have *d* orbitals that can be used in bonding.

EXAMPLE 9.9

Draw the Lewis structure for ClF_3.

Step 1: Chlorine is less electronegative than fluorine, so it occupies the central position. The skeletal structure is:

F

F Cl F

Step 2: The outer-shell electron configurations of F and Cl are $2s^22p^5$ and $3s^23p^5$, respectively. Thus, there are

$$7 + (3 \times 7) = 28 \text{ valence electrons}$$

Step 3: We draw a single covalent bond between Cl and each F, and then complete the octets for the F atoms.

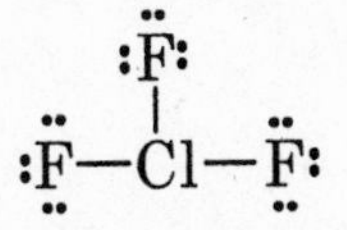

Step 4: At this point, we still have two electron pairs ($4\ e^-$) to place. Fluorine cannot exceed an octet of electrons, so the electrons must be placed as lone pairs on chlorine. The correct Lewis structure is:

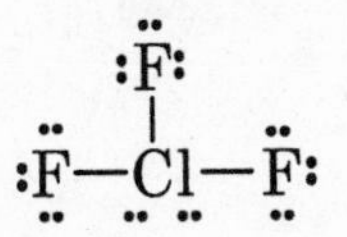

Chlorine can exceed an octet of electrons. An expanded octet can occur for atoms of elements in and beyond the third period of the periodic table. As a final check, we verify that there are 28 valence electrons in the Lewis structure of ClF_3.

PRACTICE EXERCISE

8. Write the Lewis structure for PCl_5.

Text Problem: 42

PROBLEM TYPE 6: USING BOND ENERGIES TO ESTIMATE THE ENTHALPY OF A REACTION

A quantitative measure of the stability of a molecule is its **bond dissociation energy** (or **bond energy**), which is the enthalpy change required to break a particular bond in one mole of gaseous molecules. Table 9.2 of the text lists the average bond energies of a number of bonds found in polyatomic molecules, as well as the bond energies of several diatomic molecules.

In many cases, it is possible to predict the approximate enthalpy of reaction by using the average bond energies. Energy *is always required* to break chemical bonds and chemical bond formation is always accompanied by a *release of energy*. We can estimate the enthalpy of reaction by counting the total number of bonds broken and formed in the reaction and recording all the corresponding energy changes. The enthalpy of reaction in the *gas phase* is given by:

$$\Delta H° = \text{total energy input} - \text{total energy released}$$

$$\Delta H° = \Sigma BE(\text{reactants}) - \Sigma BE(\text{products})$$

Where,

BE is the average bond energy
Σ represents summation

If the total energy input is greater than the total energy released, $\Delta H°$ *is positive* and the reaction is *endothermic*. On the other hand, if more energy is released than absorbed, $\Delta H°$ *is negative* and the reaction is *exothermic*.

EXAMPLE 9.10

Use average bond energies to estimate $\Delta H°$ for the following reaction:

$$Cl_2\,(g) + I_2\,(g) \longrightarrow 2\ ICl\,(g)$$

The average I–Cl bond energy is 210 kJ.

Recall that,

$$\Delta H° = \Sigma BE(\text{reactants}) - \Sigma BE(\text{products})$$

$$\Delta H° = BE(\text{Cl–Cl}) + BE(\text{I–I}) - 2BE(\text{I–Cl})$$

$$= 242.7 \text{ kJ} + 151.0 \text{ kJ} - 2(210 \text{ kJ})$$

$$\mathbf{\Delta H° = -26 \text{ kJ}}$$

Is the reaction exothermic or endothermic?

PRACTICE EXERCISE

9. Estimate the enthalpy change for the following reaction:

$$N_2(g) + O_2(g) \longrightarrow 2\,NO(g)$$

Hint: NO has a double bond [BE(N=O) = 630 kJ]

Text Problems: 48, 50

ANSWERS TO PRACTICE EXERCISES

1. **(a)** polar covalent **(b)** covalent **(c)** ionic

2. 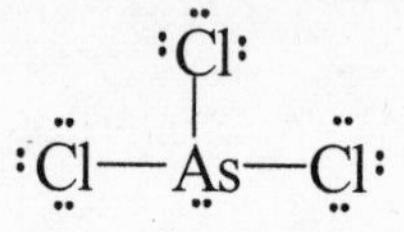

3. $[:C\equiv N:]^-$

4. **(a)** Formal charge (left N) = 5 – [4 + 1/2(4)] = –1

Formal charge (middle N) = 5 – [0 + 1/2(8)] = +1

Formal charge (O) = 6 – [4 + 1/2(4)] = 0

(b) Formal charge (O) = 6 – [6 + 1/2(2)] = –1

Formal charge (H) = 1 – [0 + 1/2(2)] = 0

5. (b) is the more plausible structure based on formal charges. There is a large positive (+2) formal charge on the S atom in structure (a). The formal charge on the S atom in structure (b) is zero.

6. :N=N=Ö: ⟷ :N≡N–Ö: ⟷ :N–N≡O:

7. 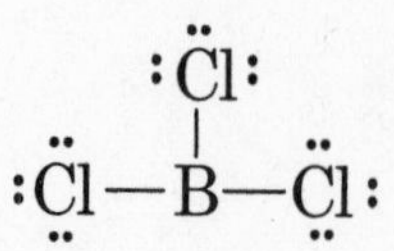

8. The lone pairs have been left off the chlorine atoms.

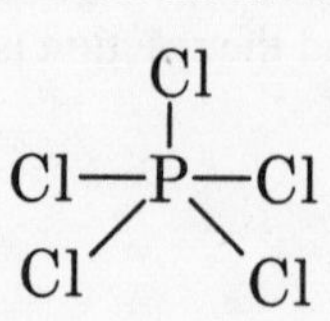

9. $\Delta H° = 180$ kJ

SOLUTIONS TO SELECTED TEXT PROBLEMS

9.14 Classifying Chemical Bonds, Problem Type 1.

The amount of ionic character is based on the electronegativity difference between the two atoms. The larger the electronegativity difference, the greater the ionic character.

Let's let ΔEN = electronegativity difference. The bonds arranged in order of increasing ionic character are:

C–H (ΔEN = 0.4) < Br–H (ΔEN = 0.7) < Na–I (ΔEN = 1.6) < F–H (ΔEN = 1.9)

< Li–Cl (ΔEN = 2.0) < K–F (ΔEN = 3.2)

9.16 The order of increasing ionic character is:

Cl–Cl (zero difference in electronegativity) < Br–Cl (difference 0.2) < Si–C (difference 0.7)

< Cs–F (difference 3.3).

9.18 **(a)** The two silicon atoms are the same. The bond is covalent.

(b) The electronegativity difference between Cl and Si is 3.0 – 1.8 = 1.2. The bond is polar covalent.

(c) The electronegativity difference between F and Ca is 4.0 – 1.0 = 3.0. The bond is ionic.

(d) The electronegativity difference between N and H is 3.0 – 2.1 = 0.9. The bond is polar covalent.

9.22 Writing Lewis Structures, Problem Type 2.

(a)

Step 1: The outer-shell electron configuration of O is $2s^2 2p^4$. Also, we must add the negative charges to the number of valence electrons, Thus, there are

$(2 \times 6) + 2 = 14$ valence electrons

Step 2: We draw a single covalent bond between each O, and then attempt to complete the octets for the O atoms.

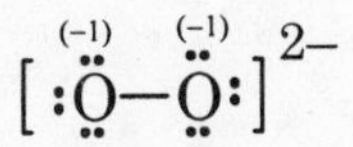

The octet rule is satisfied for both oxygen atoms. As a final check, we verify that there are 14 valence electrons in the Lewis structure of O_2^-.

Follow the same procedure as part (a) for parts (b), (c), and (d). The appropriate Lewis structures are:

(b) 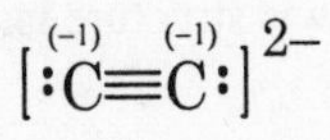 **(c)** 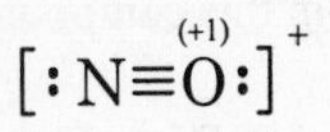 **(d)** 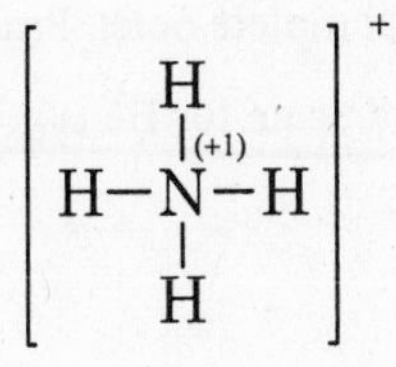

9.24 **(a)** Neither oxygen atom has a complete octet. The left-most hydrogen atom is forming two bonds (4 e^-). Hydrogen can only be surrounded by at most two electrons.

(b) The correct structure is:

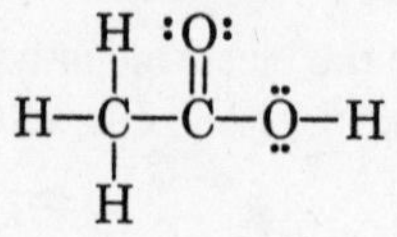

Do the two structures have the same number of electrons? Is the octet rule satisfied for all atoms other than hydrogen, which should have a duet of electrons?

9.30 Drawing Resonance Structures, Problem Type 4.

The Lewis structure for ClO_3^- is shown below.

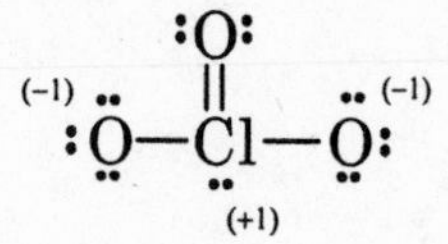

We can draw two more equivalent Lewis structures with the double bond between Cl and a different oxygen atom.

The resonance structures with formal charges are as follows:

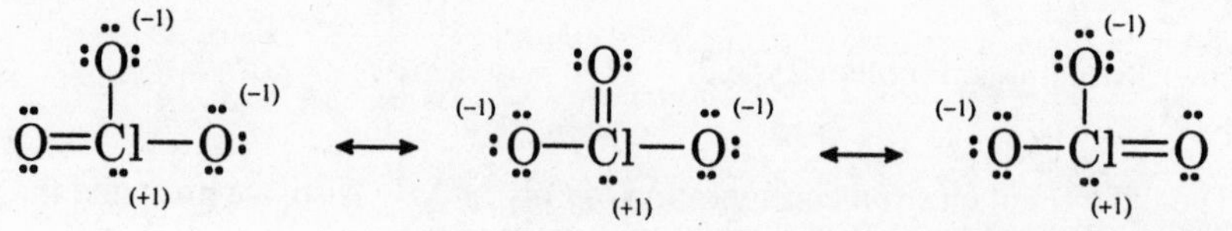

9.32 The structures of the most important resonance forms are:

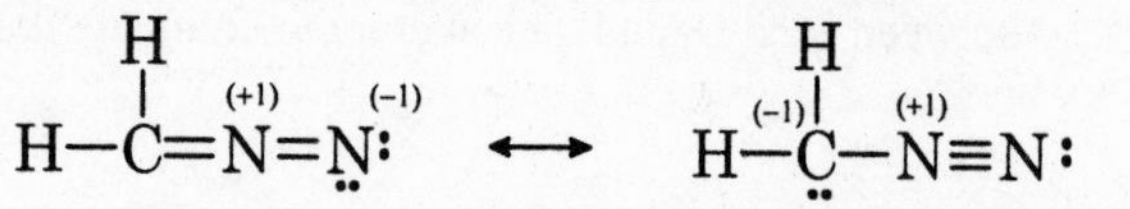

9.34 Three reasonable resonance structures with the formal charges indicated are

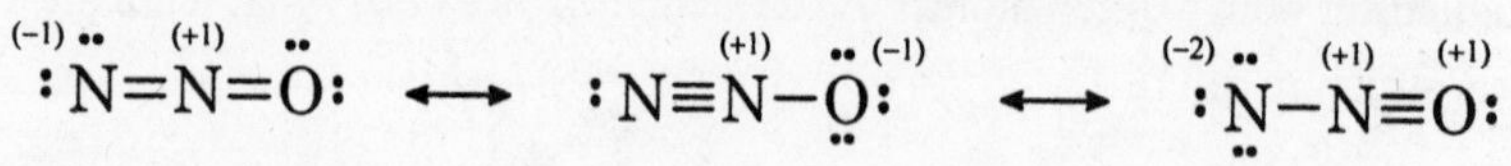

9.40 The incomplete octet, Problem Type 5A.

The octet rule for Be is *not* satisfied in this compound. The Lewis structure is:

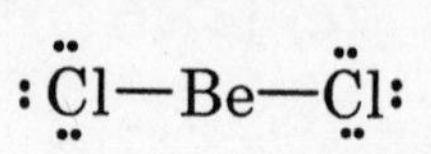

An octet of electrons on Be can only be formed by making two double bonds as shown below:

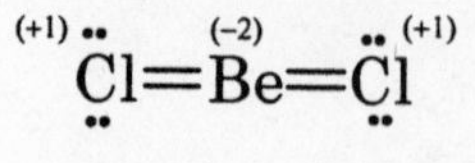

This places a high negative formal charge on Be and positive formal charges on the Cl atoms. This structure distributes the formal charges counter to the electronegativities of the elements. It is not a plausible Lewis structure.

9.42 The outer electron configuration of antimony is $5s^25p^3$. The Lewis structure is shown below. All five valence electrons are shared in the five covalent bonds. The octet rule is not obeyed. (The electrons on the chlorine atoms have been omitted for clarity.)

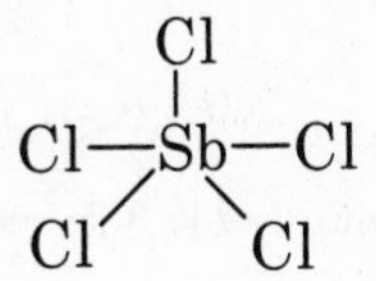

Can Sb have an expanded octet?

9.44 The reaction can be represented as:

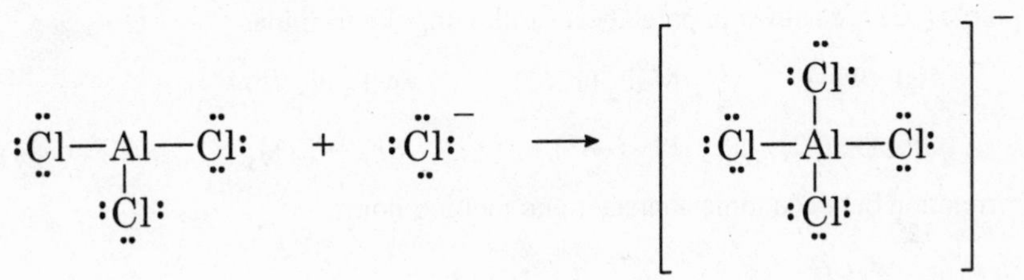

The new bond formed is called a coordinate covalent bond.

9.48 This problem is similar to Problem Type 6, Using Bond Energies to Estimate the Enthalpy of a Reaction.

There are two oxygen-to-oxygen bonds in ozone. We will represent these bonds as O–O. However, these bonds might not be true oxygen-to-oxygen single bonds.

$$\Delta H° = \Sigma BE(\text{reactants}) - \Sigma BE(\text{products})$$

$$\Delta H° = BE(\text{O=O}) - 2BE(\text{O–O})$$

In the problem, we are given $\Delta H°$ for the reaction, and we can look up the O=O bond energy in Table 9.2 of the text. Solving for the average bond energy in ozone,

$$-2BE(\text{O–O}) = \Delta H° - BE(\text{O=O})$$

$$\mathbf{\mathit{BE}(O–O)} = \frac{BE(\text{O=O}) - \Delta H°}{2} = \frac{498.7\text{ kJ/mol} + 107.2\text{ kJ/mol}}{2} = \mathbf{303.0\ kJ/mol}$$

Considering the resonance structures for ozone, is it expected that the O–O bond energy in ozone is between the single O–O bond energy (142 kJ) and the double O=O bond energy (498.7 kJ)?

9.50 **(a)**

Bonds Broken	*Number Broken*	*Bond Energy* (kJ/mol)	*Energy Change* (kJ)
C – H	12	414	4968
C – C	2	347	694
O = O	7	498.7	3491

Bonds Formed	*Number Formed*	*Bond Energy* (kJ/mol)	*Energy Change* (kJ)
C = O	8	799	6392
O − H	12	460	5520

$$\Delta H^\circ = \text{total energy input} - \text{total energy released}$$

$$= (4968 + 694 + 3491) - (6392 + 5520) = \mathbf{-2759\ kJ}$$

(b) $\Delta H^\circ = 4\,\Delta H_f^\circ(CO_2) + 6\,\Delta H_f^\circ(H_2O) - [2\,\Delta H_f^\circ(C_2H_6) + 7\,\Delta H_f^\circ(O_2)]$

$\Delta H^\circ = (4\text{ mol})(-393.5\text{ kJ/mol}) + (6\text{ mol})(-241.8\text{ kJ/mol}) - (2\text{ mol})(-84.7\text{ kJ/mol}) - (7\text{ mol})(0) = \mathbf{-2855\ kJ}$

The answers for part (a) and (b) are different, because *average* bond energies are used for part (a).

9.52 Recall that you can classify bonds as ionic or covalent based on electronegativity difference.

The melting points (°C) are shown in parentheses following the formulas.

Ionic:	NaF (993)	MgF_2 (1261)	AlF_3 (1291)	
Covalent:	SiF_4 (−90.2)	PF_5 (−83)	SF_6 (−121)	ClF_3 (−83)

Is there any correlation between ionic character and melting point?

9.54 KF is an ionic compound. It is a solid at room temperature made up of K^+ and F^- ions. It has a high melting point, and it is a strong electrolyte. Carbon dioxide, CO_2, is a covalent compound that exists as discrete molecules. It is a gas at room temperature. It has a low melting point, is only slightly soluble in water, and is a nonelectrolyte.

9.56 The resonance structures are:

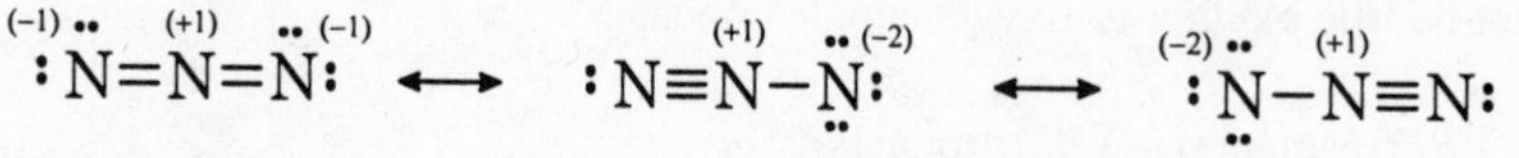

Which is the most plausible structure based on a formal charge argument?

9.58 **(a)** An example of an aluminum species that satisfies the octet rule is the anion $AlCl_4^-$. The Lewis dot structure is drawn in Problem 9.44.

(b) An example of an aluminum species containing an expanded octet is anion AlF_6^{3-}. (How many pairs of electrons surround the central atom?)

(c) An aluminum species that has an incomplete octet is the compound $AlCl_3$. The dot structure is given in Problem 9.44.

9.60 CF_2 would be very unstable because carbon does not have an octet. (How many electrons does it have?)

CH_5 does not exist because it violates the octet rule for carbon.

FH_2^- does not exist because F cannot exceed the octet rule. There are 10 valence electrons, so F would have to exceed the octet rule.

PI_5 appears to be a reasonable species (compared to PF_5 in Example 9.7 of the text). However, the iodine atoms are too large to have five of them "fit" around a single P atom.

9.62 **(a)** false **(b)** true **(c)** false **(d)** false

For question (c), what is an example of a second-period species that violates the octet rule?

9.64 The formation of CH_4 from its elements is:

$$C(s) + 2H_2(g) \longrightarrow CH_4(g)$$

The reaction could take place in two steps:

Step 1: $C(s) + 2H_2(g) \longrightarrow C(g) + 4H(g)$ $\Delta H^\circ_{rxn} = 716\text{ kJ} + 872.8\text{ kJ} = 1589\text{ kJ}$

Step 2: $C(g) + 4H(g) \longrightarrow CH_4(g)$ $\Delta H^\circ_{rxn} \cong -4 \times (\text{bond energy of C–H bond})$

$= -4 \times 414\text{ kJ} = -1656\text{ kJ}$

Therefore, $\Delta H^\circ_f(CH_4)$ would be approximately the sum of the enthalpy changes for the two steps. See Problem Type 4C, Chapter 6 (Hess's law).

$$\Delta H^\circ_f(CH_4) = \Delta H^\circ_f(1) + \Delta H^\circ_{rxn}(2)$$

$$\Delta H^\circ_f(CH_4) = 1589 - 1656\text{ kJ/mol} = \mathbf{-67\text{ kJ/mol}}$$

The actual value of $\Delta H^\circ_f(CH_4) = -74.85\text{ kJ/mol}$

9.66 Only N_2 has a triple bond. Therefore, it has the shortest bond length.

9.68 To be isoelectronic, molecules must have the same number and arrangement of valence electrons. NH_4^+ and CH_4 are isoelectronic (8 valence electrons), as are CO and N_2 (10 valence electrons), as are $B_3N_3H_6$ and C_6H_6 (30 valence electrons). Draw Lewis structures to convince yourself that the electron arrangements are the same in each isoelectronic pair.

9.70 The reaction can be represented as:

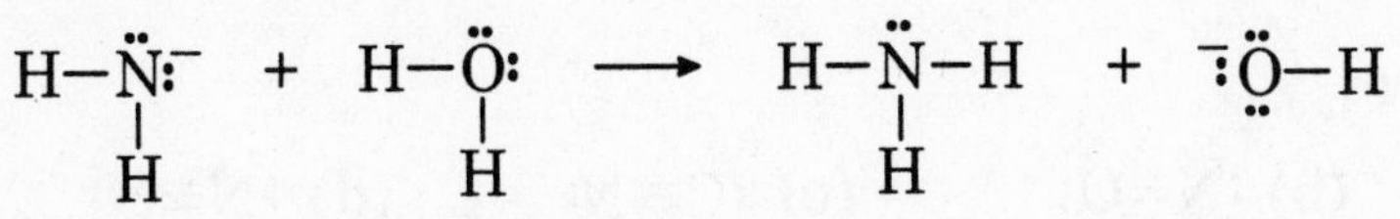

9.72 The central iodine atom in I_3^- has *ten* electrons surrounding it: two bonding pairs and three lone pairs. The central iodine has an expanded octet. Elements in the second period such as fluorine cannot have an expanded octet as would be required for F_3^-.

9.74 The skeletal structure is:

```
  H
H C N C O
  H
```

The number of valence electron is: $(1 \times 3) + (2 \times 4) + 5 + 6 = 22$ valence electrons

We can draw two resonance structures for methyl isocyanate.

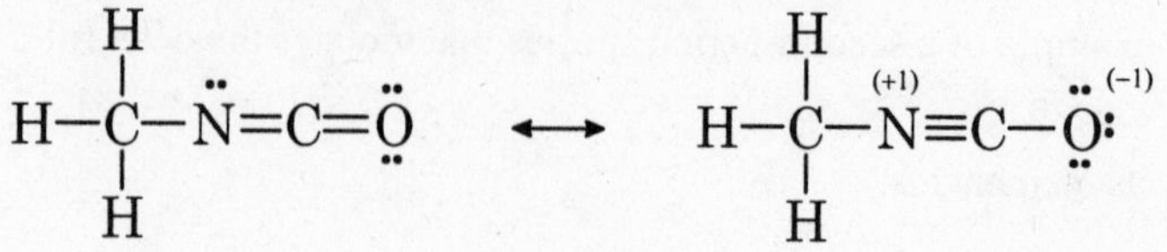

9.76 (a) This is a very good resonance form; there are no formal charges and each atom satisfies the octet rule.

(b) This is a second choice after (a) because of the positive formal charge on the oxygen (high electronegativity).

(c) This is a poor choice for several reasons. The formal charges are placed counter to the electronegativities of C and O, the oxygen atom does not have an octet, and there is no bond between the oxygen and carbon!

(d) This is a mediocre choice because of the large formal charge and lack of an octet on carbon.

9.78 The nonbonding electron pairs around Cl and F are omitted for simplicity.

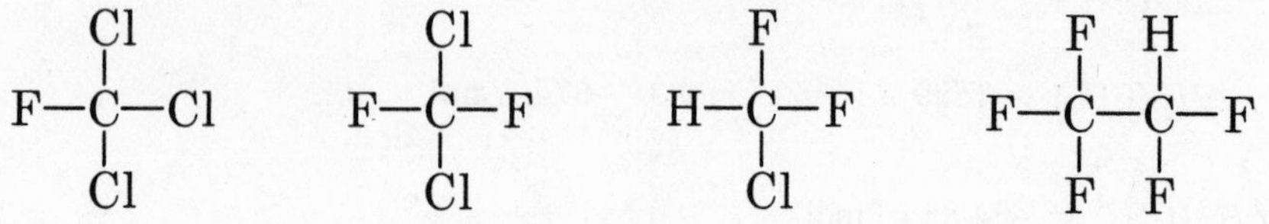

9.80 (a) Using Equation (9.4),

$$\Delta H = \sum BE(\text{reactants}) - \sum BE(\text{products})$$

$$\boldsymbol{\Delta H} = [(436.4 + 151.0) - 2(298.3)] = \mathbf{-9.2\ kJ}$$

(b) Using Equation (6.8),

$$\Delta H^\circ = 2\,\Delta H^\circ_f[\text{HI}(g)] - \{\Delta H^\circ_f[\text{H}_2(g)] + \Delta H^\circ_f[\text{I}_2(g)]\}$$

$$\boldsymbol{\Delta H^\circ} = (2\text{ mol})(25.9\text{ kJ/mol}) - [(0) + (1\text{ mol})(61.0\text{ kJ/mol})] = \mathbf{-9.2\ kJ}$$

9.82 The Lewis structures are:

9.84 True. Each noble gas atom already has completely filled *ns* and *np* subshells.

9.86 The resonance structures are:

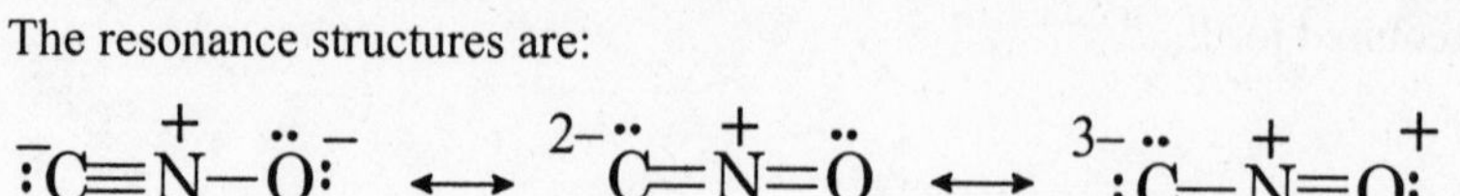

The most likely structure is on the left and the least likely structure is on the right.

9.88

(a) 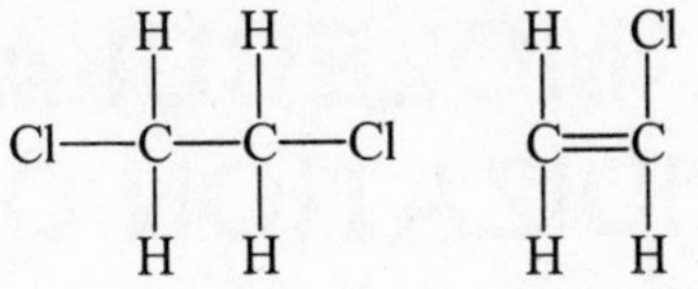

(b) 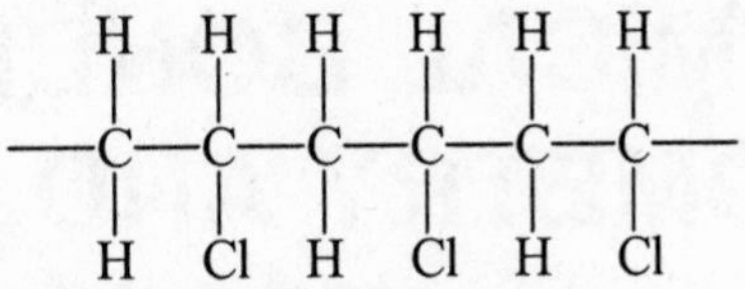

(c) In the formation of poly(vinyl chloride) form vinyl chloride, for every C=C double bond broken, 2 C–C single bonds are formed. No other bonds are broken or formed. The energy changes for 1 mole of vinyl chloride reacted are:

total energy input (breaking C=C bonds) = 620 kJ

total energy released (forming C–C bonds) = 2 × 347 kJ = 694 kJ

$\Delta H° = 620\text{ kJ} - 694\text{ kJ} = -74\text{ kJ}$

The negative sign shows that this is an exothermic reaction. To find the total heat released when 1.0×10^3 kg of vinyl chloride react, we proceed as follows:

$$\text{heat released} = 1.0\times10^6\ \cancel{\text{g}}\ C_2H_3Cl \times \frac{1\ \cancel{\text{mol}}\ C_2H_3Cl}{62.49\ \cancel{\text{g}}\ C_2H_3Cl} \times \frac{-74\text{ kJ}}{1\ \cancel{\text{mol}}\ C_2H_3Cl} = \mathbf{-1.2 \times 10^6\ kJ}$$

CHAPTER 10
CHEMICAL BONDING II: MOLECULAR GEOMETRY AND HYBRIDIZATION OF ATOMIC ORBITALS

PROBLEM-SOLVING STRATEGIES AND TUTORIAL SOLUTIONS

TYPES OF PROBLEMS

Problem Type 1: Molecular Geometry.
- **(a)** Molecules in which the central atom has *no* lone pairs.
- **(b)** Molecules in which the central atom has one or more lone pairs.
- **(c)** Geometry of molecules with more than one central atom.

Problem Type 2: Predicting Dipole Moments.

Problem Type 3: Hybridization of Atomic Orbitals.
- **(a)** Hybridization of *s* and *p* orbitals.
- **(b)** Hybridization of *s*, *p*, and *d* orbitals.
- **(c)** Hybridization in molecules containing double and triple bonds.

PROBLEM TYPE 1: MOLECULAR GEOMETRY

The model that we are going to use to study molecular geometry is called the **valence-shell electron-pair repulsion (VSEPR) model**. It accounts for the geometric arrangement of electron pairs around a central atom in terms of the repulsion between electron pairs. The geometry that a molecule ultimately adopts *minimizes* electron-pair repulsion.

Guidelines for Applying the VSEPR Model

1. Write the Lewis structure of the molecule.

2. Only consider the electron pairs around the *central atom* (the atom that is bonded to more than one other atom). Count the number of electron pairs around the central atom (bonding pairs and lone pairs). For counting purposes, treat double and triple bonds as though they were single bonds. Refer to Table 10.1 of your text to predict the overall arrangement of the electron pairs.

3. Use Tables 10.1 and 10.2 of your text to predict the *geometry* of the molecule.

4. In predicting bond angles, note that a lone pair repels another lone pair or a bonding pair more strongly than a bonding pair repels another bonding pair. There is no easy way to predict bond angles accurately when the central atom possesses one or more lone pairs.

A. Molecules in which the central atom has *no* lone pairs

For simplicity, we will only consider molecules that contain atoms of two elements, A and B, where A is the central atom. These molecules have the general formula AB_x, where x is an integer 2, 3, ... In most cases, x is between 2 and 6.

Table 10.1 of your text shows five possible arrangements of electron pairs around the central atom A. Remember, these arrangements are adopted because electron-pair repulsions are minimized.

Below is a condensed version of Table 10.1 from your text.

Arrangement of electron pairs around a central atom (A) in a molecule, and geometry of molecules if the central atom has *no* lone pairs.

Number of electron pairs around central atom	Arrangement of electron pairs	Molecular geometry
2	Linear, 180°	Linear, AB_2
3	Trigonal planar, 120°	Trigonal planar, AB_3
4	Tetrahedral, 109.5°	Tetrahedral, AB_4
5	Trigonal bipyramidal, 120°, 90°	Trigonal bipyramidal, AB_5
6	Octahedral, 90°	Octahedral, AB_6

Note: Since there are *no* lone pairs around the central atom in all the examples shown above, the molecular geometry is *always* the same as the arrangement of electron pairs around the central atom.

EXAMPLE 10.1
Use the VSEPR model to predict the geometry of the following molecules and ions: (a) $HgCl_2$, (b) $SnCl_4$, (c) NO_3^-, (d) PF_5.

(a)
Step 1: Write the Lewis structure of the molecule (see Problem Type 2, Chapter 9).

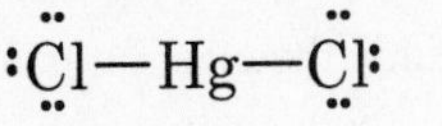

Step 2: Count the number of electron pairs around the central atom. There are two electron pairs around Hg.

Step 3: Since there are two electron pairs around Hg, the electron-pair arrangement that minimizes repulsion is **linear**.

In addition, since there are no lone pairs around the central atom, the geometry is also **linear** (AB_2).

(b)
Step 1: Write the Lewis structure of the molecule.

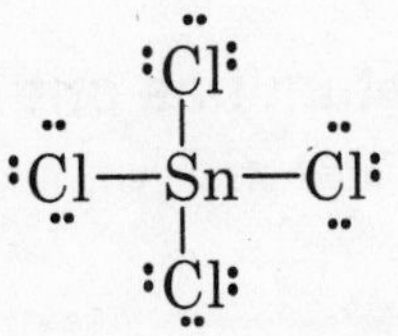

Step 2: Count the number of electron pairs around the central atom. There are four electron pairs around Sn.

Step 3: Since there are four electron pairs around Sn, the electron-pair arrangement that minimizes repulsion is **tetrahedral**.

In addition, since there are no lone pairs around the central atom, the geometry is also **tetrahedral** (AB_4).

(c)

Step 1: Write the Lewis structure of the molecule.

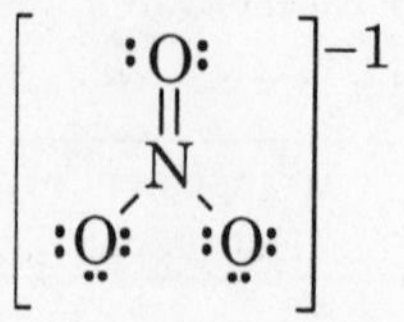

Step 2: Count the number of electron pairs around the central atom. There are three electron pairs around N. Remember, for VSEPR purposes, a double or triple bond counts the same as a single bond.

Step 3: Since there are three electron pairs around N, the electron-pair arrangement that minimizes repulsion is **trigonal planar**.

In addition, since there are no lone pairs around the central atom, the geometry is also **trigonal planar** (AB_3).

(d)

Step 1: Write the Lewis structure of the molecule.

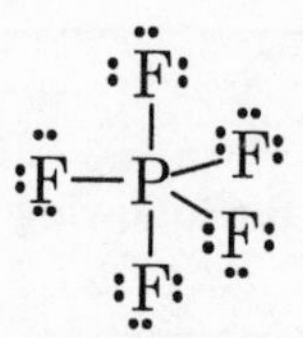

Step 2: Count the number of electron pairs around the central atom. There are five electron pairs around P.

Since there are five electron pairs around P, the electron-pair arrangement that minimizes repulsion is **trigonal bipyramidal.**

In addition, since there are no lone pairs around the central atom, the geometry is also **trigonal bipyramidal** (AB_5).

PRACTICE EXERCISE

1. Use the VSEPR model to predict the geometry of the following molecules: (a) CO_2, (b) CCl_4,(c) SO_3, (d) SF_6.

Text Problems: 8, 10, 12

B. Molecules in which the central atom has one or more lone pairs

If lone pairs are present on the central atom, the overall arrangement of electron pairs is ***not*** the same as the geometry of the molecule.

To keep track of the total number of bonding pairs and lone pairs, we will designate molecules with lone pairs as AB_xE_y, where A is the central atom, B is a surrounding atom, and E is a lone pair on the central atom, A. Both x and y are integers: x = 2, 3, . . . , and y = 1, 2, ...

When working with molecules that have lone pairs on the central atom, remember to count all electron pairs on the central atom, both bonding pairs and lone pairs. The total number of electron pairs around the central atom determines the electron arrangement around the central atom. See Table 10.1. However, the molecular geometry will not be the same as this electron arrangement. Use Table 10.2 of your text to determine the molecular geometry, or a better option is to build a model. Gum drops and toothpicks make an effective and inexpensive model kit. And besides, you can eat your model when you are finished. (Not the toothpicks!)

Below is a condensed version of Table 10.2 of your text.

Geometry of simple molecules and ions in which the central atom has one or more lone pairs.

Class of molecule	Total # of electron pairs on central atom	Number of bonding pairs	Arrangement of electron pairs	Number of lone pairs	Molecular geometry
AB_2E	3	2	Trigonal planar	1	Bent
AB_3E	4	3	Tetrahedral	1	Trigonal pyramid
AB_2E_2	4	2	Tetrahedral	2	Bent
AB_4E	5	4	Trigonal bipyramidal	1	Distorted tetrahedron
AB_3E_2	5	3	Trigonal bipyramidal	2	T-shaped
AB_2E_3	5	2	Trigonal bipyramidal	3	Linear
AB_5E	6	5	Octahedral	1	Square pyramidal
AB_4E_2	6	4	Octahedral	2	Square planar

Note: The arrangement of 3 electron-pairs is always trigonal planar; the arrangement of 4 electron-pairs is tetrahedral; the arrangement of 5 electron-pairs is trigonal bipyramidal; the arrangement of 6 electron-pairs is octahedral. However, if there are any lone pairs on the central atom, the molecular geometry will be different from the electron arrangement. **Build models!**

EXAMPLE 10.2

Use the VSEPR model to predict the geometry of the following molecules: (a) O_3, (b) XeF_2, (c) IF_5.

(a)

Step 1: Write the Lewis structure of the molecule (see Problem Type 2, Chapter 9).

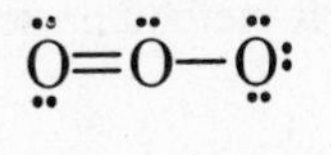

Step 2: Count the number of electron pairs around the central atom. There are three electron pairs around the central oxygen. Remember, for VSEPR purposes, a double or triple bond counts the same as a single bond.

Step 3: Since there are three electron pairs around the central O, the electron-pair arrangement that minimizes repulsion is **trigonal planar**.

However, there is one lone pair of electrons around the central atom. Consulting a model or Table 10.2 of your text, you should find that the geometry of the molecule is **bent** (AB_2E).

(b)

Step 1: Write the Lewis structure of the molecule.

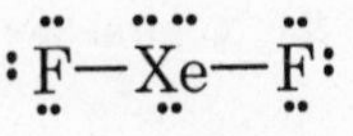

Step 2: Count the number of electron pairs around the central atom. There are five electron pairs around xenon.

Step 3: Since there are five electron pairs around xenon, the electron-pair arrangement that minimizes repulsion is **trigonal bipyramidal.**

However, there are three lone pairs of electrons around the central atom. Consulting a model or Table 10.2 of your text, you should find that the geometry of the molecule is **linear** (AB_2E_3).

(c)

Step 1: Write the Lewis structure of the molecule.

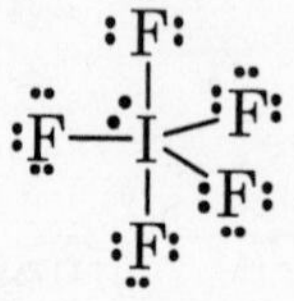

Step 2: Count the number of electron pairs around the central atom. There are six electron pairs around iodine.

Step 3: Since there are six electron pairs around iodine, the electron-pair arrangement that minimizes repulsion is **octahedral.**

However, there is one lone pair of electrons around the central atom. Consulting a model or Table 10.2 of your text, you should find that the geometry of the molecule is **square pyramidal** (AB_5E).

PRACTICE EXERCISE

2. Use the VSEPR model to predict the geometry of the following molecules: (a) SO_2, (b) XeF_4, (c) SF_4.

Text Problems: 10, 12

C. Geometry of molecules with more than one central atom

So far, we have discussed the geometry of molecules having only one central atom. (The term "central atom" means an atom that is not a terminal atom in a polyatomic molecule.) Many molecules will have more than one "central atom". In these cases, you have to apply the VSEPR method presented in parts A and B above to each of the central atoms.

EXAMPLE 10.3

Use the VSEPR model to predict the geometry of C_2H_6.

Step 1: Write the Lewis structure of the molecule. Hydrogens must be in terminal positions. The two carbons must be the "central atoms".

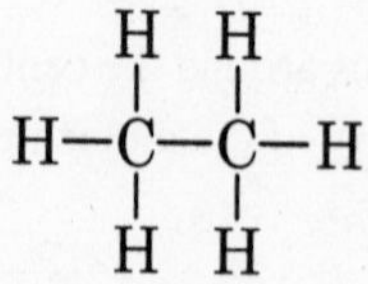

Step 2: Count the number of electron pairs around the "central atoms". There are four electron pairs around each carbon.

Step 3: Since there are four electron pairs around each carbon, the electron-pair arrangement around each carbon that minimizes repulsion is **tetrahedral.**

In addition, since there are no lone pairs around the central atoms, their geometries are also **tetrahedral.**

PRACTICE EXERCISE

3. Use the VSEPR model to predict the geometry of C_2H_2.

PROBLEM TYPE 2: PREDICTING DIPOLE MOMENTS

To determine if a molecule has a dipole moment (a measure of electrical charge separation in a molecule) you must consider two factors.

1. Are the *bonds* in the molecule polar?

   A bond will be polar if there is a difference in electronegativity between the two atoms comprising the bond (see Section 9.3 of your text). For example, in an O–H bond, there is a shift in electron density from H to O because O is more electronegative than H. The shift in electron density is symbolized by placing a crossed arrow (+⟶) above the bond to indicate the direction of the shift in electron density. This is called a *bond* moment. For example:

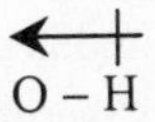

   The consequent charge separation can be represented as:

$$\overset{\delta^-}{\mathrm{O}} - \overset{\delta^+}{\mathrm{H}}$$

   where δ (delta) represents a partial charge.

2. Is the *molecule* polar?

   Diatomic molecules containing atoms of the same element do not have dipole moments and so are **nonpolar molecules.** There is no difference in electronegativity since the two elements are the same.

   However, most molecules will have polar bonds (bond moments). If a molecule has polar bonds, does this mean that it is a polar molecule? Not necessarily. The *bond moment* is a vector quantity, which means that it has both a magnitude and direction. The measured *dipole moment* of the molecule is equal to the vector sum of the bond moments. For example, in CO_2, the two bond moments are equal in magnitude and opposite in direction. The sum or resultant dipole moment will be *zero*. Hence, CO_2 is a **nonpolar molecule.**

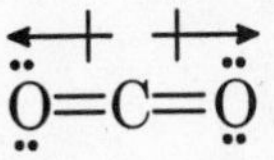

   In summary, to determine if a molecule is **polar** (i.e., does the molecule have a dipole moment), you must sum the individual bond moments to determine if there is a resultant dipole moment.

EXAMPLE 10.4

Predict whether each of the following molecules has a dipole moment: (a) CCl_4 and (b) $CHCl_3$.

(a)

Step 1: Write the Lewis structure for the molecule. The lone pairs on Cl have been omitted.

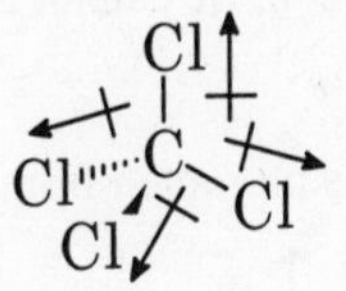

Step 2: Shown on the Lewis structure are the bond moments. Chlorine is more electronegative than C, so the arrows indicate the shift in electron density toward Cl. However, these polar bonds are arranged in a symmetric tetrahedral fashion about the central C atom. The sum or resultant dipole moment is *zero*. CCl_4 is a **nonpolar molecule**.

(b)

Step 1: Write the Lewis structure for the molecule. The lone pairs on Cl have been omitted.

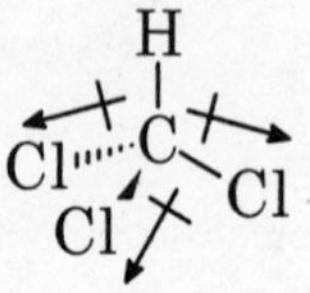

Step 2: Shown on the Lewis structure are the bond moments. Chlorine is more electronegative than C, so the arrows indicate the shift in electron density toward Cl. The electronegativity difference between C and H is very small, so a C–H bond is essentially nonpolar. In this case, the three bond moments partially reinforce each other. Thus, $CHCl_3$ has a dipole moment and is therefore a **polar molecule**.

PRACTICE EXERCISE

4. Predict whether each of the following molecules has a dipole moment: (a) CO, (b) BCl_3, and (c) XeF_4.

Text Problems: 18, 20, 22

PROBLEM TYPE 3: HYBRIDIZATION OF ATOMIC ORBITALS

The **VSEPR** model is very powerful considering its simplicity; however, it does not explain chemical bond formation in any detail. In the 1930s, **valence bond (VB) theory** was introduced to account for chemical bond formation. **VB** theory describes covalent bonding as the overlapping of atomic orbitals. This means that the orbitals share a common region in space.

VB theory uses a concept called **hybridization** to explain covalent bonding. Hybridization is the mixing of atomic orbitals in an atom (usually a central atom) to generate a set of new atomic orbitals, called *hybrid orbitals*. Hybrid orbitals are atomic orbitals obtained when two or more nonequivalent orbitals of the same atom combine. The hybrid orbitals are used to form covalent bonds.

A. Hybridization of *s* and *p* orbitals

1. *sp* hybrid orbitals

Let's consider the central atom Be in the BeH_2 molecule. Be has a ground state electron configuration of $1s^22s^2$. Only valence electrons are involved in bonding, so an orbital diagram of the valence electrons is:

$$\underset{2s}{\uparrow\downarrow} \qquad \underset{2p}{\_\_\ \ \_\_\ \ \_\_}$$

With this ground-state electron configuration, Be cannot form bonds with H because its valence electrons are paired in a 2*s* orbital.

To explain the bonding, first an electron is promoted from the 2*s* orbital to a 2*p* orbital.

$$\underset{2s}{\uparrow} \qquad \underset{2p}{\uparrow\ \ \_\_\ \ \_\_}$$

Now, we have two different orbitals that can bond to the two hydrogens, which would result in two nonequivalent Be–H bonds. However, experimental evidence shows that there are two equivalent Be–H bonds.

This is where hybridization comes in. By mixing the 2*s* orbital with one of the 2*p* orbitals, we can generate two equivalent *sp* hybrid orbitals.

↿ ↿	__ __
sp hybrid orbitals	empty *p* orbitals

Figure 10.5 of your text shows the shape and orientation of the *sp* hybrid orbitals. These two hybrid orbitals lie along the same line so that the angle between them is 180° (linear arrangement). Each of the BeH bonds is formed by the overlap of a Be *sp* hybrid and a H 1*s* orbital. The resulting BeH_2 molecule has a linear geometry.

2. ***sp*² hybrid orbitals**

Following the same type of argument for *sp* hybrids, sp^2 hybrid orbitals are formed by mixing an *s* orbital with two *p* orbitals. Three equivalent sp^2 hybrid orbitals are formed.

↿	↿ ↿ __
s	*p*

Mixing an *s* with two *p* orbitals gives:

↿ ↿ ↿	__
sp^2 hybrid orbitals	empty *p* orbital

Figure 10.7 of your text shows the shape and orientation of the sp^2 hybrid orbitals. These three hybrid orbitals lie in a plane with an angle between any two hybrids of 120°. The three sp^2 hybrid orbitals are arranged in a trigonal planar fashion.

3. ***sp*³ hybrid orbitals**

Again, following the same argument as above, sp^3 hybrid orbitals are formed by mixing an *s* orbital with three *p* orbitals. Four equivalent sp^3 hybrid orbitals are formed.

↿	↿ ↿ ↿
s	*p*

Mixing an *s* with three *p* orbitals gives:

↿ ↿ ↿ ↿
sp^3 hybrid orbitals

Figure 10.9 of your text shows the shape and orientation of the sp^3 hybrid orbitals. These four equivalent hybrid orbitals are directed toward the four corners of a regular tetrahedron with 109.5° bond angles.

Summarizing,

Type of hybrid	No. of equivalent hybrid orbitals	Arrangement of hybrid orbitals	No. of empty *p* orbitals
sp	2	linear, 180°	2
sp^2	3	trigonal planar, 120°	1
sp^3	4	tetrahedral, 109.5°	0

EXAMPLE 10.5

Determine the hybridization of the central (underlined) atom in each of the following molecules: (a) $\underline{C}Cl_4$ and (b) $\underline{B}Cl_3$.

(a)

Step 1: Write the Lewis structure of the molecule. The lone pairs on the chlorines have been omitted.

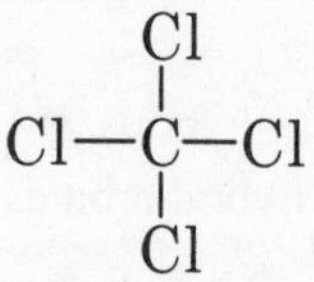

Step 2: Count the number of electron pairs around the central atom. Since there are four electron pairs around C, the electron arrangement that minimizes electron-pair repulsion is **tetrahedral.**

We conclude that C is $\boldsymbol{sp^3}$ **hybridized** because it has the electron arrangement of four sp^3 hybrid orbitals.

(b)

Step 1: Write the Lewis structure of the molecule. The lone pairs on the chlorines have been omitted.

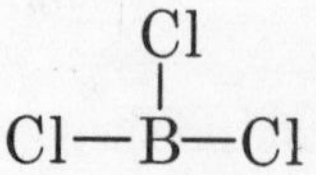

Step 2: Count the number of electron pairs around the central atom. Since there are three electron pairs around B, the electron arrangement that minimizes electron-pair repulsion is **trigonal planar**.

We conclude that B is $\boldsymbol{sp^2}$ **hybridized** because it has the electron arrangement of three sp^2 hybrid orbitals.

PRACTICE EXERCISE

5. Determine the hybridization of the central (underlined) atom in each of the following molecules or ions: (a) $\underline{C}O_2$ and (b) $\underline{C}O_3^{2-}$.

Text Problems: 32, 34, 36, **38**

B. Hybridization of *s*, *p*, and *d* orbitals

1. $\boldsymbol{sp^3d}$ hybrid orbitals

We use the same approach that we used for hybridizing *s* and *p* orbitals, but now we are also mixing in a *d* orbital. The sp^3d hybrid orbitals are formed by mixing an *s* orbital, three *p* orbitals, and a *d* orbital. Five equivalent sp^3d hybrid orbitals are formed.

Mixing an *s* orbital, three *p* orbitals, and a *d* orbital gives:

↑ ↑ ↑ ↑ ↑ — sp^3d hybrid orbitals

__ __ __ __ — empty *d* orbitals

Table 10.4 of your text shows the shape and orientation of the sp^3d hybrid orbitals. These five equivalent hybrid orbitals are directed toward the five corners of a trigonal bipyramid with bond angles of 120° and 90°.

2. sp^3d^2 hybrid orbitals

We use the same approach that we used above, but now we are mixing in two *d* orbitals. The sp^3d^2 hybrid orbitals are formed by mixing an *s* orbital, three *p* orbitals, and two *d* orbitals. Six equivalent sp^3d^2 hybrid orbitals are formed.

↑ (*s*) ↑ ↑ ↑ (*p*) ↑ ↑ __ __ __ (*d*)

Mixing an *s* orbital, three *p* orbitals, and two *d* orbitals gives:

↑ ↑ ↑ ↑ ↑ ↑ — sp^3d^2 hybrid orbitals

__ __ __ — empty *d* orbitals

Table 10.4 of your text shows the shape and orientation of the sp^3d^2 hybrid orbitals. These six equivalent hybrid orbitals are directed toward the six corners of an octahedron with bond angles of 90°.

Summarizing,

Type of hybrid	No. of equivalent hybrid orbitals	Arrangement of hybrid orbitals
sp^3d	5	trigonal bipyramid, 120°, 90°
sp^3d^2	6	octahedral, 90°

EXAMPLE 10.6

Describe the hybridization of xenon in xenon tetrafluoride (XeF_4).

Step 1: Write the Lewis structure for XeF_4. The lone pairs of electrons on F have been omitted.

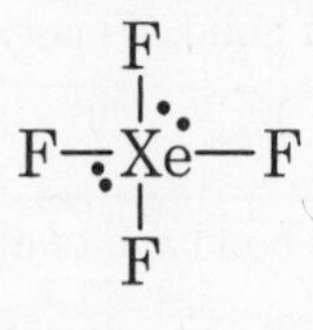

Step 2: Count the number of electron pairs around the central atom. Since there are six electron pairs around Xe, the electron arrangement that minimizes electron-pair repulsions is **octahedral**.

We conclude that Xe is **sp^3d^2 hybridized** because it has the electron arrangement of six sp^3d^2 hybrid orbitals.

> **Tip:** It is important to use the electron arrangement to determine the hybridization of the central atom and not the geometry. The two lone pairs on Xe are occupying two of the hybrid orbitals, so the lone pairs must be included to determine the correct hybridization.

PRACTICE EXERCISE

6. Describe the hybridization of sulfur in sulfur tetrafluoride, SF_4.

Text Problem: 40

C. Hybridization in molecules containing double or triple bonds

We can determine the hybridization of molecules containing double or triple bonds in the same manner as molecules with single bonds. Furthermore, we would like to determine which orbitals overlap to form the double or triple bond.

In a double or triple bond, there are two types of covalent bonds formed. One involves end-to-end overlap of orbitals in which the electron density is concentrated between the nuclei of the bonding atoms. A bond of this type is called a **sigma bond (σ bond)**. The second type involves side-to-side overlap of orbitals in which electron density is concentrated above and below the plane of the nuclei of the bonding atoms. This type of bond is called a **pi bond (π bond)**. For the molecules we will be considering, a π bond is formed from the side-to-side overlap of two *p* orbitals.

EXAMPLE 10.7

Describe the bonding in carbon dioxide, CO_2.

Step 1: Write the Lewis structure for CO_2.

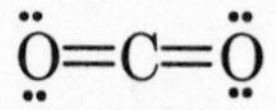

Step 2: Count the number of electron pairs around the central atom. Since there are two electron pairs around C, the electron arrangement that minimizes electron-pair repulsions is **linear.**

We conclude that C is ***sp* hybridized** because it has the electron arrangement of two *sp* hybrid orbitals.

Step 3: Count the number of electron pairs around each oxygen atom. Since there are three electron pairs around each O, the electron arrangement that minimizes repulsions is **trigonal planar**.

We conclude that each O is **sp^2 hybridized** because it has the electron arrangement of three sp^2 hybrid orbitals.

Step 4: Each of the *sp* orbitals of the C atom forms a sigma bond with an sp^2 hybrid on each of the O atoms. Carbon has two "pure" *p* orbitals that did not mix with the *s* orbital. Each of these *p* orbitals of the C atom forms a pi bond by overlapping in a side-to-side fashion with a *p* orbital on each of the oxygen atoms. Each double bond is composed of one σ bond and one π bond. Finally, the two lone pairs on each O atom are placed in its two remaining sp^2 orbitals.

Tip: A double bond is composed of one σ bond and one π bond. A triple bond is composed of one σ bond and two π bonds.

PRACTICE EXERCISE

7. Describe the bonding in a nitrogen molecule, N_2.

Text Problems: 36, **38**, 42

ANSWERS TO PRACTICE EXERCISES

1. **(a)** linear **(b)** tetrahedral **(c)** trigonal planar **(d)** octahedral

2. **(a)** bent **(b)** square planar **(c)** distorted tetrahedron (seesaw)

3. The electron arrangement around each C that minimizes electron-pair repulsion is linear. Since there are no lone pairs around each C, the geometry is also **linear**.

4. **(a)** Yes, the molecule is polar.
   (b) No, the molecule is nonpolar.
   (c) No, the molecule is nonpolar.

5. **(a)** *sp* **(b)** sp^2

6. When you draw the Lewis structure for SF_4, you will find five electron pairs around the central atom, S (four bonding pairs and one lone pair). The electron arrangement that minimizes electron-pair repulsion is trigonal bipyramidal. You should conclude that S is dsp^3 hybridized because it has the electron arrangement of five dsp^3 hybrid orbitals.

7.

:N≡N:

The structure of N_2 is **linear**. We conclude that each N is ***sp* hybridized** because it has the electron arrangement of two *sp* hybrid orbitals. An *sp* orbital of one N atom overlaps with an *sp* orbital on the other N to form a sigma bond. The other *sp* orbital of each N contains the lone pair of electrons. Each nitrogen atom has two "pure" *p* orbitals that did not mix with the *s* orbital. The two *p* orbitals on one N can form two pi bonds by overlapping side-to-side with the two *p* orbitals on the other N atom. The triple bond is composed of one σ bond and two π bonds.

SOLUTIONS TO SELECTED TEXT PROBLEMS

10.8 Molecular Geometry, Problem Type 1.

In each case, you should use the following approach.

Step 1: Write the Lewis structure of the molecule. The lone pairs on the terminal atoms have been omitted.

Step 2: Count the number of electron pairs around the central atom.

Step 3: Build a model, or consult Table 10.1 of your text to predict the *geometry* of the molecule.

	Lewis structure	Electron pairs on central atom	Electron arrangement	Lone pairs	Geometry
(a)	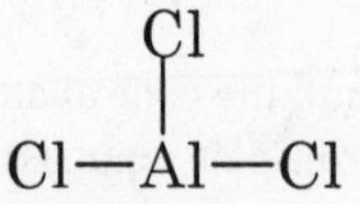	3	trigonal planar	0	trigonal planar, AB_3
(b)	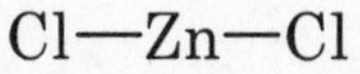	2	linear	0	linear, AB_2
(c)	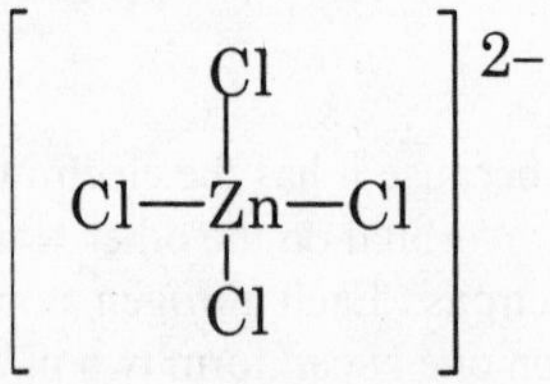	4	tetrahedral	0	tetrahedral, AB_4

10.10

(a) AB_4 tetrahedral

(b) AB_2E_2 bent

(c) AB_3 trigonal planar

(d) AB_2E_3 linear

(e) AB_4E_2 square planar

(f) AB_4 tetrahedral

(g) AB_5 trigonal bipyramid

(h) AB_3E trigonal pyramid

(i) AB_4 tetrahedral

10.12 Only molecules with four bonds to the central atom and no lone pairs are tetrahedral (AB_4).

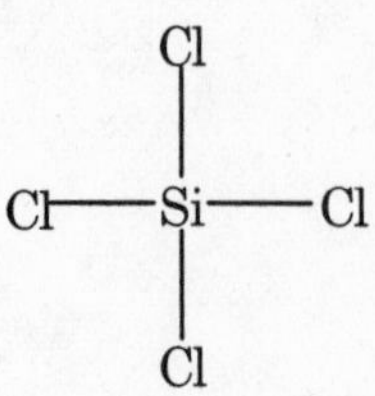

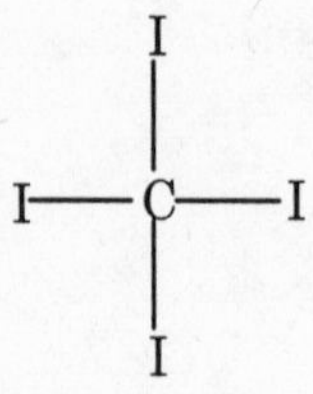

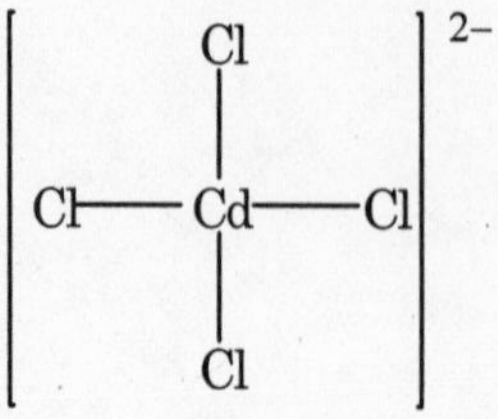

What are the Lewis structures and shapes for XeF_4 and SeF_4?

10.18 The electronegativity of the halogens decreases from F to I. Thus, the polarity of the H–X bond (where X denotes a halogen atom) also decreases from HF to HI. This difference in electronegativity accounts for the decrease in dipole moment.

10.20 Draw the Lewis structures. Both molecules are linear (AB_2). In CS_2, the two C–S bond moments are equal in magnitude and opposite in direction. The sum or resultant dipole moment will be *zero*. Hence, CS_2 is a **nonpolar molecule**. Even though OCS is linear, the C–O and C–S bond moments are not exactly equal, and there will be a small net dipole moment. Hence, OCS has a larger dipole moment than CS_2 (zero).

10.22 Predicting Dipole Moments, Problem Type 2.

Each vertex of the hexagonal structure of benzene represents the location of a C atom. Around the ring, there is no difference in electronegativity between C atoms, so the only bonds we need to consider are the polar C–Cl bonds.

The molecules shown in **(b)** and **(d)** are nonpolar. Due to the high symmetry of the molecules and the equal magnitude of the bond moments, the bond moments in each molecule cancel one another. The resultant dipole moment will be *zero*. For the molecules shown in **(a)** and **(c)**, the bond moments do not cancel and there will be net dipole moments. The dipole moment of the molecule in **(a)** is larger than that in **(c)**, because in **(a)** all the bond moments point in the same relative direction, reinforcing each other (see Lewis structure below). Therefore, the order of increasing dipole moments is:

(b) = (d) < (c) < (a).

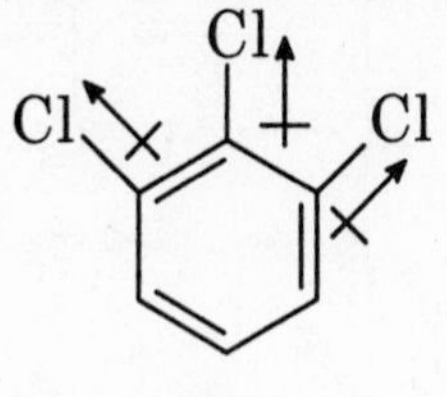

(a)

10.32 Hybridization of Atomic Orbitals, Problem Type 3.

(a)
Step 1: Write the Lewis structure of the molecule.

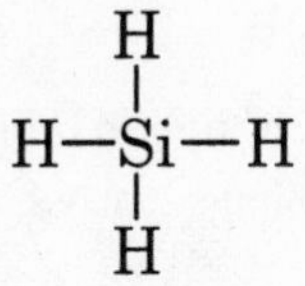

Step 2: Count the number of electron pairs around the central atom. Since there are four electron pairs around Si, the electron arrangement that minimizes electron-pair repulsion is **tetrahedral.**

We conclude that Si is ***sp***^{3} **hybridized** because it has the electron arrangement of four sp^3 hybrid orbitals.

(b)
Step 1: Write the Lewis structure of the molecule.

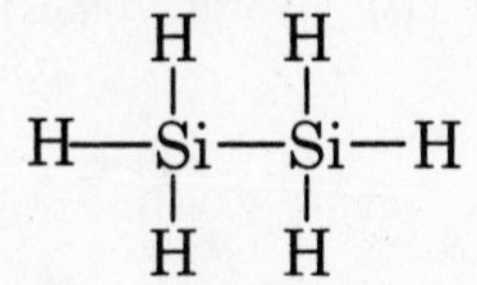

Step 2: Count the number of electron pairs around the "central atoms". Since there are four electron pairs around each Si, the electron arrangement that minimizes electron-pair repulsion for each Si is **tetrahedral.**

We conclude that each Si is ***sp***^{3} **hybridized** because it has the electron arrangement of four sp^3 hybrid orbitals.

10.34 Draw the Lewis structures. Before the reaction, boron is sp^2 hybridized (trigonal planar electron arrangement) in BF_3 and nitrogen is sp^3 hybridized (tetrahedral electron arrangement) in NH_3. After the reaction, boron and nitrogen are both sp^3 hybridized (tetrahedral electron arrangement).

10.36 **(a)** Each carbon has four bond pairs and no lone pairs and therefore has a tetrahedral electron pair arrangement. This implies sp^3 hybrid orbitals.

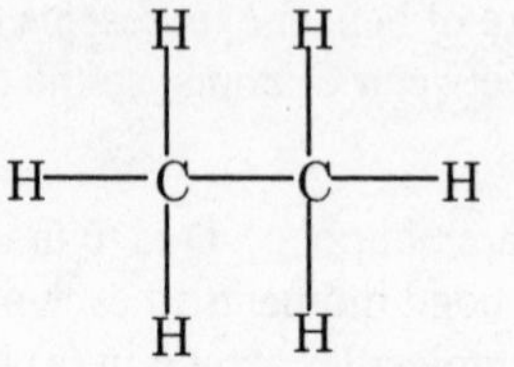

(b) The left-most carbon is tetrahedral and therefore has sp^3 hybrid orbitals. The two carbon atoms connected by the double bond are trigonal planar with sp^2 hybrid orbitals.

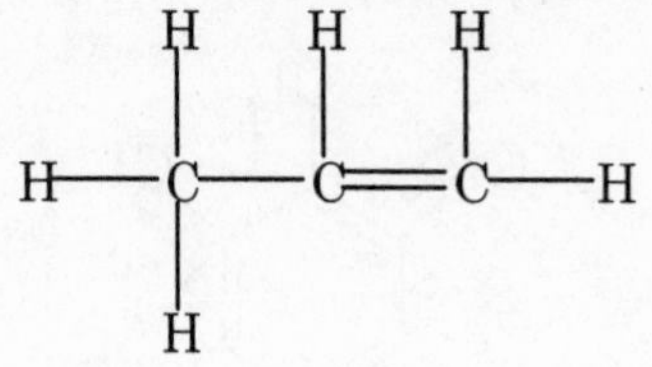

(c) This is just like (a) above. Both have sp^3 hybrid orbitals. What are the hybrid orbitals on the oxygen?

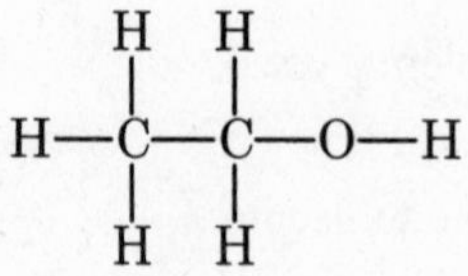

(d) The left-most carbon is tetrahedral (sp^3 hybrid orbitals). The carbon connected to oxygen is trigonal planar (why?) and has sp^2 hybrid orbitals.

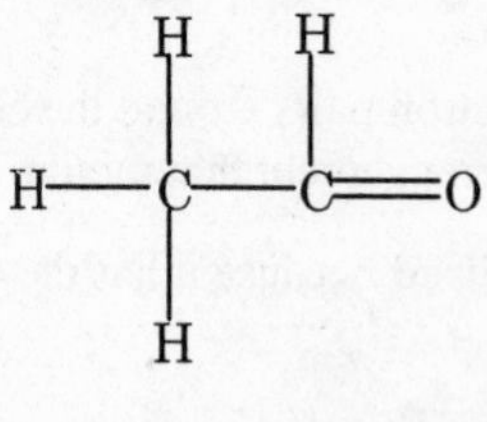

(e) The left-most carbon is tetrahedral (sp^3 hybrid orbitals). The other carbon is trigonal planar with sp^2 hybridized orbitals.

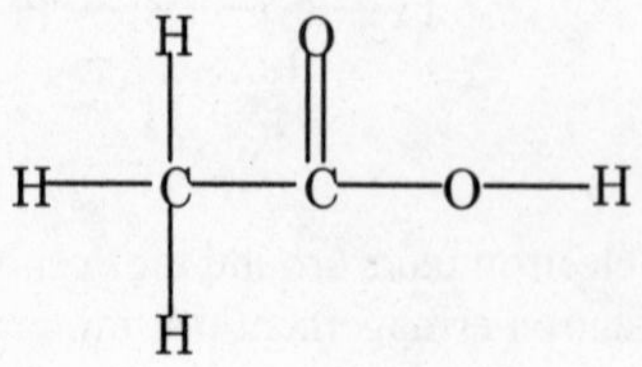

10.38 Hybridization of Atomic Orbitals, Problem Type 3.

Step 1: Write the Lewis structure of the molecule. Several resonance forms with formal charges are shown.

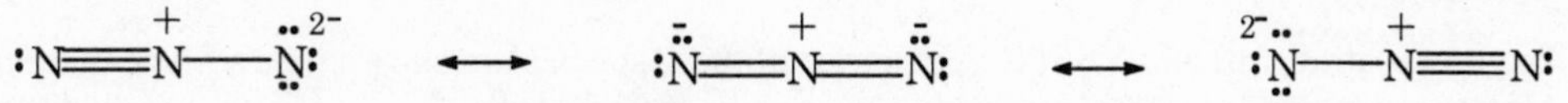

Step 2: Count the number of electron pairs around the central atom. Since there are two electron pairs around N, the electron arrangement that minimizes electron-pair repulsion is **linear** (AB_2). Remember, for VSEPR purposes a multiple bond counts the same as a single bond.

We conclude that N is ***sp* hybridized** because it has the electron arrangement of two *sp* hybrid orbitals.

10.40 Hybridization of Atomic Orbitals, Problem Type 3.

Step 1: Write the Lewis structure of the molecule.

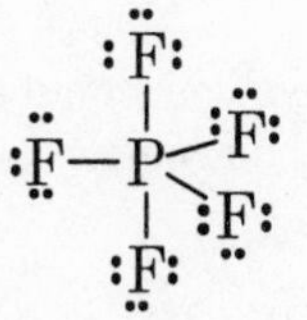

Step 2: Count the number of electron pairs around the central atom. Since there are five electron pairs around P, the electron arrangement that minimizes electron-pair repulsion is **trigonal bipyramidal** (AB_5).

We conclude that P is **sp^3d hybridized** because it has the electron arrangement of five sp^3d hybrid orbitals.

10.42 A single bond is usually a sigma bond, a double bond is usually a sigma bond and a pi bond, and a triple bond is always a sigma bond and two pi bonds. Therefore, there are nine pi bonds and nine sigma bonds in the molecule.

10.44 Molecular Geometry, Problem Type 1.

Step 1: Write the Lewis structure of the molecule (see Problem Type 2, Chapter 9).

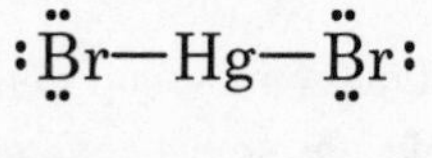

Step 2: Count the number of electron pairs around the central atom. There are two electron pairs around Hg.

Step 3: Since there are two electron pairs around Hg, the electron-pair arrangement that minimizes electron-pair repulsion is **linear**.

In addition, since there are no lone pairs around the central atom, the geometry is also **linear** (AB_2).

You could establish the geometry of $HgBr_2$ by measuring its dipole moment. If mercury(II) bromide were bent, it would have a measurable dipole moment. Experimentally, it has no dipole moment and therefore must be linear.

10.46 According to valence bond theory, a pi bond is formed through the side-to-side overlap of a pair of *p* orbitals. As atomic size increases, the distance between atoms is too large for *p* orbitals to overlap effectively in a side-to-side fashion. If two orbitals overlap poorly, that is, they share very little space in common, then the resulting bond will be very weak. This situation applies in the case of pi bonds between silicon atoms as well as between any other elements not found in the second period. It is usually far more energetically favorable for silicon, or any other heavy element, to form two single (sigma) bonds to two other atoms than to form a double bond (sigma + pi) to only one other atom.

10.48 The Lewis structures and VSEPR geometries of these species are shown below. The three nonbonding pairs of electrons on each fluorine atom have been omitted for simplicity.

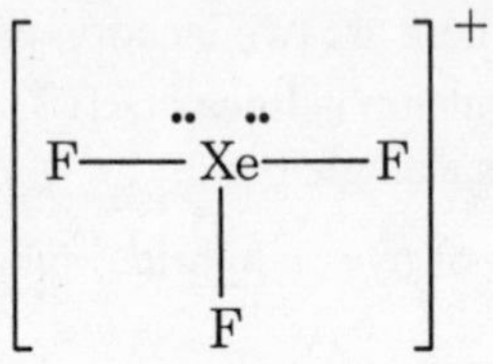

AB_3E_2
T-shaped

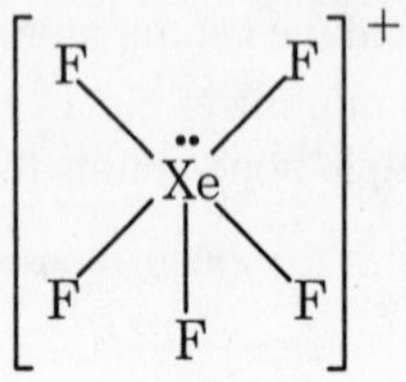

AB_5E
Square Pyramid

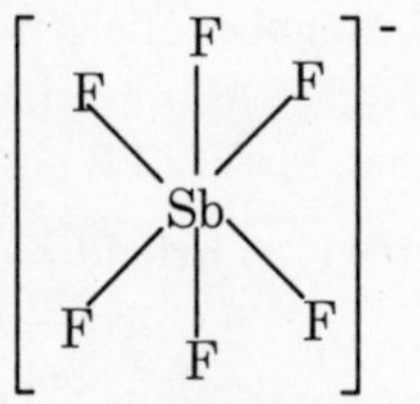

AB_6
Octahedral

10.50 To predict the bond angles for the molecules, you would have to draw the Lewis structure and determine the geometry using the VSEPR model. From the geometry, you can predict the bond angles.

(a) $BeCl_2$: AB_2 type, 180° (linear).

(b) BCl_3: AB_3 type, 120° (trigonal planar).

(c) CCl_4: AB_4 type, 109.5° (tetrahedral).

(d) CH_3Cl: AB_4 type, 109.5° (tetrahedral with a possible slight distortion resulting from the different sizes of the chlorine and hydrogen atoms).

(e) Hg_2Cl_2: Each mercury atom is of the AB_2 type. The entire molecule is linear, 180° bond angles.

(f) $SnCl_2$: AB_2E type, roughly 120° (bent).

(g) SCl_2: AB_2E_2, roughly 109.5° (bent).

(h) H_2O_2: The atom arrangement is HOOH. Each oxygen atom is of the AB_2E_2 type and the H–O–O angles will be roughly 109.5°.

(i) SnH_4: AB_4 type, 109.5° (tetrahedral).

10.52 Since arsenic and phosphorus are both in the same group of the periodic table, this problem is exactly like Problem 10.40. AsF_5 is an AB_5 type molecule, so the geometry is trigonal bipyramidal. We conclude that As is ***sp^3d* hybridized** because it has the electron arrangement of five *sp^3d* hybrid orbitals.

10.54 Only ICl_2^- and $CdBr_2$ will be linear. The rest are bent.

10.56 **(a)** Hybridization of Atomic Orbitals, Problem Type 3.

Step 1: The geometry around each nitrogen is identical. To complete an octet of electrons around N, you must add a lone pair of electrons. Count the number of electron pairs around N. There are three electron pairs around each N.

Step 2: Since there are three electron pairs around N, the electron-pair arrangement that minimizes electron-pair repulsion is **trigonal planar**.

We conclude that each N is ***sp^2* hybridized** because it has the electron arrangement of three *sp^2* hybrid orbitals.

(b) Predicting Dipole Moments, Problem Type 2.

An N–F bond is polar because F is more electronegative than N. The structure on the right has a dipole moment because the two N–F bond moments do not cancel each other out and so the molecule has a net dipole moment. On the other hand, the two N–F bond moments in the left-hand structure cancel. The sum or resultant dipole moment will be *zero*.

10.58 In 1,2-dichloroethane, the two C atoms are joined by a sigma bond. Rotation about a sigma bond does not destroy the bond, and the bond is therefore free (or relatively free) to rotate. Thus, all angles are permitted and the molecule is nonpolar because the C–Cl bond moments cancel each other because of the averaging effect brought about by rotation. In *cis*-dichloroethylene the two C–Cl bonds are locked in position. The π bond between the C atoms prevents rotation (in order to rotate, the π bond must be broken, using an energy source such as light or heat). Therefore, there is no rotation about the C=C in *cis*-dichloroethylene, and the molecule is polar.

10.60 Both sulfur–to–oxygen bonds are double bonds which occupy more space than single bonds and the repulsion between the double bonds spread out the angle close to 120°.

10.62 O_3, CO, CO_2, NO_2, and $CFCl_3$ are greenhouse gases.

10.64

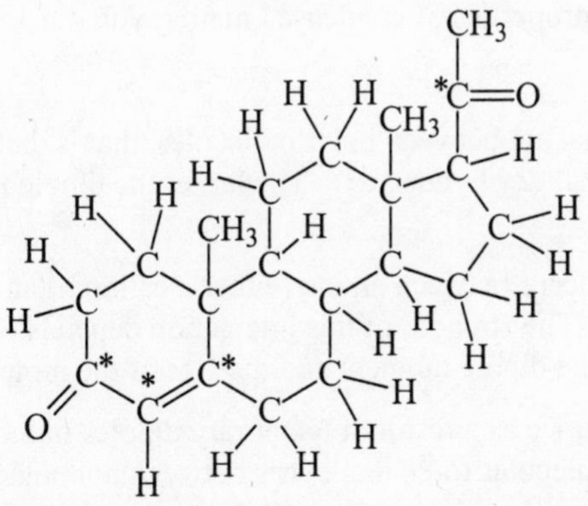

The carbons marked with the asterisks are sp^2 hybridized, the remaining carbons are sp^3 hybridized.

CHAPTER 11
INTERMOLECULAR FORCES AND LIQUIDS AND SOLIDS

PROBLEM-SOLVING STRATEGIES AND TUTORIAL SOLUTIONS

TYPES OF PROBLEMS

Problem Type 1: Identifying Intermolecular Forces.

Problem Type 2: Identifying Hydrogen Bonds.

Problem Type 3: Counting the Number of Atoms in a Unit Cell.

Problem Type 4: Calculating Density from Crystal Structure and Atomic Radius.

PROBLEM TYPE 1: IDENTIFYING INTERMOLECULAR FORCES

Intermolecular forces are attractive forces between molecules. Intermolecular forces account for the existence of the condensed states of matter--liquids and solids. To understand the properties of condensed matter, you must understand the different types of intermolecular forces.

(1) **Dipole-dipole forces:** These are attractive forces that act between polar molecules, that is, between molecules that possess dipole moments (see Section 11.2 of your text). The larger the dipole moments, the greater the attractive force.

(2) **Ion-dipole forces:** These are attractive forces that occur between an ion (either a cation or an anion) and a polar molecule (see Figure 11.2 of your text). The strength of this interaction depends on the charge and size of the ion and on the magnitude of the dipole moment and the size of the molecule.

(3) **Dispersion forces:** These are attractive forces that arise as a result of temporary dipoles induced in the atoms or molecules. This is the only type of intermolecular force that exists between nonpolar atoms or molecules. The likelihood of a dipole moment being induced depends on the polarizability of the atom or molecule. *Polarizability* is the ease with which the electron distribution in the atom or molecule can be distorted. Dispersion forces usually increase with molar mass.

For a complete discussion of intermolecular forces, see Section 11.2 of your text.

EXAMPLE 11.1
Indicate all the different types of intermolecular forces that exist in each of the following substances: (a) CCl_4 (*l*) and (b) HBr (*l*).

A good place to start with this type of problem is to determine if the molecule is polar or nonpolar. See Problem Type 2, Chapter 10 if you need to review this topic.

(a) CCl_4 is nonpolar. The only type of intermolecular forces present in nonpolar molecules is *dispersion forces*.

(b) HBr is a polar molecule. The types of intermolecular forces present are *dipole-dipole* and *dispersion forces*. There is no hydrogen bonding in HBr. The Br atom is too large and is not electronegative enough.

PRACTICE EXERCISE

1. Which of the following substances should have the strongest intermolecular attractive forces: N_2, Ar, F_2, or Cl_2?
2. The dipole moment (μ) in HCl is 1.03 D, and in HCN it is 2.99 D. Which substance should have the higher boiling point?

Text Problems: 8, 10, 16, **18**

PROBLEM TYPE 2: IDENTIFYING HYDROGEN BONDS

The boiling points of NH_3, H_2O, and HF are much higher than expected if the boiling point is solely based on molar mass (see Figure 11.7 of your text). The high boiling points in these compounds are due to extensive *hydrogen bonding*. The hydrogen bond is a special type of dipole-dipole interaction between the hydrogen atom in a polar bond, such as N–H, O–H, or F–H, and an electronegative O, N, or F atom. This interaction is written

$$A–H\cdots B \quad \text{or} \quad A–H\cdots A$$

where A and B represent O, N, or F. A–H is one molecule or part of a molecule and B is a part of another molecule. The dotted line represents the hydrogen bond.

The average energy of a hydrogen bond is quite large for a dipole-dipole interaction (up to 40 kJ/mol). Thus, hydrogen bonds are a powerful force in determining the structures and properties of many compounds.

EXAMPLE 11.2
Predict whether hydrogen bonding intermolecular attractions are present in the following substances: (a) CH_3OH (*l*) and (b) $CH_3CH_2OCH_2CH_3$ (*l*).

(a) CH_3OH (methanol) is polar and has a hydrogen atom bound to an oxygen atom. Hydrogen bonding intermolecular forces are present between CH_3OH molecules. Dipole-dipole and dispersion forces are also present.

(b) $CH_3CH_2OCH_2CH_3$ (diethyl ether) is polar and does contain both oxygen atoms and hydrogen atoms. However, all the hydrogen atoms are bonded to carbon, not oxygen. There is no hydrogen bonding in diethyl ether, because carbon is not electronegative enough. The intermolecular attractive forces present are dipole-dipole and dispersion forces.

PRACTICE EXERCISE

3. Which member of each pair has the stronger intermolecular forces of attraction: (a) H_2O or H_2S, (b) HCl or HF, and (c) NH_3 or PH_3?

Text Problems: 12, **14, 18**, 20, 32

PROBLEM TYPE 3: COUNTING THE NUMBER OF ATOMS IN A UNIT CELL

To solve this problem, you must realize that because every unit cell in a crystalline solid is adjacent to other unit cells, most of a cell's atoms are shared by neighboring cells. In cubic cells, each corner atom is shared by eight unit cells [see Figure 11.19(a) of your text]. A face-centered atom is shared by two unit cells [See Figure 11.19(b) of your text]. A center atom in a unit cell is solely contained by that unit cell and is not shared. An edge atom is shared by four unit cells. Table 11.1 summarizes this information.

TABLE 11.1

Type of atom	Amount of atom contained in unit cell
Corner	1/8
Edge	1/4
Face-centered	1/2
Center	1

EXAMPLE 11.4

If atoms of a solid occupy a face-centered cubic lattice, how many atoms are there per unit cell?

In a face-centered cubic unit cell, there are atoms at each of the eight corners, and there is one atom in each of the six faces.

(8 corner atoms)(1/8 atom per corner) + (6 face-centered atoms)(1/2 atom per face) = **4 atoms/unit cell**

PRACTICE EXERCISE

4. Atoms of polonium (Po) occupy a simple cubic lattice. How many Po atoms are there per unit cell?

Text Problems: 44, 50

PROBLEM TYPE 4: CALCULATING DENSITY FROM CRYSTAL STRUCTURE AND ATOMIC RADIUS

Step 1: To solve this type of problem, you must know how many atoms are contained in the different types of cubic unit cells. Table 11.2 summarizes the number of atoms per unit cell.

TABLE 11.2

Type of cubic unit cell	Number of atoms/unit cell
Simple cubic	1
Body-centered cubic	2
Face-centered cubic	4

Step 2: You can look up the relationship between edge length (a) and atomic radius (r). See Figure 11.20 of your text. Table 11.3 summarizes these relationships.

TABLE 11.3

Type of cubic unit cell	Relationship between a and r
Simple cubic	$a = 2r$
Body-centered cubic	$a = \frac{4}{\sqrt{3}}r$
Face-centered cubic	$a = \sqrt{8}\,r$

Step 3: Density = mass/volume. See Problem Type 1, Chapter 1. The volume of a cube is equal to the edge length cubed.

$$V = a^3$$

You will know how many atoms are in a unit cell (*Step 1* above). To calculate the mass, you need to convert from atoms/unit cell to grams/unit cell. A reasonable strategy would be

atoms/unit cell → mol/unit cell → grams/unit cell

Step 4: Substitute the mass and volume into the density equation to solve for the density.

EXAMPLE 11.3
Nickel crystallizes in a face-centered cubic lattice with an edge length of 352 pm. Calculate the density of nickel.

Step 1: A face-centered cubic unit cell contains four atoms per unit cell (see Table 11.2).

Step 2: Density = mass/volume. See Problem Type 1, Chapter 1. The volume of a cube is equal to the edge length cubed.

$$V = a^3$$

$$V = (325\text{ pm})^3 = 3.43 \times 10^7\text{ pm}^3$$

You should convert pm^3 to cm^3 because the density of a solid is typically expressed in units of g/cm^3.

$$3.43 \times 10^7\text{ pm}^3 \times \left(\frac{10^{-12}\text{ m}}{1\text{ pm}}\right)^3 \times \left(\frac{1\text{ cm}}{10^{-2}\text{ m}}\right)^3 = 3.43 \times 10^{-23}\text{ cm}^3$$

There are four atoms/unit cell. To calculate the mass, you need to convert from atoms/unit cell to grams/unit cell. A reasonable strategy would be:

atoms/unit cell → mol/unit cell → grams/unit cell

$$\frac{4\text{ Ni atoms}}{\text{unit cell}} \times \frac{1\text{ mol Ni}}{6.022 \times 10^{23}\text{ Ni atoms}} \times \frac{58.71\text{ g Ni}}{1\text{ mol Ni}} = 3.900 \times 10^{-22}\text{ g Ni/unit cell}$$

Step 3: Substitute the mass and volume into the density equation to solve for the density.

$$d = \frac{m}{V} = \frac{3.900 \times 10^{-22}\text{ g Ni}}{3.43 \times 10^{-23}\text{ cm}^3} = \mathbf{11.4\ g/cm^3}$$

PRACTICE EXERCISE

5. Potassium crystallizes in a body-centered cubic lattice and has a density of 0.856 g/cm^3 at 25°C.
   (a) How many atoms are there per unit cell?
   (b) What is the length of an edge of the unit cell?

Text Problems: 46, 48

ANSWERS TO PRACTICE EXERCISES

1. Cl_2 **2.** HCN **3.** **(a)** H_2O **(b)** HF **(c)** NH_3

4. 1 atom/unit cell **5.** **(a)** 2 atoms/unit cell **(b)** $a = 5.34 \times 10^{-8}\text{ cm} = 534\text{ pm}$

SOLUTIONS TO SELECTED TEXT PROBLEMS

11.8 Identifying Intermolecular Forces, Problem Type 1.

The three molecules are essentially nonpolar. There is little difference in electronegativity between carbon and hydrogen. Thus, the only type of intermolecular attraction in these molecules is dispersion forces. Other factors being equal, the molecule with the greater number of electrons will exert greater intermolecular attractions. By looking at the molecular formulas you can predict that the order of increasing boiling points will be $CH_4 < C_3H_8 < C_4H_{10}$.

On a very cold day, propane and butane would be liquids (boiling points −44.5°C and −0.5°C, respectively); only methane would still be a gas (boiling point −161.6°C).

11.10 **(a)** Benzene (C_6H_6) molecules are nonpolar. Only dispersion forces will be present.

(b) Chloroform (CH_3Cl) molecules are polar (why?). Dispersion and dipole-dipole forces will be present.

(c) Phosphorus trifluoride (PF_3) molecules are polar. Dispersion and dipole-dipole forces will be present.

(d) Sodium chloride (NaCl) is an ionic compound. Ion-ion (and dispersion) forces will be present.

(e) Carbon disulfide (CS_2) molecules are nonpolar. Only dispersion forces will be present.

11.12 In this problem you must identify the species capable of hydrogen bonding among themselves, not with water. In order for a molecule to be capable of hydrogen bonding with another molecule like itself, it must have at least one hydrogen atom bonded to N, O, or F. Of the choices, only **(e)** CH_3COOH (acetic acid) shows this structural feature. The others cannot form hydrogen bonds among themselves.

11.14 Identifying Hydrogen Bonds, Problem Type 2.

1-butanol has the higher boiling point because the molecules can form hydrogen bonds with each other. Diethyl ether molecules do contain both oxygen atoms and hydrogen atoms. However, all the hydrogen atoms are bonded to carbon, not oxygen. There is no hydrogen bonding in diethyl ether, because carbon is not electronegative enough.

11.16 **(a)** Xe: it has more electrons and therefore stronger dispersion forces.

(b) CS_2: it has more electrons (both molecules nonpolar) and therefore stronger dispersion forces.

(c) Cl_2: it has more electrons (both molecules nonpolar) and therefore stronger dispersion forces.

(d) LiF: it is an ionic compound, and the ion-ion attractions are much stronger than the dispersion forces between F_2 molecules.

(e) NH_3: it can form hydrogen bonds and PH_3 cannot.

11.18 Identifying Intermolecular Forces and Hydrogen Bonding, Problem Types 1 and 2.

(a) Water has O–H bonds. Therefore, water molecules can form hydrogen bonds. The attractive forces that must be overcome are hydrogen bonding and dispersion forces.

(b) Bromine (Br_2) molecules are nonpolar. Only dispersion forces must be overcome.

(c) Iodine (I_2) molecules are nonpolar. Only dispersion forces must be overcome.

(d) In this case, the F–F bond must be broken. This is an *intra*molecular force between two F atoms, not an *inter*molecular force between F_2 molecules. The attractive forces of the covalent bond must be overcome.

11.20 The lower melting compound (shown below) can form hydrogen bonds only with itself (*intra*molecular hydrogen bonds), as shown in the figure. Such bonds do not contribute to *inter*molecular attraction and do not help raise the melting point of the compound. The other compound can form *inter*molecular hydrogen bonds; therefore, it will take a higher temperature to provide molecules of the liquid with enough kinetic energy to overcome these attractive forces to escape into the gas phase.

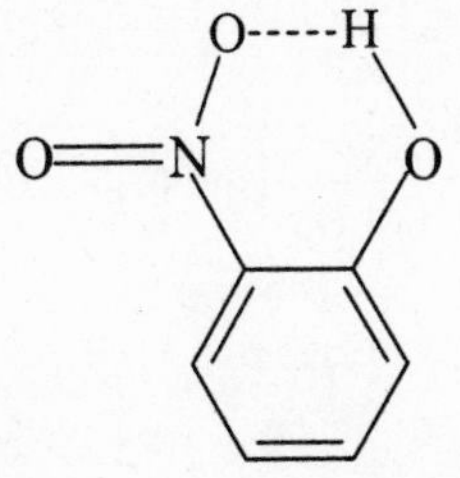

11.32 Ethylene glycol has two –OH groups, allowing it to exert strong intermolecular forces through hydrogen bonding. Its viscosity should fall between ethanol (1 OH group) and glycerol (3 OH groups).

11.44 A corner sphere is shared equally among eight unit cells, so only one-eighth of each corner sphere "belongs" to any one unit cell. A face-centered sphere is divided equally between the two unit cells sharing the face. A body-centered sphere belongs entirely to its own unit cell.

In a *simple cubic cell* there are eight corner spheres. One-eighth of each belongs to the individual cell giving a total of **one** whole sphere per cell. In a *body-centered cubic cell*, there are eight corner spheres and one body-center sphere giving a total of **two** spheres per unit cell (one from the corners and one from the body-center). In a *face-center* sphere, there are eight corner spheres and six face-centered spheres (six faces). The total number of spheres would be **four**: one from the corners and three from the faces.

11.46 Similar to Problem Type 4, Calculating the Density from Crystal Structure and Atomic Radius.

To solve this problem, we must first find the volume that a Ba atom occupies. Then, we need to calculate the volume one mole of Ba atoms occupies. Once we have these two pieces of information, we can multiply them together to end up with the number of Ba atoms per mol of Ba .

$$\frac{\text{number of Ba atoms}}{\cancel{\text{cm}^3}} \times \frac{\cancel{\text{cm}^3}}{1\text{ mol Ba}} = \frac{\text{number of Ba atoms}}{1\text{ mol Ba}}$$

Step 1: The volume that contains two barium atoms is the volume of the body-centered cubic unit cell. Some of this volume is empty space because packing is only 68.0 percent efficient. But, this will not affect our calculation.

$$V = a^3$$

Let's also convert to cm^3.

$$V = (502\ \cancel{\text{pm}})^3 \times \left(\frac{10^{-12}\ \cancel{\text{m}}}{1\ \cancel{\text{pm}}}\right)^3 \times \left(\frac{1\text{ cm}}{10^{-2}\ \cancel{\text{m}}}\right)^3 = 1.27 \times 10^{-22}\text{ cm}^3/2\text{ Ba atoms}$$

Step 2: The volume that contains one mole of barium atoms can be calculated from the density using the following strategy:

$$\frac{\text{volume}}{\text{mass of Ba}} \rightarrow \frac{\text{volume}}{\text{mol Ba}}$$

$$\frac{1\ \text{cm}^3}{3.50\ \text{g Ba}} \times \frac{137.34\ \text{g Ba}}{1\ \text{mol Ba}} = 39.2\ \text{cm}^3/\text{mol Ba}$$

Step 3: We can now calculate the number of barium atom in one mole using the strategy presented in *Step 1.*

$$\frac{\text{number of Ba atoms}}{\text{cm}^3} \times \frac{\text{cm}^3}{1\ \text{mol Ba}} = \frac{\text{number of Ba atoms}}{1\ \text{mol Ba}}$$

$$\left(\frac{2\ \text{Ba atoms}}{1.27 \times 10^{-22}\ \text{cm}^3}\right)\left(\frac{39.2\ \text{cm}^3}{1\ \text{mol Ba}}\right) = \mathbf{6.17 \times 10^{23}\ atoms / mol}$$

This is close to Avogadro's number, 6.022×10^{23} particles/mol.

11.48 The mass of the unit cell is the mass in grams of two europium atoms.

$$m = \left(\frac{2\ \text{Eu atoms}}{1\ \text{unit cell}}\right)\left(\frac{1\ \text{mol Eu}}{6.022 \times 10^{23}\ \text{Eu atoms}}\right)\left(\frac{152.0\ \text{g Eu}}{1\ \text{mol Eu}}\right) = 5.048 \times 10^{-22}\ \text{g Eu / unit cell}$$

$$V = \left(\frac{5.048 \times 10^{-22}\ \text{g}}{1\ \text{unit cell}}\right)\left(\frac{1\ \text{cm}^3}{5.26\ \text{g}}\right) = 9.60 \times 10^{-23}\ \text{cm}^3\ /\ \text{unit cell}$$

The edge length (a) is:

$$a = V^{1/3} = (9.60 \times 10^{-23}\ \text{cm}^3)^{1/3} = \mathbf{4.58 \times 10^{-8}\ cm = 458\ pm}$$

11.50 Similar to Problem Type 3, Counting the Number of Atoms in a Unit Cell.

In a face-centered cubic unit cell, there are atoms at each of the eight corners, and there is one atom in each of the six faces. Only one-half of each face-centered atom and one-eighth of each corner atom belongs to the unit cell.

X atoms/unit cell = (8 corner atoms)(1/8 atom per corner) = 1 X atom/unit cell

Y atoms/unit cell = (6 face-centered atoms)(1/2 atom per face) = 3 Y atoms/unit cell

The unit cell is the smallest repeating unit in the crystal; therefore, the empirical formula is $\mathbf{XY_3}$.

11.52 In diamond, each carbon atom is covalently bonded to four other carbon atoms. Because these bonds are strong and uniform, diamond is a very hard substance. In graphite, the carbon atoms in each layer are linked by strong bonds, but the layers are bound by weak dispersion forces. As a result, graphite may be cleaved easily between layers and is not hard.

11.72 *Step 1:* Warming ice to the melting point

$$q_1 = ms\Delta t = 866\ \text{g H}_2\text{O} \times 2.03\ \text{J/g°C} \times [0 - (-10)]\text{°C} = 17.6\ \text{kJ}$$

Step 2: Converting ice at the melting point to liquid water at 0°C (See Table 11.7 for the heat of fusion of water.)

$$q_2 = 866\,\text{g H}_2\text{O}\left(\frac{1\text{ mol}}{18.02\text{ g H}_2\text{O}}\right)\left(\frac{6.01\text{ kJ}}{1\text{ mol}}\right) = 289\text{ kJ}$$

Step 3: Heating water from 0°C to 100°C

$$q_3 = ms\Delta t = 866\text{ g H}_2\text{O} \times 4.184\text{ J/g°C} \times (100 - 0)\text{°C} = 362\text{ kJ}$$

Step 4: Converting water at 100°C to steam at 100°C (See Table 11.5 for the heat of vaporization of water.)

$$q_4 = 866\,\text{g H}_2\text{O}\left(\frac{1\text{ mol}}{18.02\text{ g H}_2\text{O}}\right)\left(\frac{40.79\text{ kJ}}{1\text{ mol}}\right) = 1.96 \times 10^3\text{ kJ}$$

Step 5: Heating steam from 100°C to 126°C

$$q_5 = ms\Delta t = 866\text{ g H}_2\text{O} \times 1.99\text{ J/g°C} \times (126 - 100)\text{°C} = 44.8\text{ kJ}$$

$$q_{\text{total}} = q_1 + q_2 + q_3 + q_4 + q_5 = \mathbf{2.67 \times 10^3\ kJ}$$

How would you set up and work this problem if you were computing the heat lost in cooling steam from 126°C to ice at −10°C?

11.74 $\Delta H_{\text{vap}} = \Delta H_{\text{sub}} - \Delta H_{\text{fus}} = 62.30\text{ kJ/mol} - 15.27\text{ kJ/mol} = \mathbf{47.03\ kJ/mol}$

11.76 Two phase changes occur in this process. First, the liquid is turned to solid (freezing), then the solid ice is turned to gas (sublimation).

11.78 When steam condenses to liquid water at 100°C, it releases a large amount of heat equal to the enthalpy of vaporization. Thus, steam at 100°C exposes one to more heat than an equal amount of water at 100°C.

11.80 We can use a modified form of the Clausius-Clapeyron equation to solve this problem. See Equation (11.4) in the text.

$P_1 = 40.1$ mmHg $\qquad$ $\mathbf{P_2 = ?}$

$T_1 = 7.6°\text{C} = 280.6\text{ K}$ $\qquad$ $T_2 = 60.6°\text{C} = 333.6\text{ K}$

$$\ln\frac{P_1}{P_2} = \frac{\Delta H_{\text{vap}}}{R}\left(\frac{1}{T_2} - \frac{1}{T_1}\right)$$

$$\ln\frac{40.1}{P_2} = \frac{31000\text{ J/mol}}{8.314\text{ J/K}\cdot\text{mol}}\left(\frac{1}{333.6\text{ K}} - \frac{1}{280.6\text{ K}}\right)$$

$$\ln\frac{40.1}{P_2} = -2.11$$

Taking the antilog of both sides, we have:

$$\frac{40.1}{P_2} = 0.121$$

$$\mathbf{P_2 = 331\ mmHg}$$

11.84 The pressure exerted by the blades on the ice lowers the melting point of the ice. A film of liquid water between the blades and the solid ice provides lubrication for the motion of the skater.

11.86 Region labels: The region containing point A is the solid region. The region containing point B is the liquid region. The region containing point C is the gas region.

(a) Raising the temperature at constant pressure beginning at A implies starting with solid ice and warming until melting occurs. If the warming continued, the liquid water would eventually boil and change to steam. Further warming would increase the temperature of the steam.

(b) At point C water is in the gas phase. Cooling without changing the pressure would eventually result in the formation of solid ice. Liquid water would never form.

(c) At B the water is in the liquid phase. Lowering the pressure without changing the temperature would eventually result in boiling and conversion to water in the gas phase.

11.88 **(a)** Boiling liquid ammonia requires breaking hydrogen bonds between molecules. Dipole–dipole and dispersion forces must also be overcome.

(b) P_4 is a nonpolar molecule, so the only intermolecular forces are of the dispersion type.

(c) CsI is an ionic solid. To dissolve in any solvent, ion–ion interparticle forces must be overcome. Also, hydrogen bonding and dispersion forces must be overcome between HF liquid molecules for them to solvate CsI solid.

(d) Metallic bonds must be broken.

11.90 The HF molecules are held together by strong intermolecular hydrogen bonds. Therefore, liquid HF has a lower vapor pressure than liquid HI. (The HI molecules do not form hydrogen bonds with each other.)

11.92 Reading directly from the graph: **(a)** solid; **(b)** vapor.

11.94 Because the critical temperature of CO_2 is only 31°C, the liquid CO_2 in the fire extinguisher is converted to a critical fluid on a hot summer day.

11.96 As the vacuum pump is turned on and the pressure is reduced, the liquid will begin to boil because the vapor pressure of the liquid is greater than the external pressure (approximately zero). The heat of vaporization is supplied by the water, and thus the water cools. Soon the water loses sufficient heat to drop the temperature below the freezing point. Finally, the ice sublimes under reduced pressure.

11.98 **(a)** Two triple points: Diamond/graphite/liquid and graphite/liquid/vapor.
(b) Diamond.
(c) Apply high pressure at high temperature.

11.100 W must be a reasonably non-reactive metal. It conducts electricity and is malleable, but doesn't react with nitric acid. Of the choices, it must be gold.

X is nonconducting (and therefore isn't a metal), is brittle, is high melting, and reacts with nitric acid. Of the choices, it must be lead sulfide.

Y doesn't conduct and is soft (and therefore is a nonmetal). It melts at a low temperature with sublimation. Of the choices, it must be iodine.

Z doesn't conduct, is chemically inert, and is high melting (network solid). Of the choices, it must be mica (SiO_2).

Would the colors of the species have been any help in determining their identity?

11.102 Sublimation temperature is −78°C or 195 K at a pressure of 1 atm.

$$\ln\frac{P_1}{P_2} = \frac{\Delta H_{\text{sub}}}{R}\left(\frac{1}{T_2} - \frac{1}{T_1}\right)$$

$$\ln\frac{1}{P_2} = \frac{25.9 \times 10^3\ \text{J/mol}}{8.314\ \text{J/mol}\cdot\text{K}}\left(\frac{1}{150\ \text{K}} - \frac{1}{195\ \text{K}}\right)$$

$$\ln\frac{1}{P_2} = 4.79$$

Taking the antilog of both sides gives:

$$\mathbf{P_2 = 8.3 \times 10^{-3}\ atm}$$

11.104 Smaller ions have more concentrated charges (charge densities) and are more effective in ion-dipole interaction. The greater the ion-dipole interaction, the larger is the heat of hydration.

11.106 $3Hg(l) + O_3(g) \rightarrow 3HgO(s)$

Conversion to solid HgO changes its surface tension.

11.108 The moles of water vapor can be calculate using the ideal gas equation.

$$n = \frac{PV}{RT} = \frac{\left(187.5\ \text{mmHg} \times \dfrac{1\ \text{atm}}{760\ \text{mmHg}}\right)(5.00\ \text{L})}{\left(0.0821\dfrac{\text{L}\cdot\text{atm}}{\text{mol}\cdot\text{K}}\right)(338\ \text{K})} = 0.0445\ \text{mol}$$

$$\text{mass of water vapor} = 0.0445\ \text{mol} \times 18.02\ \text{g/mol} = 0.802\ \text{g}$$

Now, we can calculate the percentage of the 1.20 g sample of water that is vapor.

$$\textbf{\% of H}_2\textbf{O vaporized} = \frac{0.802\ \text{g}}{1.20\ \text{g}} \times 100\% = \mathbf{66.8\%}$$

11.110 The packing efficiency is: $\dfrac{\text{volume of atoms in unit cell}}{\text{volume of unit cell}} \times 100\%$

An atom is assumed to be spherical, so the volume of an atom is $(4/3)\pi r^3$. The volume of a cubic unit cell is a^3 (a is the length of the cube edge). The packing efficiencies are calculated below:

(a) Simple cubic cell: cell edge (a) = $2r$

$$\text{Packing efficiency} = \frac{\left(\dfrac{4\pi r^3}{3}\right)\times 100\%}{(2r)^3} = \frac{4\pi r^3 \times 100\%}{24r^3} = \frac{\pi}{6}\times 100\% = \mathbf{52.4\%}$$

(b) Body-centered cubic cell: cell edge $= \dfrac{4r}{\sqrt{3}}$

$$\text{Packing efficiency} = \frac{2\times\left(\dfrac{4\pi r^3}{3}\right)\times 100\%}{\left(\dfrac{4r}{\sqrt{3}}\right)^3} = \frac{2\times\left(\dfrac{4\pi r^3}{3}\right)\times 100\%}{\left(\dfrac{64r^3}{3\sqrt{3}}\right)} = \frac{2\pi\sqrt{3}}{16}\times 100\%$$

$$= \mathbf{68.0\%}$$

Remember, there are two atoms per body-centered cubic unit cell.

(c) Face-centered cubic cell: cell edge $= \sqrt{8}r$

$$\text{Packing efficiency} = \frac{4\times\left(\dfrac{4\pi r^3}{3}\right)\times 100\%}{\left(\sqrt{8}\,r\right)^3} = \frac{\left(\dfrac{16\pi r^3}{3}\right)\times 100\%}{8r^3\sqrt{8}} = \frac{2\pi}{3\sqrt{8}}\times 100\% = \mathbf{74.0\%}$$

Remember, there are four atoms per face-centered cubic unit cell.

CHAPTER 12
PHYSICAL PROPERTIES OF SOLUTIONS

PROBLEM-SOLVING STRATEGIES AND TUTORIAL SOLUTIONS

TYPES OF PROBLEMS

Problem Type 1: Predicting Solubility Based on Intermolecular Forces.

Problem Type 2: Types of Concentration Units.
(a) Percent by mass.
(b) Molarity.
(c) Molality.

Problem Type 3: Converting between Concentration Units.
(a) Converting molality to molarity.
(b) Converting molarity to molality.
(c) Converting percent by mass to molality.

Problem Type 4: Effect of Pressure on Solubility: Henry's Law.

Problem Type 5: Colligative Properties of Nonelectrolytes.
(a) Vapor-pressure lowering, Raoult's law.
(b) Boiling-point elevation.
(c) Freezing-point depression.
(d) Osmotic pressure.

Problem Type 6: Determining Molar Mass Using Colligative Properties.
(a) Calculating molar mass from freezing-point depression.
(b) Calculating molar mass from osmotic pressure.

Problem Type 7: Colligative Properties of Electrolytes.

PROBLEM TYPE 1: PREDICTING SOLUBILITY BASED ON INTERMOLECULAR FORCES

Two substances with intermolecular forces of similar type and magnitude are likely to be soluble in each other. The saying "*like dissolves like*" will help you predict the solubility of a substance in a solvent.

EXAMPLE 12.1
Which of the following would be a better solvent for molecular I_2 (s): CCl_4 or H_2O?

I_2 is a nonpolar molecule (see Problem Type 2, Chapter 10). Using the like-dissolves-like rule, I_2 will be more soluble in the nonpolar solvent, CCl_4, than in the polar solvent, H_2O.

PRACTICE EXERCISE

1. In which solvent will NaBr be more soluble, benzene (C_6H_6) or water?

Text Problems: 8, 10

PROBLEM TYPE 2: TYPES OF CONCENTRATION UNITS

We will focus on three of the most common units of concentration: percent by mass, molarity, and molality.

A. Percent by mass

The percent by mass is defined as:

$$\textbf{percent by mass of solute} = \frac{\text{mass of solute}}{\text{mass of solute} + \text{mass of solvent}} \times 100\% \qquad (12.1)$$

$$= \frac{\text{mass of solute}}{\text{mass of soln}} \times 100\%$$

The percent by mass has no units because it is a ratio of two similar quantities.

EXAMPLE 12.2

The dehydrated form of Epsom salt is magnesium sulfate. What is the percent $MgSO_4$ by mass in a solution made from 16.0 g $MgSO_4$ and 100 mL of H_2O at 25°C? The density of water at 25°C is 0.997 g/mL.

Step 1: Calculate the mass of 100 mL of water using the density of water as a conversion factor.

$$? \text{ g of water} = 100 \text{ mL } H_2O \times \frac{0.997 \text{ g } H_2O}{1 \text{ mL } H_2O} = 99.7 \text{ g } H_2O$$

Step 2: Substitute the mass of solute and the mass of solvent into Equation (12.1) to calculate the percent by mass of $MgSO_4$.

$$\textbf{percent by mass } \mathbf{MgSO_4} = \frac{\text{mass of solute}}{\text{mass of solute} + \text{mass of solvent}} \times 100\%$$

$$= \frac{16.0 \text{ g}}{16.0 \text{ g} + 99.7 \text{ g}} \times 100\% = \mathbf{13.8\ \%}$$

PRACTICE EXERCISE

2. An aqueous solution contains 167 g $CuSO_4$ in 820 mL of solution. The density of the solution is 1.195 g/mL. Calculate the percent $CuSO_4$ by mass in the solution.

Text Problem: 14

B. Molarity (*M*)

We have already defined molarity in Chapter 4 (see Problem Type 5, Chapter 4). Molarity is defined as

$$\text{molarity} = \frac{\text{moles of solute}}{\text{liters of soln}} \qquad (12.2)$$

Molarity has units of mol/L.

EXAMPLE 12.3

What is the molarity of the $MgSO_4$ solution made in Example 12.2? Assume that the density remains unchanged upon addition of $MgSO_4$ to the water.

Step 1: Calculate the number of moles $MgSO_4$ in 16.0 g. Use the molar mass of $MgSO_4$ as a conversion factor.

$$? \text{ mol } MgSO_4 = 16.0 \text{ g } MgSO_4 \times \frac{1 \text{ mol } MgSO_4}{120.38 \text{ g } MgSO_4} = 0.133 \text{ mol } MgSO_4$$

Step 2: Calculate the volume of the solution. The total mass of the solution is:

$$16.0 \text{ g } MgSO_4 + \left(100 \text{ mL } H_2O \times \frac{0.997 \text{ g } H_2O}{1 \text{ mL } H_2O}\right) = 115.7 \text{ g}$$

Assuming that the density of the solution is the same as that of water, we can calculate the volume of the solution as follows:

$$\text{Volume of solution} = 115.7 \text{ g} \times \frac{1 \text{ mL}}{0.997 \text{ g}} = 116.0 \text{ mL soln}$$

Step 3: Calculate the molarity of the solution by substituting the mol of solute and volume of solution (in L) into Equation (12.2).

$$M = \frac{\text{moles of solute}}{\text{liters of soln}} = \frac{0.133 \text{ mol}}{0.116 \text{ L}} = \mathbf{1.15}\ \boldsymbol{M}$$

PRACTICE EXERCISE

3. An aqueous solution contains 167 g $CuSO_4$ in 820 mL of solution. The density of the solution is 1.195 g/mL. Calculate the molarity of the solution.

C. Molality (m)

Molality is defined as the number of moles of solute in 1 kg of solvent.

$$\text{molality} = \frac{\text{moles of solute}}{\text{mass of solvent (kg)}} \qquad (12.3)$$

EXAMPLE 12.4

What is the molality of the $MgSO_4$ solution made in Example 12.2?

Step 1: Calculate the number of moles $MgSO_4$ in 16.0 g. Use the molar mass of $MgSO_4$ as a conversion factor.

$$? \text{ mol } MgSO_4 = 16.0 \text{ g } MgSO_4 \times \frac{1 \text{ mol } MgSO_4}{120.38 \text{ g } MgSO_4} = 0.133 \text{ mol } MgSO_4$$

Step 2: Calculate the mass (in kg) of 100 mL of H_2O. Use the density of water as a conversion factor.

$$? \text{ mass of } H_2O = 100 \text{ mL } H_2O \times \frac{0.997 \text{ g } H_2O}{1 \text{ mL } H_2O} \times \frac{1 \text{ kg}}{1000 \text{ g}} = 0.0997 \text{ kg } H_2O$$

Step 3: Calculate the molality by substituting the mol of solute and the mass of solvent (in kg) into Equation (12.3).

$$m = \frac{\text{moles of solute}}{\text{mass of solvent (kg)}} = \frac{0.133 \text{ mol}}{0.0997 \text{ kg}} = \mathbf{1.33}\ \boldsymbol{m}$$

PRACTICE EXERCISE

4. An aqueous solution contains 167 g $CuSO_4$ in 820 mL of solution. The density of the solution is 1.195 g/mL. Calculate the molality of the solution.

Text Problems: 16, 18

PROBLEM TYPE 3: CONVERTING BETWEEN CONCENTRATION UNITS

There are advantages and disadvantages to each type of concentration unit. An advantage of molality and percent by mass is that the concentration is temperature independent. On the other hand, molarity changes with temperature, because solution volume typically increases with increasing temperature. However, the advantage of molarity is that it is generally easier to measure the volume of solution than to weigh the solvent or solution.

A. Converting molality to molarity

EXAMPLE 12.5
Calculate the molarity of a 2.44 *m* NaCl solution given that its density is 1.089 g/mL.

Step 1: The mass of the solution must be converted to volume of solution to convert from molality to molarity. To calculate the mass of the solution, you must calculate the mass of the solute and then add that to the mass of water (1 kg = 1000 g). Remember that 2.44 *m* means 2.44 mol of solute per 1 kg (1000 g) of solvent.

Use the molar mass of the solute as a conversion factor to convert from moles of solute to mass of solute.

$$? \text{ mass of solute} = 2.44 \text{ mol NaCl} \times \frac{58.44 \text{ g NaCl}}{1 \text{ mol NaCl}} = 143 \text{ g NaCl}$$

$$? \text{ mass of solution} = \text{mass of solute} + \text{mass of solvent}$$

$$= 143 \text{ g} + 1000 \text{ g} = 1143 \text{ g}$$

Step 2: From the mass of the solution, you can calculate the volume of solution using the solution density as a conversion factor.

$$? \text{ volume of solution} = 1143 \text{ g} \times \frac{1 \text{ mL}}{1.089 \text{ g}} = 1.050 \times 10^3 \text{ mL} = 1.050 \text{ L}$$

Step 3: The moles of solute are given in the molality (2.44 mol), and the volume of solution was calculated in *Step 2*. Divide moles of solute by liters of solution to calculate the molarity of the solution.

$$M = \frac{\text{mol of solute}}{\text{L of soln}} = \frac{2.44 \text{ mol}}{1.050 \text{ L}} = \mathbf{2.32}\ \boldsymbol{M}$$

PRACTICE EXERCISE

5. Concentrated hydrochloric acid is 15.7 *m*. Calculate the molarity of concentrated HCl given that its density is 1.18 g/mL.

Text Problem: 20

B. Converting molarity to molality

EXAMPLE 12.6
Calculate the molality of a 2.55 *M* NaCl solution given that its density is 1.089 g/mL.

Step 1: From the volume of solution, you can calculate the mass of the solution using the solution density as a conversion factor. Remember that 1.55 *M* means 1.55 mol of solute per 1 L (1000 mL) of solution.

$$? \text{ mass of solution} = 1000 \text{ mL soln} \times \frac{1.089 \text{ g}}{1 \text{ mL soln}} = 1089 \text{ g}$$

Step 2: To calculate the mass of the *solvent* (water), you need to subtract the mass due to the *solute* from the mass of solution calculated in *Step 1*. You can calculate the mass of the solute from the moles of solute using molar mass as a conversion factor.

$$? \text{ mass of solute} = 2.55 \text{ mol NaCl} \times \frac{58.44 \text{ g NaCl}}{1 \text{ mol NaCl}} = 149 \text{ g NaCl}$$

$$? \text{ mass of solvent} = \text{mass of soln} - \text{mass of solute}$$

$$= 1089 \text{ g} - 149 \text{ g} = 9.40 \times 10^2 \text{ g} = 0.940 \text{ kg solvent}$$

Step 3: The moles of solute are given in the molarity (2.55 mol). Divide moles of solute by mass of solvent in kg to calculate the molality of the solution.

$$m = \frac{\text{mol of solute}}{\text{kg of solvent}} = \frac{2.55 \text{ mol}}{0.940 \text{ kg}} = \mathbf{2.71}\ \boldsymbol{m}$$

PRACTICE EXERCISE

6. Concentrated sulfuric acid, H_2SO_4 is 18.0 *M*. Calculate the molality of concentrated H_2SO_4 given that its density is 1.83 g/mL.

C. Converting percent by mass to molality

EXAMPLE 12.7

Concentrated hydrochloric acid is 36.5 percent HCl by mass. Its density is 1.18 g/mL. Calculate the molality of concentrated HCl.

Step 1: Assume that the mass of the solution is 100.0 g. The percent by mass of the solute can then be converted directly to grams. Then, the mass of solute can be converted to moles of solute using molar mass as a conversion factor.

Assuming 100.0 g solution,

$$36.5 \text{ g HCl} \times \frac{1 \text{ mol HCl}}{36.46 \text{ g HCl}} = 1.00 \text{ mol HCl}$$

Step 2: The mass of the solvent (water) can then be calculated by subtracting the mass of solute from the mass of the solution (100.0 g). Remember, the mass of the solution is equal to the mass of solute + mass of solvent.

$$? \text{ mass of solvent} = \text{mass of solution} - \text{mass of solute}$$

$$= 100.0 \text{ g} - 36.5 \text{ g} = 63.5 \text{ g} = 0.0635 \text{ kg solvent (water)}$$

Step 3: Moles of solute was calculated in *Step 1*, and mass of solvent was calculated in *Step 2*. Divide moles of solute by mass of solvent in kg to calculate the molality of the solution.

$$m = \frac{\text{mol of solute}}{\text{kg of solvent}} = \frac{1.00 \text{ mol}}{0.0635 \text{ kg}} = \mathbf{15.7}\ \boldsymbol{m}$$

PRACTICE EXERCISE

7. What is the molality of a 3.0 percent hydrogen peroxide (H_2O_2) aqueous solution? The density of the solution is 1.0 g/mL.

Text Problem: 20

PROBLEM TYPE 4: EFFECT OF PRESSURE ON SOLUBILITY: HENRY'S LAW

Solubility is defined as the maximum amount of a solute that will dissolve in a given quantity of solvent at a specific temperature. For all practical purposes, external pressure has no influence on the solubilities of liquids and solids, but it does greatly affect the solubility of gases. There is a quantitative relationship between gas solubility and pressure called **Henry's law**, which states that the solubility of a gas in a liquid is proportional to the pressure of the gas over the solution:

$$c = kP$$

where,

- c is the molar concentration (mol/L) of the dissolved gas.
- P is the pressure, in atmospheres, of the gas over the solution.
- k is a constant for a given gas that depends only on the temperature. k has units of mol/L·atm.

As the pressure of the gas over the solution increases, more gas molecules strike the surface of the liquid increasing the number of gas molecules that dissolve in the liquid. For further discussion, see Section 12.5 of your text.

EXAMPLE 12.8

What is the concentration of O_2 at 25°C in water that is saturated with *air* at an atmospheric pressure of 645 mmHg? The Henry's law constant (k) for oxygen is 3.5×10^{-4} mol/L·atm. Assume that the mole fraction of oxygen in air is 0.209.

Step 1: We want to calculate the molar concentration (c) of O_2 in water. Using Henry's law, $c_{O_2} = kP_{O_2}$. k is given in the problem. To calculate the molar concentration, we must first calculate the partial pressure of oxygen in air. The partial pressure of O_2 is found using Dalton's law of partial pressures (see Problem Type 3, Chapter 5).

$$P_{O_2} = X_{O_2} P_T = (0.209)(645 \text{ mmHg}) \times \frac{1 \text{ atm}}{760 \text{ mmHg}} = 0.177 \text{ atm}$$

Step 2: Substitute the partial pressure of O_2 and k into the Henry's law expression to solve for the molar concentration of O_2 in water.

$$c_{O_2} = kP_{O_2} = (3.5 \times 10^{-4} \text{ mol/L·atm})(0.177 \text{ atm}) = \mathbf{6.2 \times 10^{-5} \text{ mol/L}}$$

PRACTICE EXERCISE

8. The Henry's law constant for CO is 9.73×10^{-4} mol/L·atm at 25°C. What is the concentration of dissolved CO in water if the partial pressure of CO in the air is 0.015 mmHg?

Text Problems: 34, 36

PROBLEM TYPE 5: COLLIGATIVE PROPERTIES OF NONELECTROLYTES

Several important properties of solutions depend on the number of solute particles in solution and not on the nature of the solute particles. These properties are called **colligative properties.**

A. Vapor pressure lowering, Raoult's law

If a solute is nonvolatile, the vapor pressure of its solution is always less than that of the pure solvent. Raoult's law quantifies this relationship by stating that the partial pressure of a solvent over a solution, P_1, is given by the vapor pressure of the pure solvent, P_1°, times the mole fraction of the solvent in the solution, X_1:

$$P_1 = X_1 P_1^\circ \qquad (12.4)$$

If the solution contains only one solute, $X_1 = 1 - X_2$, where X_2 is the mole fraction of the solute. Substituting for X_1 in Equation (12.4) gives:

$$P_1 = (1 - X_2) P_1^\circ$$

$$P_1 = P_1^\circ - X_2 P_1^\circ$$

$$P_1^\circ - P_1 = \Delta P = X_2 P_1^\circ \qquad (12.5)$$

Equation (12.5) shows that a decrease in vapor pressure, ΔP, is directly proportional to the concentration of the solute in solution, X_2.

EXAMPLE 12.9

Calculate the vapor pressure of an aqueous solution at 30°C made from 1.00×10^2 g of sucrose ($C_{12}H_{22}O_{11}$) and 1.00×10^2 g of water. The vapor pressure of pure water at 30°C is 31.8 mmHg.

Step 1: Equation (12.4) above gives a relationship between the vapor pressure of the solvent over a solution and the mole fraction of the solvent. If you can calculate the mole fraction of the solvent, you can calculate the vapor pressure of the solution. Problem Type 3, Chapter 5 defines mole fraction.

$$\text{mol water} = 1.00 \times 10^2 \text{ g water} \times \frac{1 \text{ mol water}}{18.02 \text{ g water}} = 5.55 \text{ mol}$$

$$\text{mol sucrose} = 1.00 \times 10^2 \text{ g sucrose} \times \frac{1 \text{ mol sucrose}}{342.3 \text{ g sucrose}} = 0.292 \text{ mol}$$

$$X_{\text{water}} = \frac{\text{mol}_{\text{water}}}{\text{mol}_{\text{water}} + \text{mol}_{\text{sucrose}}} = \frac{5.55 \text{ mol}}{5.55 \text{ mol} + 0.292 \text{ mol}} = 0.950$$

Step 2: Substitute X_{water} and the vapor pressure of pure water into Equation (12.4) to solve for the vapor pressure of the solution.

$$P_{\text{soln}} = X_{\text{water}} P_{\text{water}}^\circ = (0.950)(31.8 \text{ mmHg}) = \mathbf{30.2\ mmHg}$$

Tip: You could also have solved this problem using Equation (12.5). You could calculate the change in vapor pressure (ΔP) from the mole fraction of solute. Then, you could calculate the vapor pressure of the solution by subtracting ΔP from the vapor pressure of the pure solvent (water).

PRACTICE EXERCISE

9. At 25°C, the vapor pressure of pure water is 23.76 mmHg and that of an aqueous sucrose ($C_{12}H_{22}O_{11}$) is 23.28 mmHg. Calculate the molality of the solution.

Text Problems: 50, 52, 54

B. Boiling-point elevation

The **boiling point** of a solution is the temperature at which its vapor pressure equals the external atmospheric pressure (see Section 11.6 of the text). We just saw in Part (a) that a nonvolatile solute always decreases the vapor pressure of the solution relative to the pure solvent. Consequently, the boiling point of the solution is *higher* than the pure solvent, because more energy in the form of heat must be added to raise the vapor pressure of the solution to the external atmospheric pressure. The change in boiling point is proportional to the concentration of solute.

$$\Delta T_b = K_b m \qquad (12.6)$$

where,

$\Delta T_b = T_b - T_b^\circ$ (where T_b is the boiling point of the solution and T_b° is the boiling point of the pure solvent)

m is the molal concentration of the solute

K_b is the molal boiling-point elevation constant of the solvent with units of °C/m

Do you know why molality is used for the concentration instead of molarity? Because we are dealing with a system that is not kept at constant temperature. We cannot express the concentration in molarity because molarity changes with temperature.

EXAMPLE 12.10

What is the boiling point of an "antifreeze/coolant" solution made from a 50-50 mixture (by volume) of ethylene glycol, $C_2H_6O_2$ and water? Assume the density of water is 1.00 g/mL and the density of ethylene glycol is 1.11 g/mL.

Using Equation (12.6), you can calculate the change in boiling point, ΔT_b, by calculating the molality of the solution and then multiplying by K_b for water (see Table 12.2 of your text).

Step 1: For simplicity, assume that you have 100.0 mL of solution. Since the mixture is 50-50 by volume, there are 50.0 mL of ethylene glycol and 50.0 mL of water. To calculate the molality of the solution, you need the moles of solute (ethylene glycol) and the mass of solvent (water) in kg.

$$\text{mol of ethylene glycol} = 50.0 \text{ mL } C_2H_6O_2 \times \frac{1.11 \text{ g } C_2H_6O_2}{1 \text{ mL } C_2H_6O_2} \times \frac{1 \text{ mol } C_2H_6O_2}{62.07 \text{ g } C_2H_6O_2} = 0.894 \text{ mol}$$

$$\text{mass of water} = 50.0 \text{ mL } H_2O \times \frac{1.00 \text{ g } H_2O}{1 \text{ mL } H_2O} \times \frac{1 \text{ kg}}{1000 \text{ g}} = 0.0500 \text{ kg}$$

$$\text{molality} = \frac{\text{mol solute}}{\text{kg solvent}} = \frac{0.894 \text{ mol}}{0.0500 \text{ kg}} = 17.9\ m$$

Step 2: Substitute the molality of the solution and K_b into Equation (12.6) to solve for the change in boiling point. Then, add the change in boiling point to the normal boiling point of water (100.0°C) to calculate the boiling point of the solution.

$$\Delta T_b = K_b m = (0.52°\text{C}/m)(17.9\ m) = 9.3°\text{C}$$

$$\textbf{b.p. of soln} = 100.0°\text{C} + 9.3°\text{C} = \mathbf{109.3°C}$$

PRACTICE EXERCISE

10. What is the boiling point of an aqueous solution of a nonvolatile solute that freezes at −3.0°C?

C. Freezing-point depression

The **freezing point** of a liquid (or the melting point of a solid) is the temperature at which the solid and liquid phases coexist in equilibrium. It might be easier to understand freezing-point depression by looking at the opposite of freezing, melting. To melt a solid, the intermolecular forces holding the solid molecules together must be overcome. Adding another solid substance to a pure solid disrupts the intermolecular forces of the formerly pure solid. Hence, it is easier to overcome the intermolecular forces and the mixture melts at a lower temperature than the pure solid. The melting point (or the freezing point) is depressed. The depression of freezing point can be represented by the following equation.

$$\Delta T_f = K_f m \qquad (12.7)$$

where,

$\Delta T_f = T_f^\circ - T_f$ (where T_f is the freezing point of the solution and T_f° is the freezing point of the pure solvent)

m is the molal concentration of the solute

K_f is the molal freezing-point depression constant with units of °C/m

EXAMPLE 12.11

How many grams of isopropyl alcohol, C_3H_7OH, should be added to 1.0 L of water to give a solution that will not freeze above −16°C?

We want to lower the freezing point of 1.0 L of water by 16°C. Using Equation (12.7), we can calculate the molality of the solution needed. Then, from the molality, we can calculate the grams of isopropyl alcohol needed.

Step 1: Rearrange Equation (12.7) to solve for the molality. You can look up K_f in Table 12.2 of your text.

$$m = \frac{\Delta T_f}{K_f} = \frac{16\ \cancel{^\circ C}}{1.86\ \cancel{^\circ C}/m} = 8.6\ m$$

Step 2: 8.6 m means that the solution contains 8.6 mol of solute per 1 kg of solvent. Assume that the density of water is 1.0 g/mL; thus, 1 kg (1000 g) of water has a volume of 1 L. Convert 8.6 mol of isopropyl alcohol to grams of isopropyl alcohol using the molar mass as a conversion factor.

$$\textbf{? g isopropyl alcohol} = 8.6\ \cancel{mol}\ C_3H_7OH \times \frac{60.09\ g\ C_3H_7OH}{1\ \cancel{mol}\ C_3H_7OH} = \mathbf{5.2 \times 10^2\ g}$$

PRACTICE EXERCISE

11. Benzene melts at 5.50°C. When 2.11 g of naphthalene, $C_{10}H_8$, is added to 100 g of benzene, the solution freezes at 4.65°C. Calculate the freezing-point depression constant (K_f) for benzene.

Text Problems: 56, 60

D. Osmotic pressure

Osmosis is the net movement of solvent molecules through a semipermeable membrane from a pure solvent or from a dilute solution to a more concentrated solution. The **osmotic pressure (π)** of a solution is the pressure required to stop osmosis. The osmotic pressure of a solution is given by:

$$\pi = MRT \qquad (12.8)$$

where,

M is the molarity of the solution

R is the gas constant (0.0821 L·atm/K·mol)

T is the absolute temperature in K

EXAMPLE 12.12

The average osmotic pressure of seawater is about 30.0 atm at 25°C. Calculate the molar concentration of an aqueous solution of urea (NH_2CONH_2) that is isotonic with seawater.

A solution of urea that is isotonic with seawater must have the same osmotic pressure, 30.0 atm. Solve Equation (12.8) algebraically for the molar concentration. Then, substitute π, R, and T (in K) into the equation to solve for the molar concentration of the urea solution.

$$M = \frac{\pi}{RT} = \frac{30.0\ \cancel{\text{atm}}}{298\ \cancel{\text{K}}} \times \frac{\text{mol}\cdot\cancel{\text{K}}}{0.0821\ \text{L}\cdot\cancel{\text{atm}}} = \mathbf{1.23}\ \boldsymbol{M}$$

PRACTICE EXERCISE

12. The walls of red blood cells are semipermeable membranes, and the solution of NaCl within those walls exerts an osmotic pressure of 7.82 atm at 37°C. What concentration of NaCl must a *surrounding* solution have so that this pressure is balanced and cell rupture (hemolysis) is prevented?

PROBLEM TYPE 6: DETERMINING MOLAR MASS USING COLLIGATIVE PROPERTIES

Any of the four colligative properties discussed in Problem Type 5 can be used to calculate the molar mass of the solute. However, in practice, only freezing-point depression and osmotic pressure are used because they show the most pronounced changes.

A. Calculating molar mass from freezing-point depression

You can solve this type of problem by first calculating the molality of the solute using Equation (12.7) and then calculating the molar mass from the molality.

EXAMPLE 12.13

Benzene has a normal freezing point of 5.51°C. The addition of 1.25 g of an unknown compound to 85.0 g of benzene produces a solution with a freezing point of 4.52°C. What is the molar mass of the unknown compound?

Step 1: Solve Equation (12.7) algebraically for molality (m), then substitute ΔT_f and K_f into the equation to calculate the molality.

$$\Delta T_f = 5.51°\text{C} - 4.52°\text{C} = 0.99°\text{C}$$

$$m = \frac{\Delta T_f}{K_f} = \frac{0.99\ °\cancel{\text{C}}}{5.12\ °\cancel{\text{C}}/m} = 0.19\ m$$

Step 2: Multiplying the molality by the mass of solvent (in kg) gives moles of unknown solute. Then, dividing the mass of solute (in g) by the moles of solute, gives the molar mass of the unknown solute.

$$?\ \text{mol of unknown solute} = \frac{0.19\ \text{mol solute}}{1\ \cancel{\text{kg}}\ \text{benzene}} \times 0.085\ \cancel{\text{kg}}\ \text{benzene}$$

$$= 0.016\ \text{mol solute}$$

$$\textbf{molar mass of unknown} = \frac{1.25\ \text{g}}{0.016\ \text{mol}} = \mathbf{78\ g/mol}$$

PRACTICE EXERCISE

13. When 48 g of glucose (a nonelectrolyte) is dissolved in 500 g of H_2O, the solution has a freezing point of −0.94°C. What is the molar mass of glucose?

14. In the course of research, a chemist isolates a new compound. An elemental analysis shows the following: C, 50.7 percent; H, 4.25 percent; O, 45.1 percent. If 5.01 g of the compound is dissolved in 100 g of water, a solution with a freezing point of −0.65°C is produced. What is the molecular formula of the compound?

Text Problems: 58, 62

B. Calculating the molar mass from osmotic pressure

You can solve this type of problem by first calculating the molarity of the solution using Equation (12.8) and then calculating the molar mass from the molarity.

EXAMPLE 12.14

30.0 g of sucrose is dissolved in water making 1.00 × 10² mL of solution. The solution has an osmotic pressure of 20.8 atm at 16.0°C. What is the molar mass of sucrose?

Step 1: Solve Equation (12.8) algebraically for molarity (M), then substitute π, R, and T (in K) into the equation to calculate the molarity.

$$M = \frac{\pi}{RT} = \frac{20.8\ \cancel{\text{atm}}}{289\ \cancel{\text{K}}} \times \frac{\text{mol}\cdot\cancel{\text{K}}}{0.0821\ \text{L}\cdot\cancel{\text{atm}}} = 0.877\ M$$

Step 2: Multiplying the molarity by the volume of solution (in L) gives moles of solute. Then, dividing the mass of solute (in g) by the moles of solute, gives the molar mass of the solute.

$$?\ \text{mol of sucrose} = (0.877\ \text{mol}/\cancel{\text{L}})(0.100\ \cancel{\text{L}}) = 0.0877\ \text{mol sucrose}$$

$$\textbf{molar mass of sucrose} = \frac{30.0\ \text{g sucrose}}{0.0877\ \text{mol sucrose}} = \textbf{342 g/mol}$$

PRACTICE EXERCISE

15. Peruvian Indians use a dart poison from root extracts called curare. It is a nonelectrolyte. The osmotic pressure at 20.0°C of an aqueous solution containing 0.200 g of curare in 1.00 × 10² mL of solution is 56.2 mmHg. Calculate the molar mass of curare.

Text Problems: 64, 66

PROBLEM TYPE 7: COLLIGATIVE PROPERTIES OF ELECTROLYTES

Electrolytes dissociate into ions in solution, so this requires us to take a slightly different approach than that used for the colligative properties of nonelectrolytes. Remember, it is the number of solute particles that determines the colligative properties of a solution. To account for the dissociation of an electrolyte into ions, the equations for colligative properties must be modified as follows:

$$\Delta T_b = iK_bm \qquad (12.9)$$
$$\Delta T_f = iK_fm$$
$$\pi = iMRT$$

where,

i is the van't Hoff factor which is defined as:

$$i = \frac{\text{actual number of particles in soln after dissociation}}{\text{number of formula units initially dissolved in soln}}$$

Thus, *i* should be 1 for all nonelectrolytes. For strong electrolytes such as KCl and $BaSO_4$, *i* should be 2, and for strong electrolytes such as $CaCl_2$ and Na_2CO_3, *i* should be 3. In reality, the colligative properties of *electrolyte* solutions are usually smaller than anticipated. At higher concentrations, electrostatic forces come into play, drawing cations and anions together. The ion pairs that are formed decrease the number of solute particles in solution. Table 12.3 of your text lists the van't Hoff factor (*i*) for various solutes.

EXAMPLE 12.15

Calculate the value of *i* for an electrolyte that should dissociate into 2 ions, if a 1.0 *m* aqueous solution of the electrolyte freezes at −3.28°C. Why is the value of *i* less than 2?

Solve Equation (12.9) algebraically for *i*, then substitute the values of ΔT_f, K_f, and *m* into the equation to solve for *i*.

$\Delta T_f = 3.28°C$

$$i = \frac{\Delta T_f}{K_f m} = \frac{3.28\ \cancel{°C}}{(1.86\ \cancel{°C}/\cancel{m})(1.0\ \cancel{m})} = 1.76$$

The value if *i* is less than 2 because ion pairing reduces the number of particles in solution. Remember, colligative properties depend only on the number of solute particles in solution and *not* on the nature of the solute particles.

PRACTICE EXERCISE

16. Arrange the following aqueous solutions in order of increasing boiling points: 0.100 *m* ethanol, 0.050 *m* $Ca(NO_3)_2$, 0.100 *m* NaBr, 0.050 *m* HCl.

Text Problems: 74, 76, **78**, 80

ANSWERS TO PRACTICE EXERCISES

1. Remember that "like dissolves like". Therefore, the ionic solid NaBr will be more soluble in the polar solvent, H_2O.

2. 17.0 percent $CuSO_4$

3. 1.28 *M* $CuSO_4$

4. 1.29 *m* $CuSO_4$

5. 11.8 *M* HCl

6. $m = 3 \times 10^2\ m$

7. 0.91 *m* H_2O_2

8. 1.9×10^{-8} *M* CO

9. $m = 1.1\ m$

10. b.p. of soln = 100.84°C

11. K_f (benzene) = 5.15°C/*m*

12. 0.15 *M* NaCl

13. molar mass = 1.9×10^2 g/mol

14. $C_6H_6O_4$

15. molar mass of curare = 6.50×10^2 g/mol

16. ethanol ≅ HCl < $Ca(NO_3)_2$ < NaBr

SOLUTIONS TO SELECTED TEXT PROBLEMS

12.8 Predicting Solubility Based on Intermolecular Forces, Problem Type 1.

Recall that "like dissolves like". Strong hydrogen bonding (dipole-dipole attraction) is the principal intermolecular attraction in liquid ethanol, but in liquid cyclohexane the intermolecular forces are dispersion forces because cyclohexane is nonpolar. Cyclohexane cannot form hydrogen bonds with ethanol, and therefore cannot attract ethanol molecules strongly enough to form a solution.

12.10 The longer the C–C chain, the more the molecule "looks like" a hydrocarbon and the less important the –OH group becomes. Hence, as the C–C chain length increases, the molecule becomes less polar. Since "like dissolves like", as the molecules become more nonpolar, the solubility in polar water decreases. The –OH group of the alcohols can form strong hydrogen bonds with water molecules, but this property decreases as the chain length increases.

12.14 Percent by Mass Calculation, Problem Type 2A.

(a) The percent by mass is defined as

$$\textbf{percent by mass of solute} = \frac{\text{mass of solute}}{\text{mass of solute} + \text{mass of solvent}} \times 100\%$$

We are given percent by mass of solute and the mass of solute in the problem. Substituting those two quantities into the above equation, we can solve for the mass of solvent (water).

$$16.2\% = \frac{5.00\text{ g urea}}{5.00\text{ g urea} + \text{mass of water}} \times 100\%$$

$$(0.162)(\text{mass of water}) = 5.00\text{ g} - (0.162)(5.00\text{g})$$

$$\textbf{mass of water} = \textbf{25.9 g}$$

(b) Similar to part (a),

$$1.5\% = \frac{26.2\text{ g MgCl}_2}{26.2\text{ g MgCl}_2 + \text{mass of water}} \times 100\%$$

$$\textbf{mass of water} = \mathbf{1.72 \times 10^3\ g}$$

12.16 $$\text{molality} = \frac{\text{moles of solute}}{\text{mass of solvent (kg)}}$$

(a) $$\text{mass of 1 L soln} = 1000\ \cancel{\text{mL}}\left(\frac{1.08\text{ g}}{1\ \cancel{\text{mL}}}\right) = 1080\text{ g}$$

$$\text{mass of water} = 1080\text{ g} - 2.50\ \cancel{\text{mol}}\text{ NaCl}\left(\frac{58.44\text{ g NaCl}}{1\ \cancel{\text{mol}}\text{ NaCl}}\right) = 934\text{ g} = 0.934\text{ kg}$$

$$m = \frac{2.50\text{ mol NaCl}}{0.934\text{ kg H}_2\text{O}} = \mathbf{2.68}\ \boldsymbol{m}$$

(b) 100 g of the solution contains 48.2 g KBr and 51.8 g H_2O.

$$\text{mol of KBr} = 48.2\ \text{g KBr} \times \left(\frac{1\ \text{mol KBr}}{119.0\ \text{g KBr}}\right) = 0.405\ \text{mol KBr}$$

$$\text{mass of } H_2O \text{ (in kg)} = 51.8\ \text{g } H_2O \times \frac{1\ \text{kg } H_2O}{1000\ \text{g } H_2O} = 0.0518\ \text{kg } H_2O$$

$$m = \frac{0.405\ \text{mol KBr}}{0.0518\ \text{kg } H_2O} = \mathbf{7.82}\ \boldsymbol{m}$$

12.18 Let's assume that we have 1.0 L of a 0.010 *M* solution.

Assuming a solution density of 1.0 g/mL, the mass of 1.0 L of the solution is:

$$\text{Mass of 1.0 L} = 1.0\ \text{L}\left(\frac{1000\ \text{mL}}{1\ \text{L}}\right)\left(\frac{1.0\ \text{g}}{1\ \text{mL}}\right) = 1.0 \times 10^3\ \text{g}$$

The mass of 0.010 mole of urea is:

$$0.010\ \text{mol urea}\left(\frac{60.06\ \text{g urea}}{1\ \text{mol urea}}\right) = 0.60\ \text{g urea}$$

The mass of the solvent is:

$$(\text{solution mass}) - (\text{solute mass}) = (1.0 \times 10^3\ \text{g}) - (0.60\ \text{g}) = 1.0 \times 10^3\ \text{g} = 1.0\ \text{kg}$$

$$m = \frac{\text{moles solute}}{\text{mass solvent}} = \frac{0.010\ \text{mol}}{1.0\ \text{kg}} = \mathbf{0.010}\ \boldsymbol{m}$$

12.20 Converting between Concentration Units, Problem Type 3.

(a) Converting percent by mass to molality

Step 1: Assume that the mass of the solution is 100.0 g. The percent by mass of the solute can then be converted directly to grams. Then, the mass of solute can be converted to moles of solute using molar mass as a conversion factor. Assuming 100.0 g solution,

$$98.0\ \text{g } H_2SO_4 \times \frac{1\ \text{mol } H_2SO_4}{98.09\ \text{g } H_2SO_4} = 0.999\ \text{mol } H_2SO_4$$

Step 2: The mass of the solvent (water) can then be calculated by subtracting the mass of solute from the mass of the solution (100.0 g). Remember, the mass of the solution is equal to the mass of solute + mass of solvent.

$$?\ \text{mass of solvent} = \text{mass of solution} - \text{mass of solute}$$

$$= 100.0\ \text{g} - 98.0\ \text{g} = 2.0\ \text{g} = 2.0 \times 10^{-3}\ \text{kg solvent (water)}$$

Step 3: Moles of solute was calculated in *Step 1*, and mass of solvent was calculated in *Step 2*. Divide moles of solute by mass of solvent in kg to calculate the molality of the solution.

$$m = \frac{\text{mol of solute}}{\text{kg of solvent}} = \frac{0.999\ \text{mol}}{2.0 \times 10^{-3}\ \text{kg}} = \mathbf{5.0 \times 10^2}\ \boldsymbol{m}$$

(b) Converting molality to molarity.

Step 1: The mass of the solution must be converted to volume of solution to convert from molality to molarity. We assumed 100.0 g of solution in part (a). Use the solution density as a conversion factor to convert to volume of solution.

$$? \text{ volume of solution} = 100.0 \not{g} \times \frac{1 \text{ mL}}{1.83 \not{g}} = 54.6 \text{ mL} = 0.0546 \text{ L}$$

Step 2: We calculated moles of solute in part (a), 0.999 mol H_2SO_4. Divide moles of solute by liters of solution to calculate the molarity of the solution.

$$M = \frac{\text{mol of solute}}{\text{L of soln}} = \frac{0.999 \text{ mol}}{0.0546 \text{ L}} = \mathbf{18.3}\ \boldsymbol{M}$$

12.22 Assume 100.0 g of solution.

(a) The mass of ethanol in the solution is 0.100 × 100.0 g = 10.0 g. The mass of the water is 100.0 g − 10.0 g = 90.0 g = 0.0900 kg. The amount of ethanol in moles is:

$$10.0 \not{g} \text{ ethanol} \left(\frac{1 \text{ mol}}{46.07 \not{g}}\right) = 0.217 \text{ mol ethanol}$$

$$m = \frac{\text{mol solute}}{\text{kg solvent}} = \frac{0.217 \text{ mol}}{0.0900 \text{ kg}} = \mathbf{2.41}\ \boldsymbol{m}$$

(b) The volume of the solution is:

$$100.0 \not{g} \left(\frac{1 \text{ mL}}{0.984 \not{g}}\right) = 102 \text{ mL} = 0.102 \text{ L}$$

The amount of ethanol in moles is 0.217 mol [part (a)].

$$M = \frac{\text{mol solute}}{\text{liters of soln}} = \frac{0.217 \text{ mol}}{0.102 \text{ L}} = \mathbf{2.13}\ \boldsymbol{M}$$

(c) $\textbf{Solution volume} = 0.125 \not{\text{mol}} \left(\frac{1 \text{ L}}{2.13 \not{\text{mol}}}\right) = \mathbf{0.0587\ L} = \mathbf{58.7\ mL}$

12.26 At 75°C, 155 g of KNO_3 dissolves in 100 g of water to form 255 g of solution. When cooled to 25°C, only 38.0 g of KNO_3 remain dissolved. This means that (155 − 38.0) g = 117 g of KNO_3 will crystallize.

The amount of KNO_3 formed when 100 g of saturated solution at 75°C is cooled to 25°C can be found by a simple unit conversion.

$$(100 \not{g} \text{ saturated soln}) \times \left(\frac{117 \text{ g } KNO_3 \text{ crystallized}}{255 \not{g} \text{ saturated soln}}\right) = \mathbf{45.9\ g\ KNO_3}$$

12.34 According to Henry's law, the solubility of a gas in a liquid increases as the pressure increases (c = kP). The soft drink tastes flat at the bottom of the mine because the carbon dioxide pressure is greater and the dissolved gas is not released from the solution. As the miner goes up in the elevator, the atmospheric carbon dioxide pressure decreases and dissolved gas is released from his stomach.

12.36 Effect of Pressure on Solubility, Problem Type 4. To solve this problem, you need to compare the solubilities of N_2 in blood under normal pressure (0.80 atm) and under a greater pressure that a deep-sea diver might experience (4.0 atm).

Step 1: Calculate the Henry's law constant, *k*, using the concentration of N_2 in blood at 0.80 atm.

$$k = \frac{c}{P}$$

$$k = \frac{5.6 \times 10^{-4}\ \text{mol/L}}{0.80\ \text{atm}} = 7.0 \times 10^{-4}\ \text{mol/L·atm}$$

Step 2: Calculate the concentration of N_2 in blood at 4.0 atm using *k* calculated above.

$$c = kP$$

$$c = (7.0 \times 10^{-4}\ \text{mol/L·}\cancel{\text{atm}})(4.0\ \cancel{\text{atm}}) = 2.8 \times 10^{-3}\ \text{mol/L}$$

Step 3: From each of the concentrations of N_2 in blood, we can calculate the number of moles of N_2 dissolved by multiplying by the total blood volume of 5.0 L. Then, we can calculate the number of moles of N_2 released when the diver returns to the surface.

The number of moles of N_2 in 5.0 L of blood at 0.80 atm is:

$$(5.6 \times 10^{-4}\ \text{mol/}\cancel{\text{L}}) \times 5.0\ \cancel{\text{L}} = 2.8 \times 10^{-3}\ \text{mol}$$

The number of moles of N_2 in 5.0 L of blood at 4.0 atm is:

$$(2.8 \times 10^{-3}\ \text{mol/}\cancel{\text{L}}) \times 5.0\ \cancel{\text{L}} = 1.4 \times 10^{-2}\ \text{mol}$$

The amount of N_2 released in moles when the diver returns to the surface is:

$$1.4 \times 10^{-2}\ \text{mol} - 2.8 \times 10^{-3}\ \text{mol} = 1.1 \times 10^{-2}\ \text{mol}$$

Step 4: We can now calculate the volume of N_2 released using the ideal gas equation. The total pressure pushing on the N_2 that is released is atmospheric pressure. You should assume that the atmospheric pressure is 1 atm.

The volume of N_2 released is:

$$V_{N_2} = \frac{nRT}{P}$$

$$V_{N_2} = \frac{(1.1 \times 10^{-2}\ \cancel{\text{mol}})(310\ \cancel{\text{K}})}{(1.0\ \cancel{\text{atm}})} \times \frac{0.0821\ \text{L}\cdot\cancel{\text{atm}}}{\cancel{\text{mol}}\cdot\cancel{\text{K}}} = \mathbf{0.28\ L}$$

12.50 Vapor-pressure lowering, Raoult's law, Problem Type 5A.

Step 1: Using Equation (12.5) from Problem Type 5A, we can calculate the mole fraction of sucrose that causes a 2.0 mmHg drop in vapor pressure.

$$\Delta P = X_2 P_1^\circ \qquad (12.5)$$

$$\Delta P = X_{\text{sucrose}} P^{\circ}_{\text{water}}$$

$$X_{\text{sucrose}} = \frac{\Delta P}{P^{\circ}_{\text{water}}} = \frac{2.0\ \text{mmHg}}{17.5\ \text{mmHg}} = 0.11$$

Step 2: From the definition of mole fraction, we can calculate the number of moles of sucrose needed to cause a 2.0 mmHg drop in vapor pressure.

$$X_{\text{sucrose}} = \frac{n_{\text{sucrose}}}{n_{\text{water}} + n_{\text{sucrose}}}$$

$$\text{moles of water} = 552\ \text{g} \times \left(\frac{1\ \text{mol}}{18.02\ \text{g}}\right) = 30.6\ \text{mol H}_2\text{O}$$

$$X_{\text{sucrose}} = 0.11 = \frac{n_{\text{sucrose}}}{30.6 + n_{\text{sucrose}}}$$

$$n_{\text{sucrose}} = 3.8\ \text{mol sucrose}$$

Step 3: Using the molar mass of sucrose as a conversion factor, we can calculate the mass of sucrose.

$$\textbf{mass of sucrose} = 3.8\ \text{mol sucrose} \times \left(\frac{342.3\ \text{g sucrose}}{1\ \text{mol sucrose}}\right) = \mathbf{1.3 \times 10^3\ g\ sucrose}$$

12.52 For any solution the sum of the mole fractions of the components is always 1.00, so the mole fraction of 1–propanol is 0.700. The partial pressures are:

$$P_{\text{ethanol}} = X_{\text{ethanol}} \times P^{\circ}_{\text{ethanol}} = 0.300 \times 100\ \text{mmHg} = \mathbf{30.0\ mmHg}$$

$$P_{\text{1-propanol}} = X_{\text{1-propanol}} \times P^{\circ}_{\text{1-propanol}} = 0.700 \times 37.6\ \text{mmHg} = \mathbf{26.3\ mmHg}$$

Is the vapor phase richer in one of the components than the solution? Which component? Should this always be true for ideal solutions?

12.54 This problem is very similar to Problem 12.50.

$$\Delta P = X_{\text{urea}} P^{\circ}_{\text{water}}$$

$$2.50\ \text{mmHg} = X_{\text{urea}}(31.8\ \text{mmHg})$$

$$X_{\text{urea}} = 0.0786$$

The number of moles of water is:

$$n_{\text{water}} = 450\ \text{g H}_2\text{O}\left(\frac{1\ \text{mol H}_2\text{O}}{18.02\ \text{g H}_2\text{O}}\right) = 25.0\ \text{mol H}_2\text{O}$$

$$X_{\text{urea}} = \frac{n_{\text{urea}}}{n_{\text{water}} + n_{\text{urea}}}$$

$$0.0786 = \frac{n_{\text{urea}}}{25.0 + n_{\text{urea}}}$$

$$n_{\text{urea}} = 2.13\text{ mol}$$

$$\textbf{mass of urea} = 2.13\text{ mol urea} \times \left(\frac{60.06\text{ g urea}}{1\text{ mol urea}}\right) = \textbf{128 g of urea}$$

12.56 $$m = \frac{\Delta T_f}{K_f} = \frac{1.1^\circ\text{C}}{1.86^\circ\text{C}/m} = \mathbf{0.59}\ \boldsymbol{m}$$

12.58 This is a combination of Problem Type 4B, Chapter 3, and Problem Type 6A, Chapter 12. First, we need to find the empirical formula from mass percent data. Then, we can determine the molar mass from the freezing-point depression. Finally, from the empirical formula and the molar mass, we can find the molecular formula.

Step 1: Assume you have exactly 100 g of substance. 100 g is a convenient amount, because all the percentages sum to 100 percent. In 100 g of the solid extracted from gum arabic there will be 40.0 g C, 6.7 g H, and 53.3 g O.

Step 2: Calculate the number of moles of each element in the compound. Remember, an *empirical formula* tells us which elements are present and the simplest whole-number ratio of their atoms. This ratio is also a mole ratio. Let n_C, n_H, and n_O be the number of moles of elements present. Use the molar masses of these elements as conversion factors to convert to moles.

$$n_C = 40.0\text{ g C} \times \frac{1\text{ mol C}}{12.01\text{ g C}} = \textbf{3.33 mol C}$$

$$n_H = 6.7\text{ g H} \times \frac{1\text{ mol H}}{1.008\text{ g H}} = \textbf{6.6 mol H}$$

$$n_O = 53.3\text{ g O} \times \frac{1\text{ mol O}}{16.00\text{ g O}} = \textbf{3.33 mol O}$$

Thus, we arrive at the formula $C_{3.33}H_{6.6}O_{3.33}$, which gives the identity and the ratios of atoms present. However, chemical formulas are written with whole numbers.

Step 3: Try to convert to whole numbers by dividing all the subscripts by the smallest subscript.

$$\textbf{C:}\ \frac{3.33}{3.33} = 1.00 \qquad \textbf{H:}\ \frac{6.6}{3.33} = 2.0 \qquad \textbf{O:}\ \frac{3.33}{3.33} = 1.00$$

This gives us the empirical, $\mathbf{CH_2O}$.

Step 4: Use the freezing point data to determine the molar mass. First, calculate the molality of the solution.

$$m = \frac{\Delta T_f}{K_f} = \frac{1.56^\circ\text{C}}{8.00^\circ\text{C}/m} = 0.195\ m$$

Step 5: Multiplying the molality by the mass of solvent (in kg) gives moles of unknown solute. Then, dividing the mass of solute (in g) by the moles of solute, gives the molar mass of the unknown solute.

$$?\text{ mol of unknown solute} = \frac{0.195\text{ mol solute}}{1\text{ kg diphenyl}} \times 0.0278\text{ kg diphenyl}$$

$$= 0.00542\text{ mol solute}$$

$$\textbf{molar mass of unknown} = \frac{0.650\text{ g}}{0.00542\text{ mol}} = \mathbf{1.20 \times 10^2}\ \textbf{g/mol}$$

Step 6: Calculate the empirical molar mass. Then, determine the number of CH_2O units present in the molecular formula by taking the ratio of molar mass to empirical mass.

$$\text{empirical molar mass} = 12.01\text{ g} + 2(1.008\text{ g}) + 16.00\text{ g} = 30.03\text{ g/mol}$$

The number of (CH_2O) units present in the molecular formula is:

$$\frac{\text{molar mass}}{\text{empirical molar mass}} = \frac{1.20 \times 10^2 \cancel{\text{g}}}{30.03 \cancel{\text{g}}} = 4.00$$

Thus, there are four CH_2O units in each molecule of the compound, so the molecular formula is $(CH_2O)_4$, or $\mathbf{C_4H_8O_4}$.

12.60 We first find the number of moles of gas using the ideal gas equation.

$$n = \frac{PV}{RT} = \frac{(748/760)\cancel{\text{atm}}(4.00\cancel{\text{L}})}{(27+273)\cancel{\text{K}}} \times \frac{\text{mol}\cdot\cancel{\text{K}}}{0.0821\cancel{\text{L}}\cdot\cancel{\text{atm}}} = 0.160\text{ mol}$$

$$\text{molality} = \frac{0.160\text{ mol}}{0.0580\text{ kg benzene}} = 2.76\ m$$

See Table 12.2 of the text for the normal freezing point and K_f value for benzene.

$$\Delta T_f = K_f m = (5.12°\text{C}/\cancel{m})(2.76\cancel{m}) = 14.1°\text{C}$$

$$\textbf{freezing point} = 5.5°\text{C} - 14.1°\text{C} = \mathbf{-8.6°C}$$

12.62 First, from the freezing point depression we can calculate the molality of the solution. See Table 12.2 of the text for the normal freezing point and K_f value for benzene.

$$\Delta T_f = (5.5 - 4.3)°\text{C} = 1.2°\text{C}$$

$$m = \frac{\Delta T_f}{K_f} = \frac{1.2°\cancel{\text{C}}}{5.12°\cancel{\text{C}}/m} = 0.23\ m$$

Multiplying the molality by the mass of solvent (in kg) gives moles of unknown solute. Then, dividing the mass of solute (in g) by the moles of solute, gives the molar mass of the unknown solute.

$$?\text{ mol of unknown solute} = \frac{0.23\text{ mol solute}}{1\cancel{\text{kg}}\text{ benzene}} \times 0.0250\cancel{\text{kg}}\text{ benzene}$$

$$= 0.0058\text{ mol solute}$$

$$\textbf{molar mass of unknown} = \frac{2.50\text{ g}}{0.0058\text{ mol}} = \mathbf{4.3 \times 10^2\ g/mol}$$

The empirical molar mass of C_6H_5P is 108.1 g/mol. Therefore, the molecular formula is $(C_6H_5P)_4$ or $\mathbf{C_{24}H_{20}P_4}$.

12.64 Calculating molar mass from osmotic pressure, Problem Type 6B.

Step 1: Solve Equation (12.8) from Problem Type 6B algebraically for molarity (M), then substitute π, R, and T (in K) into the equation to calculate the molarity.

$$\pi = MRT \qquad (12.8)$$

$$M = \frac{\pi}{RT} = \frac{\left(5.20\ \text{mmHg} \times \frac{1\ \text{atm}}{760\ \text{mmHg}}\right)}{298\ \text{K}} \times \frac{\text{mol}\cdot\text{K}}{0.0821\ \text{L}\cdot\text{atm}} = 2.80 \times 10^{-4}\ M$$

Step 2: Multiplying the molarity by the volume of solution (in L) gives moles of solute. Then, dividing the mass of solute (in g) by the moles of solute, gives the molar mass of the solute.

$$\text{? mol of polymer} = (2.80 \times 10^{-4}\ \text{mol/L})(0.170\ \text{L}) = 4.76 \times 10^{-5}\ \text{mol polymer}$$

$$\textbf{molar mass of polymer} = \frac{0.8330\ \text{g polymer}}{4.76 \times 10^{-5}\ \text{mol polymer}} = \mathbf{1.75 \times 10^{4}\ g/mol}$$

12.66 We use the osmotic pressure data to determine the molarity.

$$M = \frac{\pi}{RT} = \frac{4.61\ \text{atm}}{(20+273)\ \text{K}} \times \frac{\text{mol}\cdot\text{K}}{0.0821\ \text{L}\cdot\text{atm}} = 0.192\ \text{mol / L}$$

Next we use the density and the solution mass to find the volume of the solution.

$$\text{mass of soln} = 6.85\ \text{g} + 100.0\ \text{g} = 106.9\ \text{g soln}$$

$$\text{volume of soln} = 106.9\ \text{g soln} \times \frac{1\ \text{mL}}{1.024\ \text{g}} = 104.4\ \text{mL} = 0.1044\ \text{L}$$

Multiplying the molarity by the volume (in L) gives moles of solute (carbohydrate).

$$\text{mol of solute} = M \times \text{L} = (0.192\ \text{mol/L})(0.1044\ \text{L}) = 0.0200\ \text{mol solute}$$

Finally, dividing mass of carbohydrate by moles of carbohydrate gives the molar mass of the carbohydrate.

$$\textbf{molar mass} = \frac{6.85\ \text{g carbohydrate}}{0.0200\ \text{mol carbohydrate}} = \mathbf{343\ g/mol}$$

12.72 Boiling point, vapor pressure, and osmotic pressure all depend on particle concentration. Therefore, these solutions also have the same boiling point, osmotic pressure, and vapor pressure.

12.74 The freezing point will be depressed most by the solution that contains the most solute particles. You should try to classify each solute as a strong electrolyte, a weak electrolyte, or a nonelectrolyte. All three solutions have the same concentration, so comparing the solutions is straightforward. HCl is a strong electrolyte, so under ideal conditions it will completely dissociate into two particles per molecule. The concentration of particles will be 1.00 m. Acetic acid is a weak electrolyte, so it will only dissociate to a small extent. The concentration of particles will be greater than 0.50 m, but less than 1.00 m. Glucose is a nonelectrolyte, so glucose molecules remain as glucose molecules in solution. The concentration of particles will be 0.50 m. For these solutions, the order in which the freezing points become *lower* is:

$$0.50\ m\ \text{glucose} > 0.50\ m\ \text{acetic acid} > 0.50\ m\ \text{HCl}$$

In other words, the HCl solution will have the lowest freezing point (greatest freezing point depression).

12.76 Using Equation (12.5) from Problem Type 5A, we can find the mole fraction of the NaCl. We use subscript 1 for H_2O and subscript 2 for NaCl.

$$\Delta P = X_2 P_1^\circ$$

$$X_2 = \frac{\Delta P}{P_1^\circ}$$

$$X_2 = \frac{23.76 \text{ mmHg} - 22.98 \text{ mmHg}}{23.76 \text{ mmHg}} = 0.03283$$

Let's assume that we have 1000 g (1 kg) of water as the solvent, because the definition of molality is moles of solute per kg of solvent. We can find the number of moles of particles dissolved in the water using the definition of mole fraction.

$$X_2 = \frac{n_2}{n_1 + n_2}$$

$$n_1 = 1000 \text{ g H}_2\text{O}\left(\frac{1 \text{ mol H}_2\text{O}}{18.02 \text{ g H}_2\text{O}}\right) = 55.49 \text{ mol H}_2\text{O}$$

$$\frac{n_2}{55.49 + n_2} = 0.03283$$

$$n_2 = 1.884 \text{ mol}$$

Since NaCl dissociates to form two particles (ions), the number of moles of NaCl is half of the above result.

$$\text{Moles NaCl} = 1.884 \text{ mol particles} \times \frac{1 \text{ mol NaCl}}{2 \text{ mol particles}} = 0.9420 \text{ mol}$$

The molality of the solution is:

$$\frac{0.9420 \text{ mol}}{1.000 \text{ kg}} = \mathbf{0.9420\ \mathit{m}}$$

12.78 Colligative Properties of Electrolytes, Problem Type 7.

Step 1: To calculate the osmotic pressure of "physiological saline", we must first calculate the molarity of the solution since $\pi = iMRT$.

To calculate molarity, let's assume that we have 1.000 L of solution (1.000×10^3 mL). We can use the solution density as a conversion factor to calculate the mass of 1.000×10^3 mL of solution.

$$1.000 \times 10^3 \text{ mL soln} \times \frac{1.005 \text{ g soln}}{1 \text{ mL soln}} = 1005 \text{ g of soln}$$

Since the solution is 0.86% by mass NaCl, the mass of NaCl in the solution is:

$$1005 \text{ g} \times \frac{0.86\%}{100\%} = 8.6 \text{ g NaCl}$$

The molarity of the solution is:

$$\left(\frac{8.6\ \text{g NaCl}}{1.000\ \text{L}}\right)\left(\frac{1\ \text{mol NaCl}}{58.44\ \text{g NaCl}}\right) = 0.15\ M$$

Step 2: Since NaCl is a strong electrolyte, let's assume that the van't Hoff factor is 2. Substituting *i*, *M*, *R*, and *T* into the equation for osmotic pressure gives:

$$\pi = iMRT = 2\left(\frac{0.15\ \text{mol}}{\text{L}}\right) \times \left(\frac{0.0821\ \text{L}\cdot\text{atm}}{\text{mol}\cdot\text{K}}\right) \times (310\ \text{K}) = \mathbf{7.6\ atm}$$

12.80 From Table 12.3, $i = 1.3$

$$\pi = iMRT$$

$$\pi = (1.3)\left(\frac{0.0500\ \text{mol}}{\text{L}}\right) \times \left(\frac{0.0821\ \text{L}\cdot\text{atm}}{\text{mol}\cdot\text{K}}\right) \times (295\ \text{K})$$

$$\pi = \mathbf{1.6\ atm}$$

12.82 At constant temperature, the osmotic pressure of a solution is proportional to the molarity. When equal volumes of the two solutions are mixed, the molarity will just be the mean of the molarities of the two solutions (assuming additive volumes). Since the osmotic pressure is proportional to the molarity, the osmotic pressure of the solution will be the mean of the osmotic pressure of the two solutions.

$$\pi = \frac{2.4\ \text{atm} + 4.6\ \text{atm}}{2} = \mathbf{3.5\ atm}$$

12.84 **(a)** You can use Equation (12.4) from Problem Type 5A to calculate the vapor pressure of each component.

$$P_1 = X_1 P_1^\circ \qquad (12.4)$$

First, you must calculate the mole fraction of each component.

$$X_A = \frac{n_A}{n_A + n_B} = \frac{1.00\ \text{mol}}{1.00\ \text{mol} + 1.00\ \text{mol}} = 0.500$$

Similarly,

$$X_B = 0.500$$

Substitute the mole fraction calculated above and the vapor pressure of the pure solvent into Equation (12.4) to calculate the vapor pressure of each component of the solution.

$$P_A = X_A P_A^\circ = 0.500 \times 76\ \text{mmHg} = 38\ \text{mmHg}$$

$$P_B = X_B P_B^\circ = 0.500 \times 132\ \text{mmHg} = 66\ \text{mmHg}$$

The total vapor pressure is the sum of the vapor pressures of the two components.

$$P_{\text{Total}} = P_A + P_B = 38\ \text{mmHg} + 66\ \text{mmHg} = \mathbf{104\ mmHg}$$

(b) This problem is solved similarly to part (a).

$$X_A = \frac{n_A}{n_A + n_B} = \frac{2.00\ \text{mol}}{2.00\ \text{mol} + 5.00\ \text{mol}} = 0.286$$

Similarly,

$$X_B = 0.714$$

$$P_A = X_A P_A^\circ = 0.286 \times 76 \text{ mmHg} = 22 \text{ mmHg}$$

$$P_B = X_B P_B^\circ = 0.714 \times 132 \text{ mmHg} = 94 \text{ mmHg}$$

$$\mathbf{P_{Total}} = P_A + P_B = 22 \text{ mmHg} + 94 \text{ mmHg} = \mathbf{116\ mmHg}$$

12.86 From the osmotic pressure, you can calculate the molarity of the solution.

$$M = \frac{\pi}{RT} = \frac{\left(30.3 \text{ mmHg} \times \frac{1 \text{ atm}}{760 \text{ mmHg}}\right)}{308 \text{ K}} \times \frac{\text{mol} \cdot \text{K}}{0.0821 \text{ L} \cdot \text{atm}} = 1.58 \times 10^{-3} \text{ mol/L}$$

Multiplying molarity by the volume of solution in liters gives the moles of solute.

$$(1.58 \times 10^{-3} \text{ mol solute/L soln}) \times (0.262 \text{ L soln}) = 4.14 \times 10^{-4} \text{ mol solute}$$

Divide the grams of solute by the moles of solute to calculate the molar mass.

$$\textbf{molar mass of solute} = \frac{1.22 \text{ g}}{4.14 \times 10^{-4} \text{ mol}} = \mathbf{2.95 \times 10^3\ g/mol}$$

12.88 Solve Equation (12.7) from Problem Type 6A algebraically for molality (m), then substitute ΔT_f and K_f into the equation to calculate the molality. You can find the normal freezing point for benzene and K_f for benzene in Table 12.2 of your text.

$$\Delta T_f = 5.5°\text{C} - 3.9°\text{C} = 1.6°\text{C}$$

$$m = \frac{\Delta T_f}{K_f} = \frac{1.6°\text{C}}{5.12°\text{C}/m} = 0.31\ m$$

Multiplying the molality by the mass of solvent (in kg) gives moles of unknown solute. Then, dividing the mass of solute (in g) by the moles of solute, gives the molar mass of the unknown solute.

$$? \text{ mol of unknown solute} = \frac{0.31 \text{ mol solute}}{1 \text{ kg benzene}} \times 8.0 \times 10^{-3} \text{ kg benzene}$$

$$= 2.5 \times 10^{-3} \text{ mol solute}$$

$$\text{molar mass of unknown} = \frac{0.50 \text{ g}}{2.5 \times 10^{-3} \text{ mol}} = \mathbf{2.0 \times 10^2\ g/mol}$$

The molar mass for cocaine $C_{17}H_{21}NO_4$ = 303 g/mol, so the compound is not cocaine. We assume in our analysis that the compound is a pure, monomeric, nonelectrolyte.

12.90 First, we calculate the number of moles of HCl in 100 g of solution.

$$n_{HCl} = 100 \text{ g soln} \times \frac{37.7 \text{ g HCl}}{100 \text{ g soln}} \times \frac{1 \text{ mol HCl}}{36.46 \text{ g HCl}} = 1.03 \text{ mol HCl}$$

Next, we calculate the volume of 100 g of solution.

$$V = 100\,\text{g} \times \frac{1\text{ mL}}{1.19\text{ g}} \times \frac{1\text{ L}}{1000\text{ mL}} = 0.0840\text{ L}$$

Finally, the molarity of the solution is:

$$\frac{1.03\text{ mol}}{0.0840\text{ L}} = \mathbf{12.3}\ \boldsymbol{M}$$

12.92 Solution A: Let molar mass be $\mathcal{M}$.

$$\Delta P = X_A P_A^\circ$$

$$(760 - 754.5) = X_A(760)$$

$$X_A = 7.237 \times 10^{-3}$$

$$n = \frac{\text{mass}}{\text{molar mass}}$$

$$X_A = \frac{n_A}{n_A + n_{\text{water}}} = \frac{5.00/\mathcal{M}}{5.00/\mathcal{M} + 100/18.02} = 7.237 \times 10^{-3}$$

$$\mathcal{M} = \mathbf{124\ g/mol}$$

Solution B: Let molar mass be $\mathcal{M}$

$$\Delta P = X_B P_B^\circ$$

$$X_B = 7.237 \times 10^{-3}$$

$$n = \frac{\text{mass}}{\text{molar mass}}$$

$$X_B = \frac{n_B}{n_B + n_{\text{benzene}}} = \frac{2.31/\mathcal{M}}{2.31/\mathcal{M} + 100/78.11} = 7.237 \times 10^{-3}$$

$$\mathcal{M} = \mathbf{248\ g/mol}$$

The molar mass in benzene is about twice that in water. This suggests some sort of dimerization is occurring in a nonpolar solvent such as benzene.

12.94 **(a)** Boiling under reduced pressure.

(b) CO_2 boils off, expands and cools, condensing water vapor to form fog.

12.96 **(a)** Seawater has a larger number of ionic compounds dissolved in it; thus the boiling point is elevated.

(b) Carbon dioxide escapes from an opened soft drink bottle because gases are less soluble in liquids at lower pressure (Henry's law).

(c) As you proved in Problem 12.18, at dilute concentrations molality and molarity are almost the same because the density of the solution is almost equal to that of the pure solvent.

(d) For colligative properties we are concerned with the number of solute particles in solution relative to the number of solvent particles. Since in colligative particle measurements we frequently are dealing with changes in temperature (and since density varies with temperature), we need a concentration unit that is temperature invariant. We use units of moles per kilograms of solvent (molality) rather than moles per liter of solution (molarity).

(e) Methanol is very water soluble (why?) and effectively lowers the freezing point of water. However in the summer, the temperatures are sufficiently high so that most of the methanol would be lost to vaporization.

12.98 $\Delta T_f = 5.5 - 2.2 = 3.3°C$ $C_{10}H_8$: 128.2 g/mol

$$m = \frac{\Delta T_f}{K_f} = \frac{3.3}{5.12} = 0.645\ m$$ C_6H_{12}: 84.16 g/mol

Let x = mass of C_6H_{12} (in grams).

Using,

$$m = \frac{\text{mol solute}}{\text{kg solvent}} \quad \text{and} \quad \text{mol} = \frac{\text{mass}}{\text{molar mass}}$$

$$0.645 = \frac{\dfrac{x}{84.16} + \dfrac{1.32 - x}{128.2}}{0.0189\text{ kg}}$$

$$0.0122 = \frac{128.2x + 111.1 - 84.16x}{84.16 \times 128.2}$$

$$x = 0.47\text{ g}$$

$$\%\ \mathbf{C_6H_{12}} = \frac{0.47}{1.32} \times 100\% = \mathbf{36\%}$$

$$\%\ \mathbf{C_{10}H_8} = \frac{0.86}{1.32} \times 100\% = \mathbf{65\%}$$

The percentages don't add up to 100% because of rounding procedures.

12.100 The completed table is shown below:

Attractive Forces	Deviation from Raoult's	$\Delta H_{solution}$
A ↔ A, B ↔ B > A ↔ B	Positive	Positive (endothermic)
A ↔ A, B ↔ B < A ↔ B	Negative	Negative (exothermic)
A ↔ A, B ↔ B = A ↔ B	Zero	Zero

The first row represents a Case 1 situation in which A's attract A's and B's attract B's more strongly than A's attract B's. As described in Section 12.6 of the text, this results in positive deviation from Raoult's law (higher vapor pressure than calculated) and positive heat of solution (endothermic).

In the second row a negative deviation from Raoult's law (lower than calculated vapor pressure) means A's attract B's better than A's attract A's and B's attract B's. This causes a negative (exothermic) heat of solution.

In the third row a zero heat of solution means that A–A, B–B, and A–B interparticle attractions are all the same. This corresponds to an ideal solution which obeys Raoult's law exactly.

What sorts of substances form ideal solutions with each other?

12.102 Since the total volume is less than the sum of the two volumes, the ethanol and water must have an intermolecular attraction that results in an overall smaller volume.

12.104 **(a)** Looking at Figure 12.9(b) of the text, you would have to apply a pressure greater than the osmotic pressure to the right compartment to force pure solvent (water) through to the left compartment.

(b) Reverse osmosis uses high pressure to force water from a more concentrated solution to a less concentrated one through a semipermeable membrane. Desalination by reverse osmosis is considerably cheaper than by distillation and avoids the technical difficulties associated with desalination by freezing. Reverse osmosis does *not* involve a phase transition.

(c) To reverse the osmotic migration of water across a semipermeable membrane, an external pressure exceeding the osmotic pressure must be applied. To find the osmotic pressure of 0.70 *M* NaCl solution, we must use the van't Hoff factor because NaCl is a strong electrolyte (see Table 12.3 of the text).

The osmotic pressure of sea water is:

$$\pi = iMRT = 1.9(0.70\ \text{mol/L})(0.0821\ \text{L·atm/K·mol})(298\ \text{K}) = \mathbf{33\ atm}$$

To cause reverse osmosis a pressure in excess of 33 atm must be applied.

CHAPTER 13
INTRODUCTION TO ORGANIC CHEMISTRY

PROBLEM-SOLVING STRATEGIES AND TUTORIAL SOLUTIONS

TYPES OF PROBLEMS

Problem Type 1: Determining the Number of Structural Isomers.

Problem Type 2: Organic Nomenclature.
(a) Alkanes.
(b) Alkenes.
(c) Alkynes.
(d) Aromatic Compounds.

Problem Type 3: Addition Reactions.

Problem Type 4: Distinguishing between Structural and Geometric Isomers.

Problem Type 5: Functional Groups.

Problem Type 6: Chirality.

PROBLEM TYPE 1: DETERMINING THE NUMBER OF STRUCTURAL ISOMERS

Structural isomers are molecules that have the same molecular formula but a different order of linking the atoms. For small hydrocarbon molecules (eight or fewer carbon atoms), it is relatively easy to determine the number of structural isomers by trial and error.

EXAMPLE 13.1
How many structural isomers can be identified for hexane, C_6H_{14}?

Step 1: Write down the straight-chain structure.

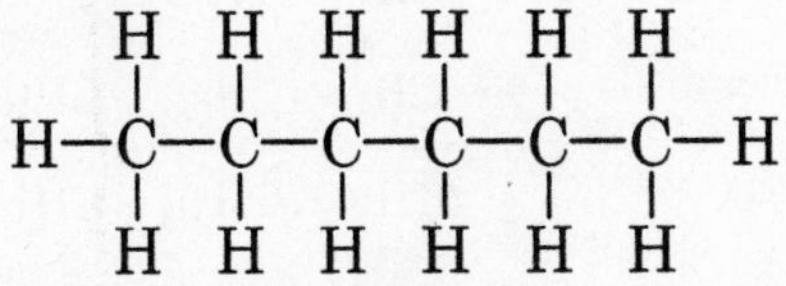

n-hexane

Step 2: By necessity, the other structures must have branched chains. First, try single methyl substituents.

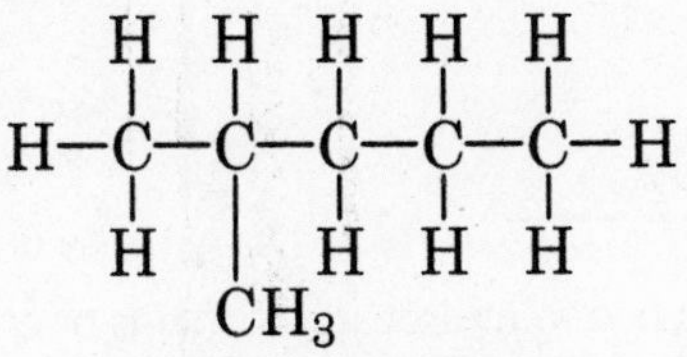

2-methylpentane

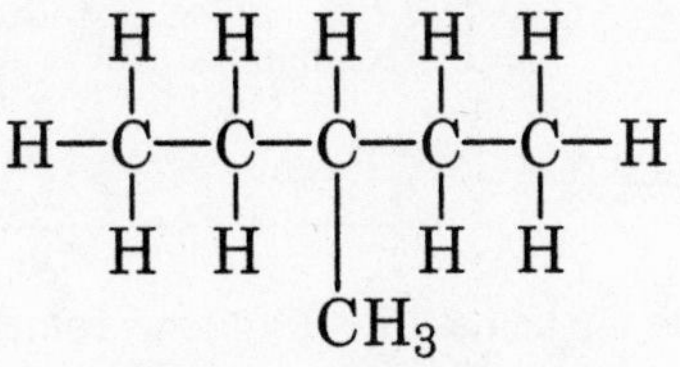

3-methylpentane

Then, try structures that have two methyl groups.

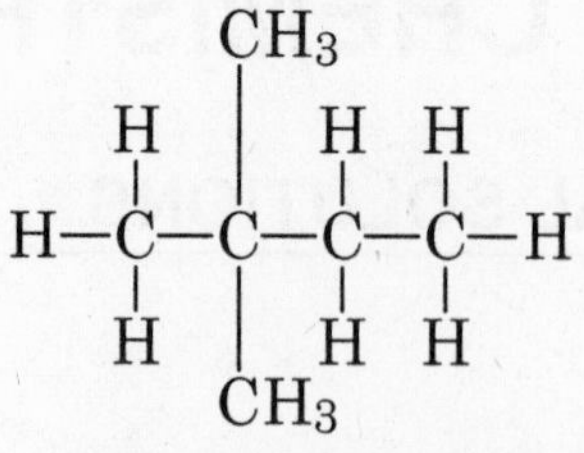

2,2-dimethylbutane

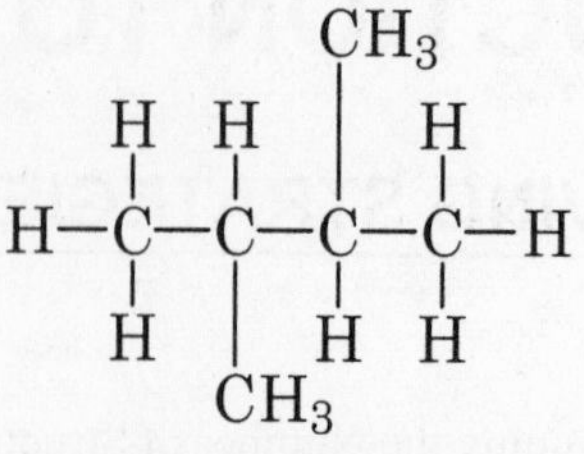

2,3-dimethylbutane

Hexane has five structural isomers, in which the numbers of carbon and hydrogen atoms remain unchanged despite the differences in structure.

PRACTICE EXERCISE

1. How many structural isomers are there of $C_4H_{10}O$? **Hint:** Consider both alcohols and ethers.

Text Problems: 8, 10

PROBLEM TYPE 2: ORGANIC NOMENCLATURE

The first step in learning the nomenclature of hydrocarbons is to know the names of the first ten straight-chain alkanes. Except for the first four members, the number of carbon atoms in each alkane is identified by the Greek prefix. See Table 13.1 below.

TABLE 13.1
The First Ten Straight-Chain Alkanes

Name of hydrocarbon	Molecular formula
Methane	CH_4
Ethane	CH_3-CH_3
Propane	$CH_3-CH_2-CH_3$
Butane	$CH_3-(CH_2)_2-CH_3$
Pentane	$CH_3-(CH_2)_3-CH_3$
Hexane	$CH_3-(CH_2)_4-CH_3$
Heptane	$CH_3-(CH_2)_5-CH_3$
Octane	$CH_3-(CH_2)_6-CH_3$
Nonane	$CH_3-(CH_2)_7-CH_3$
Decane	$CH_3-(CH_2)_8-CH_3$

Next, you need to learn the names of substituents (other than hydrogen) attached to hydrocarbon chains or aromatic compounds. For example, when a hydrogen atom is removed from methane, a $-CH_3$ fragment is left, which is called a *methyl* group. Similarly, removing a hydrogen atom from an ethane molecule gives an *ethyl* group, $-CH_2CH_3$.

These groups or substituents that are derived from alkanes are called *alkyl* groups. See Table 13.2 of your text for other common alkyl groups.

A. Alkanes

Alkanes have the general formula C_nH_{2n+2}, where $n = 1, 2, \ldots$. Alkanes contain only single covalent bonds. The bonds are said to be saturated because no more hydrogen atoms can be added to the carbon atoms. Thus, alkanes are also called *saturated hydrocarbons*.

There are other rules to follow when naming hydrocarbons.

1. Find the longest carbon chain in the molecule. The parent name of the compound is based on the longest carbon chain. For example, if the longest carbon chain contains five C atoms, the parent name of the compound is *pentane*. See Table 13.1 above.
2. Next, you must specify the name and location of any groups attached to the longest carbon chain. See Table 13.2 of your text for naming *alkyl* groups. To specify the location of the group or groups, start numbering the longest chain from the end that is closer to the carbon atom bearing the substituent group.
3. If there is more than one of a particular group attached to the longest carbon chain, you must specify the number of groups with a prefix. The prefixes are di– (2 groups), tri– (3 groups), tetra– (4 groups), and so on.
4. There can be many different types of substituents other than alkyl groups. Table 13.3 of your text lists the names of some common functional groups.

EXAMPLE 13.2
Give the correct name for the following structure:

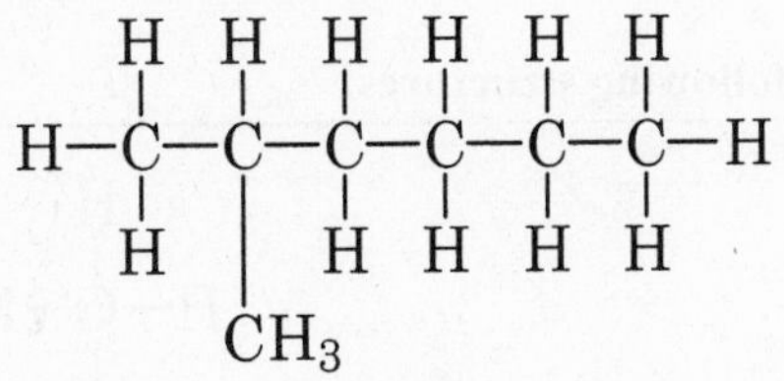

Step 1: Find the longest carbon chain. The longest chain has *six* carbons. Therefore, the parent name of the compound is *hexane*.

Step 2: Specify the name and location of any groups attached to the longest carbon chain. You should see that there is a methyl group attached to the second carbon from the left of the chain. You should number the carbon chain from the end that is closer to the carbon atom bearing the substituent group. If you start numbering from the left, the methyl group is on the second carbon in the chain. However, if you start numbering from the right, the methyl group is on the fifth carbon in the chain. Numbering from the left is correct.

The correct name for this compound is **2-methylhexane**.

Tip: You should always put a dash (–) between a number and a "word" when naming an organic compound.

EXAMPLE 13.3
Give the correct name for the following structure:

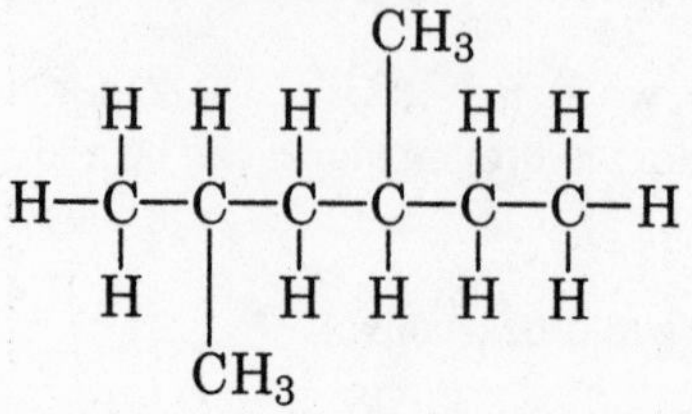

Step 1: You should find that the longest carbon chain has six carbons. The parent name is *hexane*.

Step 2: Numbering from the left, there are methyl groups on the second and fourth carbons. Since there are two methyl groups, we must specify this by using the prefix di–.

The correct name for the compound is **2,4-dimethylhexane**.

Let's examine the name of the compound if we had numbered the longest chain from right to left. The methyl groups would be on the third and fifth carbons. Hence, the name would be 3,5-dimethylhexane, which is incorrect. The correct name should always have the lowest numbering scheme as possible.

Tip: Always use commas to separate numbers when naming organic compounds.

EXAMPLE 13.4
Give the correct name for each of the following structures:

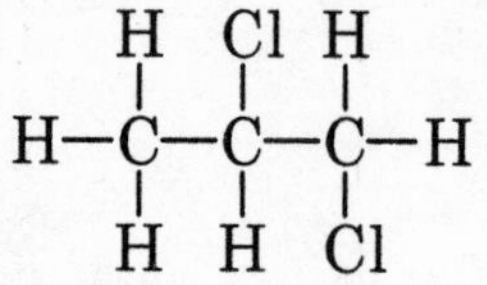

(a)

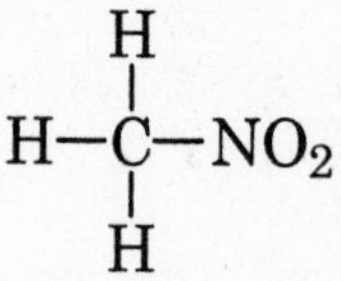

(b)

(a)
Step 1: You should find that the longest carbon chain in the molecule is *three* carbons. The parent name is *propane*.

Step 2: Referring to Table 13.3 of your text, you should find that a –Cl group is called a chloro group. There are chloro groups on the 1 and 2 carbons, numbering from right to left. (Why not number from left to right?)

Hence, the correct name for the compound is **1,2-dichloropropane**.

(b)
Step 1: There is only one carbon in the molecule, so the parent name is *methane*.

Step 2: Referring to Table 13.3 of your text, you should find that a $-NO_2$ group is called a nitro group. The correct name for the compound is **nitromethane**.

Why isn't the correct name 1-nitromethane?

PRACTICE EXERCISE

2. Give the systematic name for each of the following structural formulas:

(a) $CH_3(CH_2)_7CH_3$

(b) 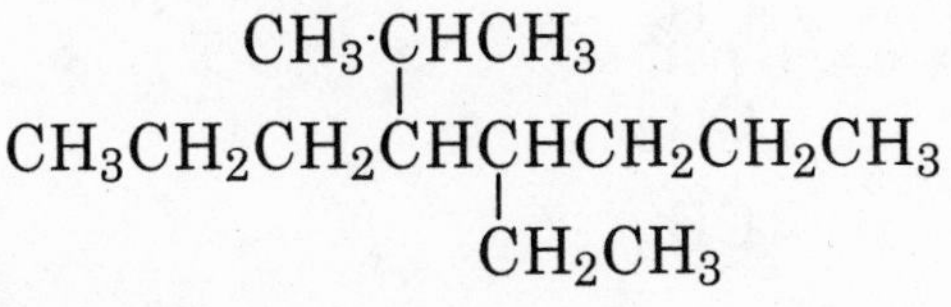

(c) $BrCH_2CH_2CHBrCH_2Br$

Text Problem: 34

B. Alkenes

Alkenes are molecules that contain at least one carbon-carbon double bond. Alkenes have the general formula C_nH_{2n}, where $n = 2, 3, \ldots$.

To name alkenes, you follow the same rules as outlined for alkanes; except, you must specify the position(s) of the carbon-carbon double bond(s), and the name of alkenes ends with an "–ene" suffix.

EXAMPLE 13.5
Give the correct name for the following structure:

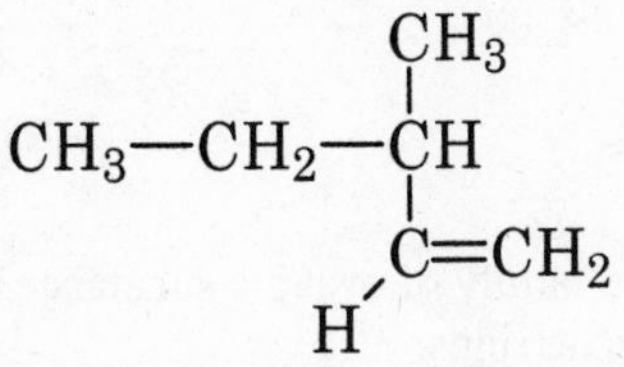

Step 1: The longest carbon chain in the molecule is five carbons. The parent name is pent*ene*, because the molecule has a double bond.

Step 2: For this molecule, we number from right to left, because we want the double bond to be at the lowest number possible. In this case, the double bond starts on the first carbon if we number from right to left. There is also a methyl substituent on the third carbon.

The correct name for this compound is **3-methyl-1-pentene**. The number 1 before pentene specifies that the double bond starts on the first carbon.

PRACTICE EXERCISE

3. Draw the structural formula for 4-methyl-2-hexene.

C. Alkynes

Alkynes contain at least one carbon-carbon triple bond. They have the general formula C_nH_{2n-2}, where $n = 2, 3, \ldots$.

To name alkynes, you follow the same rules as outlined above for alkanes; except, you must specify the position(s) of the carbon-carbon triple bond(s), and the name of alkynes ends with an "–yne" suffix.

EXAMPLE 13.6
Give the correct name for the following structure:

$$
\begin{array}{l}
CH_3-CH_2-C\equiv C-CH_2-\underset{|}{C}H-CH_3 \\
\qquad\qquad\qquad\qquad\qquad\quad \underset{|}{C}H_2 \\
\qquad\qquad\qquad\qquad\qquad\quad \underset{|}{C}H_2 \\
\qquad\qquad\qquad\qquad\qquad\quad CH_3
\end{array}
$$

Step 1: The longest carbon chain in the molecule contains nine carbons. The parent name is non*yne* because there is a triple bond.

Step 2: You should number the carbon chain from left to right, placing the triple bond on the third carbon. If you numbered from right to left, the triple bond would be on the sixth carbon. You should also notice that there is a methyl substituent on the sixth carbon.

The correct name for this molecule is **6-methyl-3-nonyne.**

PRACTICE EXERCISE

4. Give an acceptable name for each of the following structures:

(a) $(CH_3)_3CC\equiv CCH_2CH_3$

(b)
$$
\begin{array}{l}
CH_3CH_2CH_2\underset{|}{C}HC\equiv CCH_3 \\
\qquad\qquad\quad\; CH_2CH_2CH_3
\end{array}
$$

D. Aromatic Compounds

Benzene is the parent compound of this large family of organic substances. In naming aromatic compounds, we will consider both mono- and di-substituted benzene rings.

The naming of mono-substituted benzene rings (benzenes in which one H atom has been replaced by another atom or groups of atoms) is straightforward. You simply name the substituent followed by the name "benzene".

EXAMPLE 13.7
Give the correct name for each of the following structures:

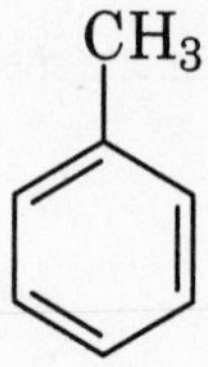

(a)

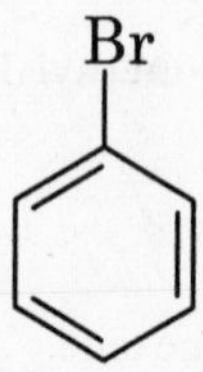

(b)

(a) The substituent is a methyl group, so the correct name is **methylbenzene**. However, there are older names that are still in common use for many compounds. Methylbenzene is usually called **toluene**.

(b) The substituent is a bromo group, so the correct name is **bromobenzene**.

For di-substituted benzenes, we must indicate the location of the second group relative to the first. For example, three different dichlorobenzenes are possible.

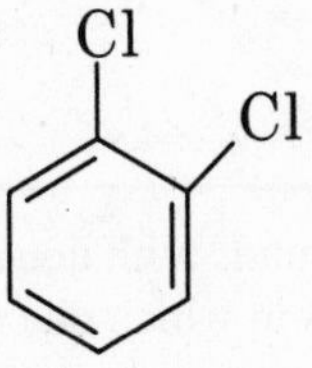

(a)

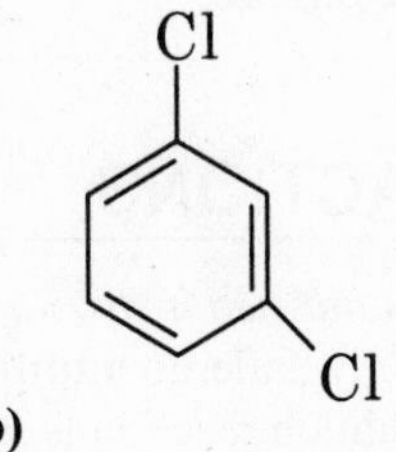

(b)

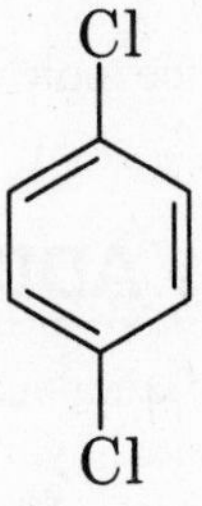

(c)

The systematic way to name these molecules is to number the carbon atoms of the benzene ring as follows:

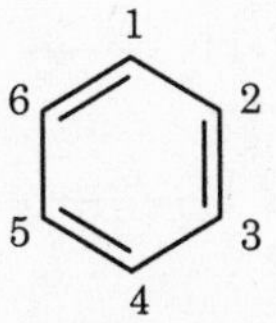

Thus, (a) would be named 1,2-dichlorobenzene, (b) 1,3-dichlorobenzene, and (c) 1,4-dichlorobenzene. However, the prefixes *o*- (*ortho*-), *m*- (*meta*), and *p*- (*para*-) are used more often to denote the relative positions of the two substituted groups. *Ortho*- designates 1,2 substituents, *meta*- designates 1,3 substituents, and *para*- designates 1,4 substituents. Thus, (a) is named *o*-dichlorobenzene, (b) *m*-dichlorobenzene, and (c) *p*-dichlorobenzene.

In compounds with two different substituted groups, the positions of the substituents can be specified with numbers or with the *o*-, *m*-, or *p*- prefixes.

PRACTICE EXERCISE

5. Draw structural formulas for:

(a) ethylbenzene (b) *m*-bromochlorobenzene (c) *p*-nitrotoluene

Text Problem: 36

Finally in some molecules, benzene is named as a substituent. A benzene molecule minus a hydrogen atom (C_6H_5) is called a *phenyl* group.

EXAMPLE 13.8

Give the correct name for the following structure:

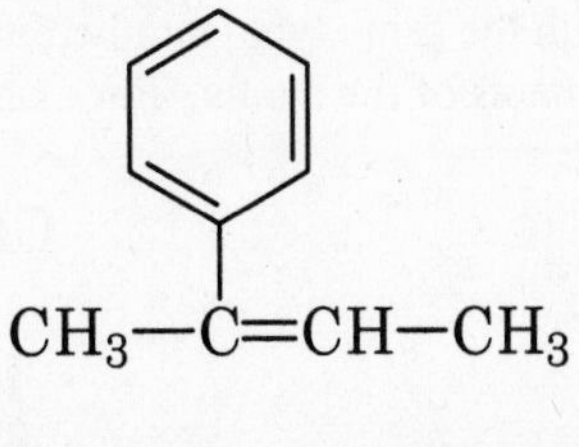

For this molecule, it would be difficult to name the carbon chain as a substituent on the benzene ring. Therefore, we name the benzene ring as a substituent on the carbon chain.

Step 1: The longest carbon chain contains four carbons. The parent name is but*ene* because the molecule contains a double bond.

Step 2: You should number the carbon chain from left to right so that both the double bond and the phenyl group are on the second carbon.

The correct name for the molecule is **2-phenyl-2-butene**.

PROBLEM TYPE 3: ADDITION REACTIONS

Alkenes, alkynes, and aromatic compounds are called *unsaturated hydrocarbons*, compounds with double or triple carbon-carbon bonds. Unsaturated hydrocarbons commonly undergo **addition reactions** in which one molecule adds to another to form a single product. An example of an addition reaction is *hydrogenation*, which is the addition of molecular hydrogen to compounds containing C=C and C≡C bonds.

$$CH_2{=}CH_2 + H_2 \longrightarrow CH_3{-}CH_3$$

$$CH{\equiv}CH + 2\,H_2 \longrightarrow CH_3{-}CH_3$$

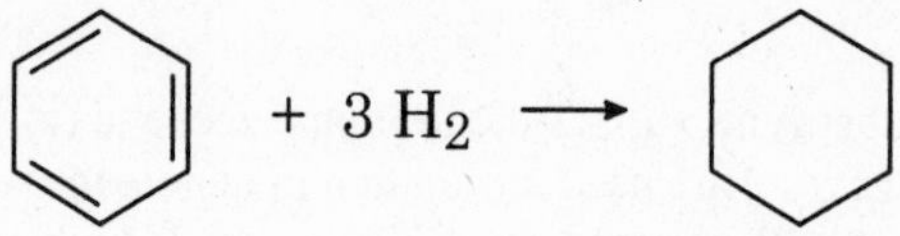

Alkanes are called *saturated hydrocarbons* because no more hydrogen can be added to the carbon atoms. The carbon atoms in a saturated hydrocarbon are already bonded to the maximum number of H atoms.

$$CH_3{-}CH_3 + H_2 \longrightarrow \text{no reaction}$$

Other addition reactions involve an addition of HX or X_2 to the multiple bond, where X represents a halogen (Cl, Br, or I). Examples of these addition reactions are:

$$CH_2{=}CH_2 + HCl \longrightarrow CH_3{-}CH_2Cl$$

$$CH_2{=}CH_2 + Cl_2 \longrightarrow CH_2Cl{-}CH_2Cl$$

PROBLEM TYPE 4: DISTINGUISHING BETWEEN STRUCTURAL AND GEOMETRIC ISOMERS

Recall that **structural isomers** are molecules that have the same molecular formula but a different order of linking the atoms. **Geometric isomers** are molecules with the same type, number, and order of attachment of atoms and the same chemical bonds, but different spatial arrangements of the atoms. For example, 1,2-dichloroethene exists as two geometric isomers.

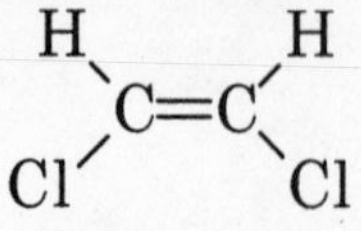

cis-1,2-dichloroethene

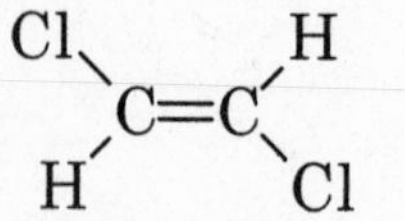

trans-1,2-dichloroethene

Note that in the *cis* isomer, the two Cl atoms (and the two H atoms) are adjacent to each other, whereas in the *trans* isomer, the Cl atoms are on the opposite side of the C=C bond.

EXAMPLE 13.9

Draw Lewis structures for all compounds, not including cyclic compounds, with the molecular formula C_5H_{10} and determine which are geometric isomers.

Alkenes have the general formula C_nH_{2n}. Thus the structures with molecular formula C_5H_{10} should be alkenes.

To draw the correct Lewis structures, follow the procedure outlined in Chapter 9. We can draw six Lewis structures as follows:

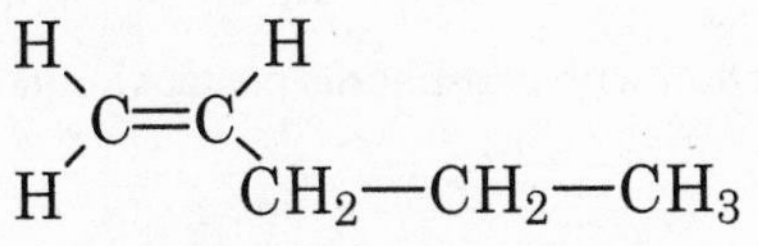

1-pentene

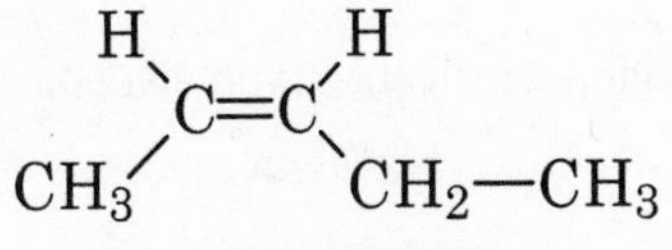

cis-2-pentene

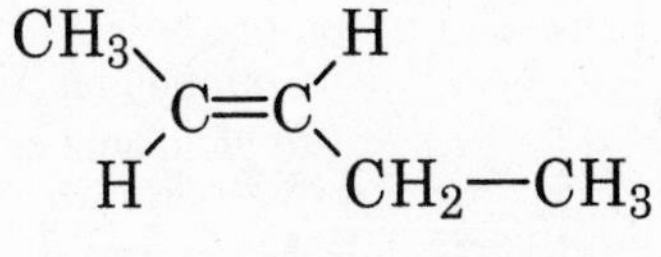

trans-2-pentene

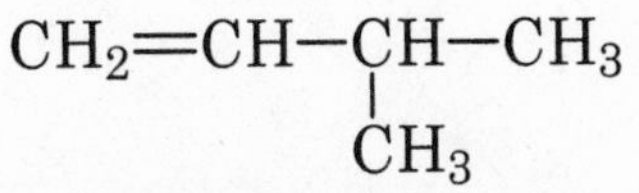

3-methyl-1-butene

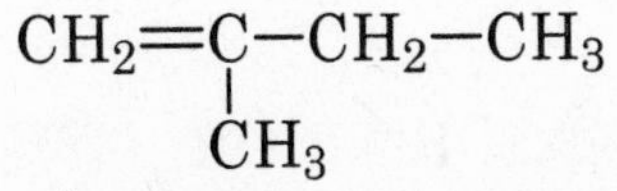

2-methyl-1-butene

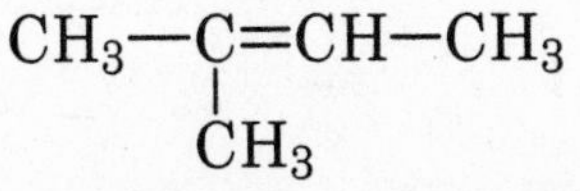

2-methyl-2-butene

The geometric isomers are *cis*- and *trans*-2-pentene. 1-pentene, 3-methyl-1-butene, 2-methyl-1-butene, and 2-methyl-2-butene are structural isomers to the geometric isomers because the atoms are attached in a different order.

Why don't 1-pentene, 3-methyl-1-butene, 2-methyl-1-butene, and 2-methyl-2-butene have geometric isomers?

PRACTICE EXERCISE

6. Draw the structural formula for 3-bromo-2,5-dimethyl-*trans*-3-hexene.

Text Problem: 18

PROBLEM TYPE 5: FUNCTIONAL GROUPS

We will discuss the most common organic functional groups, with emphasis on classes of compounds in which the functional groups include oxygen or nitrogen.

1. **Alcohols.** All alcohols contain the hydroxyl group, –OH. Figure 13.7 of your text shows some common alcohols. Alcohols typically undergo esterification (formation of an ester) reactions with carboxylic acids. Oxidation to aldehydes, ketones, and carboxylic acids are also common reactions.

2. **Ethers.** Ethers contain the R–O–R' linkage, where R and R' are either an alkyl group or a group derived from an aromatic hydrocarbon.

3. **Aldehydes and Ketones.** These compounds contain the carbonyl functional group.

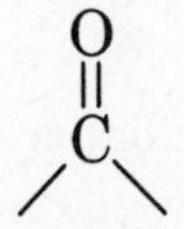

The difference between aldehydes and ketones is that in aldehydes at least one hydrogen atom is bonded to the carbon atom of the carbonyl group. In ketones, no hydrogen atoms are bonded to the carbonyl carbon atom. Common reactions include reduction to yield alcohols. Oxidation of aldehydes yields carboxylic acids.

4. **Carboxylic Acids.** These compounds contain the carboxyl group, –COOH.

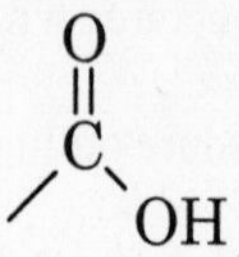

Common reactions include esterification with alcohols and reaction with phosphorus pentachloride to yield acid chlorides.

5. **Esters.** Esters have the general formula R'COOR.

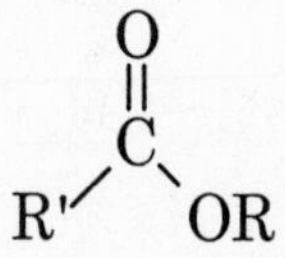

R' can be H, an alkyl, or an aromatic hydrocarbon group, and R is an alkyl or an aromatic hydrocarbon group. A common reaction is hydrolysis to yield acids and alcohols.

6. **Amines.** Amines are organic bases. They have the general formula R_3N, where one of the R groups must be an alkyl group or an aromatic hydrocarbon group. A common reaction is formation of ammonium salts with acids.

Table 13.4 of your text summarizes important functional groups. For more detailed discussions of reactions, see Section 13.3 of your text.

EXAMPLE 13.12

Identify the functional groups in the following molecules: (a) $C_5H_{11}OH$, (b) CH_3CHO, (c) $C_3H_7OCH_3$, (d) $CH_3COC_2H_5$, (e) CH_3COOCH_3.

Learning to recognize functional groups requires memorization of their structural formulas. Table 13.4 of your text shows a number of the important functional groups.

(a) $C_5H_{11}OH$ contains a *hydroxyl* group. It is an *alcohol.*

(b) CH_3CHO is a way to represent

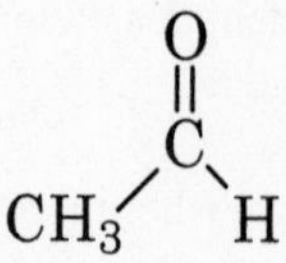

on a single line of type. C=O is a carbonyl group. Since there is a hydrogen atom bonded to the carbonyl carbon, CH_3CHO is an *aldehyde.*

(c) $C_3H_7OC_2H_5$ contains a C–O–C group and is therefore an *ether.*

(d) $CH_3COC_2H_5$ contains a carbonyl group. Since a hydrogen atom is not bonded to the carbonyl carbon, this molecule is a *ketone.*

(e) CH_3COOCH_3 is a condensed structural formula for an *ester*.

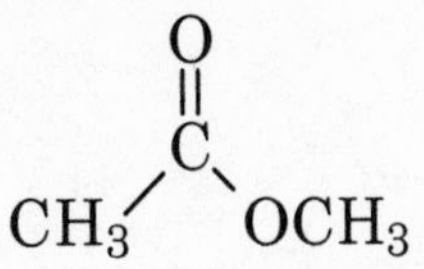

PRACTICE EXERCISE

7. Indicate the functional groups by name that are in the following molecules.

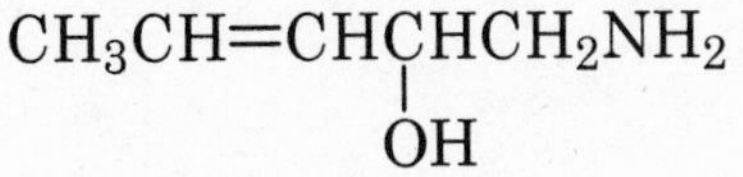

(a)

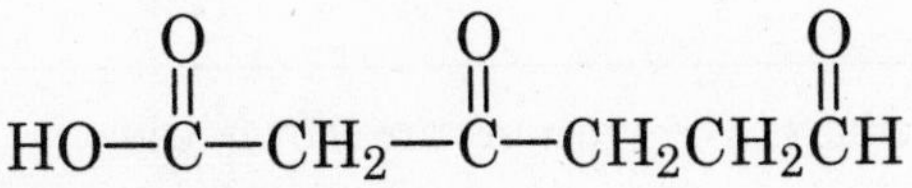

(b)

Text Problems: 22, 24, 26, 28

PROBLEM TYPE 6: CHIRALITY

Compounds that come as mirror-image pairs can be compared with left-handed and right-handed gloves and are thus referred to as **chiral**, or handed, molecules. While every molecule can have a mirror image, the *chiral* mirror-image pairs are *nonsuperimposable*. Conversely, achiral (nonchiral) pairs are superimposable. See Figure 13.11 of your text for an illustration of superimposable compared to nonsuperimposable.

Observations show that most simple chiral molecules contain at least one *asymmetric* carbon atom; that is, a carbon atom bonded to four different atoms or groups of atoms.

> **Tip:** Consider your hands when thinking about chirality. If you view your left hand in a mirror, the mirror image of your left hand is a right hand. However, your left and right hands are nonsuperimposable. To verify this, try putting a "left-handed" glove on your right hand.

EXAMPLE 13.13

Classify the following objects as chiral or achiral:

(a) shoe **(b) screw** **(c) fork** **(d) coffee cup**

A shoe and a screw are chiral. The mirror image of a right shoe is a left shoe. A left shoe is not superimposable on a right shoe. A mirror image of a screw with clockwise threads will have counterclockwise threads. The screws would not be superimposable.

A fork and a coffee cup are achiral. Their mirror images are superimposable.

PRACTICE EXERCISE

8. Are the following molecules chiral?

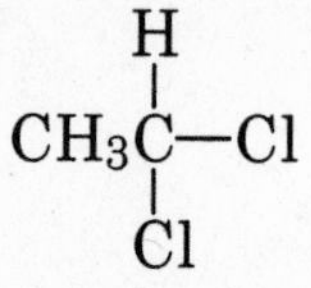

(a) 1,1-dichloroethane

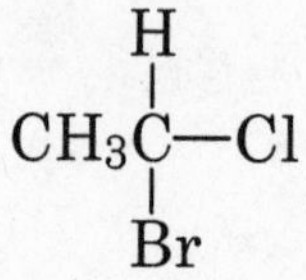

(b) 1-bromo-1-chloroethane

Text Problem: 32

ANSWERS TO PRACTICE EXERCISES

1. There are seven structural isomers—four alcohols and three ethers. The structures are shown below.

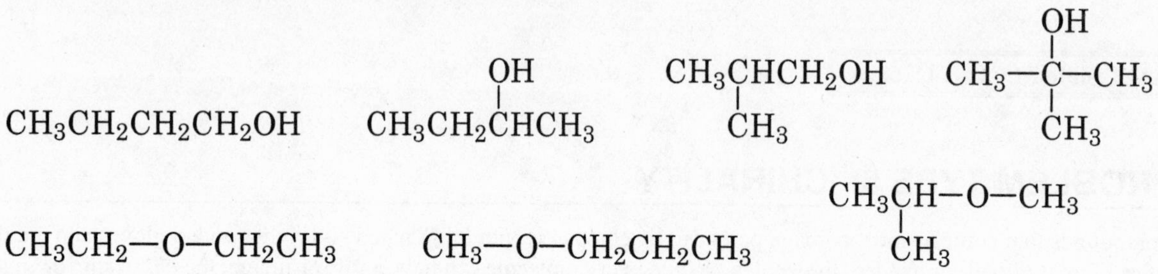

2. **(a)** nonane **(b)** 4-ethyl-5-isopropyloctane **(c)** 1,2,4-tribromobutane

3.

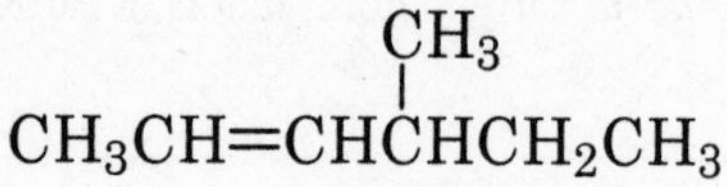

4. **(a)** 2,2-dimethyl-3-hexyne **(b)** 4-propyl-2-heptyne

5. The structures are:

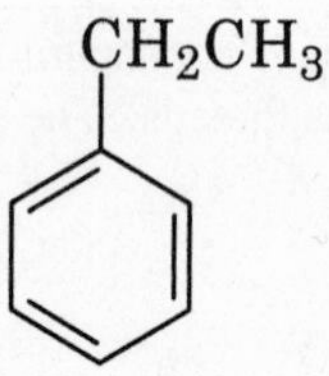

(a)

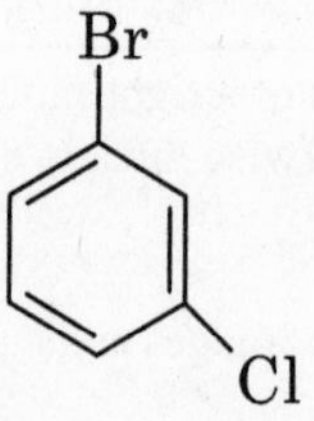

(b)

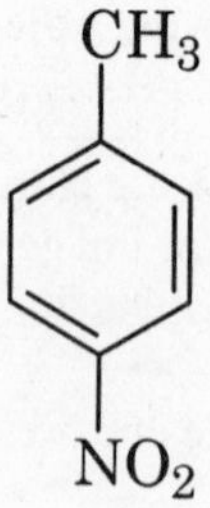

(c)

6. The structure of 3-bromo-2,5-dimethyl-*trans*-3-hexene is:

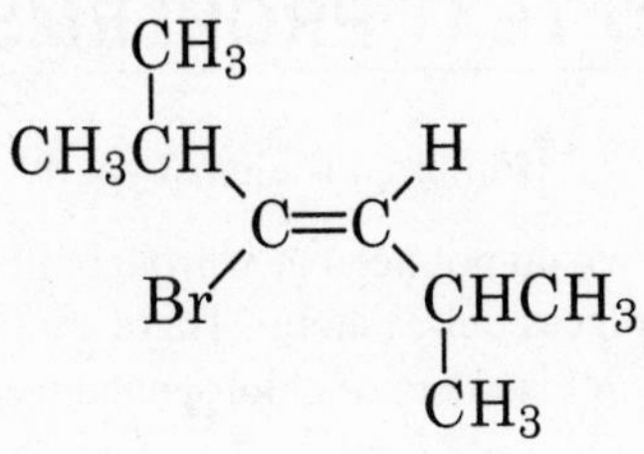

7. **(a)** carbon-carbon double bond, hydroxyl (–OH), and amine ($R–NH_2$).

(b) carboxyl (–COOH), carbonyl (ketone), and carbonyl (aldehyde).

8. **(a)** There are only three different groups bonded to the second carbon atom. This molecule is *achiral.*

(b) Replacing one of the chloro groups with a bromo group places four different groups on the second carbon atom. This molecule is *chiral.*

SOLUTIONS TO SELECTED TEXT PROBLEMS.

13.8 Determining the Number of Structural Isomers, Problem Type 1.

We are starting with *n*-pentane, so we do not need to worry about any branched chain structures. In the chlorination reaction, a Cl atom replaces one H atom. There are three different carbons on which the Cl atom can be placed. Hence, *three* structural isomers of chloropentane can be derived from *n*–pentane:

$CH_3CH_2CH_2CH_2CH_2Cl$ $CH_3CH_2CH_2CHClCH_3$ $CH_3CH_2CHClCH_2CH_3$

13.10 Both alkenes and cycloalkanes have the general formula C_nH_{2n}. Let's start with C_3H_6. It could be an alkene or a cycloalkane.

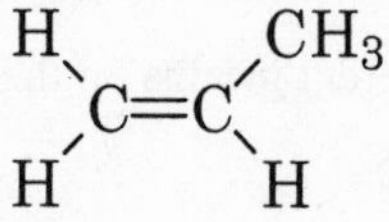
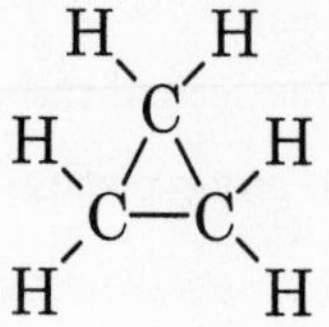

Now, let's replace one H with a Br atom to form C_3H_5Br. *Four* isomers are possible.

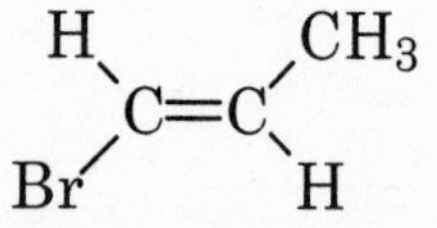
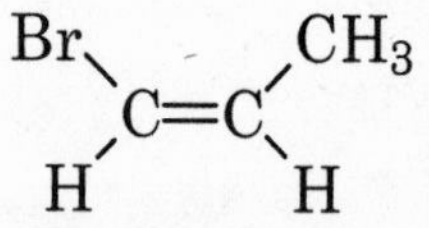
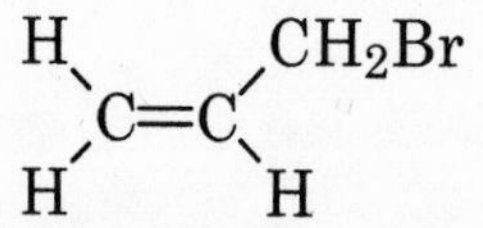
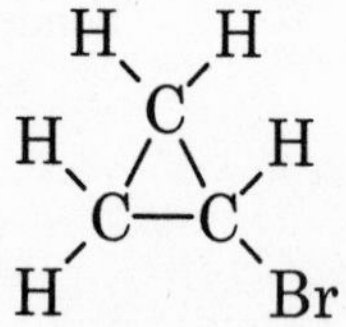

There is only one isomer for the cycloalkane. Note that all three carbons are equivalent in this structure.

13.12 **(a)** This compound could be an alkene or a cycloalkane; both have the general formula, C_nH_{2n}.

(b) This could be an alkyne with general formula, C_nH_{2n-2}. It could also be a hydrocarbon with two double bonds (a diene). It could be a cyclic hydrocarbon with one double bond (a cycloalkene).

(c) This must be an alkane; the formula is of the C_nH_{2n+2} type.

(d) This compound could be an alkene or a cycloalkane; both have the general formula, C_nH_{2n}.

(e) This compound could be an alkyne with one triple bond, or it could be a cyclic alkene (unlikely because of ring strain).

13.14 One compound is an alkane; the other is an alkene. Alkenes characteristically undergo addition reactions with hydrogen, with halogens (Cl_2, Br_2, I_2) and with hydrogen halides (HCl, HBr, HI). Alkanes do not react with these substances under ordinary conditions.

13.16 In this problem you are asked to calculate the standard enthalpy of reaction. This type of problem was covered in Chapter 6, Problem Type 4A.

$$\Delta H^\circ_{rxn} = \Sigma n\, \Delta H^\circ_f\,(\text{products}) - \Sigma m\, \Delta H^\circ_f\,(\text{reactants})$$

$$\Delta H^\circ_{rxn} = \Delta H^\circ_f\,(C_6H_6) - 3\,\Delta H^\circ_f\,(C_2H_2)$$

You can look up ΔH_f° values in Appendix 2 of your textbook.

$$\Delta H^\circ_{\mathbf{rxn}} = (1\ \text{mol})(49.04\ \text{kJ/mol}) - (3\ \text{mol})(226.6\ \text{kJ/mol}) = \mathbf{-630.8\ kJ}$$

13.18 In this problem you must distinguish between *cis* and *trans* isomers. Recall that *cis* means that two particular atoms (or groups of atoms) are adjacent to each other, and *trans* means that the atoms (or groups of atoms) are on opposite sides in the structural formula.

In (a), the Cl atoms are adjacent to each other. This is the *cis* isomer. In (b), the Cl atoms are on opposite sides of the structure. This is the *trans* isomer.

The names are: **(a)** *cis*-1,2-dichlorocyclopropane; and **(b)** *trans*-1,2-dichlorocyclopropane.

Are any other dichlorocyclopropane isomers possible?

13.22 Functional Groups, Problem Type 5.

(a) $H_3C–O–CH_2–CH_3$ contains a C–O–C group and is therefore an *ether*.

(b) This molecule contains an RNH_2 group and is therefore an *amine*.

(c) This molecule is an *aldehyde*. It contains a carbonyl group in which one of the atoms bonded to the carbonyl carbon is a hydrogen atom.

(d) This molecule also contains a carbonyl group. However, in this case there are no hydrogen atoms bonded to the carbonyl carbon. This molecule is a *ketone*.

(e) This molecule contains a carboxyl group. It is a *carboxylic acid*.

(f) This molecule contains a hydroxyl group (–OH). It is an *alcohol*.

(g) This molecule has both an RNH_2 group and a carboxyl group. It is therefore both an *amine* and a *carboxylic acid*, commonly called an *amino acid*.

13.24 Alcohols react with carboxylic acids to form esters. The reaction is:

$$HCOOH + CH_3OH \longrightarrow HCOOCH_3 + H_2O$$

The structure of the product is:

$$H-\overset{\overset{\displaystyle O}{\|}}{C}-O-CH_3$$ (methyl formate)

13.26 Alcohols can be oxidized to ketones under controlled conditions. The possible starting compounds are:

$$CH_3CH_2CH_2\overset{\overset{\displaystyle OH}{|}}{C}HCH_3 \qquad CH_3CH_2\overset{\overset{\displaystyle OH}{|}}{C}HCH_2CH_3 \qquad (CH_3)_2CH\overset{\overset{\displaystyle OH}{|}}{C}HCH_3$$

The corresponding products are:

$$CH_3CH_2CH_2\overset{\overset{\displaystyle O}{\|}}{C}CH_3 \qquad CH_3CH_2\overset{\overset{\displaystyle O}{\|}}{C}CH_2CH_3 \qquad (CH_3)_2CH\overset{\overset{\displaystyle O}{\|}}{C}CH_3$$

Why isn't the alcohol $CH_3CH_2CH_2CH_2CH_2OH$ a possible starting compound?

13.28 **(a)** ketone **(b)** ester **(c)** ether

13.32 Chirality, Problem Type 6.

Recall that most simple chiral molecules contain at least one asymmetric carbon atom; that is, a carbon atom bonded to four different atoms or groups of atoms. You should find that the second carbon of (a) and also the second carbon of (c) are bonded to four different atoms or groups of atoms. Therefore, the amino acids **(a)** and **(c)** are chiral.

13.34 **(a)** This is a six-carbon chain with a methyl group on the third carbon.

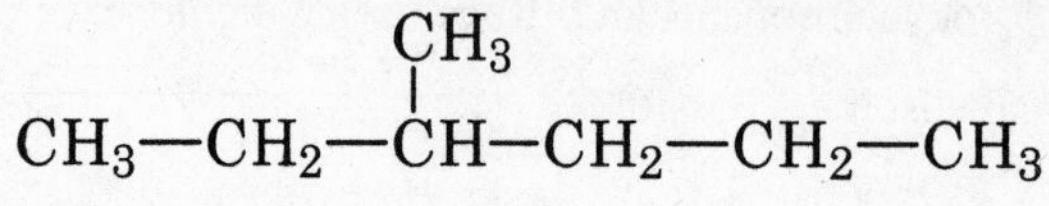

(b) This is a six carbon ring with chlorine atoms on the 1,3, and 5 carbons.

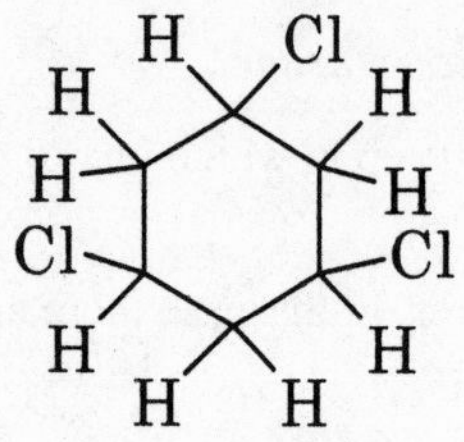

Note: The carbon atoms in the ring have been omitted for simplicity.

(c) This is a five carbon chain with methyl groups on the 2 and 3 carbons.

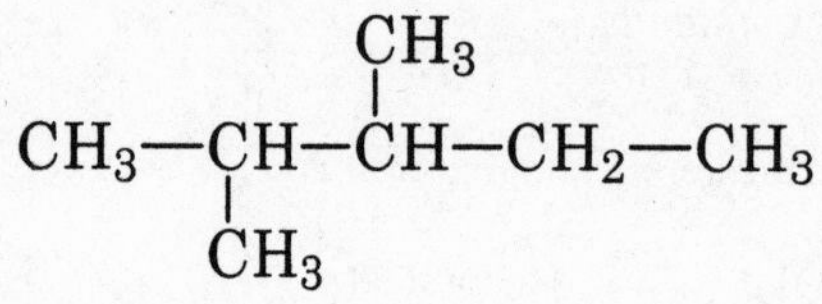

(d) This is a five carbon chain with a bromine atom on the fourth carbon and a phenyl group (a benzene molecule minus a hydrogen atom, C_6H_5) on the second carbon.

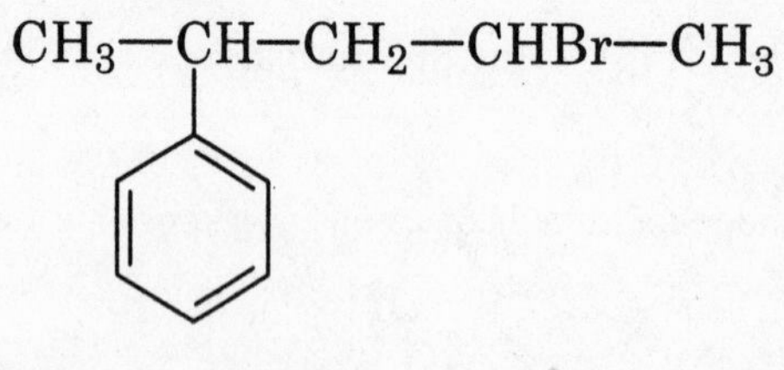

(e) This is an eight carbon chain with methyl groups on the 3, 4, and 5 carbons.

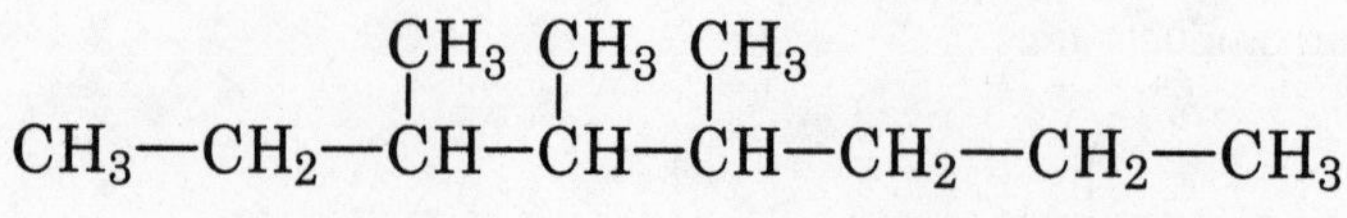

13.36 Organic Nomenclature, Problem Type 2.

When a benzene ring has more than *two* substituents, you must specify the location of the substituents with numbers. Remember to number the ring so that you end up with the lowest numbering scheme as possible, giving preference to alphabetical order.

(a) Since a chloro group comes alphabetically before a methyl group, let's start by numbering the top carbon of the ring as 1. If we number clockwise, this places the second chloro group on carbon 3 and a methyl group on carbon 4.

This compound is **1,3–dichloro–4–methylbenzene**.

(b) If we start numbering counterclockwise from the bottom carbon of the ring, the name is 2–ethyl–1,4–dinitrobenzene. Numbering clockwise from the top carbon gives 3–ethyl–1,4–dinitrobenzene.

Numbering as low as possible, the correct name is **2–ethyl–1,4–dinitrobenzene**.

(c) Again, keeping the numbers as low as possible, the correct name for this compound is **1,2,4,5–tetramethylbenzene**. You should number clockwise from the top carbon of the ring.

13.38 See Problem Type 4C, Chapter 6.

If we rearrange the equations given and multiply times the necessary factors, we have:

$$2\,CO_2\,(g) + 2\,H_2O\,(l) \longrightarrow C_2H_4\,(g) + 3\,O_2\,(g) \qquad \Delta H° = 1411\text{ kJ}$$

$$C_2H_2\,(g) + \frac{5}{2}\,O_2\,(g) \longrightarrow 2\,CO_2\,(g) + H_2O\,(l) \qquad \Delta H° = -1299.5\text{ kJ}$$

$$H_2\,(g) + \frac{1}{2}\,O_2\,(g) \longrightarrow H_2O\,(l) \qquad \Delta H° = -285.8\text{ kJ}$$

$$C_2H_2\,(g) + H_2\,(g) \longrightarrow C_2H_4\,(g) \qquad \mathbf{\Delta H° = -174\text{ kJ}}$$

The heat of hydrogenation for acetylene is **–174 kJ**.

13.40 To form a hydrogen bond *with water* a molecule must have at least one H–F, H–O, or H–N bond, *or* must contain an O, N, or F atom. The following can form hydrogen bonds with water:

(a) carboxylic acids **(c)** ethers **(d)** aldehydes **(f)** amines

13.42 In any stoichiometry problem, you must start with a balanced equation. The balanced equation for the combustion reaction is:

$$2\,C_8H_{18}\,(l) + 25\,O_2\,(g) \rightarrow 16\,CO_2\,(g) + 18\,H_2O\,(l)$$

To find the number of moles of octane in one liter, use density as a conversion factor to find grams of octane, then use the molar mass of octane to convert to moles of octane. The strategy is:

L octane → mL octane → g octane → mol octane

$$1.0\,\cancel{L} \times \left(\frac{1000\,\cancel{mL}}{1\,\cancel{L}}\right) \times \left(\frac{0.70\,\cancel{g}\,C_8H_{18}}{1\,\cancel{mL}\,C_8H_{18}}\right) \times \left(\frac{1\text{ mol }C_8H_{18}}{114.2\,\cancel{g}\,C_8H_{18}}\right) = 6.13\text{ mol }C_8H_{18}$$

Using the mole ratio from the balanced equation, the number of moles of oxygen used is:

$$6.13\,\cancel{mol}\,C_8H_{18} \times \left(\frac{25\text{ mol }O_2}{2\,\cancel{mol}\,C_8H_{18}}\right) = 76.6\text{ mol }O_2$$

From the ideal gas equation, we can calculate the volume of oxygen.

$$V_{O_2} = \frac{n_{O_2}RT}{P} = \frac{(76.6\ \text{mol})(293\ \text{K})}{1.00\ \text{atm}} \times \frac{0.0821\ \text{L}\cdot\text{atm}}{\text{mol}\cdot\text{K}} = 1.84 \times 10^3\ \text{L}$$

Air is only 22% O_2 by volume. Thus, 100 L of air will contain 22 L of O_2. Setting up the appropriate conversion factor, we find that the volume of air is:

$$\textbf{? vol of air} = 1.84 \times 10^3\ \text{L}\ O_2 \times \left(\frac{100\ \text{L air}}{22\ \text{L}\ O_2}\right) = \mathbf{8.4 \times 10^3\ L\ air}$$

13.44 **(a)** A benzene ring has six carbon-carbon bonds; hence, benzene has **six** C–C sigma bonds.

(b) Cyclobutane has four carbon-carbon bonds; hence, cyclobutane has **four** sigma bonds.

(c) Looking at the carbon skeleton of 2–methyl–3–ethylpentane, you should find **seven** C–C sigma bonds.

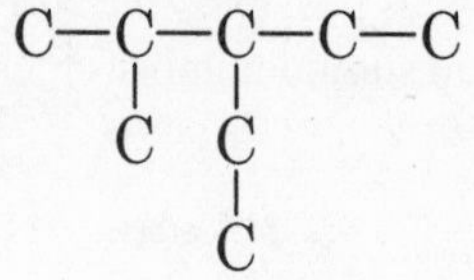

13.46 This problem relates back to Problem Type 5, Chapter 3—The Experimental Determination of Empirical Formulas.

(a) The easiest way to calculate the mg of C in CO_2 is by mass ratio. There are 12.01 g of C in 44.01 g CO_2 or 12.01 mg C in 44.01 mg CO_2.

$$\textbf{? mg C} = 57.94\ \text{mg}\ CO_2 \times \left(\frac{12.01\ \text{mg C}}{44.01\ \text{mg}\ CO_2}\right) = \mathbf{15.81\ mg\ C}$$

Similarly,

$$\textbf{? mg H} = 11.85\ \text{mg}\ H_2O \times \left(\frac{2.016\ \text{mg H}}{18.02\ \text{mg}\ H_2O}\right) = \mathbf{1.33\ mg\ H}$$

The mg of oxygen can be found by difference.

$$\textbf{? mg O} = 20.63\ \text{mg Y} - 15.81\ \text{mg C} - 1.33\ \text{mg H} = \mathbf{3.49\ mg\ O}$$

(b) *Step 1:* Calculate the number of moles of each element present in the sample. Use molar mass as a conversion factor.

$$?\ \text{mol C} = 15.81\ \text{mg C} \times \frac{1\ \text{g}}{1000\ \text{mg}} \times \frac{1\ \text{mol C}}{12.01\ \text{g C}} = 1.316 \times 10^{-3}\ \text{mol C}$$

Similarly,

$$?\ \text{mol H} = 1.33\ \text{mg H} \times \frac{1\ \text{g}}{1000\ \text{mg}} \times \frac{1\ \text{mol H}}{1.008\ \text{g H}} = 1.319 \times 10^{-3}\ \text{mol H}$$

$$?\ \text{mol O} = 3.49\ \text{mg O} \times \frac{1\ \text{g}}{1000\ \text{mg}} \times \frac{1\ \text{mol O}}{16.00\ \text{g O}} = 2.18 \times 10^{-4}\ \text{mol O}$$

Thus, we arrive at the formula $C_{1.316 \times 10^{-3}}H_{1.319 \times 10^{-3}}O_{2.18 \times 10^{-4}}$, which gives the identity and the ratios of atoms present. However, chemical formulas are written with whole numbers.

Step 2: Try to convert to whole numbers by dividing all the subscripts by the smallest subscript.

$$\text{C: } \frac{1.316 \times 10^{-3}}{2.18 \times 10^{-4}} = 6.04 \approx 6 \qquad \text{H: } \frac{1.319 \times 10^{-3}}{2.18 \times 10^{-4}} = 6.05 \approx 6 \qquad \text{O: } \frac{2.18 \times 10^{-4}}{2.18 \times 10^{-4}} = 1.00$$

This gives us the empirical formula, $\mathbf{C_6H_6O}$.

(c) The presence of six carbons and a corresponding number of hydrogens suggests a benzene derivative. A plausible structure is shown below.

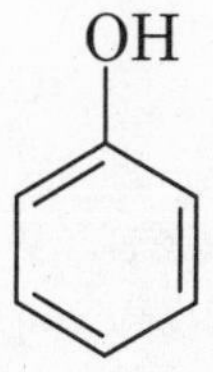

13.48 In this problem, you must consider the *cis/trans* possibilities about each of the two double bonds. Three geometric isomers are possible.

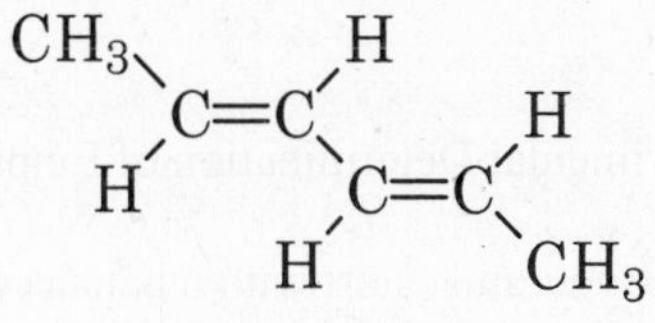

trans, trans

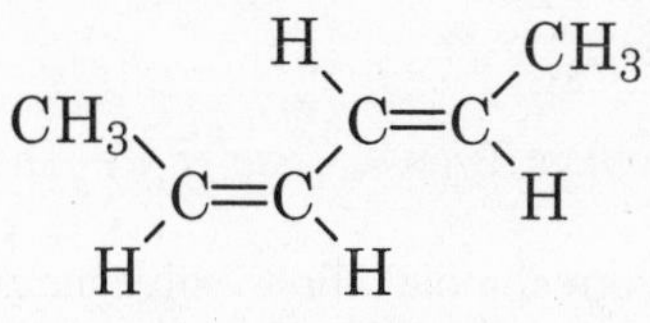

cis, trans

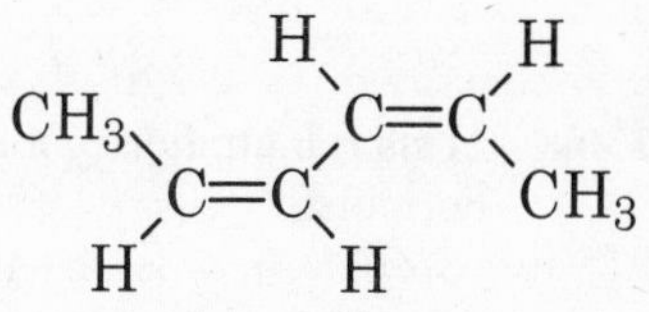

cis, cis

13.50 **(a)** In comparing the compound in part (a) with the starting alkyne, it is clear that a molecule of HBr has been added to the triple bond. The reaction is:

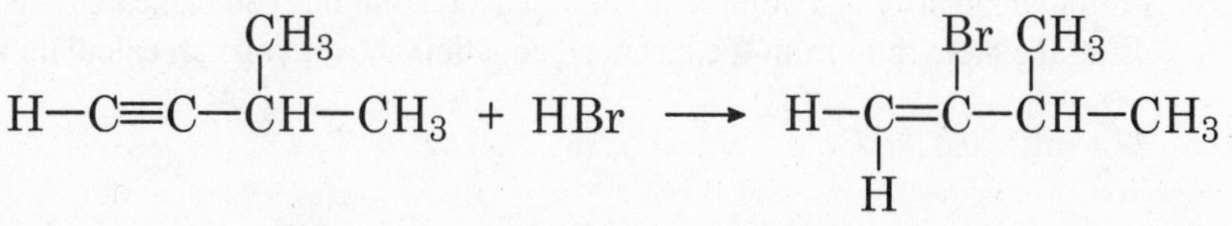

(b) This compound can be made from the product formed in part (a) by addition of bromine to the double bond.

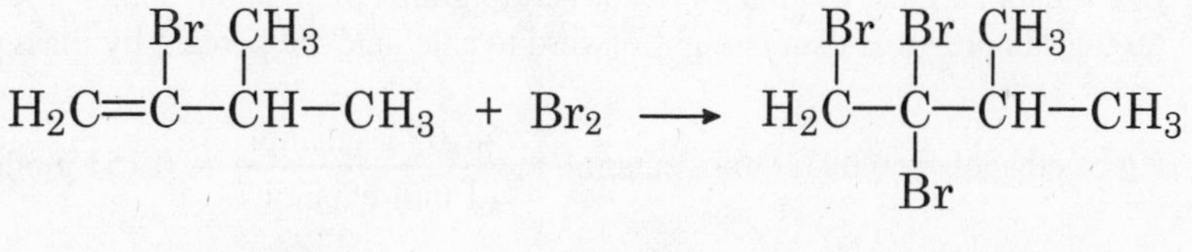

(c) This compound can be made from the product of part (a) by addition of hydrogen to the double bond.

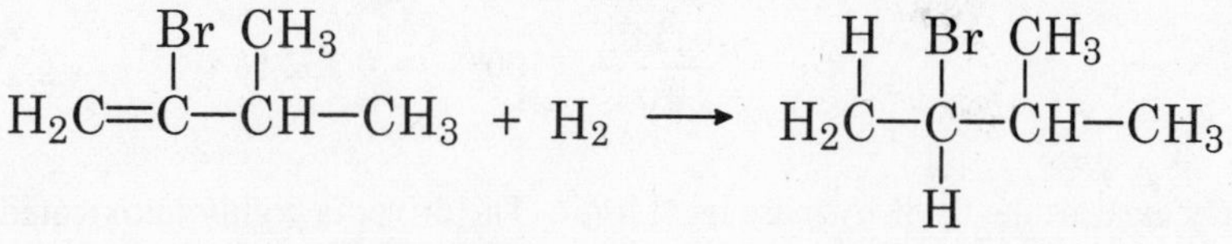

13.52 **(a)** alkyne **(b)** alcohol **(c)** ether **(d)** aldehyde **(e)** carboxylic acid **(f)** amine

13.54 In Chapter 11, we found that salts with their electrostatic intermolecular attractions had low vapor pressures and thus high boiling points. Ammonia and its derivatives (amines) are molecules with dipole–dipole attractions. If the nitrogen has one direct N–H bond, the molecule will have hydrogen bonding. Even so, these molecules will have much weaker intermolecular attractions than ionic species and hence higher vapor pressures. Thus, if we could convert the neutral ammonia–type molecules into salts, their vapor pressures, and thus associated odors, would decrease. Lemon juice contains acids which can react with ammonia–type (amine) molecules to form ammonium salts.

$$NH_3 + H^+ \longrightarrow NH_4^+ \qquad\qquad RNH_2 + H^+ \longrightarrow RNH_3^+$$

13.56 Acetone is a ketone with the formula, CH_3COCH_3. We must write the structure of an aldehyde that has the same number and types of atoms (C_3H_6O). Removing the aldehyde functional group (–CHO) from the formula leaves C_2H_5. This is the formula of an ethyl group. The aldehyde that is a structural isomer of acetone is:

$$CH_3CH_2\overset{\overset{\displaystyle O}{\|}}{C}\text{—}H$$

13.58 This is a titration problem, Problem Type 8, Chapter 4.

Step 1: In order to have the correct mole ratio to solve the problem, you must start with a balanced chemical equation. Dichromate ($Cr_2O_7^{2-}$) is an oxidizing agent that oxidizes ethanol to acetic acid. The balanced equation is given in Section 13.3 of your text.

$$3CH_3CH_2OH + 2K_2Cr_2O_7 + 8H_2SO_4 \rightarrow 3CH_3COOH + 2Cr_2(SO_4)_3 + 2K_2SO_4 + 11H_2O$$

Step 2: From the molarity and volume of the $K_2Cr_2O_7$ solution, you can calculate moles of $K_2Cr_2O_7$. Then, using the mole ratio from the balanced equation above, you can calculate moles of ethanol.

28.6 mL = 0.0286 L

$$?\ \text{mol ethanol} = \frac{0.0765\ \text{mol}\ K_2Cr_2O_7}{1\ \text{L of soln}} \times (0.0286\ \text{L soln}) \times \frac{3\ \text{mol ethanol}}{2\ \text{mol}\ K_2Cr_2O_7} = 0.00328\ \text{mol}$$

Step 3: Use the molar mass of ethanol to convert to grams of ethanol. This is the mass of ethanol in the 60.0 g sample. It is then straightforward to calculate the percent by mass of ethanol.

$$?\ \text{g of ethanol} = 0.00328\ \text{mol ethanol} \times \frac{46.07\ \text{g ethanol}}{1\ \text{mol ethanol}} = 0.151\ \text{g ethanol}$$

$$\textbf{percent by mass ethanol} = \frac{\text{mass of ethanol}}{\text{mass of sample}} \times 100\%$$

$$= \frac{0.151\ \text{g}}{60.0\ \text{g}} \times 100\% = \mathbf{0.252\ \%}$$

This clearly exceeds the 0.1% by mass legal limit. The driver is highly intoxicated and should be prosecuted.

13.60 Marsh gas (methane, CH_4); grain alcohol (ethanol, C_2H_5OH); wood alcohol (methanol, CH_3OH); rubbing alcohol (isopropyl alcohol, $(CH_3)_2CHOH$); antifreeze (ethylene glycol, CH_2OHCH_2OH); mothballs (naphthalene, $C_{10}H_8$); vinegar (acetic acid, CH_3COOH).

13.62 The red bromine vapor absorbs photons of blue light and dissociates to form bromine atoms.

$$Br_2 \rightarrow 2Br\bullet$$

The bromine atoms collide with methane molecules and abstract hydrogen atoms.

$$Br\bullet + CH_4 \rightarrow HBr + \bullet CH_3$$

The methyl radical then reacts with Br_2, giving the observed product and regenerating a bromine atom to start the process over again:

$$\bullet CH_3 + Br_2 \rightarrow CH_3Br + Br\bullet$$

$$Br\bullet + CH_4 \rightarrow HBr + \bullet CH_3 \qquad \text{and so on...}$$

13.64 **(a)** Aromatic hydrocarbons have the highest octane rating and branched-chain aliphatic hydrocarbons have higher octane numbers than straight-chain aliphatic hydrocarbons. Toluene is an aromatic compound. 2,2,4-trimethylpentane contains more branching than 2-methylhexane. *n*-heptane is a straight-chain aliphatic hydrocarbon. The correct order in decreasing octane number is:

toluene > 2,2,4-trimethylpentane > 2-methylhexane > *n*-heptane

(b) $CH_3CH_2CH_2CH_2CH_2CH_2CH_3 \xrightarrow{\text{catalyst}}$

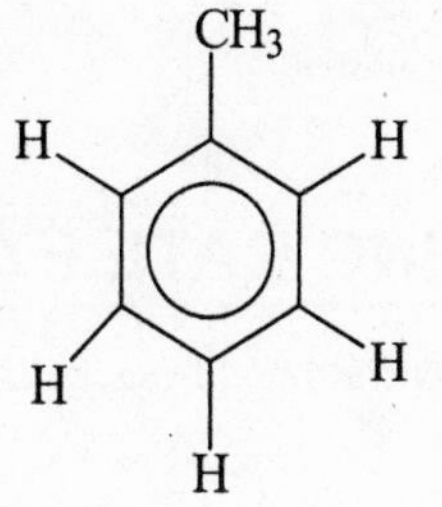

$+\ 4\,H_2(g)$

(c)

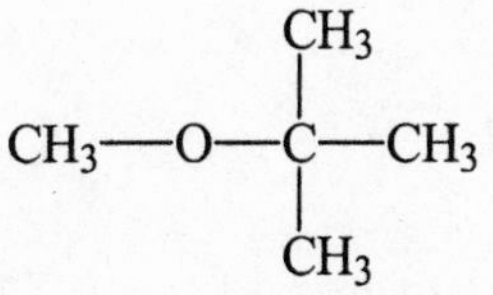

The two substituents of the ether are a methyl group on the left and a *tert*-butyl group on the right.

CHAPTER 14
CHEMICAL KINETICS

PROBLEM-SOLVING STRATEGIES AND TUTORIAL SOLUTIONS

TYPES OF PROBLEMS

Problem Type 1: Writing Rate Expressions.

Problem Type 2: Determining the Rate Law of a Reaction.

Problem Type 3: First-Order Reactions.
(a) Analyzing a first-order reaction.
(b) Determining the half-life of a first-order reaction.
(c) Analyzing first-order kinetics.

Problem Type 4: Second-Order Reactions.
(a) Analyzing a second-order reaction.
(b) Determining the half-life of a second-order reaction.
(c) Analyzing second-order kinetics.

Problem Type 5: The Arrhenius Equation.
(a) Applying the Arrhenius equation.
(b) Applying a modified form of the Arrhenius equation that relates the rate constants at two different temperatures.

Problem Type 6: Studying Reaction Mechanisms.

PROBLEM TYPE 1: WRITING RATE EXPRESSIONS

As a chemical reaction proceeds, the concentrations of reactants and products change with time. As the reaction

$$A + B \longrightarrow C$$

progresses, the concentration of C increases. The rate can be expressed as the change in concentration of C during the time interval Δt.

$$\text{rate} = \frac{\Delta[C]}{\Delta t}$$

For a specific reaction, we need to take into account the stoichiometry of the balanced equation. For example, let's express the rate of the following reaction in terms of the concentrations of the individual reactants and products.

$$2\,NO\,(g) + O_2\,(g) \longrightarrow 2\,NO_2\,(g)$$

Notice from the balanced equation that the concentration of NO will decrease twice as fast as that of O_2. We can write the rate as

$$\text{rate} = -\frac{1}{2}\frac{\Delta[NO]}{\Delta t} \quad \text{or} \quad \text{rate} = -\frac{\Delta[O_2]}{\Delta t} \quad \text{or} \quad \text{rate} = \frac{1}{2}\frac{\Delta[NO_2]}{\Delta t}$$

Division of each concentration by the coefficient from the balanced equation makes all of the above rates equal. Notice also that a negative sign is inserted before terms involving reactants. The Δ[NO] is a negative quantity because the concentration of NO *decreases* with time. Therefore, multiplying Δ[NO] by a negative sign makes the rate of reaction a positive quantity.

For a general reaction:

$$aA + bB \longrightarrow cC + dD$$

the rate of reaction can be expressed in terms of any reactant or product.

$$\text{rate} = -\frac{1}{a}\frac{\Delta[A]}{\Delta t} = -\frac{1}{b}\frac{\Delta[B]}{\Delta t} = \frac{1}{c}\frac{\Delta[C]}{\Delta t} = \frac{1}{d}\frac{\Delta[D]}{\Delta t}$$

EXAMPLE 14.1

Write expressions for the rate of the following reaction in terms of each of the reactants and products.

$$2\,N_2O_5\,(g) \longrightarrow 4\,NO_2\,(g) + O_2\,(g)$$

Recall that the rate is defined as the change in concentration of a reactant or product with time. Each "change in concentration" term is divided by the corresponding stoichiometric coefficient. Terms involving reactants are preceded by a minus sign.

$$\textbf{Rate} = -\frac{1}{2}\frac{\Delta[N_2O_5]}{\Delta t} = \frac{1}{4}\frac{\Delta[NO_2]}{\Delta t} = \frac{\Delta[O_2]}{\Delta t}$$

PRACTICE EXERCISE

1. Oxygen gas can be formed by the decomposition of nitrogen monoxide (nitric oxide).

$$2\,NO\,(g) \longrightarrow O_2\,(g) + N_2\,(g)$$

If the rate of formation of O_2 is 0.054 *M*/s, what is the rate of change of the NO concentration?

Text Problem: 6

PROBLEM TYPE 2: DETERMINING THE RATE LAW OF A REACTION

The **rate law** is an expression relating the rate of a reaction to the rate constant and the concentrations of reactants. For a general reaction of the type

$$aA + bB \longrightarrow cC + dD$$

the rate law takes the form

$$\text{rate} = k[A]^x[B]^y$$

The term k is the **rate constant**, a constant of proportionality between the reaction rate and the concentrations of the reactants. The sum of the powers to which all reactant concentrations appearing in the rate law are raised is called the overall **reaction order**. In the rate law expression shown above, the overall reaction order is given by x + y.

k, x, and y must be determined experimentally. One method to determine x and y is to keep the concentrations of all but one reactant fixed. Then, the rate of reaction is measured as a function of the concentration of the one reactant whose concentration is varied. Any variation in rate is due solely to the variation in this reactant's concentration.

The example below will show you how to determine the rate law and how to calculate the rate constant.

EXAMPLE 14.2

The following rate data were collected for the reaction

$$2\,NO + 2H_2 \longrightarrow N_2 + 2H_2O$$

(a) Determine the rate law.

(b) Calculate the rate constant.

Experiment	$[NO]_0$ (M)	$[H_2]_0$ (M)	$\Delta[N_2]/\Delta t$ (M/h)
1	0.60	0.15	0.076
2	0.60	0.30	0.15
3	0.60	0.60	0.30
4	1.20	0.60	1.21

(a) To determine the rate law, we must determine the exponents in the equation

$$\text{rate} = k[NO]^x[H_2]^y$$

To determine the order of the reaction with respect to H_2, find two experiments in which the [NO] is held constant. Compare the data from experiments 1 and 2. When the concentration of H_2 is doubled, the reaction rate doubles. Thus, the reaction is *first-order* in H_2.

To determine the order with respect to NO, compare experiments 3 and 4. When the NO concentration is doubled, the reaction rate quadruples. Thus, the reaction is *second-order* in NO.

The rate law is:

$$\textbf{rate} = \boldsymbol{k[NO]^2[H_2]}$$

(b)

Step 1: Rearrange the rate law from part (a), solving for *k*.

$$k = \frac{\text{rate}}{[NO]^2[H_2]}$$

Step 2: Substitute the data from any one of the experiments to calculate *k*. Using the data from Experiment 1,

$$k = \frac{0.076\ \cancel{M}/\text{h}}{(0.60\ M)^2(0.15\ \cancel{M})} = \mathbf{1.4}\ \boldsymbol{M^{-2}h^{-1}}$$

PRACTICE EXERCISE

2. The following experimental data were obtained for the reaction

$$2A + B \longrightarrow \text{products}$$

What is the rate law for this reaction?

Experiment	$[A]_0$ (M)	$[B]_0$ (M)	Rate (M/s)
1	0.40	0.20	5.6×10^{-3}
2	0.80	0.20	5.5×10^{-3}
3	0.40	0.40	22.3×10^{-3}

Text Problems: 16, 18, 48 a,b

PROBLEM TYPE 3: FIRST-ORDER REACTIONS

One of the most widely encountered kinetic forms is the first-order rate equation. Consider the reaction,

$$\text{A} \longrightarrow \text{products}$$

For a first-order reaction, the exponent of [A] in the rate law is 1. We can write:

$$\text{rate} = k[\text{A}]$$

We also know that

$$\text{rate} = \frac{-\Delta[\text{A}]}{\Delta t}$$

Combining the two equations gives:

$$k[\text{A}] = \frac{-\Delta[\text{A}]}{\Delta t}$$

Using calculus, we can integrate both sides of the above equation to give:

$$\ln\frac{[\text{A}]_0}{[\text{A}]} = kt \qquad (14.1)$$

where ln is the natural logarithm, and $[\text{A}]_0$ and [A] are the concentrations of A at times $t = 0$ and $t = t$, respectively.

A. Analyzing a first-order reaction

EXAMPLE 14.3

Methyl isocyanide undergoes a first-order isomerization to from methyl cyanide.

$$CH_3NC\ (g) \longrightarrow CH_3CN\ (g)$$

The reaction was studied at 199°C. The initial concentration of CH_3NC was 0.0258 mol/L and after 11.4 min, analysis showed the concentration of the product to be 1.30×10^{-3} mol/L.

(a) What is the first-order rate constant?

(b) How long will it take for 90.0 percent of the CH_3NC to react?

(a) This problem illustrates the use of the first-order rate equation

$$\ln\frac{[CH_3NC]_0}{[CH_3NC]} = kt$$

where k is the rate constant and $[CH_3NC]$ is the reactant concentration at time t.

Step 1: We can calculate $[CH_3NC]$ by realizing that the amount of product formed equals the amount of reactant lost due to the 1:1 mole ratio between reactant and product. Thus,

$$[CH_3NC] = [CH_3NC]_0 - 1.30 \times 10^{-3}\ M$$

$$= 0.0258\ M - 1.30 \times 10^{-3}\ M = 0.0245\ M$$

Step 2: Substitute the concentrations and the time into the first-order rate equation to calculate the rate constant k.

$$\ln\frac{[CH_3NC]_0}{[CH_3NC]} = kt$$

$$\ln \frac{0.0258\text{ M}}{0.0245\text{ M}} = k(11.4\text{ min})$$

$$\boldsymbol{k} = \frac{\ln 1.05}{11.4\text{ min}} = \mathbf{4.54 \times 10^{-3}\ min^{-1}}$$

(b) If 90 percent of the initial CH_3NC is consumed, then 10 percent remains. We can write,

$$[CH_3NC] = 0.10[CH_3NC]_0$$

Step 1: Substitution into the first-order rate equation gives:

$$\ln \frac{[CH_3NC]_0}{0.10[CH_3NC]_0} = kt$$

Step 2: Substitute in the rate constant k calculated in part (a) to solve for t.

$$\ln \frac{[CH_3NC]_0}{0.10[CH_3NC]_0} = (4.54 \times 10^{-3}\text{ min}^{-1})t$$

$$t = \frac{\ln 10}{4.54 \times 10^{-3}\text{ min}^{-1}} = \mathbf{507\ min}$$

PRACTICE EXERCISE

3. The hydrolysis of sucrose ($C_{12}H_{22}O_{11}$) yields the simple sugars glucose ($C_6H_{12}O_6$) and fructose ($C_6H_{12}O_6$), which happen to be isomers.

$$C_{12}H_{22}O_{11} + H_2O \longrightarrow C_6H_{12}O_6 + C_6H_{12}O_6$$

The reaction is first-order in glucose concentration.

$$\text{rate} = k[C_{12}H_{22}O_{11}]$$

At 27°C, the rate constant is $2.1 \times 10^{-6}\ s^{-1}$. Starting with a sucrose solution with a concentration of 0.10 M at 27°C, what would the concentration of sucrose be 24 hours later? (The solution is maintained at 27°C.)

Text Problem: 24

B. Determining the half-life of a first-order reaction

The **half-life** of a reaction, $t_{\frac{1}{2}}$, is the time required for the concentration of a reactant to decrease to half of its initial concentration. From Equation (14.1), we can write

$$\ln \frac{[A]_0}{[A]} = kt \qquad (14.1)$$

$$t = \frac{1}{k}\ln \frac{[A]_0}{[A]} \qquad (14.2)$$

From the definition of half-life, when $t = t_{\frac{1}{2}}$

$$[A] = \frac{[A]_0}{2}$$

Substituting into Equation (14.2) gives:

$$t_{\frac{1}{2}} = \frac{1}{k}\ln\frac{[A]_0}{[A]_0/2}$$

or

$$t_{\frac{1}{2}} = \frac{1}{k}\ln 2 = \frac{0.693}{k} \qquad (14.3)$$

Equation (14.3) shows that the half-life of a first-order reaction is *independent* of the initial concentration of the reactant. Thus, it would take the same time for the concentration of the reactant to decrease from 1.0 *M* to 0.50 *M* as it would to decrease from 0.10 *M* to 0.050 *M*.

The half-life can also be used to determine the rate constant of a first-order reaction.

EXAMPLE 14.4

Methyl isocyanide undergoes a first-order isomerization to form methyl cyanide.

$$CH_3NC\ (g) \longrightarrow CH_3CN\ (g)$$

The reaction was studied at 199°C. The initial concentration of CH_3NC was 0.0258 mol/L and after 11.4 min, analysis showed the concentration of the product to be 1.30×10^{-3} mol/L. Using the rate constant calculated in Example 14.3, calculate the half-life of methyl isocyanide.

The rate constant calculated in Example 14.3 is $4.54 \times 10^{-3}\ \text{min}^{-1}$. Substituting the rate constant into Equation (14.3) gives,

$$t_{\frac{1}{2}} = \frac{0.693}{k} = \frac{0.693}{4.54 \times 10^{-3}\ \text{min}^{-1}} = \mathbf{153\ min}$$

PRACTICE EXERCISE

4. The hydrolysis of sucrose ($C_{12}H_{22}O_{11}$) yields the simple sugars glucose ($C_6H_{12}O_6$) and fructose ($C_6H_{12}O_6$), which happen to be isomers.

$$C_{12}H_{22}O_{11} + H_2O \longrightarrow C_6H_{12}O_6 + C_6H_{12}O_6$$

The reaction is first-order in sucrose concentration.

$$\text{rate} = k[C_{12}H_{22}O_{11}]$$

At 27°C, the rate constant is $2.1 \times 10^{-6}\ s^{-1}$. What is the half-life of sucrose at 27°C?

Text Problem: 24

C. Analyzing first-order kinetics

Equation (14.1) can be rearranged as follows:

$$\ln\frac{[A]_0}{[A]} = kt \qquad (14.1)$$

$$\ln[A]_0 - \ln[A] = kt$$

or

$$\ln[A] = -kt + \ln[A]_0 \qquad (14.4)$$

Equation (14.4) has the form of a linear equation.

$$y = mx + b$$

where,

m is the slope of the line
b is the y-intercept

A plot of ln[A] versus t (y vs. x) gives a straight line with a slope of $-k$ (m). Thus, we can calculate the rate constant k from the slope of the plot.

EXAMPLE 14.5
At 500 K, butadiene gas converts to cyclobutene gas:

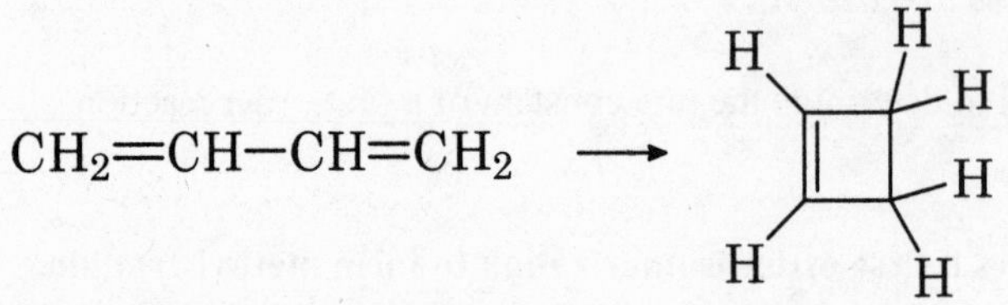

Given the following data, is the reaction first-order in butadiene concentration?

Time from start (s)	Concentration of butadiene (M)
195	0.0162
604	0.0147
1246	0.0129
2180	0.0110
4140	0.0084
8135	0.0057

If the reaction is first-order in butadiene, then a plot of ln[butadiene] versus t will be a straight line.

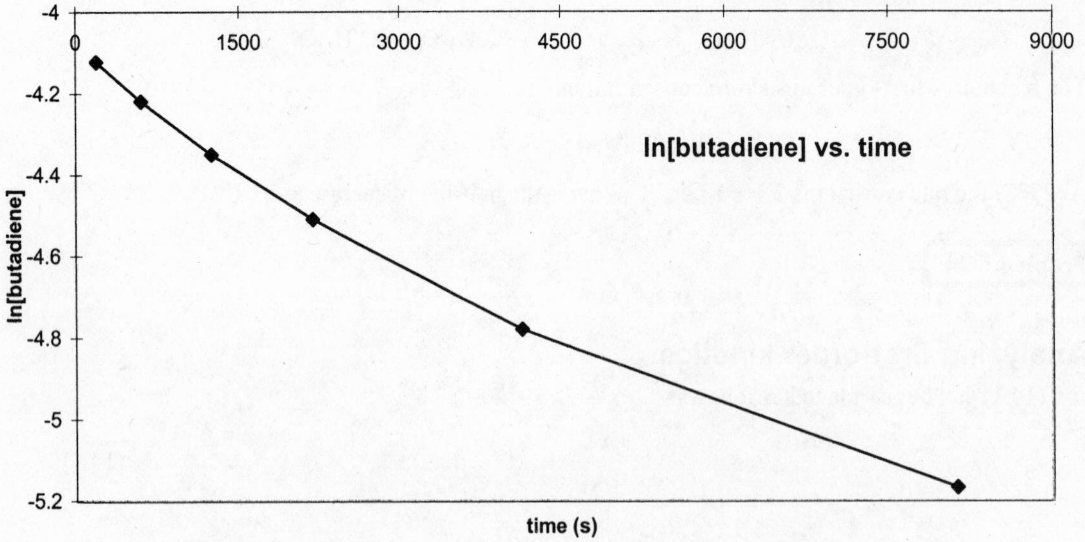

As you can see, a plot of ln[butadiene] versus t does *not* give a straight line. Hence, the reaction is *not* first-order in butadiene.

PRACTICE EXERCISE

5. In a certain experiment, the rate of hydrogen peroxide decomposition,

$$2\ H_2O_2 \rightarrow 2\ H_2O + O_2$$

is followed by titration against a potassium permanganate solution. At regular intervals, an equal volume of H_2O_2 is withdrawn to give the following data:

Time (min)	0	10.0	20.0	30.0
Volume of $KMnO_4$ used (mL)	22.8	13.8	8.25	5.00

Confirm that the reaction is first-order in hydrogen peroxide and calculate the rate constant.

Hint: The concentration of hydrogen peroxide is directly proportional to the volume (in mL) of $KMnO_4$ used in each titration.

PROBLEM TYPE 4: SECOND-ORDER REACTIONS

Second-order reactions are also encountered quite often. Consider the reaction,

$$A \longrightarrow \text{products}$$

For a second-order reaction, the exponent of [A] in the rate law is 2. We can write:

$$\text{rate} = k[A]^2$$

We also know that

$$\text{rate} = \frac{-\Delta[A]}{\Delta t}$$

Combining the two equations gives:

$$k[A]^2 = \frac{-\Delta[A]}{\Delta t}$$

Using calculus, we can integrate both sides of the above equation to give:

$$\frac{1}{[A]} = \frac{1}{[A]_0} + kt \qquad (14.5)$$

where $[A]_0$ and [A] are the concentrations of A at times $t = 0$ and $t = t$, respectively.

A. Analyzing a second-order reaction

EXAMPLE 14.6
At 500 K, butadiene gas converts to cyclobutene gas:

$CH_2{=}CH{-}CH{=}CH_2 \longrightarrow$ 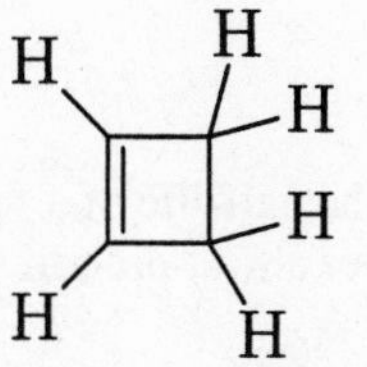

At 500 K, the rate constant for the reaction is 0.0143/*M*·s. If the initial concentration of butadiene is 0.272 *M*, calculate the concentration of butadiene after 30.0 min.

This problem illustrates the use of the second-order rate equation:

$$\frac{1}{[\text{butadiene}]} = \frac{1}{[\text{butadiene}]_0} + kt$$

where k is the rate constant and [butadiene] is the reactant concentration at time t.

We can calculate [butadiene] by substituting k, t, and the initial concentration of butadiene into the above equation.

$$\frac{1}{[\text{butadiene}]} = \frac{1}{0.272\ M} + (0.0143\ /\ M\cdot\text{s})(1.80 \times 10^3\ \text{s})$$

$$\frac{1}{[\text{butadiene}]} = 29.4\ M^{-1}$$

$$\mathbf{[butadiene]\ =\ 0.0340}\ \boldsymbol{M}$$

PRACTICE EXERCISE

6. For the reaction shown in Example 14.6 above, how long will it take for the concentration of butadiene to decrease from its initial concentration of 0.272 *M* to 0.100 *M*?

Text Problem: 26

B. Determining the half-life of a second-order reaction

The **half-life** of a reaction, $t_{\frac{1}{2}}$, is the time required for the concentration of a reactant to decrease to half of its initial concentration. From Equation (14.5), we can write:

$$\frac{1}{[\text{A}]} = \frac{1}{[\text{A}]_0} + kt \qquad (14.5)$$

$$t = \left(\frac{1}{k}\right)\left(\frac{1}{[\text{A}]} - \frac{1}{[\text{A}]_0}\right) \qquad (14.6)$$

From the definition of half-life, when $t = t_{\frac{1}{2}}$

$$[\text{A}] = \frac{[\text{A}]_0}{2}$$

Substituting into Equation (14.6) gives:

$$t_{\frac{1}{2}} = \left(\frac{1}{k}\right)\left(\frac{1}{[\text{A}]_0/2} - \frac{1}{[\text{A}]_0}\right)$$

or

$$t_{\frac{1}{2}} = \frac{1}{k[\text{A}]_0} \qquad (14.7)$$

Equation (14.7) shows that the half-life of a second-order reaction is *dependent* on the initial concentration of the reactant, unlike the half-life of a first-order reaction.

EXAMPLE 14.7
The following reaction follows second-order kinetics:

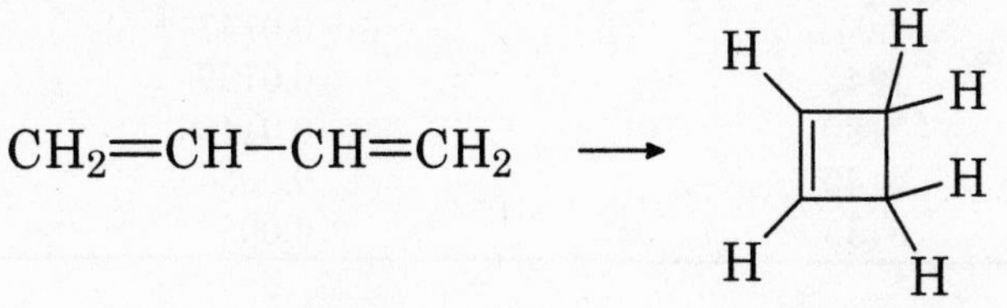

butadiene (*g*) cyclobutene (*g*)

At 500 K, the rate constant is 0.0143/*M*·s. The initial concentration of butadiene is 0.272 *M*. What is the half-life for this reaction?

The rate constant is 0.0143 /*M*·s. Substituting the rate constant and the initial concentration into the Equation (14.7) gives:

$$t_{\frac{1}{2}} = \frac{1}{k[\text{A}]_0} = \frac{1}{(0.0143/\text{ M}\cdot\text{s})(0.272\text{ M})} = \mathbf{257\ s}$$

PRACTICE EXERCISE

7. For the reaction shown in Example 14.7 above, the half-life of the reaction is determined to be 66.6 s at 500 K. What is the initial concentration of butadiene? ($k = 0.0143$ /m·s at 500 K)

C. Analyzing second-order kinetics

Equation (14.5) has the form of a linear equation.

$$\frac{1}{[\text{A}]} = kt + \frac{1}{[\text{A}]_0} \qquad (14.5)$$

$$y = mx + b$$

where,

m is the slope of the line
b is the *y*-intercept

A plot of $\frac{1}{[\text{A}]}$ versus *t* (*y* vs. *x*) gives a straight line with a slope of *k* (*m*). Thus, we can calculate the rate constant *k* from the slope of the plot.

EXAMPLE 14.8
At 500 K, butadiene gas converts to cyclobutene gas:

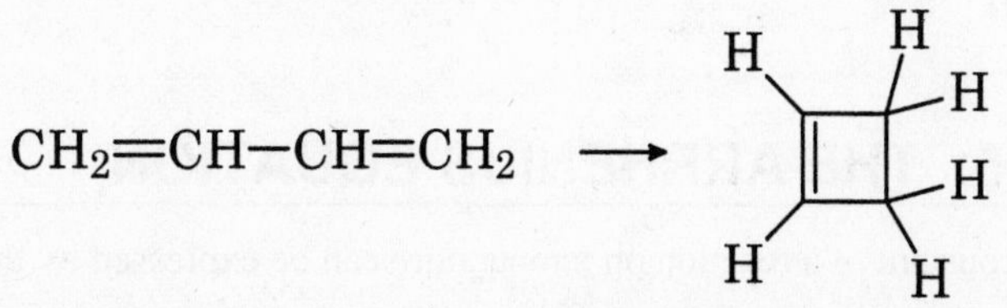

Given the following data, is the reaction second-order in butadiene concentration?

Time from start (s)	Concentration of butadiene (M)
195	0.0162
604	0.0147
1246	0.0129
2180	0.0110
4140	0.0084
8135	0.0057

If the reaction is second-order in butadiene, then a plot of $\frac{1}{[\text{butadiene}]}$ versus t will be a straight line.

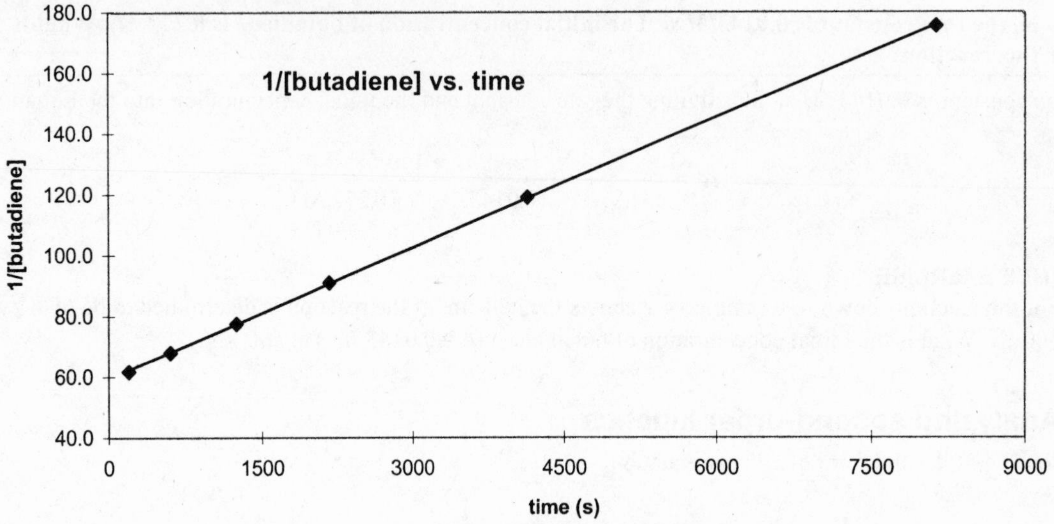

As you can see, a plot of $\frac{1}{[\text{butadiene}]}$ versus t *does* give a straight line. Hence, the reaction is second-order in butadiene.

PRACTICE EXERCISE

8. If a plot of $\frac{1}{[\text{A}]}$ versus time produces a straight line, which of the following must be *true*?

a. The reaction is first-order in A.
b. The reaction is second-order in A.
c. The reaction is first-order in two reactants.
d. The rate of the reaction does not depend on the concentration of A.
e. None of the above

PROBLEM TYPE 5: THE ARRHENIUS EQUATION

The dependence of the rate constant of a reaction on temperature can be expressed by the following equation, called the **Arrhenius equation**.

$$k = Ae^{-E_a/RT}$$

where, E_a is the activation energy of the reaction (in kJ/mol)
R is the gas constant (8.314 J/mol·K)
T is the absolute temperature (in K)
e is the base of the natural logarithm scale (see Appendix 3)

The quantity A represents the collision frequency and is called the *frequency factor*. It can be treated as a constant for a given reacting system over a fairly wide temperature range. The Arrhenius equation shows that the rate constant is directly proportional to A. Therefore, as the number of collisions increase, the rate increases.

The minus sign associated with the exponent E_a/RT indicates that the rate constant decreases with increasing activation energy and increases with increasing temperature.

The Arrhenius equation can be expressed in a more useful form by taking the natúral logarithm of both sides.

$$\ln k = \ln[\, Ae^{-E_a/RT}\,]$$

$$\ln k = \ln A - \frac{E_a}{RT} \qquad (14.8)$$

or

$$\ln k = \left(-\frac{E_a}{R}\right)\left(\frac{1}{T}\right) + \ln A \qquad (14.9)$$

Equation (14.9) has the form of the linear equation

$$y = mx + b$$

where,

m is the slope of the line
b is the y-intercept

A plot of $\ln[k]$ versus $\frac{1}{T}$ (y vs. x) gives a straight line with a slope of $\frac{-E_a}{R}$ (m). Thus, we can calculate the activation energy (E_a) from the slope of the plot.

A. Applying the Arrhenius Equation

EXAMPLE 14.9
Variation of the rate constant with temperature for the reaction

$$\mathbf{NO + O_3 \rightarrow NO_2 + O_2}$$

is given in the following table. Determine graphically the activation energy for the reaction.

Temperature (K)	k ($M^{-1}s^{-1}$)
283	9.30×10^6
293	1.08×10^7
303	1.25×10^7
313	1.43×10^7
323	1.62×10^7

A plot of $\ln[k]$ versus $\frac{1}{T}$ should give a straight line with a slope of $\frac{-E_a}{R}$. Thus, we can calculate the activation energy (E_a) from the slope of the plot.

From the given data we obtain:

$1/T$ (K^{-1})	ln k
3.53×10^{-3}	16.05
3.41×10^{-3}	16.20
3.30×10^{-3}	16.34
3.19×10^{-3}	16.48
3.10×10^{-3}	16.60

These data, when plotted, yield the graph shown below.

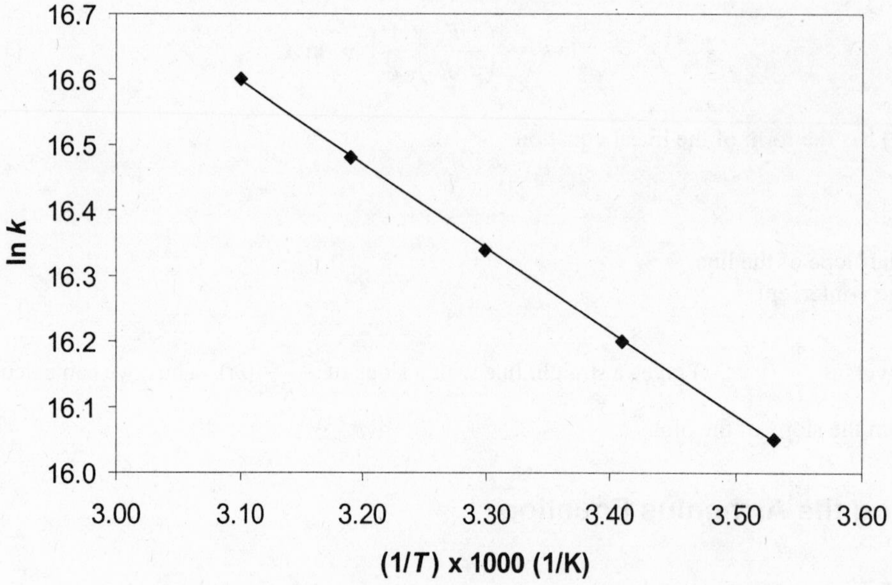

Calculating the slope from the first and last points gives:

$$\text{slope} = \frac{16.05 - 16.60}{(3.53 \times 10^{-3}) - (3.10 \times 10^{-3})} = -1.28 \times 10^3 \text{ K}$$

The slope is equal to $-E_a/R$ or

$$E_a = -\text{slope} \times R$$

$$\boldsymbol{E_a} = -(-1.28 \times 10^3 \cancel{\text{K}})(8.314 \text{ J/mol}\cdot\cancel{\text{K}}) = \mathbf{1.06 \times 10^4 \textbf{ J/mol} = 10.6 \textbf{ kJ/mol}}$$

B. Applying a modified form of the Arrhenius equation that relates the rate constants at two different temperatures

Starting with Equation (14.8), we can write,

$$\ln k_1 = \ln A - \frac{E_a}{RT_1}$$

and

$$\ln k_2 = \ln A - \frac{E_a}{RT_2}$$

Subtracting $\ln k_2$ from $\ln k_1$ gives:

$$\ln k_1 - \ln k_2 = -\frac{E_a}{RT_1} + \frac{E_a}{RT_2}$$

$$\ln\frac{k_1}{k_2} = \frac{E_a}{R}\left(\frac{1}{T_2} - \frac{1}{T_1}\right)$$

or

$$\ln\frac{k_1}{k_2} = \frac{E_a}{R}\left(\frac{T_1 - T_2}{T_1 T_2}\right) \qquad (14.10)$$

Using Equation (14.10), we can calculate the activation energy if we know the rate constants at two temperatures or find the rate constant at another temperature if the activation energy is known.

EXAMPLE 14.10
For the reaction

$$\mathbf{NO + O_3 \longrightarrow NO_2 + O_2}$$

the following rate constants have been obtained. Calculate the activation energy for this reaction.

Temperature (°C)	k ($M^{-1}s^{-1}$)
10.0	9.3×10^6
30.0	1.25×10^7

We can calculate the activation energy using Equation (14.10).

$$\ln\frac{k_1}{k_2} = \frac{E_a}{R}\left(\frac{T_1 - T_2}{T_1 T_2}\right) \qquad (14.10)$$

Let,

$k_1 = 9.3 \times 10^6\ M^{-1}s^{-1}$ at $T_1 = 273 + 10.0 = 283$ K

$k_2 = 1.25 \times 10^7\ M^{-1}s^{-1}$ at $T_2 = 273 + 30.0 = 303$ K

Recall that $R = 8.314$ J/mol·K. Substitute the rate constants, the temperatures, and R into Equation (14.10) to solve for E_a.

$$\ln\left(\frac{9.3 \times 10^6\ \text{M}^{-1}\text{s}^{-1}}{1.25 \times 10^7\ \text{M}^{-1}\text{s}^{-1}}\right) = \frac{E_a}{8.314\ \text{J/mol}\cdot\text{K}}\left(\frac{283\ \text{K} - 303\ \text{K}}{(283\ \text{K})(303\ \text{K})}\right)$$

$$(\ln 0.744)(8.314\ \text{J/mol}\cdot\text{K}) = E_a\left(\frac{-20\ \cancel{\text{K}}}{8.57 \times 10^4\ \text{K}^{\cancel{2}}}\right)$$

$$(-0.296)\left(8.314\ \frac{\text{J}}{\text{mol}\cdot\cancel{\text{K}}}\right) = E_a\left(-2.33 \times 10^{-4}\ \frac{1}{\cancel{\text{K}}}\right)$$

$$\mathbf{E_a = 1.06 \times 10^4\ J/mol = 10.6\ kJ/mol}$$

Compare the answer in this example with that obtained in Example 14.9.

PRACTICE EXERCISE

9. At 300 K, the rate constant is $1.5 \times 10^{-5}\ M^{-1}\cdot s^{-1}$ for the following reaction:

$$2\ NOCl \longrightarrow 2\ NO + Cl_2$$

The activation energy is 90.2 kJ/mol. Calculate the value of the rate constant at 310 K.

Text Problems: 34, 36, 38

PROBLEM TYPE 6: STUDYING REACTION MECHANISMS

For a complete discussion of reaction mechanisms, see Section 14.5 of your text. Experimental studies of reaction mechanisms begin with the collection of data (rate measurements). Next the data are analyzed to determine the rate constant and the order of reaction, so that the rate law can be written. Finally, a plausible mechanism is suggested for the reaction in terms of elementary steps. This sequence of steps is summarized in Figure 14.14 of your text.

The elementary steps of the proposed mechanism must satisfy two requirements:

- The sum of the elementary steps must give the overall balanced equation for the reaction.
- The **rate-determining step**, which is the slowest step in the sequence of steps leading to product formation, should predict the same rate law as is determined experimentally.

EXAMPLE 14.11

The rate law for the substitution of NH_3 for H_2O in the following reaction is first order in $Ni(H_2O)_6^{2+}$ and zero order in NH_3.

$$\mathbf{Ni(H_2O)_6^{2+}\ (\mathit{aq}) + NH_3\ (\mathit{aq}) \longrightarrow Ni(H_2O)_5(NH_3)^{2+}\ (\mathit{aq}) + H_2O\ (\mathit{l})}$$

Show that the following mechanism is consistent with the experimental rate law.

$$Ni(H_2O)_6^{2+}\ (aq) \longrightarrow Ni(H_2O)_5^{2+}\ (aq) + H_2O\ (l) \qquad \text{(slow)}$$

$$Ni(H_2O)_5^{2+}\ (aq) + NH_3\ (aq) \longrightarrow Ni(H_2O)_5(NH_3)^{2+}\ (aq) \qquad \text{(fast)}$$

First, does the sum of the elementary steps give the overall balanced equation. Yes, the sum of the steps gives:

$$Ni(H_2O)_6^{2+}\ (aq) + NH_3\ (aq) \longrightarrow Ni(H_2O)_5(NH_3)^{2+}\ (aq) + H_2O\ (l)$$

Next, the reaction was experimentally determined to be first order in $Ni(H_2O)_6^{2+}$ and zero order in NH_3. We can write the rate law for the reaction.

$$\text{rate} = k[Ni(H_2O)_6^{2+}]$$

Does the rate law match the rate law of the proposed mechanism? We can write the rate law from the rate determining step of the proposed mechanism. Step 1 is the slow step, so we can write:

$$\text{rate} = k[Ni(H_2O)_6^{2+}]$$

which matches the experimentally determined rate law.

The elementary steps of the proposed mechanism satisfy the two requirements outlined above; therefore, it is a valid mechanism.

PRACTICE EXERCISE

10. The reaction of nitric oxide and chlorine,

$$2\ NO\ (g) + Cl_2\ (g) \rightarrow 2\ NOCl\ (g)$$

has been proposed to proceed by the following two-step mechanism:

$$NO\ (g) + Cl_2\ (g) \rightarrow NOCl_2\ (g)$$

$$NO\ (g) + NOCl_2\ (g) \rightarrow 2\ NOCl\ (g)$$

What rate law is predicted if the first step of the proposed mechanism is the rate-determining step?

Text Problems: 48, 56

ANSWERS TO PRACTICE EXERCISES

1. Rate of change of NO = −0.11 M/s

2. Rate = $k[B]^2$

3. [sucrose] = 0.083 M after 24 h

4. $t_{\frac{1}{2}} = 3.3 \times 10^5$ s

5. The slope of the straight line plot equals $-k$. You should find $\boldsymbol{k}$ **= 0.0504 min^{-1}**.

6. 442 s

7. 1.05 M

8. **(b)**

9. $k = 4.8 \times 10^{-5}\ M^{-1}s^{-1}$

10. rate = $k[NO][Cl_2]$

SOLUTIONS TO SELECTED TEXT PROBLEMS

14.6 Writing Rate Expressions, Problem Type 1.

The rate is defined as the change in concentration of a reactant or product with time. Each "change in concentration" term is divided by the corresponding stoichiometric coefficient. Terms involving reactants are proceeded by a minus sign.

$$\text{rate} = -\frac{\Delta[N_2]}{\Delta t} = -\frac{1}{3}\frac{\Delta[H_2]}{\Delta t} = \frac{1}{2}\frac{\Delta[NH_3]}{\Delta t}$$

(a) If hydrogen is reacting at the rate of −0.074 *M*/s, the rate at which ammonia is being formed is

$$\frac{1}{2}\frac{\Delta[NH_3]}{\Delta t} = -\frac{1}{3}\frac{\Delta[H_2]}{\Delta t}$$

or

$$\frac{\Delta[NH_3]}{\Delta t} = -\frac{2}{3}\frac{\Delta[H_2]}{\Delta t}$$

$$\frac{\Delta[NH_3]}{\Delta t} = -\frac{2}{3}(-0.074\ M/\text{s}) = \mathbf{0.049\ \mathit{M}/s}$$

(b) The rate at which nitrogen is reacting must be:

$$\frac{\Delta[N_2]}{\Delta t} = \frac{1}{3}\frac{\Delta[H_2]}{\Delta t} = \frac{1}{3}(-0.074\ M/\text{s}) = \mathbf{-0.025\ \mathit{M}/s}$$

Will the rate at which ammonia forms always be twice the rate of reaction of nitrogen, or is this true only at the instant described in this problem?

14.16 **(a)** Assume the rate law has the form:

$$\text{rate} = k[F_2]^x[ClO_2]^y$$

To determine the order of the reaction with respect to F_2, find two experiments in which the $[ClO_2]$ is held constant. Compare the data from experiments 1 and 3. When the concentration of F_2 is doubled, the reaction rate doubles. Thus, the reaction is *first-order* in F_2.

To determine the order with respect to ClO_2, compare experiments 1 and 2. When the ClO_2 concentration is quadrupled, the reaction rate quadruples. Thus, the reaction is *first-order* in ClO_2.

The rate law is:

$$\text{rate} = k[F_2][ClO_2]$$

(b) The value of *k* can be found using the data from any of the experiments. If we take the numbers from the second experiment we have:

$$k = \frac{\text{rate}}{[F_2][ClO_2]} = \frac{4.8\times10^{-3}\ \cancel{M}/\text{s}}{(0.10\ \cancel{M})(0.040\ M)} = 1.2\ M^{-1}\text{s}^{-1}$$

Verify that the same value of *k* can be obtained from the other sets of data.

(c) Since we now know the rate law and the value of the rate constant, we can calculate the rate at any concentration of reactants.

$$\textbf{rate} = k[F_2][ClO_2] = (1.2\,\cancel{M}^{-1}s^{-1})(0.010\,\cancel{M})(0.020\,M) = \mathbf{2.4 \times 10^{-4}}\ \boldsymbol{M}\textbf{/s}$$

14.18 Determining the Rate Law of a Reaction, Problem Type 2.

(a) Compare the second and fifth set of data. Doubling [X] at constant [Y] increases the rate by a factor of 0.508/0.127 or 4. Therefore, the reaction is second-order in X. Next, compare the second and fourth set of data. Doubling [Y] at constant [X] increases the rate by a factor of 1.016/0.127 or 8 times. Therefore, the reaction is third order in Y. The rate law is:

$$\text{rate} = k[X]^2[Y]^3$$

The order of the reaction is (2 + 3) = **5**. The reaction is *5th-order.*

(b)

Step 1: First, use any set of data to calculate the rate constant *k*. Using the first set of data,

$$k = \frac{\text{rate}}{[X]^2[Y]^3} = \frac{0.147\ \cancel{M}/s}{(0.10\ M)^{\cancel{2}}(0.50\ M)^3} = 1.2 \times 10^2\ M^{-4}s^{-1}$$

Step 2: Substitute the concentrations of X and Y into the rate law to calculate the initial rate of disappearance of X.

$$\textbf{rate} = (1.2 \times 10^2\ M^{-4}s^{-1})(0.30\ M)^2(0.40\ M)^3 = \mathbf{0.69}\ \boldsymbol{M}\textbf{/s}$$

14.20 **(a)** For a reaction first-order in A,

$$\text{Rate} = k[A]$$

$$1.6 \times 10^{-2}\ \cancel{M}/s = k(0.35\ \cancel{M})$$

$$\boldsymbol{k} = \mathbf{0.046\ s^{-1}}$$

(b) For a reaction second-order in A,

$$\text{Rate} = k[A]^2$$

$$1.6 \times 10^{-2}\ M/s = k(0.35\ M)^2$$

$$\boldsymbol{k} = \mathbf{0.13}\ \boldsymbol{M}^{-1}\mathbf{s^{-1}}$$

14.24 Determining the half-life of a first-order reaction, Problem Type 3B.

(a) For any first order reaction the rate constant can be found from the half–life using the following equation.

$$k = \frac{0.693}{t_{\frac{1}{2}}}$$

$$\boldsymbol{k} = \frac{0.693}{t_{\frac{1}{2}}} = \frac{0.693}{35.0\ s} = \mathbf{0.0198\ s^{-1}}$$

(b) The time required for 95% of the phosphine to decompose can be found using the following equation.

$$\ln\frac{[A]_0}{[A]} = kt$$

or

$$t = \frac{1}{k}\ln\frac{[A]_0}{[A]}$$

The value of [A] after 95% of the reactant has decomposed is $0.0500[A]_0$. Substitute the rate constant calculated in part (a) and $[A] = 0.0500[A]_0$ into the above equation to calculate the time required for 95% of the phosphine to decompose.

$$t = \frac{1}{k}\ln\frac{[A]_0}{[A]} = \left(\frac{1}{0.0198\ \text{s}^{-1}}\right)\ln\frac{[A]_0}{0.0500[A]_0}$$

$$t = \frac{\ln 20.0}{0.0198\ \text{s}^{-1}} = \mathbf{151\ s}$$

14.26 $$\frac{1}{[A]} = \frac{1}{[A]_0} + kt$$

$$\frac{1}{0.28} = \frac{1}{0.62} + 0.54t$$

t = **3.6 s**

14.34 Applying a modified form of the Arrhenius equation, Problem Type 5B.

The ratio of the rates at two different temperatures is equivalent to the ratio of the rate constants at the two temperatures.

Use the modified form of the Arrhenius equation shown below to calculate the energy of activation.

Let T_1 = 250°C = 523 K and T_2 = 150°C = 423 K.

$$\ln\frac{k_1}{k_2} = \frac{E_a}{R}\left(\frac{T_1 - T_2}{T_1T_2}\right)$$

$$\ln(1.50 \times 10^3) = \frac{E_a}{8.314\ \text{J/mol}\cdot\text{K}}\left(\frac{523\ \text{K} - 423\ \text{K}}{(523\ \text{K})(423\ \text{K})}\right)$$

$$7.31 = \frac{E_a}{8.314\frac{\text{J}}{\text{mol}\cdot\cancel{\text{K}}}}\left(4.52 \times 10^{-4}\ \frac{1}{\cancel{\text{K}}}\right)$$

$$\mathbf{E_a = 1.35 \times 10^5\ J/mol = 135\ kJ/mol}$$

14.36 Use the modified form of the Arrhenius equation to calculate the temperature at which the rate constant is $8.80 \times 10^{-4}\ \text{s}^{-1}$.

$$\ln\frac{k_1}{k_2} = \frac{E_a}{R}\left(\frac{1}{T_2} - \frac{1}{T_1}\right)$$

$$\ln\left(\frac{4.60 \times 10^{-4}\ \cancel{s^{-1}}}{8.80 \times 10^{-4}\ \cancel{s^{-1}}}\right) = \frac{1.04 \times 10^5\ \cancel{J}/\cancel{mol}}{8.314\ \cancel{J}/\cancel{mol}\cdot K}\left(\frac{1}{T_2} - \frac{1}{623\ K}\right)$$

$$(\ln 0.523) = 1.25 \times 10^4\ K\left(\frac{1}{T_2} - \frac{1}{623\ K}\right)$$

$$-0.649 + 20.1 = \frac{1.25 \times 10^4\ K}{T_2}$$

$$19.5\ T_2 = 1.25 \times 10^4\ K$$

$$\mathbf{T_2 = 641\ K = 368°C}$$

14.38 Since the ratio of rates is equal to the ratio of rate constants, we can write:

$$\ln\frac{\text{rate}_1}{\text{rate}_2} = \ln\frac{k_1}{k_2}$$

$$\ln\frac{k_1}{k_2} = \ln\left(\frac{2.0 \times 10^2}{39.6}\right) = \frac{E_a}{8.314\ J/K\cdot mol}\left(\frac{(300\ K - 278\ K)}{(300\ K)(278\ K)}\right)$$

$$\mathbf{E_a = 5.10 \times 10^4\ J/mol = 51.0\ kJ/mol}$$

14.48 (a) Determining the Rate Law of a Reaction, Problem Type 2.

Since the reaction rate doubles when the X_2 concentration is doubled, the reaction is first-order in X. The reaction rate triples when the concentration of Y is tripled, so the reaction is also first-order in Y. The concentration of Z has no effect on the rate, so the reaction is zero-order in Z.

The rate law is:

$$\mathbf{rate = k[X_2][Y]}$$

(b) If a change in the concentration of Z has no effect on the rate, the concentration of Z is not a term in the rate law. This implies that Z does not participate in the rate-determining step of the reaction mechanism.

(c) Studying Reaction Mechanisms, Problem Type 6.

The rate law shows that the slow step involves reaction of a molecule of X_2 with a molecule of Y. Since Z is not present in the rate law, it does not take part in the slow step and must appear in a fast step at a later time. (If the fast step involving Z happened before the rate-determining step, the rate law would involve Z in a more complex way.) A mechanism that is consistent with the rate law could be:

$$X_2 + Y \longrightarrow XY + X \quad \text{(slow)}$$

$$X + Z \longrightarrow XZ \quad \text{(fast)}$$

The rate law only tells us about the slow step. Other mechanisms with different subsequent fast steps are possible. Try to invent one.

14.56 The rate-determining step involves the breakdown of ES to E and P. The rate law for this step is:

$$\text{rate} = k_2[\text{ES}]$$

In the first elementary step, the intermediate ES is in equilibrium with E and S. The equilibrium relationship is:

$$\frac{[\text{ES}]}{[\text{E}][\text{S}]} = \frac{k_1}{k_{-1}}$$

or

$$[\text{ES}] = \frac{k_1}{k_{-1}}[\text{E}][\text{S}]$$

Substitute [ES] into the rate law expression.

$$\textbf{rate} = k_2[\text{ES}] = \boldsymbol{\frac{k_1 k_2}{k_{-1}}[\text{E}][\text{S}]}$$

14.58 Temperature, energy of activation, concentration of reactants, and a catalyst.

14.60 First, calculate the radius of the 10.0 cm^3 sphere.

$$V = \frac{4}{3}\pi r^3$$

$$10.0\ \text{cm}^3 = \frac{4}{3}\pi r^3$$

$$r = 1.34\ \text{cm}$$

The surface area of the sphere is:

$$\textbf{area} = 4\pi r^2 = 4\pi(1.34\ \text{cm})^2 = \mathbf{22.6\ cm^2}$$

Next, calculate the radius of the 1.25 cm^3 sphere.

$$V = \frac{4}{3}\pi r^3$$

$$1.25\ \text{cm}^3 = \frac{4}{3}\pi r^3$$

$$r = 0.668\ \text{cm}$$

The surface area of one sphere is:

$$\text{area} = 4\pi r^2 = 4\pi(0.668\ \text{cm})^2 = 5.61\ \text{cm}^2$$

$$\textbf{The total area of 8 spheres} = 5.61\ \text{cm}^2 \times 8 = \mathbf{44.9\ cm^2}$$

Obviously, the surface area of the eight spheres (44.9 cm^2) is greater than that of one larger sphere (22.6 cm^2). A greater surface area promotes the catalyzed reaction more effectively.

14.62 If water is also the solvent in this reaction, it is present in vast excess over the other reactants and products. Throughout the course of the reaction, the concentration of the water will not change by a measurable amount. As a result, the reaction rate will not appear to depend on the concentration of water.

14.64 **(a)** To determine the rate law, we must determine the exponents in the equation

$$\text{rate} = k[CH_3COCH_3]^x[Br_2]^y[H^+]^z$$

To determine the order of the reaction with respect to CH_3COCH_3, find two experiments in which the $[Br_2]$ and $[H^+]$ are held constant. Compare the data from experiments (a) and (e). When the concentration of CH_3COCH_3 is increased by a factor of 1.33, the reaction rate increases by a factor of 1.33. Thus, the reaction is first-order in CH_3COCH_3.

To determine the order with respect to Br_2, compare experiments (a) and (b). When the Br_2 concentration is doubled, the reaction rate does not change. Thus, the reaction is zero-order in Br_2.

To determine the order with respect to H^+, compare experiments (a) and (c). When the H^+ concentration is doubled, the reaction rate doubles. Thus, the reaction is first-order in H^+.

The rate law is:

$$\textbf{rate} = \boldsymbol{k[CH_3COCH_3][H^+]}$$

(b) Rearrange the rate law from part (a), solving for k.

$$k = \frac{\text{rate}}{[CH_3COCH_3][H^+]}$$

Substitute the data from any one of the experiments to calculate k. Using the data from Experiment (a),

$$\boldsymbol{k} = \frac{5.7\times10^{-5}\ M/\text{s}}{(0.30\ M)(0.050\ M)} = \mathbf{3.8\times10^{-3}\ /}\boldsymbol{M}\mathbf{\cdot s}$$

14.66 Recall that the pressure of a gas is directly proportional to the number of moles of gas. This comes from the ideal gas equation.

$$P = \frac{nRT}{V}$$

The balanced equation is:

$$2\ N_2O\ (g) \longrightarrow 2\ N_2\ (g) + O_2\ (g)$$

From the stoichiometry of the balanced equation, for every one mole of N_2O that decomposes, one mole of N_2 and 0.5 moles of O_2 will be formed. Let's assume that we had 2 moles of N_2O at $t = 0$. After one half-life there will be one mole of N_2O remaining and one mole of N_2 and 0.5 moles of O_2 will be formed. The total number of moles of gas after one half-life will be:

$$n_T = n_{N_2O} + n_{N_2} + n_{O_2} = 1\text{ mol} + 1\text{ mol} + 0.5\text{ mol} = 2.5\text{ moles}$$

At $t = 0$, there were 2 mol of gas. Now, at $t_{\frac{1}{2}}$, there are 2.5 mol of gas. Since the pressure of a gas is directly proportional to the number of moles of gas, we can write:

$$\frac{2.10\text{ atm}}{2\text{ mol gas }(t = 0)} \times 2.5\text{ mol gas}\left(\text{at } t_{\frac{1}{2}}\right) = \mathbf{2.63\ atm\ after\ one\ half\text{-}life}$$

14.68 The rate expression for a third order reaction is:

$$\text{rate} = -\frac{\Delta[\text{A}]}{\Delta t} = k[\text{A}]^3$$

The units for the rate law are:

$$\frac{M}{\text{s}} = kM^3$$

$$\mathbf{k = M^{-2}s^{-1}}$$

14.70 Both compounds, A and B, decompose by first-order kinetics. Therefore, we can write a first-order rate equation for A and also one for B.

$$[\text{A}] = [\text{A}]_0 e^{-k_\text{A}t} \qquad [\text{B}] = [\text{B}]_0 e^{-k_\text{B}t}$$

We can calculate each of the rate constants, k_A and k_B, from their respective half-lives.

$$k_\text{A} = \frac{0.693}{50.0\text{ min}} = 0.0139\text{ min}^{-1} \qquad k_\text{B} = \frac{0.693}{18.0\text{ min}} = 0.0385\text{ min}^{-1}$$

The initial concentration of A and B are equal. $[\text{A}]_0 = [\text{B}]_0$. Therefore, from the first-order rate equations, we can write:

$$\frac{[\text{A}]}{[\text{B}]} = 4 = \frac{[\text{A}]_0 e^{-k_\text{A}t}}{[\text{B}]_0 e^{-k_\text{B}t}} = \frac{e^{-k_\text{A}t}}{e^{-k_\text{B}t}} = e^{(k_\text{B} - k_\text{A})t} = e^{(0.0385-0.0139)t}$$

$$4 = e^{0.0246t}$$

$$t = \mathbf{56.4\text{ min}}$$

14.72 **(a)** Changing the concentration of a reactant has no effect on *k*.

(b) If a reaction is run in a solvent rather than in the gas phase, then the reaction mechanism will probably change and will thus change *k*.

(c) Doubling the pressure simply changes the concentration. No effect on *k*, as in (a).

(d) The rate constant *k* changes with temperature.

(e) A catalyst changes the reaction mechanism and therefore changes *k*.

14.74 Mathematically, the amount left after ten half–lives is:

$$\left(\frac{1}{2}\right)^{10} = 9.8 \times 10^{-4} = \mathbf{0.098\%}$$

14.76 The net ionic equation is:

$$\text{Zn}\,(s) + 2\,\text{H}^+\,(aq) \longrightarrow \text{Zn}^{2+}\,(aq) + \text{H}_2\,(g)$$

(a) Changing from the same mass of granulated zinc to powdered zinc **increases** the rate because the surface area of the zinc (and thus its concentration) has increased.

(b) Decreasing the mass of zinc (in the same granulated form) will **decrease** the rate because the total surface area of zinc has decreased.

(c) The concentration of protons has decreased in changing from the strong acid (hydrochloric) to the weak acid (acetic); the rate will **decrease**.

(d) An increase in temperature will increase the rate constant k; therefore, the rate of reaction **increases**.

14.78 In terms of the initial concentration of A, the concentration of A remaining after 4.90 min is 0.645 $[A]_0$. Using the first-order rate equation,

$$\ln\frac{[A]_0}{[A]} = kt$$

or

$$\ln\frac{[A]_0}{0.645[A]_0} = kt$$

$$0.439 = k(4.90\text{ min})$$

$$k = \frac{0.439}{4.90\text{ min}} = \mathbf{0.0896\ min^{-1}}$$

14.80 The first-order rate equation can be arranged to take the form of a straight line.

$$\ln[A] = -kt + \ln[A]_0$$

If a reaction obeys first-order kinetics, a plot of ln[A] vs. t will be a straight line with a slope of $-k$.

The slope of a plot of $\ln[N_2O_5]$ vs. t is $-6.18 \times 10^{-4}\text{ min}^{-1}$. Thus,

$$k = 6.18 \times 10^{-4}\text{ min}^{-1}$$

The equation for the half-life of a first-order reaction is:

$$t_{\frac{1}{2}} = \frac{0.693}{k}$$

$$t_{\frac{1}{2}} = \frac{0.693}{6.18 \times 10^{-4}\text{ min}^{-1}} = \mathbf{1.12 \times 10^3\ min}$$

14.82 **(a)** We can write the rate law for an elementary step directly from the stoichiometry of the balanced reaction. In this rate-determining elementary step three molecules must collide simultaneously (one X and two Y's). This makes the reaction termolecular, and consequently the rate law must be third order: first order in X and second order in Y.

The rate law is:

$$\textbf{rate} = \boldsymbol{k}\textbf{[X][Y]}^2$$

(b) The value of the rate constant can be found by solving algebraically for k.

$$k = \frac{\text{rate}}{[X][Y]^2} = \frac{3.8 \times 10^{-3}\ \cancel{M}/\text{s}}{(0.26\ \cancel{M})(0.88\ M)^2} = \mathbf{1.9 \times 10^{-2}\ \boldsymbol{M}^{-2}s^{-1}}$$

Could you write the rate law if the reaction shown were the overall balanced equation and not an elementary step?

14.84 We can calculate the ratio of k_1/k_2 at 40°C using the Arrhenius equation.

$$\frac{k_1}{k_2} = \frac{Ae^{-E_{a_1}/RT}}{Ae^{-E_{a_2}/RT}} = e^{-(E_{a_1} - E_{a_2})/RT}$$

$$8.0 = e^{-\Delta E_a/(8.314\ \text{J/K·mol})(313\ \text{K})}$$

$$\Delta E_a = -5.4 \times 10^3\ \text{J/mol}$$

Having calculate ΔE_a, we can substitute back into the equation to calculate the ratio k_1/k_2 at 300°C (573 K).

$$\frac{\boldsymbol{k_1}}{\boldsymbol{k_2}} = e^{-(-5.4\times10^3\text{J/mol})/(8.314\ \text{J/K·mol})(573\ \text{K})} = \mathbf{3.1}$$

14.86 During the first five minutes or so the engine is relatively cold, so the exhaust gases will not fully react with the components of the catalytic converter. Remember, for almost all reactions, the rate of reaction increases with temperature.

14.88 Let k_2 be the rate constant for the slow step:

$$\text{rate} = k_2[\text{CH}_3\text{–C}(=\text{OH}^+)\text{—CH}_3][\text{H}_2\text{O}] \qquad (1)$$

Let k_1 and k_{-1} be the rate constants for the forward and reverse steps in the fast equilibrium.

$$k_1[\text{CH}_3\text{COCH}_3][\text{H}_3\text{O}^+] = k_{-1}[\text{CH}_3\text{-C}(=\text{OH}^+)\text{—CH}_3][\text{H}_2\text{O}] \qquad (2)$$

Therefore, Equation (1) becomes

$$\text{rate} = \frac{k_1k_2}{k_{-1}}[\text{CH}_3\text{COCH}_3][\text{H}_3\text{O}^+]$$

which is the same as (a) in Problem 14.64.

14.90 At very high $[H_2]$,

$$k_2[\text{H}_2] >> 1$$

$$\textbf{rate} = \frac{k_1[\text{NO}]^2[\text{H}_2]}{k_2[\text{H}_2]} = \frac{\boldsymbol{k_1}}{\boldsymbol{k_2}}[\textbf{NO}]^2$$

At very low $[H_2]$,

$$k_2[\text{H}_2] << 1$$

$$\textbf{rate} = \frac{k_1[\text{NO}]^2[\text{H}_2]}{1} = \boldsymbol{k_1}[\textbf{NO}]^2[\textbf{H}_2]$$

The result from Problem 14.77 agrees with the rate law determined for low $[H_2]$.

14.92 A plausible two-step mechanism is:

$$NO_2 + NO_2 \rightarrow NO_3 + NO \quad \text{slow}$$
$$NO_3 + CO \rightarrow NO_2 + CO_2 \quad \text{fast}$$
$$\overline{CO + NO_2 \rightarrow CO_2 + NO}$$

14.94 At high pressure of PH_3, all the sites on W are occupied, so the rate is independent of $[PH_3]$.

14.96 **(a)** $\frac{\Delta[B]}{\Delta t} = k_1[A] - k_2[B]$

(b) If $\frac{\Delta[B]}{\Delta t} = 0$

Then, from part (a) of this problem:

$$k_1[A] = k_2[B]$$

$$\mathbf{[B] = \frac{k_1}{k_2}[A]}$$

14.98 **(a)** Drinking too much alcohol too fast means all the alcohol dehydrogenase (ADH) active sites are tied up and the excess alcohol will damage the central nervous system.

(b) Both ethanol and methanol will compete for the same site at ADH. An excess of ethanol will replace methanol at the active site, leading to methanol's discharge from the body.

CHAPTER 15
CHEMICAL EQUILIBRIUM

PROBLEM-SOLVING STRATEGIES AND TUTORIAL SOLUTIONS

TYPES OF PROBLEMS

Problem Type 1: Homogeneous Equilibria.
- **(a)** Writing expressions for K_c and K_p.
- **(b)** Calculating equilibrium partial pressures.
- **(c)** Converting between K_c and K_p.

Problem Type 2: Heterogeneous Equilibria, Calculating K_p and K_c.

Problem Type 3: The Form of K and the Equilibrium Equation.

Problem Type 4: Using the Reaction Quotient (Q_c) to Predict the Direction of a Reaction.

Problem Type 5: Calculating Equilibrium Concentrations.

Problem Type 6: Factors that Affect Chemical Equilibrium.
- **(a)** Changes in concentration.
- **(b)** Changes in pressure and volume.
- **(c)** Changes in temperature.
- **(d)** The effect of a catalyst.

PROBLEM TYPE 1: HOMOGENEOUS EQUILIBRIA

Let's start by considering the following reversible reaction.

$$a\text{A} + b\text{B} \rightleftharpoons c\text{C} + d\text{D}$$

where *a, b, c,* and *d* are the stoichiometric coefficients for the reacting species A, B, C, and D. The equilibrium constant for the reaction at a particular temperature is:

$$K = \frac{[\text{C}]^c[\text{D}]^d}{[\text{A}]^a[\text{B}]^b}$$

This equation is the mathematical form of the **law of mass action**. It relates the concentrations of reactants and products at equilibrium in terms of a quantity called the **equilibrium constant (*K*)**.

The magnitude of the equilibrium constant, K, is important. If the equilibrium constant is much greater than one ($K >> 1$), we say that the equilibrium lies to the right and favors the products. Conversely, if $K << 1$, the equilibrium lies to the left and favors the reactants.

A. Writing expressions for K_c and K_p

The term **homogeneous equilibrium** applies to reactions in which all reacting species are in the same phase. An example of a homogeneous gas-phase equilibrium is the reaction between sulfur dioxide and oxygen to form sulfur trioxide.

$$2\,SO_2\,(g) + O_2\,(g) \rightleftharpoons 2\,SO_3\,(g)$$

The equilibrium constant as given in Equation (15.2) of your text is:

$$K_c = \frac{[SO_3]^2}{[SO_2]^2[O_2]}$$

The subscript c of K_c denotes that the concentrations of all species are expressed in mol/L.

The concentrations of reactants and products in gas-phase reactions can also be expressed in terms of their partial pressures. Starting with the ideal gas equation, we can write:

$$P = \left(\frac{n}{V}\right)RT$$

At constant temperature, the pressure (P) of a gas is directly related to the concentration of the gas in mol/L, $\left(\frac{n}{V}\right)$.

Thus, for the equilibrium shown above, we can write:

$$K_p = \frac{P_{SO_3}^2}{P_{SO_2}^2 P_{O_2}}$$

where P_{SO_3}, P_{SO_2}, and P_{O_2} are the equilibrium partial pressures (in atmospheres) of SO_3, SO_2, and O_2, respectively. The subscript p of K_p indicates the equilibrium concentrations are expressed in terms of pressure.

EXAMPLE 15.1

Write the equilibrium constant expression for the following reversible reaction:

$$4\,NH_3\,(g) + 5\,O_2\,(g) \rightleftharpoons 4\,NO\,(g) + 6\,H_2O\,(g)$$

Remember that the concentration of each component in the equilibrium constant expression is raised to a power equal to its coefficient in the balanced equation.

$$K_c = \frac{[NO]^4[H_2O]^6}{[NH_3]^4[O_2]^5}$$

You could also write the equilibrium constant expression in terms of the partial pressures of the gaseous components.

$$K_p = \frac{P_{NO}^4 P_{H_2O}^6}{P_{NH_3}^4 P_{O_2}^5}$$

PRACTICE EXERCISE

1. Write the equilibrium constant expression for the following reaction.

$$2\,N_2O\,(g) + 3\,O_2\,(g) \rightleftharpoons 2\,N_2O_4\,(g)$$

Text Problem: 15.8

B. Calculating equilibrium partial pressures

EXAMPLE 15.2

At 400°C, K_p = 64 for the following reaction:

$$H_2\,(g) + I_2\,(g) \rightleftharpoons 2\,HI\,(g)$$

At equilibrium, the partial pressures of H_2 and I_2 in a closed container are 0.20 atm and 0.50 atm respectively. What is the partial pressure of HI in the mixture?

Step 1: Write K_p in terms of the partial pressures of the reacting species.

$$K_p = \frac{P_{HI}^2}{P_{H_2}P_{I_2}}$$

Step 2: Solve the above equation algebraically for the unknown partial pressure. Then, substitute the known values into the equation to solve for the unknown.

$$P_{HI}^2 = K_p P_{H_2} P_{I_2}$$

$$P_{HI} = \sqrt{K_p P_{H_2} P_{I_2}}$$

$$\mathbf{P_{HI}} = \sqrt{(64)(0.20)(0.50)} = \mathbf{2.5\ atm}$$

PRACTICE EXERCISE

2. Consider the following reaction at a temperature of 250°C.

$$PCl_5\,(g) \rightleftharpoons PCl_3\,(g) + Cl_2\,(g)$$

At equilibrium, the partial pressures of PCl_5, PCl_3, and Cl_2 are 0.0704 atm, 0.340 atm, and 0.218 atm, respectively. What is the value of K_p?

C. Converting K_c to K_p

In general, K_c is not equal to K_p, since the partial pressures of reactants and products are not equal to their concentrations expressed in mol/L. A simple relationship between K_p and K_c can be derived. See Section 15.2 of your text for the derivation. The relationship is:

$$K_p = K_c(RT)^{\Delta n}$$

where,

Δn = mol of gaseous product in balanced equation − mol of gaseous reactants in balanced eq.

If we express pressure in atmospheres, the gas constant R is given by 0.0821 L·atm/mol·K. We can write the relationship between K_p and K_c as

$$K_p = K_c(0.0821\ T)^{\Delta n} \qquad (15.1)$$

In general, $K_p \neq K_c$ except in the special case when $\Delta n = 0$. In this case, Equation (15.1) can be written as:

$$K_p = K_c(0.0821\ T)^0$$

and

$$K_p = K_c$$

EXAMPLE 15.3

In the decomposition of carbon dioxide at 2000°C,

$$\mathbf{2\,CO_2\,(g) \rightleftharpoons 2\,CO\,(g) + O_2\,(g)}$$

$\mathbf{K_p = 1.2 \times 10^{-4}}$**. Calculate K_c for this reaction.**

Step 1: Rearrange Equation (15.1), which relates K_p and K_c, solving for K_c.

$$K_p = K_c(0.0821\ T)^{\Delta n}$$

$$K_c = \frac{K_p}{(0.0821\ T)^{\Delta n}}$$

Step 2: Substitute the given values into the above equation to solve for K_c. Temperature must be in units of Kelvin. **Note:** In the balanced equation, there are a total of three moles of gaseous products and two moles of gaseous reactants, so $\Delta n = 3 - 2$.

$$\mathbf{K_c} = \frac{K_p}{(0.0821\ T)^{\Delta n}} = \frac{1.2 \times 10^{-4}}{[(0.0821)(2273\text{K})]^{(3-2)}} = \mathbf{6.4 \times 10^{-7}}$$

PRACTICE EXERCISE

3. At 400°C, $K_p = 64$ for the following reaction:

$$H_2\,(g) + I_2\,(g) \rightleftharpoons 2\,HI\,(g)$$

What is the value of K_c for this reaction?

Text Problems: 14, 18

PROBLEM TYPE 2: HETEROGENEOUS EQUILIBRIA, CALCULATING K_p AND K_c

Whenever a reaction involves reactants and products that exist in different phases, it is called a *heterogeneous reaction.* If the reaction is carried out in a closed container, a **heterogeneous equilibrium** will result. For example, when steam is brought into contact with charcoal in a closed container, the following equilibrium is established:

$$C\,(s) + H_2O\,(g) \rightleftharpoons H_2\,(g) + CO\,(g)$$

At equilibrium, we might write the equilibrium constant as:

$$K_c' = \frac{[H_2][CO]}{[C][H_2O]} \qquad (15.2)$$

However, the "concentration" of a solid, like its density, is an intensive property and thus does not depend on the amount of substance present. For this reason, the term [C] is a constant and can be combined with the equilibrium constant. We can simplify Equation (15.2) by writing

$$[C]\,K_c' = K_c = \frac{[H_2][CO]}{[H_2O]}$$

Keep in mind that the value of K_c does not depend on how much carbon is present, as long as some amount is present at equilibrium.

The argument presented above for solids also applies to pure liquids. Thus, if a liquid is a reactant or product, we can treat its concentration as a constant and omit it from the equilibrium constant expression.

EXAMPLE 15.4

What are the values of K_p and K_c at 1000°C for the reaction

$$\mathbf{CaCO_3\,(s) \rightleftharpoons CaO\,(s) + CO_2\,(g)}$$

if the pressure of CO_2 in equilibrium with $CaCO_3$ and CaO is 3.87 atm?

Step 1: As stated above, if a solid or a liquid is a reactant or product, we can treat its concentration as a constant and omit it from the equilibrium constant expression. Thus, enough information is given to calculate K_p for this heterogeneous equilibrium.

$$K_p = P_{CO_2} = 3.87$$

Step 2: To calculate K_c, rearrange the equation relating K_p and K_c, solving for K_c. Then, substitute the given values into the equation to obtain the answer.

$$K_p = K_c(0.0821\ T)^{\Delta n}$$

$$K_c = \frac{K_p}{(0.0821\ T)^{\Delta n}}$$

$$\mathbf{K_c} = \frac{3.87}{[(0.0821)(1273\text{K})]^{(1-0)}} = \mathbf{0.0370}$$

Text Problems: 18, **20**

PROBLEM TYPE 3: THE FORM OF *K* AND THE EQUILIBRIUM EQUATION

The equilibrium constant expression and its value depend on how an equation is balanced. Often an equation can be balanced with more than one set of coefficients. For example,

$$2\ SO_2\ (g) + O_2\ (g) \rightleftharpoons 2\ SO_3\ (g) \qquad K_c = \frac{[SO_3]^2}{[SO_2]^2[O_2]}$$

and

$$SO_2\ (g) + 1/2\ O_2\ (g) \rightleftharpoons SO_3\ (g) \qquad K_c' = \frac{[SO_3]}{[SO_2][O_2]^{1/2}}$$

Is there a relationship between the equilibrium constants for the two reactions? Note that K_c' is the square root of the equilibrium constant for the first reaction, K_c.

$$K_c' = \frac{[SO_3]}{[SO_2][O_2]^{1/2}} = \sqrt{\frac{[SO_3]^2}{[SO_2]^2[O_2]}} = \sqrt{K_c}$$

The general relationship is that you raise K_c to the power by which the equation was multiplied. To come up with the second balanced equation above, we had to multiply the first equation by 1/2. Thus,

$$K_c' = (K_c)^{1/2} = \sqrt{K_c}$$

What if the reaction is written in the reverse direction?

$$2SO_3\ (g) \rightleftharpoons 2\ SO_2\ (g) + O_2\ (g) \qquad K_c'' = \frac{[SO_2]^2[O_2]}{[SO_3]^2}$$

Is there a relationship between K_c'' and K_c? By inspection, you should find that K_c'' is the reciprocal of K_c for the forward reaction.

$$K_c'' = \frac{[SO_2]^2[O_2]}{[SO_3]^2} = \frac{1}{K_c}$$

Remember, always use the K_c expression and value that are consistent with the way in which the balanced equation is written.

EXAMPLE 15.5

For the reaction, $2\,HBr\,(g) \rightleftharpoons H_2\,(g) + Br_2\,(g)$, $K_p = 1.4 \times 10^{-5}$ at 700 K. What are the values of K_p for the following reactions at the same temperature?

(a) $4\,HBr\,(g) \rightleftharpoons 2\,H_2\,(g) + 2\,Br_2\,(g)$
(b) $H_2\,(g) + Br_2\,(g) \rightleftharpoons 2\,HBr\,(g)$

(a) In Equation (a), the original equation has been multiplied by *two*. The general relationship is that you raise K to the power by which the equation was multiplied by. Thus,

$$K_p' = (K_p)^2 = (1.4 \times 10^{-5})^2 = \mathbf{2.0 \times 10^{-10}}$$

(b) Equation (b) is the reverse of the original equation. By inspection, you should find that K_p'' is the reciprocal of K_p.

$$K_p'' = \frac{1}{K_p} = \frac{1}{1.4 \times 10^{-5}} = \mathbf{7.1 \times 10^4}$$

PRACTICE EXERCISE

4. For the reaction, $2\,HBr\,(g) \rightleftharpoons H_2\,(g) + Br_2\,(g)$, $K_p = 1.4 \times 10^{-5}$ at 700 K. What is the value of K_p for the following reaction at the same temperature?

$$HBr\,(g) \rightleftharpoons 1/2\,H_2\,(g) + 1/2\,Br_2\,(g)$$

Text Problem: 16

PROBLEM TYPE 4: USING THE REACTION QUOTIENT (*Q*) TO PREDICT THE DIRECTION OF A REACTION

Equilibrium constants provide useful information about chemical reaction systems. For instance, equilibrium constants can be used to predict the direction in which a reaction will proceed to establish equilibrium.

The reaction quotient, Q_c, is a useful aid in predicting whether or not a reaction system is at equilibrium. Again, consider the reaction

$$2\,SO_2\,(g) + O_2\,(g) \rightleftharpoons 2\,SO_3\,(g)$$

The reaction quotient is:

$$Q_c = \frac{[SO_3]_0^2}{[SO_2]_0^2[O_2]_0}$$

You should notice that Q_c has the same algebraic form as K_c. However, the concentrations are not necessarily equilibrium concentrations. We will call them initial concentrations, represented by a subscript 0 after the square brackets, $[\;]_0$. Substituting initial concentrations into the reaction quotient, gives a value for Q_c. In order to predict whether the system is at equilibrium, the magnitude of Q_c must be compared with that of K_c.

If $Q_c = K_c$, The initial concentrations are equilibrium concentrations. The system is at equilibrium.

If $Q_c > K_c$, The ratio of initial concentrations of products to reactants is too large. To reach equilibrium, products must be converted to reactants. The system proceeds from right to left (consuming products, forming reactants) to reach equilibrium.

If $Q_c < K_c$, The ratio of initial concentrations of products to reactants is too small. To reach equilibrium, reactants must be converted to products. The system proceeds from left to right (consuming reactants, forming products) to reach equilibrium.

EXAMPLE 15.6

At a certain temperature, the reaction $CO(g) + Cl_2(g) \rightleftharpoons COCl_2(g)$, has an equilibrium constant $K_c = 13.8$. Is the following mixture an equilibrium mixture? If not, in which direction (right or left) will the reaction proceed to reach equilibrium?

$$[CO]_0 = 2.5\ M,\ [Cl_2]_0 = 1.2\ M,\ \text{and}\ [COCl_2]_0 = 5.0\ M$$

Step 1: Recall that for a system to be at equilibrium, $Q_c = K_c$. Substitute the given concentrations into the equation for the reaction quotient to calculate Q_c.

$$Q_c = \frac{[COCl_2]_0}{[CO]_0[Cl_2]_0} = \frac{5.0}{(2.5)(1.2)} = 1.7$$

Step 2: Compare Q_c to K_c. Since $Q_c < K_c$, the ratio of initial concentrations of products to reactants is too small. To reach equilibrium, reactants must be converted to products. The system proceeds from **left to right** (consuming reactants, forming products) to reach equilibrium.

PRACTICE EXERCISE

5. Given the reaction,

$$N_2(g) + O_2(g) \rightleftharpoons 2\,NO(g) \qquad K_c = 2.5 \times 10^{-3} \text{ at } 2130°C$$

decide whether the following mixture is at equilibrium or if a net forward or reverse reaction will occur. $[NO] = 0.0050\ M$, $[O_2] = 0.25\ M$, and $[N_2] = 0.020\ M$

Text Problem: 28

PROBLEM TYPE 5: CALCULATING EQUILIBRIUM CONCENTRATIONS

The expected concentrations at equilibrium can be calculated from a knowledge of the initial concentrations and the equilibrium constant. In these types of problems, it will be very helpful to recall that

$$\text{equilibrium concentration} = \text{initial concentration} \pm \text{the change due to reaction.}$$

The next *two* examples illustrate this important type of calculation.

EXAMPLE 15.7

A 0.25 mol sample of N_2O_4 dissociates and comes to equilibrium in a 1.5 L flask at 100°C. The reaction is

$$N_2O_4(g) \rightleftharpoons 2\,NO_2(g) \qquad K_c = 0.36 \text{ at } 100°C$$

What are the equilibrium concentrations of NO_2 and N_2O_4?

Step 1: Express the equilibrium concentrations of all species in terms of initial concentrations and a single unknown x, that represents the change in concentration. Let $(-x)$ be the depletion in concentration (mol/L) of N_2O_4.

From the stoichiometry of the reaction, it follows that the increase in concentration for NO_2 must be $2x$. Complete a table that lists the initial concentrations, the change in concentrations, and the equilibrium concentrations.

	$N_2O_4\ (g)$ $\rightleftharpoons$	$2\ NO_2\ (g)$
Initial (M):	0.25 mol/1.5 L = 0.17	0
Change (M):	$-x$	$+2x$
Equilibrium (M):	$0.17 - x$	$2x$

Step 2: Write the equilibrium constant expression in terms of the equilibrium concentrations. Knowing the value of the equilibrium constant, solve for x.

$$K_c = \frac{[NO_2]^2}{[N_2O_4]}$$

$$0.36 = \frac{(2x)^2}{0.17 - x}$$

$$0.061 - 0.36x = 4x^2$$

$$4x^2 + 0.36x - 0.061 = 0$$

The above equation is a quadratic equation of the form $ax^2 + bx + c = 0$. The solution for a quadratic equation is :

$$x = \frac{-b \pm \sqrt{b^2 - 4ac}}{2a}$$

Here, we have a = 4, b = 0.36, and c = −0.061. Substituting into the above equation,

$$x = \frac{-0.36 \pm \sqrt{(0.36)^2 - 4(4)(-0.061)}}{2(4)}$$

$$x = \frac{-0.36 \pm 1.05}{8}$$

$$x = 0.086\ M \quad \text{or} \quad x = -0.18\ M$$

The second solution is physically impossible because you cannot have a negative concentration. The first solution is the correct answer.

> **Tip:** In solving a quadratic equation of this type, one answer is always physically impossible, so the choice of which value to use for x is easy to make.

Step 3: Having solved for x, calculate the equilibrium concentrations of all species.

$$[N_2O_4] = (0.17 - 0.086)M = \mathbf{0.08\ M}$$

$$[NO_2] = 2(0.086\ M) = \mathbf{0.17\ M}$$

EXAMPLE 15.8

A 1.00 L vessel initially contains 0.777 mol of SO_3 (*g*) at 1100 K. What is the value of K_c for the following reaction if 0.520 mol of SO_3 remain at equilibrium?

$$2\ SO_3\ (g) \rightleftharpoons 2\ SO_2\ (g) + O_2\ (g)$$

If we can calculate the equilibrium concentrations for all species, we can calculate the equilibrium constant K_c.

Step 1: In this problem, we are given both the initial and equilibrium concentrations of SO_3. Recalling that

$$\text{equilibrium concentration} = \text{initial concentration} \pm \text{the change due to reaction}$$

we can calculate the change in concentration of SO_3 due to reaction. Let's call this change, $2x$, because of the coefficient of 2 for SO_3 in the balanced equation.

$$\text{change in concentration of } SO_3 = 2x = 0.777\ M - 0.520\ M = 0.257\ M$$

$$x = 0.129\ M$$

Complete a table that lists the initial concentrations, the change in concentrations, and the equilibrium concentrations.

	$2\,SO_3\,(g)$	$\rightleftharpoons$	$2\,SO_2\,(g)$	+	$O_2\,(g)$
Initial (*M*):	0.777		0		0
Change (*M*):	$-2x = -0.257$		$+2x$		$+x$
Equilibrium (*M*):	0.520		$2x$		x

So, at equilibrium,

$$[SO_2] = 2x = 0.257\ M$$
$$[O_2] = x = 0.129\ M$$

Tip: You probably could have come up with the equilibrium concentrations of SO_2 and O_2 without the use of a table. However, a table is a simple way to keep all the data organized.

Step 2: Substitute the equilibrium concentrations into the equilibrium constant expression to solve for K_c.

$$K_c = \frac{[SO_2]^2[O_2]}{[SO_3]^2}$$

$$\mathbf{K_c} = \frac{(0.257)^2(0.129)}{(0.520)^2} = \mathbf{0.0315}$$

PRACTICE EXERCISE

6. For the reaction,

$$N_2\,(g) + O_2\,(g) \rightleftharpoons 2\,NO\,(g) \qquad K_p = 3.80 \times 10^{-4} \text{ at } 2000°C$$

what equilibrium pressures of N_2, O_2, and NO will result when a 10.0 L reactor vessel is filled with 2.00 atm of N_2 and 0.400 atm of O_2 and the reaction is allowed to come to equilibrium?

7. Initially a 1.0 L vessel contains 10.0 mol of NO and 6.0 mol of O_2 at a certain temperature.

$$2\,NO\,(g) + O_2\,(g) \rightleftharpoons 2\,NO_2\,(g)$$

At equilibrium, the vessel contains 8.8 mol of NO_2. Determine the value of K_c at this temperature.

Text Problems: 30, 32, 34, 36

PROBLEM TYPE 6: FACTORS THAT AFFECT CHEMICAL EQUILIBRIUM

What effect does a change in concentration, pressure, volume, or temperature have on a system at equilibrium? This question can be answered qualitatively by using **Le Chatelier's principle**. It states that when an external stress is applied to a system at equilibrium, the system adjusts in such a way that the stress is partially offset. The word "stress" here means a change in concentration, pressure, volume, or temperature that removes a system from the equilibrium state.

A. Changes in concentration

EXAMPLE 15.9

For the following reaction at equilibrium in a closed container,

$$2\ NaHCO_3\ (s) \rightleftharpoons Na_2CO_3\ (s) + H_2O\ (g) + CO_2\ (g)$$

state the effects (increase, decrease, or no change) of the following stresses on the number of moles of sodium carbonate, Na_2CO_3. Note that Na_2CO_3 is a solid (this is a heterogeneous equilibrium); its concentration will remain constant, but its amount can change.

(a) Removing CO_2 (*g*).
(b) Adding H_2O (*g*).
(c) Adding $NaHCO_3$ (*s*).

Applying Le Chatelier's principle,

(a) If CO_2 concentration is lowered, the system will react to offset the change. That is, a shift to the right will replace some of the removed CO_2. Moles of Na_2CO_3 will **increase**.

(b) The system will respond to the stress of added H_2O by shifting to the left to remove some of the water. Moles of Na_2CO_3 will **decrease**.

(c) The position of a heterogeneous equilibrium does not depend on the amounts of pure solids or liquids present. Remember that the concentrations of pure solids and liquids are constant and thus do not enter into the equilibrium constant expression. Hence, there is no shift in the equilibrium, and the amount of Na_2CO_3 is **unchanged**.

B. Changes in pressure and volume

The pressure of a system of gases in chemical equilibrium can be increased by decreasing the available volume. The ideal gas equation shows this inverse relationship between pressure and volume.

$$PV = nRT$$

A decrease in volume causes the concentrations of all *gas-phase* components to increase. Remember that pressure is directly proportional to the concentration of a gas.

$$P = \left(\frac{n}{V}\right)RT$$

The stress caused by the increased pressure will be partially offset by a net reaction that will lower the total pressure. In other words, the system will adjust by lowering the total concentration of gas molecules to reestablish equilibrium. Again, consider the following equilibrium:

$$2\ SO_2\ (g) + O_2\ (g) \rightleftharpoons 2\ SO_3\ (g)$$

When the molecules of the above gases are compressed into a smaller volume, the total pressure increases and hence the total concentration increases (this is a stress). A net forward reaction (shift to the right) will bring the system to a new state of equilibrium, in which the *total concentration* of all molecules will be lowered somewhat. This partially offsets the initial stress on the system. The total concentration is lowered somewhat because when 2 moles of SO_2 react with 1 mole of O_2 (a total of 3 moles), only 2 moles of SO_3 are produced.

In general, an increase in pressure (decrease in volume) favors the net reaction that decreases the total number of moles of gases. On the other hand, a decrease in pressure (increase in volume) favors the net reaction that increases the total number of moles of gases.

EXAMPLE 15.10

For the following reaction at equilibrium in a closed container,

$$2\ NaHCO_3\ (s) \rightleftharpoons Na_2CO_3\ (s) + H_2O\ (g) + CO_2\ (g)$$

state the effect (increase, decrease, or no change) of increasing the volume of the container on the number of moles of sodium carbonate, Na_2CO_3. Note that Na_2CO_3 is a solid (this is a heterogeneous equilibrium); its concentration will remain constant, but its amount can change.

Pressure and volume are inversely related. Thus, an increase in the volume of the container corresponds to a decrease in pressure. The system will shift to offset the decreased pressure by undergoing the reaction that increases the total number of moles of gases. Looking at the above reaction, there are zero moles of gas on the reactants' side, and two moles of gas on the products' side.

The reaction will shift to the right to establish equilibrium, and the amount of Na_2CO_3 will **increase**.

C. Changes in temperature

A change in concentration, pressure, or volume may alter the equilibrium position, but it does not change the value of the equilibrium constant. However, a change in temperature can alter the equilibrium constant.

To decide how a temperature stress will affect a system at equilibrium, you must look at whether the reaction is endothermic or exothermic. The following reaction is endothermic (positive ΔH).

$$PCl_5\ (g) \rightleftharpoons PCl_3\ (g) + Cl_2\ (g) \qquad \Delta H° = 92.9\ \text{kJ}$$

What happens if we heat the above system at equilibrium? We have added heat (a stress). The system will shift in the direction that will partially offset this stress by removing some heat. Since endothermic processes absorb heat from the surroundings (reducing the stress), heating favors dissociation of PCl_5 into PCl_3 and Cl_2 molecules (the forward reaction). Consequently, the equilibrium constant for the above reaction, given by

$$K_c = \frac{[PCl_3][Cl_2]}{[PCl_5]}$$

increases with temperature.

In general, a temperature *increase* favors an *endothermic* reaction, and a temperature *decrease* favors an *exothermic* reaction.

EXAMPLE 15.11

For the following reaction at equilibrium in a closed container,

$$2\ NaHCO_3\ (s) \rightleftharpoons Na_2CO_3\ (s) + H_2O\ (g) + CO_2\ (g) \qquad \Delta H = 128\ \text{kJ}$$

state the effect (increase, decrease, or no change) of decreasing the temperature on the number of moles of sodium carbonate, Na_2CO_3. Note that Na_2CO_3 is a solid (this is a heterogeneous equilibrium); its concentration will remain constant, but its amount can change.

Recall that a temperature decrease favors an exothermic reaction. The above equation, as written, is endothermic (positive ΔH). However, the reverse reaction is exothermic with $\Delta H = -128$ kJ.

$$Na_2CO_3 + H_2O + CO_2 \rightleftharpoons 2\ NaHCO_3 \qquad \Delta H = -128\text{ kJ}$$

Thus, the reverse reaction is favored, and the amount of Na_2CO_3 will **decrease**.

PRACTICE EXERCISE

8. For the decomposition of calcium carbonate

$$CaCO_3\,(s) \rightleftharpoons CaO\,(s) + CO_2\,(g) \qquad \Delta H° = 175\text{ kJ}$$

how will the amount (not concentration) of $CaCO_3$ (s) change with the following stresses to the system at equilibrium?

(a) CO_2 (*g*) is removed.
(b) CaO (*s*) is added.
(c) The temperature is raised.
(d) The volume of the container is decreased.

Text Problems: 42, **44**, **46**, 48, 50

D. The effect of a catalyst

A catalyst increases the rate at which a reaction occurs. For a reversible reaction, a catalyst affects the rate in the forward and reverse directions to the same extent. Therefore, the presence of a catalyst does not alter the equilibrium constant, nor does it shift the position of an equilibrium system.

ANSWERS TO PRACTICE EXERCISES

1. If concentration is expressed in mol/L, $K_c = \dfrac{[N_2O_4]^2}{[N_2O]^2[O_2]^3}$

or, in terms of partial pressures, $K_p = \dfrac{P_{N_2O_4}^2}{P_{N_2O}^2 P_{O_2}^3}$

2. $K_p = 1.05$

3. $K_p = K_c = 64$

4. $K_p' = 3.7 \times 10^{-3}$

5. Since $Q_c > K_c$, the ratio of initial concentrations of products to reactants is too large. To reach equilibrium, products must be converted to reactants. The system proceeds from **right to left** (consuming products, forming reactants) to reach equilibrium.

6. $P_{NO_2} = 1.74 \times 10^{-2}$ atm

$P_{N_2} = 1.99$ atm

$P_{O_2} = 0.391$ atm

7. $K_c = 34$

8. **(a)** The amount of $CaCO_3$ will **decrease**.
(b) The amount of $CaCO_3$ will **not change**.
(c) The amount of $CaCO_3$ will **decrease**.
(d) The amount of $CaCO_3$ will **increase**.

SOLUTIONS TO SELECTED TEXT PROBLEMS

15.8 (a) $K_c = \dfrac{[CO]^2[O_2]}{[CO_2]^2}$ $\quad K_p = \dfrac{P_{CO}^2 \cdot P_{O_2}}{P_{CO_2}^2}$

(b) $K_c = \dfrac{[O_3]^2}{[O_2]^3}$ $\quad K_p = \dfrac{P_{O_3}^2}{P_{O_2}^3}$

(c) $K_c = \dfrac{[COCl_2]}{[CO][Cl_2]}$ $\quad K_p = \dfrac{P_{COCl_2}}{P_{CO} \cdot P_{Cl_2}}$

(d) $K_c = \dfrac{[CO][H_2]}{[H_2O]}$ $\quad K_p = \dfrac{P_{CO} \cdot P_{H_2}}{P_{H_2O}}$

(e) $K_c = \dfrac{[H^+][HCOO^-]}{[HCOOH]}$

(f) $K_c = [O_2]$ $\quad K_p = P_{O_2}$

15.12 The problem states that the system is at equilibrium, so we simply substitute the equilibrium concentrations into the equilibrium constant expression to calculate K_c.

Step 1: Calculate the concentrations of the components in units of mol/L. The molarities can be calculated by simply dividing the number of moles by the volume of the flask.

$$[H_2] = \frac{2.50 \text{ mol}}{12.0 \text{ L}} = 0.208\ M$$

$$[S_2] = \frac{1.35 \times 10^{-5} \text{ mol}}{12.0 \text{ L}} = 1.13 \times 10^{-6}\ M$$

$$[H_2S] = \frac{8.70 \text{ mol}}{12.0 \text{ L}} = 0.725\ M$$

Step 2: Once the molarities are known, K_c can be found by substituting the molarities into the equilibrium constant expression.

$$\mathbf{K_c} = \frac{[H_2S]^2}{[H_2]^2[S_2]} = \frac{(0.725)^2}{(0.208)^2(1.13 \times 10^{-6})} = \mathbf{1.08 \times 10^7}$$

If you forget to convert moles to moles/liter, will you get a different answer? Under what circumstances will the two answers be the same?

15.14 Converting between K_c and K_p, Problem Type 1C.

Step 1: Rearrange the equation relating K_p and K_c, solving for K_c.

$$K_p = K_c(0.0821\ T)^{\Delta n}$$

$$K_c = \frac{K_p}{(0.0821T)^{\Delta n}}$$

Step 2: Substitute the given values into the above equation to solve for K_c. Temperature must be in units of Kelvin.

$$\boldsymbol{K_c} = \frac{K_p}{(0.0821T)^{\Delta n}} = \frac{5.0 \times 10^{-4}}{[(0.0821)(575\ \text{K})]^{(3-2)}} = \mathbf{1.1 \times 10^{-5}}$$

15.16 The equilibrium constant expressions are:

(a) $$K_c = \frac{[NH_3]^2}{[N_2][H_2]^3}$$

(b) $$K_c = \frac{[NH_3]}{[N_2]^{1/2}[H_2]^{3/2}}$$

Substituting the given equilibrium concentration gives:

(a) $$\boldsymbol{K_c} = \frac{(0.25)^2}{(0.11)(1.91)^3} = \mathbf{0.082}$$

(b) $$\boldsymbol{K_c} = \frac{(0.25)}{(0.11)^{1/2}(1.91)^{3/2}} = \mathbf{0.29}$$

Is there a relationship between the K_c values from parts (a) and (b)?

15.18 Because pure solids do not enter into an equilibrium constant expression, we can calculate K_p directly from the pressure that is due solely to CO_2 (*g*).

$$\boldsymbol{K_p} = P_{CO_2} = \mathbf{0.105}$$

Now, we can convert K_p to K_c using the following equation.

$$K_p = K_c(0.0821\ T)^{\Delta n}$$

$$K_c = \frac{K_p}{(0.0821T)^{\Delta n}}$$

$$\boldsymbol{K_c} = \frac{0.105}{(0.0821 \times 623)^{(1-0)}} = \mathbf{2.05 \times 10^{-3}}$$

15.20 Heterogeneous Equilibria, Calculating K_p. Problem Type 2.

Step 1: If a solid or a liquid is a reactant or product, we can treat its concentration as a constant and omit it from the equilibrium constant expression. Therefore, the equilibrium constant expression for the reaction is

$$K_p = P_{NH_3}^2 P_{CO_2}$$

Step 2: The total pressure in the flask (0.363 atm) is a sum of the partial pressures of NH_3 and CO_2.

$$P_T = P_{NH_3} + P_{CO_2} = 0.363 \text{ atm}$$

Let's let the partial pressure of $CO_2 = x$. From the stoichiometry of the balanced equation, you should find that $P_{NH_3} = 2P_{CO_2}$. Therefore, the partial pressure of $NH_3 = 2x$. Substituting into the equation for total pressure gives:

$$P_T = P_{NH_3} + P_{CO_2} = 2x + x = 3x$$

$$3x = 0.363 \text{ atm}$$

$$x = P_{CO_2} = 0.121 \text{ atm}$$

$$P_{NH_3} = 2x = 0.242 \text{ atm}$$

Step 3: Substitute the equilibrium pressures into the equilibrium constant expression to solve for K_p.

$$\boldsymbol{K_p} = P_{NH_3}^2 P_{CO_2} = (0.242)^2(0.121) = \mathbf{7.09 \times 10^{-3}}$$

15.22 If the CO pressure at equilibrium is 0.497 atm, the balanced equation requires the chlorine pressure to have the same value. The initial pressure of phosgene gas can be found from the ideal gas equation.

$$P = \frac{nRT}{V} = \frac{(3.00 \times 10^{-2} \text{ mol})(0.0821 \text{ L} \cdot \text{atm / mol} \cdot \text{K})(800 \text{ K})}{(1.50 \text{ L})} = 1.31 \text{ atm}$$

Since there is a 1:1 mole ratio between phosgene and CO, the partial pressure of CO formed equals the partial pressure of phosgene reacted. The phosgene pressure at equilibrium is:

$$P_{COCl_2} = (1.31 - 0.497) \text{ atm} = 0.81 \text{ atm}$$

The value of K_p is then found by substitution.

$$\boldsymbol{K_p} = \frac{P_{COCl_2}}{P_{CO}P_{Cl_2}} = \frac{0.81}{(0.497)^2} = \mathbf{3.3}$$

15.24 In this problem, you are asked to calculate K_c.

Step 1: Calculate the initial concentration of NOCl

$$[NOCl]_0 = \frac{2.50 \text{ mol}}{1.50 \text{ L}} = 1.67\ M$$

Step 2: If 28.0 percent of the NOCl has dissociated at equilibrium, 72.0 percent remains at equilibrium. The concentration of NOCl is 72.0 percent of its initial value.

$$[NOCl] = (0.720)(1.67\ M) = 1.20\ M$$

Since there is a 1:1 mole ratio between NOCl and NO, the concentration of NO at equilibrium is equal to the 28 percent of NOCl that dissociated.

$$[NO] = (0.280)(1.67\ M) = 0.468\ M$$

The mole ratio between NOCl and Cl_2 is 2:1. This means that the concentration of Cl_2 is equal to half of the 28 percent NOCl that dissociated.

$$[Cl_2] = \frac{1}{2}(0.280)(1.67\ M) = 0.234\ M$$

Step 3: The equilibrium constant K_c can be calculated by substituting the above concentrations into the equilibrium constant expression.

$$\mathbf{K_c} = \frac{[NO]^2[Cl_2]}{[NOCl]^2} = \frac{(0.468)^2(0.234)}{(1.20)^2} = \mathbf{0.0356}$$

15.28 Using the Reaction Quotient (Q_c) to Predict the Direction of a Reaction, Problem Type 4.

Step 1: Recall that for a system to be at equilibrium, $Q_c = K_c$. Substitute the given concentrations into the equation for the reaction quotient to calculate Q_c.

$$Q_c = \frac{[NH_3]^2}{[N_2][H_2]^3} = \frac{[0.48]^2}{[0.60][0.76]^3} = 0.87$$

Step 2: Compare Q_c to K_c. Since $Q_c < K_c$ (0.87 < 1.2), the ratio of initial concentrations of products to reactants is too small. To reach equilibrium, reactants must be converted to products. The system proceeds from left to right (consuming reactants, forming products) to reach equilibrium.

Therefore, **$[NH_3]$ will increase** and **$[N_2]$ and $[H_2]$ will decrease** at equilibrium.

15.30 Calculating Equilibrium Concentrations, Problem Type 5.

Step 1: Since the reaction started with only pure NO_2, the equilibrium concentration of NO must be twice the equilibrium concentration of O_2, due to the 2:1 mole ratio of the balanced equation. Therefore, the equilibrium partial pressure of **NO is 0.50 atm.**

Step 2: We can find the equilibrium NO_2 pressure by rearranging the equilibrium constant expression, then substituting in the known values.

$$K_p = \frac{P_{NO}^2 P_{O_2}}{P_{NO_2}^2}$$

$$\mathbf{P_{NO_2}} = \sqrt{\frac{(P_{NO})^2 P_{O_2}}{K_p}} = \sqrt{\frac{(0.50)^2(0.25)}{158}} = \mathbf{0.020\ atm}$$

15.32 Calculating Equilibrium Concentrations, Problem Type 5.

Step 1: Complete a table that lists the initial concentrations, the change in concentrations, and the equilibrium concentrations. The initial concentration of I_2 is 0.0456 mol/2.30 L = 0.0198 *M*. Let *x* be the amount (in mol/L) of I_2 dissociated.

	$I_2(g)$	$\rightleftharpoons$	$2\,I(g)$
Initial (*M*):	0.0198		0.000
Change (*M*):	$-x$		$+2x$
Equilibrium (*M*):	$(0.0198 - x)$		$2x$

Step 2: Write the equilibrium constant expression in terms of the equilibrium concentrations. Knowing the value of the equilibrium constant, solve for x.

$$K_c = \frac{[I]^2}{[I_2]} = \frac{(2x)^2}{(0.0198 - x)} = 3.80 \times 10^{-5}$$

$$4x^2 + (3.80 \times 10^{-5})x - (7.52 \times 10^{-7}) = 0$$

The above equation is a quadratic equation of the form $ax^2 + bx + c = 0$. The solution for a quadratic equation is

$$x = \frac{-b \pm \sqrt{b^2 - 4ac}}{2a}$$

Here, we have a = 4, b = 3.80×10^{-5}, and c = -7.52×10^{-7}. Substituting into the above equation,

$$x = \frac{-3.80 \times 10^{-5} \pm \sqrt{(3.80 \times 10^{-5})^2 - 4(4)(-7.52 \times 10^{-7})}}{2(4)}$$

$$x = \frac{-3.80 \times 10^{-5} \pm 3.47 \times 10^{-3}}{8}$$

$$x = 4.29 \times 10^{-4}\,M \quad \text{or} \quad x = -4.39 \times 10^{-4}\,M$$

The second solution is physically impossible because you cannot have a negative concentration. The first solution is the correct answer.

Step 3: Having solved for x, calculate the equilibrium concentrations of all species.

$$[I] = 2x = (2)(4.29 \times 10^{-4}\,M) = \mathbf{8.58 \times 10^{-4}\,M}$$

$$[I_2] = (0.0198 - x) = (0.0198 - 4.29 \times 10^{-4})\,M = \mathbf{0.0194\,M}$$

Tip: We could have simplified this problem by assuming that x was small compared to 0.0198. We could then assume that $0.0198 - x \cong 0.0198$. By making this assumption, we could have avoided solving a quadratic equation.

15.34 **(a)** The equilibrium constant, K_c, can be found by simple substitution.

$$K_c = \frac{[H_2O][CO]}{[CO_2][H_2]} = \frac{(0.040)(0.050)}{(0.086)(0.045)} = \mathbf{0.52}$$

(b) The magnitude of the reaction quotient Q_c for the system after the concentration of CO_2 becomes 0.50 mol/L, but before equilibrium is reestablished, is:

$$Q_c = \frac{(0.040)(0.050)}{(0.50)(0.045)} = 0.089$$

The value of Q_c is smaller than K_c; therefore, the system will shift to the right, increasing the concentrations of CO and H_2O and decreasing the concentrations of CO_2 and H_2. Let x be the depletion in the concentration of CO_2 at equilibrium. The stoichiometry of the balanced equation then requires that the decrease in the concentration of H_2 must also be x, and that the concentration increases of CO and H_2O be equal to x as well. The changes in the original concentrations are shown in the table.

	CO_2 +	H_2 ⇌	CO +	H_2O
Initial (M):	0.50	0.045	0.050	0.040
Change (M):	$-x$	$-x$	$+x$	$+x$
Equilibrium (M):	$(0.50 - x)$	$(0.045 - x)$	$(0.050 + x)$	$(0.040 + x)$

The equilibrium constant expression is:

$$K_c = \frac{[H_2O][CO]}{[CO_2][H_2]} = \frac{(0.040+x)(0.050+x)}{(0.50-x)(0.045-x)} = 0.52$$

$$0.52(x^2 - 0.545x + 0.0225) = x^2 + 0.090x + 0.0020$$

$$0.48x^2 + 0.373x - (9.7 \times 10^{-3}) = 0$$

The positive root of the equation is $x = 0.025$.

The equilibrium concentrations are:

$$\mathbf{[CO_2]} = (0.50 - 0.025)\ M = \mathbf{0.48\ M}$$
$$\mathbf{[H_2]} = (0.045 - 0.025)\ M = \mathbf{0.020\ M}$$
$$\mathbf{[CO]} = (0.050 + 0.025)\ M = \mathbf{0.075\ M}$$
$$\mathbf{[H_2O]} = (0.040 + 0.025)\ M = \mathbf{0.065\ M}$$

15.36 The initial concentrations are $[H_2] = 0.80\ \text{mol}/5.0\ \text{L} = 0.16\ M$ and $[CO_2] = 0.80\ \text{mol}/5.0\ \text{L} = 0.16\ M$.

	$H_2(g)$ +	$CO_2(g)$	⇌	$H_2O(g)$ +	$CO_2(g)$
Initial (M):	0.16	0.16		0.00	0.00
Change (M):	$-x$	$-x$		$+x$	$+x$
Equilibrium (M):	$(0.16 - x)$	$(0.16 - x)$		x	x

$$K_c = \frac{[H_2O][CO]}{[H_2][CO_2]} = 4.2 = \frac{x^2}{(0.16-x)^2}$$

Taking the square root of both sides, we obtain:

$$\frac{x}{0.16 - x} = 2.0$$

$$x = 0.11\ M$$

The equilibrium concentrations are:

$$\mathbf{[H_2]} = \mathbf{[CO_2]} = (0.16 - 0.11)\ M = \mathbf{0.05\ M}$$

$$\mathbf{[H_2O]} = \mathbf{[CO]} = \mathbf{0.11\ M}$$

15.42 **(a)** Removal of $CO_2(g)$ from the system would shift the position of equilibrium to the **right**.

(b) Addition of more solid Na_2CO_3 would have **no effect**. $[Na_2CO_3]$ does not appear in the equilibrium constant expression.

(c) Removal of some of the solid $NaHCO_3$ would have **no effect**. Same reason as (b).

15.44 Factors that Affect Chemical Equilibrium, Problem Type 6.

(a) Changes in pressure ordinarily do not affect the concentrations of reacting species in condensed phases because liquids and solids are virtually incompressible. Pressure change should have **no effect** on this system.

(b) Same situation as (a).

(c) Only the product is in the gas phase. An increase in pressure should favor the reaction (forward reaction or reverse reaction) that decreases the total number of moles of gas. The equilibrium should shift to the **left**, that is, the amount of B should decrease and that of A should increase.

(d) In this equation there are equal moles of gaseous reactants and products. A shift in either direction will have no effect on the total number of moles of gas present. There will be **no change** when the pressure is increased.

(e) A shift in the direction of the reverse reaction (**left**) will have the result of decreasing the total number of moles of gas present.

15.46 Factors that Affect Chemical Equilibrium, Problem Type 6.

(a) A temperature *increase* favors an *endothermic* reaction. The reaction is endothermic in the forward direction, so the equilibrium will shift to the **right**.

(b) The system will shift in the direction to remove some of the added Cl_2. The equilibrium will shift to the **left**.

(c) The system will shift to replace some of the PCl_3 that was removed. The equilibrium will shift to the **right**.

(d) An increase in pressure should favor the reaction that decreases the total number of moles of gas. The equilibrium will shift to the **left**.

(e) Adding a catalyst will not affect the position of the equilibrium.

15.48 There will be no change in the pressures. A catalyst has no effect on the position of the equilibrium.

15.50 For this system, $K_p = [CO_2]$.

This means that to remain at equilibrium, the pressure of carbon dioxide must stay at a fixed value as long as the temperature remains the same.

(a) If the volume is increased, the pressure of CO_2 will drop (Boyle's law, pressure and volume are inversely proportional). Some $CaCO_3$ will break down to form more CO_2 and CaO. (Shift right)

(b) Assuming that the amount of added solid CaO is not so large that the volume of the system is altered significantly, there should be no change at all. If a huge amount of CaO were added, this would have the effect of reducing the volume of the container. What would happen then?

(c) Assuming that the amount of $CaCO_3$ removed doesn't alter the container volume significantly, there should be no change. Removing a huge amount of $CaCO_3$ will have the effect of increasing the container volume. The result in that case will be the same as in part (a).

(d) The pressure of CO_2 will be greater and will exceed the value of K_p. Some CO_2 will combine with CaO to form more $CaCO_3$. (Shift left)

(e) Carbon dioxide combines with aqueous NaOH according to the equation

$$CO_2\,(g) + NaOH\,(aq) \rightarrow NaHCO_3\,(aq)$$

This will have the effect of reducing the CO_2 pressure and causing more $CaCO_3$ to break down to CO_2 and CaO. (Shift right)

(f) Carbon dioxide does not react with hydrochloric acid, but $CaCO_3$ does.

$$CaCO_3\,(s) + 2\,HCl\,(aq) \rightarrow CaCl_2\,(aq) + CO_2\,(g) + H_2O\,(l)$$

The CO_2 produced by the action of the acid will combine with CaO as discussed in (d) above. (Shift left)

(g) This is a decomposition reaction. Decomposition reactions are endothermic. Increasing the temperature will favor this reaction and produce more CO_2 and CaO. (Shift right)

15.52 (a) Since the total pressure is 1.00 atm, the sum of the partial pressures of NO and Cl_2 is

$$1.00 \text{ atm} - \text{partial pressure of NOCl} = 1.00 \text{ atm} - 0.64 \text{ atm} = 0.36 \text{ atm}$$

The stoichiometry of the reaction requires that the partial pressure of NO be twice that of Cl_2. Hence, the partial pressure of NO is **0.24 atm** and the partial pressure of Cl_2 is **0.12 atm**.

(b) The equilibrium constant K_p is found by substituting the partial pressures calculated in part (a) into the equilibrium constant expression.

$$\mathbf{K_p} = \frac{P_{NO}^2 P_{Cl_2}}{P_{NOCl}^2} = \frac{(0.24)^2(0.12)}{(0.64)^2} = \mathbf{0.017}$$

15.54 The equilibrium expression for this system is given by:

$$K_p = P_{CO_2} P_{H_2O}$$

(a) In a closed vessel the decomposition will stop when the product of the partial pressures of CO_2 and H_2O equals K_p. Adding more sodium bicarbonate will have no effect.

(b) In an open vessel, CO_2 (g) and H_2O (g) will escape from the vessel, and the partial pressures of CO_2 and H_2O will never become large enough for their product to equal K_p. Therefore, equilibrium will never be established. Adding more sodium bicarbonate will result in the production of more CO_2 and H_2O.

15.56 (a) The equation that relates K_p and K_c is:

$$K_p = K_c(0.0821\,T)^{\Delta n}$$

For this reaction, $\Delta n = 3 - 2 = 1$

$$\mathbf{K_c} = \frac{K_p}{(0.0821\,T)} = \frac{2 \times 10^{-42}}{(0.0821 \times 298)} = \mathbf{8 \times 10^{-44}}$$

(b) Because of a very large activation energy, the reaction of hydrogen with oxygen is infinitely slow without a catalyst or an initiator. The action of a single spark on a mixture of these gases results in the explosive formation of water.

15.58 (a) Calculate the value of K_p by substituting the equilibrium partial pressures into the equilibrium constant expression.

$$\mathbf{K_p} = \frac{P_B}{P_A^2} = \frac{(0.60)}{(0.60)^2} = \mathbf{1.7}$$

(b) The total pressure is the sum of the partial pressures for the two gaseous components, A and B. We can write:

$$P_A + P_B = 1.5 \text{ atm}$$

and

$$P_B = 1.5 - P_A$$

Substituting into the expression for K_p gives:

$$K_p = \frac{(1.5 - P_A)}{P_A^2} = 1.7$$

$$1.7P_A^2 + P_A - 1.5 = 0$$

Solving the quadratic equation, we obtain:

$$\mathbf{P_A = 0.69\ atm}$$

and by difference,

$$\mathbf{P_B = 0.81\ atm}$$

15.60 Total number of moles of gas is:

$$0.020 + 0.040 + 0.96 = 1.02 \text{ mol of gas}$$

You can calculate the partial pressure of each gaseous component from the mole fraction and the total pressure (See Dalton's Law of Partial Pressures, Problem Type 3, Chapter 5).

$$P_{NO} = X_{NO}P_T = \frac{0.040}{1.02} \times 0.20 = 0.0078 \text{ atm}$$

$$P_{O_2} = X_{O_2}P_T = \frac{0.020}{1.02} \times 0.20 = 0.0039 \text{ atm}$$

$$P_{NO_2} = X_{NO_2}P_T = \frac{0.96}{1.02} \times 0.20 = 0.19 \text{ atm}$$

Calculate K_p by substituting the partial pressures into the equilibrium constant expression.

$$\mathbf{K_p} = \frac{P_{NO_2}^2}{P_{NO}^2 P_{O_2}} = \frac{(0.19)^2}{(0.0078)^2(0.0039)} = \mathbf{1.5 \times 10^5}$$

15.62 Set up a table that contains the initial concentrations, the change in concentrations, and the equilibrium concentration. Assume that the vessel has a volume of 1 L.

The reaction is:	H_2	+	Cl_2	$\rightleftharpoons$	2 HCl
Initial (M):	0.47		0		3.59
Change (M):	$+x$		$+x$		$-2x$
Equilibrium (M):	$(0.47 + x)$		x		$(3.59 - 2x)$

Substitute the equilibrium concentrations into the equilibrium constant expression, then solve for x. Since $\Delta n = 0$, $K_c = K_p$

$$K_c = \frac{[HCl]^2}{[H_2][Cl_2]} = \frac{(3.59 - 2x)^2}{(0.47 + x)x} = 193$$

Solving the quadratic equation,

$$x = 0.10$$

Having solved for x, calculate the equilibrium concentrations of all species.

$$[H_2] = 0.57\,M \qquad [Cl_2] = 0.10\,M \qquad [HCl] = 3.39\,M$$

Since we assumed that the vessel had a volume of 1 L, the above molarities also correspond to the number of moles of each component.

From the mole fraction of each component and the total pressure, we can calculate the partial pressure of each component (See Dalton's Law of Partial Pressures, Problem Type 3, Chapter 5).

$$\text{Total number of moles} = 0.57 + 0.10 + 3.39 = 4.06\text{ mol}$$

$$P_{H_2} = \frac{0.57}{4.06} \times 2.00 = \mathbf{0.28\ atm}$$

$$P_{Cl_2} = \frac{0.10}{4.06} \times 2.00 = \mathbf{0.049\ atm}$$

$$P_{HCl} = \frac{3.39}{4.06} \times 2.00 = \mathbf{1.67\ atm}$$

15.64 This is a difficult problem. Express the equilibrium number of moles in terms of the initial moles and the change in number of moles (x). Next, calculate the mole fraction of each component. Using the mole fraction, you should come up with a relationship between partial pressure and total pressure for each component. Substitute the partial pressures into the equilibrium constant expression to solve for the total pressure, P_T.

The reaction is:

	N_2	+	$3\,H_2$	$\rightleftharpoons$	$2\,NH_3$
Initial (mol):	1		3		0
Change (mol):	$-x$		$-3x$		$2x$
Equilibrium (mol):	$(1 - x)$		$(3 - 3x)$		$2x$

$$\text{Mole fraction of } NH_3 = \frac{\text{mol of } NH_3}{\text{total number of moles}}$$

$$X_{NH_3} = \frac{2x}{(1-x)+(3-3x)+2x} = \frac{2x}{4-2x}$$

$$0.21 = \frac{2x}{4-2x}$$

$$x = 0.35\text{ mol}$$

Substituting x into the following mole fraction equations, the mole fractions of N_2 and H_2 can be calculated.

$$X_{N_2} = \frac{1-x}{4-2x} = \frac{1 - 0.35}{4 - 2(0.35)} = 0.20$$

$$X_{H_2} = \frac{3-3x}{4-2x} = \frac{3 - 3(0.35)}{4 - 2(0.35)} = 0.59$$

The partial pressures of each component are equal to the mole fraction multiplied by the total pressure.

$$P_{NH_3} = 0.21P_T \qquad P_{N_2} = 0.20P_T \qquad P_{H_2} = 0.59P_T$$

Substitute the partial pressures above (in terms of P_T) into the equilibrium constant expression, and solve for P_T.

$$K_p = \frac{P_{NH_3}^2}{P_{H_2}^3 P_{N_2}}$$

$$4.31 \times 10^{-4} = \frac{(0.21)^2 P_T^2}{(0.59P_T)^3(0.20P_T)}$$

$$4.31 \times 10^{-4} = \frac{1.07}{P_T^2}$$

$$\mathbf{P_T = 5.0 \times 10^1\ atm}$$

15.66 The initial number of moles of SO_2Cl_2 is:

$$\text{mol } SO_2Cl_2 = 6.75\ \text{g } SO_2Cl_2 \times \frac{1\ \text{mol } SO_2Cl_2}{135.0\ \text{g } SO_2Cl_2} = 0.0500\ \text{mol}$$

At equilibrium, there is 0.0345 mol of SO_2 (and therefore also 0.0345 mol of Cl_2). The amount of SO_2Cl_2 left = (0.0500 mol – 0.0345 mol) = (0.0155 mol). Thus, the equilibrium concentrations are:

$$[SO_2Cl_2] = 0.0155\ \text{mol}/2.00\ \text{L} = 0.00775\ M$$

$$[SO_2] = [Cl_2] = 0.0345\ \text{mol}/2.00\ \text{L} = 0.0173\ M$$

Substitute the equilibrium concentrations into the equilibrium constant expression to calculate K_c.

$$\mathbf{K_c} = \frac{[SO_2][Cl_2]}{[SO_2Cl_2]} = \frac{(0.0173)(0.0173)}{(0.00775)} = \mathbf{3.86 \times 10^{-2}}$$

15.68 $I_2(g) \rightleftharpoons 2\,I(g)$

Assuming 1 mole of I_2 is present originally and α moles reacts, at equilibrium: $[I_2] = 1 - \alpha$, $[I] = 2\alpha$. The total number of moles present in the system = $(1 - \alpha) + 2\alpha = 1 + \alpha$. From Problem 15.63(a) in your text, we know that K_p is equal to:

$$K_p = \frac{4\alpha^2}{1-\alpha^2}P \qquad (1)$$

If there were no dissociation, then the pressure would be:

$$P = \frac{nRT}{V} = \frac{\left(1.00\ \text{g} \times \frac{1\ \text{mol}}{253.8\ \text{g}}\right)\left(0.0821\ \frac{\text{L}\cdot\text{atm}}{\text{mol}\cdot\text{K}}\right)(1473\ \text{K})}{0.500\ \text{L}} = 0.953\ \text{atm}$$

$$\frac{\text{observed pressure}}{\text{calculated pressure}} = \frac{1.51\ \text{atm}}{0.953\ \text{atm}} = \frac{1+\alpha}{1}$$

$$\alpha = 0.584$$

Substituting in equation (1) above:

$$\mathbf{K_p} = \frac{4\alpha^2}{1-\alpha^2}P = \frac{4 \times (0.584)^2}{1-(0.584)^2} \times 1.51 = \mathbf{3.13}$$

15.70 According to the ideal gas law, pressure is directly proportional to the concentration of a gas in mol/L if the reaction is at constant volume and temperature. Therefore, pressure may be used as a concentration unit. The reaction is:

	N_2	+	$3\ H_2$	$\rightleftharpoons$	$2\ NH_3$
Initial (atm):	0.862		0.373		0
Change (atm):	$-x$		$-3x$		$+2x$
Equilibrium (atm):	$(0.862 - x)$		$(0.373 - 3x)$		$2x$

$$K_p = \frac{P_{NH_3}^2}{P_{NH_2}^3 P_{N_2}}$$

$$4.31 \times 10^{-4} = \frac{(2x)^2}{(0.373 - 3x)^3(0.862 - x)}$$

At this point, we need to make two assumptions that $3x$ is very small compared to 0.373 and that x is very small compared to 0.862. Hence,

$$0.373 - 3x \approx 0.373$$

and

$$0.862 - x \approx 0.862$$

$$4.31 \times 10^{-4} \approx \frac{(2x)^2}{(0.373)^3(0.862)}$$

Solving for x.

$$x = 2.20 \times 10^{-3} \text{ atm}$$

The equilibrium pressures are:

$$P_{N_2} = (0.862 - 2.20 \times 10^{-3}) \text{ atm} = \mathbf{0.860\ atm}$$

$$P_{H_2} = (0.373 - 3 \times 2.20 \times 10^{-3}) \text{ atm} = \mathbf{0.366\ atm}$$

$$P_{NH_3} = 2(2.20 \times 10^{-3}) \text{ atm} = \mathbf{4.40 \times 10^{-3}\ atm}$$

Was the assumption valid that we made above? Typically, the assumption is considered valid if x is less than 5 percent of the number that we said it was very small compared to. Is this the case?

15.72 (a) The equation is:

	fructose $\rightleftharpoons$	glucose
Initial (M):	0.244	0
Change (M):	−0.131	(0.244 − 0.113)
Equilibrium (M):	0.113	0.131

Calculating the equilibrium constant,

$$\mathbf{K_c} = \frac{[\text{glucose}]}{[\text{fructose}]} = \frac{0.131}{0.113} = \mathbf{1.16}$$

(b) $$\textbf{The percent converted} = \frac{\text{amount of fructose converted}}{\text{original amount of fructose}} \times 100\%$$

$$= \frac{0.244 - 0.113}{0.244} \times 100\% = \mathbf{53.7\%}$$

15.74 **(a)** There is only one gas phase component, O_2. The equilibrium constant is simply

$$K_p = P_{O_2} = \mathbf{0.49\ atm}$$

(b) From the ideal gas equation, we can calculate the moles of O_2 produced by the decomposition of CuO.

$$n_{O_2} = \frac{PV}{RT} = \frac{0.49\ \text{atm} \times 2.0\ \text{L}}{0.0821\ \text{L}\cdot\text{atm/K}\cdot\text{mol} \times 1297\ \text{K}} = 9.2 \times 10^{-3}\ \text{mol}\ O_2$$

From the balanced equation,

$$9.2 \times 10^{-3}\ \text{mol}\ O_2 \times \frac{4\ \text{mol CuO}}{1\ \text{mol}\ O_2} = 3.7 \times 10^{-2}\ \text{mol CuO decomposed}$$

$$\textbf{Fraction of CuO decomposed} = \frac{\text{amount of CuO lost}}{\text{original amount of CuO}}$$

$$= \frac{3.7 \times 10^{-2}\ \text{mol}}{0.16\ \text{mol}} = \mathbf{0.23}$$

(c) If a 1.0 mol sample were used, the pressure of oxygen would still be the same (0.49 atm) and it would be due to the same quantity of O_2. Remember, a pure solid does not affect the equilibrium position. The moles of CuO lost would still be 3.7×10^{-2} mol. Thus the fraction decomposed would be:

$$\frac{0.037}{1.0} = \mathbf{0.037}$$

(d) If the number of moles of CuO were less than 3.7×10^{-2} mol, the equilibrium could not be established because the pressure of O_2 would be less than 0.49 atm. Therefore, the smallest number of moles of CuO needed to establish equilibrium must be slightly greater than 3.7×10^{-2} mol.

15.76 We first must find the initial concentrations of all the species in the system.

$$[H_2]_0 = \frac{0.714\ \text{mol}}{2.40\ \text{L}} = 0.298\ M$$

$$[I_2]_0 = \frac{0.984\ \text{mol}}{2.40\ \text{L}} = 0.410\ M$$

$$[HI]_0 = \frac{0.886\ \text{mol}}{2.40\ \text{L}} = 0.369\ M$$

Calculate the reaction quotient by substituting the initial concentrations into the appropriate equation.

$$Q_c = \frac{[HI]_0^2}{[H_2]_0[I_2]_0} = \frac{(0.369)^2}{(0.298)(0.410)} = 1.11$$

We find that Q_c is less than K_c. The equilibrium will shift to the right, decreasing the concentrations of H_2 and I_2 and increasing the concentration of HI.

We set up the usual table. Let x be the decrease in concentration of H_2 and I_2.

	H_2	+	I_2	$\rightleftharpoons$	2 HI
Initial (M):	0.298		0.410		0.369
Change (M):	$-x$		$-x$		$+2x$
Equilibrium (M):	$(0.298 - x)$		$(0.410 - x)$		$(0.369 + 2x)$

The equilibrium constant expression is:

$$K_c = \frac{[HI]^2}{[H_2][I_2]} = \frac{(0.369 + 2x)^2}{(0.298 - x)(0.410 - x)} = 54.3$$

This becomes the quadratic equation

$$50.3x^2 - 39.9x + 6.48 = 0$$

The smaller root is $x = 0.228\ M$. (The larger root is physically impossible.)
Having solved for x, calculate the equilibrium concentrations.

$$[\mathbf{H_2}] = (0.298 - 0.228)\ M = \mathbf{0.070\ M}$$
$$[\mathbf{I_2}] = (0.410 - 0.228)\ M = \mathbf{0.182\ M}$$
$$[\mathbf{HI}] = 0.369 + 2(0.228)\ M = \mathbf{0.825\ M}$$

15.78 The gas cannot be **(a)** because the color became lighter with heating. Heating **(a)** to 150°C would produce some HBr, which is colorless and would lighten rather than darken the gas.

The gas cannot be **(b)** because Br_2 doesn't dissociate into Br atoms at 150°C, so the color shouldn't change.

The gas must be **(c)**. From 25°C to 150°C, heating causes N_2O_4 to dissociate into NO_2, thus darkening the color (NO_2 is a brown gas).

$$N_2O_4\ (g) \rightarrow 2\ NO_2\ (g)$$

Above 150°C, the NO_2 breaks up into colorless NO and O_2.

$$2\ NO_2\ (g) \rightarrow 2\ NO\ (g) + O_2\ (g)$$

An increase in pressure shifts the equilibrium back to the left, forming NO_2, thus darkening the gas again.

$$2\ NO\ (g) + O_2\ (g) \rightarrow 2\ NO_2\ (g)$$

15.80 Decomposition of N_2O_4 is an endothermic process, $N_2O_4 \rightleftharpoons 2\ NO_2$. As the temperature is increased, the system shifts to the right to reestablish equilibrium.

(a) Color deepens, **(b)** increases, **(c)** decreases
(d) increases, **(e)** unchanged (conservation of mass)

15.82 Potassium is more volatile than sodium. Therefore, its removal shifts the equilibrium from left to right.

15.84 **(a)** Since both reactions are endothermic ($\Delta H°$ is positive), according to LeChatelier's principle the products would be favored at high temperatures. Indeed, the steam-reforming process is carried out at very high temperatures (between 800°C and 1000°C). It is interesting to note that in a plant that uses natural gas (methane) for both hydrogen generation and heating, about one-third of the gas is burned to maintain the high temperatures.

In each reaction there are more moles of products than reactants; therefore, we expect products to be favored at low pressures. In reality, the reactions are carried out at high pressures. The reason is that when the hydrogen gas produced is used captively (usually in the synthesis of ammonia), high pressure leads to higher yields of ammonia.

(b) (i) The relation between K_c and K_p is given by Equation 15.5 of the text:

$$K_p = K_c(RT)^{\Delta n}$$

Since $\Delta n = 4 - 2 = 2$, we write:

$$K_p = (18)(0.0821 \times 1073)^2 = \mathbf{1.4 \times 10^5}$$

(ii) Let x be the amount of CH_4 and H_2O (in atm) reacted. We write:

	CH_4	+	H_2O	$\rightleftharpoons$	CO	+	$3H_2$
Initial (atm):	15		15		0		0
Change (atm):	$-x$		$-x$		$+x$		$+3x$
Equilibrium (atm):	$15 - x$		$15 - x$		x		$3x$

The equilibrium constant is given by:

$$K_p = \frac{P_{CO}P_{H_2}^3}{P_{CH_4}P_{H_2O}}$$

$$1.4 \times 10^5 = \frac{(x)(3x)^3}{(15 - x)(15 - x)} = \frac{27x^4}{(15 - x)^2}$$

Taking the square root of both sides, we obtain:

$$3.7 \times 10^2 = \frac{5.2x^2}{15 - x}$$

which can be expressed as

$$5.2x^2 + 3.7 \times 10^2x - 5.6 \times 10^3 = 0$$

Solving the quadratic equation, we obtain

$$x = 13 \text{ atm}$$

(The other solution for x is negative and is physically impossible.)

At equilibrium, the pressures are:

$$P_{CH_4} = (15 - 13) = \mathbf{2\ atm}$$

$$P_{H_2O} = (15 - 13) = \mathbf{2\ atm}$$

$$P_{CO} = \mathbf{13\ atm}$$

$$P_{H_2} = 3(13 \text{ atm}) = \mathbf{39\ atm}$$

15.86 $K_p = P_{NH_3}P_{HCl}$

$$P_{NH_3} = P_{HCl} = \frac{2.2}{2} = 1.1 \text{ atm}$$

$$K_p = (1.1)(1.1) = \mathbf{1.2}$$

15.88 **(a)** React Ni with CO above 50°C. Pump away the $Ni(CO)_4$ vapor (shift equilibrium to right), leaving the solid impurities behind.

(b) $Ni\,(s) + 4\,CO\,(g) \rightarrow Ni(CO)_4\,(g)$

$$\Delta H° = \Delta H_f°[Ni(CO)_4] - 4\,\Delta H_f°(CO)$$

$$= (1 \text{ mol})(-602.9 \text{ kJ/mol}) - (4 \text{ mol})(-110.5 \text{ kJ/mol})$$

$$= -160.9 \text{ kJ}$$

The reaction is exothermic — formation of $Ni(CO)_4$ is less favorable at high temperatures. Heat $Ni(CO)_4$ above 200°C to convert it back to Ni.

15.90 **(a)** $\mathbf{K_p = P_{Hg} = 0.0020\ mmHg = 2.6 \times 10^{-6}\ atm}$

$$K_c = \frac{K_p}{(0.0821T)^{\Delta n}} = \frac{2.6 \times 10^{-6}}{(0.0821 \times 299)^1} = \mathbf{1.1 \times 10^{-7}}$$

(b) Volume of lab $= (6.1 \text{ m})(5.3 \text{ m})(3.1 \text{ m}) = 100 \text{ m}^3$

$[Hg] = K_c$

$$\textbf{Total mass of Hg vapor} = \frac{1.1 \times 10^{-7} \text{ mol}}{1 \text{ L}} \times \frac{200.6 \text{ g}}{1 \text{ mol}} \times \frac{1 \text{ L}}{1 \text{ dm}^3} \times \left(\frac{1 \text{ dm}}{0.1 \text{ m}}\right)^3 \times 100 \text{ m}^3 = \mathbf{2.2\ g}$$

The concentration of mercury vapor in the room is:

$$\frac{2.2 \text{ g}}{100 \text{ m}^3} = 0.22 \text{ g/m}^3 = \mathbf{22\ mg/m^3}$$

Yes! This concentration exceeds the safety limit of 0.05 mg/m^3. Better clean up the spill!

15.92 Initially, at equilibrium: $[NO_2] = 0.0475\,M$ and $[N_2O_4] = 0.491\,M$. At the instant the volume is halved, the concentrations double.

$[NO_2] = 2(0.0475\,M) = 0.0950\,M$ and $[N_2O_4] = 2(0.491\,M) = 0.982\,M$. The system is no longer at equilibrium. The system will shift to the left to offset the increase in pressure when the volume is halved. When a new equilibrium position is established, we write:

$$N_2O_4 \rightleftharpoons 2\,NO_2$$
$$0.982\,M + x \qquad 0.0950\,M - 2x$$

$$K_c = \frac{[NO_2]^2}{[N_2O_4]} = \frac{(0.0950 - 2x)^2}{(0.982 + x)}$$

$$4x^2 - 0.3846x + 4.48 \times 10^{-3} = 0$$

Solving $x = 0.0826\ M$ (impossible) and $x = 0.0136\ M$

At the new equilibrium,

$$[N_2O_4] = 0.982 + 0.0136 = 0.996\ M$$
$$[NO_2] = 0.0950 - (2 \times 0.0136) = 0.0678\ M$$

As we can see, the new equilibrium concentration of NO_2 is *greater* than the initial equilibrium concentration (0.0475 *M*). Therefore, the gases should look *darker*!

CHAPTER 16
ACIDS AND BASES

PROBLEM-SOLVING STRATEGIES AND TUTORIAL SOLUTIONS

TYPES OF PROBLEMS

Problem Type 1: Identifying Conjugate Acid-Base Pairs.

Problem Type 2: The Ion-Product Constant (K_w), Calculating [H^+] from [OH^-].

Problem Type 3: pH Calculations.
(a) Calculating pH from [H^+].
(b) Calculating [H^+] from pH.

Problem Type 4: Calculating the pH of a Strong Acid and a Strong Base.

Problem Type 5: Weak Acids.
(a) Ionization of a weak monoprotic acid.
(b) Determining K_a from a pH measurement.

Problem Type 6: Calculating the pH of a Diprotic Acid.

Problem Type 7: Calculating the pH of a Weak Base.

Problem Type 8: Predicting the Acid-Base Properties of Salt Solutions.

PROBLEM TYPE 1: IDENTIFYING CONJUGATE ACID-BASE PAIRS

A **conjugate acid-base pair** can be defined as *an acid and its conjugate base or a base and its conjugate acid.* The conjugate base of a Brønsted acid is the species that remains when *one* proton has been removed from the acid. Conversely, a conjugate acid results from the addition of *one* proton to a Brønsted base.

EXAMPLE 16.1
Consider the reaction

$$HSO_4^-\,(aq) + HCO_3^-\,(aq) \rightleftharpoons SO_4^{2-}\,(aq) + H_2CO_3\,(aq)$$

(a) Identify the acids and bases for the forward and reverse reactions.
(b) Identify the conjugate acid-base pairs.

(a) In the forward reaction, HSO_4^- is the proton donor, which makes it an acid. The proton acceptor, HCO_3^-, is a base. In the reverse reaction, the proton donor (acid) is H_2CO_3, and SO_4^{2-} is the proton acceptor (base).

(b)
$$\underset{\text{acid}_1}{HSO_4^-\,(aq)} + \underset{\text{base}_2}{HCO_3^-\,(aq)} \rightleftharpoons \underset{\text{base}_1}{SO_4^{2-}\,(aq)} + \underset{\text{acid}_2}{H_2CO_3\,(aq)}$$

The subscripts 1 and 2 designate the two conjugate acid-base pairs.

PRACTICE EXERCISE

1. Identify the conjugate bases of the following acids:

   (a) CH_3COOH (b) H_2S (c) HSO_3^- (d) HClO

Text Problems: 4, 6, 8

PROBLEM TYPE 2: THE ION-PRODUCT CONSTANT (K_w), CALCULATING [H^+] FROM [OH^-]

Water is a very weak electrolyte and therefore a poor conductor of electricity, but it does ionize to a small extent:

$$H_2O\,(l) \rightleftharpoons H^+\,(aq) + OH^-\,(aq)$$

We can write the equilibrium constant for the autoionization of water as

$$K_c = \frac{[H^+][OH^-]}{[H_2O]}$$

Since a very small fraction of water molecules are ionized, the concentration of water, [H_2O], remains virtually unchanged. Therefore, we assume that the concentration of water is constant, and we write:

$$K_c[H_2O] = K_w = [H^+][OH^-] \qquad (16.1)$$

In pure water at 25°C, the concentrations of H^+ and OH^- ions are equal and found to be $[H^+] = [OH^-] = 1.0 \times 10^{-7}\,M$. Substituting these concentrations into Equation (16.1),

$$K_w = [H^+][OH^-] = (1 \times 10^{-7})(1 \times 10^{-7}) = 1 \times 10^{-14}$$

Thus, knowing K_w and one of the concentrations, either [H^+] or [OH^-], we can easily calculate the other concentration.

EXAMPLE 16.2

The OH^- ion concentration in a certain solution at 25°C is 5.0×10^{-5} *M*. What is the H^+ concentration?

The ion product of water is applicable to all aqueous solutions. At 25°C,

$$K_w = 1.0 \times 10^{-14} = [H^+][OH^-]$$

Rearrange the equation to solve for [H^+].

$$[H^+] = \frac{1.0 \times 10^{-14}}{[OH^-]} = \frac{1.0 \times 10^{-14}}{5.0 \times 10^{-5}} = \mathbf{2.0 \times 10^{-10}\,M}$$

PRACTICE EXERCISE

2. The H^+ concentration in a certain solution is $6.6 \times 10^{-4}\,M$. What is the OH^- concentration?

Text Problem: 16c

PROBLEM TYPE 3: pH CALCULATIONS

Because the concentrations of H^+ and OH^- ions in aqueous solutions are frequently very small numbers making them inconvenient to work with, the Danish biochemist Soren Sorensen in 1909 proposed a more practical measure called pH. The **pH** of a solution is defined as *the negative log of the hydrogen ion concentration* (in moles per liter).

$$pH = -\log[H^+]$$

The pOH of a solution is defined in a similar manner.

$$pOH = -\log[OH^-]$$

A useful relationship between pH and pOH is:

$$pH + pOH = 14.$$

See Section 16.3 of your text for a complete discussion.

A. Calculating pH from $[H^+]$

EXAMPLE 16.3

The $[H^+]$ of a solution is 0.015 *M*. What is the pH of the solution?

The relationship between $[H^+]$ and pH is

$$pH = -\log[H^+]$$

Substitute the H^+ ion concentration into the above equation to calculate the pH of the solution.

$$\mathbf{pH} = -\log[H^+] = -\log[0.015] = \mathbf{1.82}$$

PRACTICE EXERCISE

3. The OH^- concentration of a certain ammonia solution is 7.2×10^{-4} *M*. What is the pOH and pH?

Text Problem: 18

B. Calculating $[H^+]$ from pH

EXAMPLE 16.4

What is the H^+ concentration in a solution with a pOH of 3.9?

Step 1: Recall that pH + pOH = 14. Since pOH is given in the problem, we can calculate the pH.

$$pH = 14.0 - pOH$$

$$pH = 14.0 - 3.9 = 10.1$$

Step 2: The H^+ concentration can now be determined from the pH.

$$pH = -\log[H^+]$$

or,

$$-pH = \log[H^+]$$

Taking the antilog of both sides of the equation,

$$10^{-pH} = [H^+]$$

$$[H^+] = 10^{-10.1} = \mathbf{8 \times 10^{-11}}\ M$$

PRACTICE EXERCISE

4. The pH of a solution is 4.45. What is the H^+ concentration?

Text Problem: 16a,b

PROBLEM TYPE 4: CALCULATING THE pH OF A STRONG ACID AND A STRONG BASE

Strong acids and strong bases are strong electrolytes that are assumed to ionize completely in water. For example, consider the strong acid, hydrochloric acid [HCl (*aq*)]. We assume that it ionizes completely in water.

$$HCl\ (aq) \rightarrow H^+\ (aq) + Cl^-\ (aq)$$

Since strong acids ionize completely in water, we can easily calculate the $[H^+]$ in solution. For example, consider a 0.10 *M* HCl solution. Let's set up a table to determine the $[H^+]$ of the solution.

	HCl (*aq*) →	H^+ (*aq*) +	Cl^- (*aq*)
initial conc.	0.10 *M*	0	0
conc. after ionization	0	0.10 *M*	0.10 *M*

Since there is a one:one mole ratio between HCl and H^+, the H^+ concentration after ionization equals the initial concentration of HCl.

Note that the above reaction can be written more accurately as:

$$HCl\ (aq) + H_2O\ (l) \rightarrow H_3O^+\ (aq) + Cl^-\ (aq)$$

HCl is an acid and donates a proton (H^+) to the weak base, H_2O. H^+ in aqueous solution is really shorthand notation for H_3O^+.

You need to know the six strong acids and the six strong bases. Half the battle in pH calculations is recognizing the type of species in solution (i.e., strong acid, strong base, weak acid, weak base, or salt).

The six strong acids

$HClO_4$	perchloric acid
HI	hydroiodic acid
HBr	hydrobromic acid
HCl	hydrochloric acid
H_2SO_4	sulfuric acid
HNO_3	nitric acid

The six **strong bases** include the five alkali metal hydroxides (LiOH, NaOH, KOH, RbOH, and CsOH), plus barium hydroxide, $Ba(OH)_2$.

EXAMPLE 16.5

Calculate the pH of a 0.10 *M* $Ba(OH)_2$ solution.

Step 1: First, you must recognize that barium hydroxide is one of the six strong bases. Strong bases ionize completely in solution. Let's write the reaction and set up a table to calculate the OH^- concentration in solution.

	$Ba(OH)_2\ (aq)$ →	$Ba^{2+}\ (aq)$ +	$2\ OH^-\ (aq)$
initial conc.	0.10 *M*	0	0
conc. after ionization	0	0.10 *M*	2(0.10 *M*) = 0.20 *M*

Since there is a 2:1 mole ratio between OH^- and $Ba(OH)_2$, the OH^- concentration is double the $Ba(OH)_2$ concentration.

Step 2: Calculate the pOH from the OH^- concentration.

$$\text{pOH} = -\log[\text{OH}^-]$$
$$= -\log[0.20\ M] = 0.70$$

Step 3: Use the relationship, pH + pOH = 14.0 to calculate the pH of the $Ba(OH)_2$ solution.

$$\textbf{pH} = 14.00 - 0.70 = \mathbf{13.30}$$

PRACTICE EXERCISE

5. What is the pH of a 0.025 *M* HCl solution?

Text Problems: 22, 24

PROBLEM TYPE 5: WEAK ACIDS

Most acids ionize only to a limited extent in water. Such acids are classified as **weak acids**. For example, HNO_2 (nitrous acid), is a weak acid. Its ionization in water is represented by

$$HNO_2\ (aq) + H_2O\ (l) \rightleftharpoons H_3O^+\ (aq) + NO_2^-\ (aq)$$

or simply

$$HNO_2\ (aq) \rightleftharpoons H^+\ (aq) + NO_2^-\ (aq)$$

The equilibrium constant for this acid ionization is called the **acid ionization constant, K_a**. It is given by:

$$K_a = \frac{[H^+][NO_2^-]}{[HNO_2]}$$

In general,

$$K_a = \frac{[H^+][A^-]}{[HA]}$$

where,

A^- is the conjugate base of the weak acid, HA.

A. Ionization of a weak monoprotic acid

This problem is very similar to Problem Type 5, Chapter 15: Calculating Equilibrium Concentrations. Try to think of this problem as just another equilibrium calculation. The only difference is that in Chapter 15, we dealt mostly with gas phase reactions. Now, we are dealing with weak acids in aqueous solution. Some problems will ask you to calculate the equilibrium concentrations of all species, others will ask for the pH of the solution.

EXAMPLE 16.6

Calculate the pH of a 0.20 *M* acetic acid, CH_3COOH, solution.

First, recognize that acetic acid is a weak acid. It is not one of the six strong acids, so it must be a weak acid.

Step 1: Express the equilibrium concentrations of all species in terms of initial concentrations and a single unknown x, that represents the change in concentration. Let $(-x)$ be the depletion in concentration (mol/L) of CH_3COOH. From the stoichiometry of the reaction, it follows that the increase in concentration for both H^+ and CH_3COO^- must be x. Complete a table that lists the initial concentrations, the change in concentrations, and the equilibrium concentrations.

	$CH_3COOH\ (aq)$	$\rightleftharpoons$	$H^+\ (aq)$	+	$CH_3COO^-\ (aq)$
Initial (*M*):	0.20		0		0
Change (*M*):	$-x$		$+x$		$+x$
Equilibrium (*M*):	$0.20 - x$		x		x

Step 2: Write the ionization constant expression in terms of the equilibrium concentrations. Knowing the value of the equilibrium constant (K_a), solve for x.

$$K_a = \frac{[H^+][CH_3COO^-]}{[CH_3COOH]}$$

You can look up the K_a value for acetic acid in Table 16.3 of your text.

$$1.8 \times 10^{-5} = \frac{(x)(x)}{(0.20 - x)}$$

At this point, we can make an assumption that x is very small compared to 0.20. Hence,

$$0.20 - x \cong 0.20$$

Oftentimes, assumptions such as these are valid if K is very small. A very small value of K means that a very small amount of reactants go to products. Hence, x is small. If we did not make this assumption, we would have to solve a quadratic equation.

$$1.8 \times 10^{-5} \cong \frac{(x)(x)}{0.20}$$

Solving for x.

$$x = 1.9 \times 10^{-3}\ M = [H^+]$$

Checking the validity of the assumption,

$$\frac{1.9 \times 10^{-3}}{0.20} \times 100\% = 0.95\% < 5\%$$

The assumption was valid.

Step 3: Having solved for the $[H^+]$, calculate the pH of the solution.

$$\mathbf{pH} = -\log[H^+] = -\log(1.9 \times 10^{-3}) = \mathbf{2.72}$$

PRACTICE EXERCISE

6. Calculate the pH of a 0.50 *M* nitrous acid (HNO_2) solution. The K_a value for HNO_2 is 4.5×10^{-4}.

Text Problems: 42, 44, 46, 48

B. Determining K_a from a pH measurement

This problem is the reverse of Example 16.6 above. From the pH of the solution, you can calculate the $[H^+]$. Then from the $[H^+]$, you can calculate the equilibrium concentrations of the other species in solution. Substitute these equilibrium concentrations into the ionization constant expression to solve for K_a.

EXAMPLE 16.7

The pH of a 0.10 *M* solution of a weak monoprotic acid is 5.15. What is the K_a of the acid?

Step 1: Calculate the $[H^+]$ from the pH.

$$pH = -\log[H^+]$$

or,

$$-pH = \log[H^+]$$

Taking the antilog of both sides of the equation,

$$10^{-pH} = [H^+]$$

$$[H^+] = 10^{-5.15} = 7.1 \times 10^{-6}\ M$$

Step 2: Complete a table that lists the initial concentrations, the change in concentrations, and the equilibrium concentrations.

	$HA\ (aq) \rightleftharpoons$	$H^+\ (aq)$ +	$A^-\ (aq)$
Initial (*M*):	0.10	0	0
Change (*M*):	$-x$	$+x$	$+x$
Equilibrium (*M*):	$0.10 - x$	x	x

In this case, x is known.

$$x = [H^+] = [A^-] = 7.1 \times 10^{-6}\ M$$

$$[HA] = 0.10 - x = (0.10 - 7.1 \times 10^{-6})M = 0.10\ M$$

Step 3: Substitute the equilibrium concentrations into the ionization constant expression to solve for K_a.

$$K_a = \frac{[H^+][A^-]}{[HA]}$$

$$\mathbf{K_a} = \frac{(7.1 \times 10^{-6})^2}{(0.10)} = \mathbf{5.0 \times 10^{-10}}$$

PRACTICE EXERCISE

7. The pH of a 0.50 *M* monoprotic weak acid is 2.24. What is the K_a of the acid?

Text Problem: 56

PROBLEM TYPE 6: CALCULATING THE pH OF A DIPROTIC ACID

Diprotic acids may yield more than one hydrogen ion per molecule. These acids ionize in a stepwise manner, that is, they lose one proton at a time. An ionization constant expression can be written for each ionization stage. Consequently, two equilibrium constant expressions must be used to calculate the concentrations of species in the acid solution. For the generic diprotic acid, H_2A, we can write:

$$H_2A\,(aq) \rightleftharpoons H^+\,(aq) + HA^-\,(aq) \qquad K_{a_1} = \frac{[H^+][HA^-]}{[H_2A]}$$

$$HA^-\,(aq) \rightleftharpoons H^+\,(aq) + A^{2-}\,(aq) \qquad K_{a_2} = \frac{[H^+][A^{2-}]}{[HA^-]}$$

Note that the conjugate base in the first ionization becomes the acid in the second ionization.

EXAMPLE 16.8

Calculate the concentrations of H_2A, HA^-, H^+, and A^{2-} in a 1.0 *M* H_2A solution. The first and second ionization constants for H_2A are 1.3×10^{-2} and 6.3×10^{-8}, respectively.

Step 1: Complete a table showing the concentrations for the first ionization stage

	$H_2A\,(aq)$	$\rightleftharpoons$	$H^+\,(aq)$	+	$HA^-\,(aq)$
Initial (*M*):	1.0		0		0
Change (*M*):	$-x$		$+x$		$+x$
Equilibrium (*M*):	$1.0 - x$		x		x

Step 2: Write the ionization constant expression for K_{a_1}. Then, solve for x.

$$K_{a_1} = \frac{[H^+][HA^-]}{[H_2A]}$$

$$1.3 \times 10^{-2} = \frac{(x)(x)}{(1.0 - x)}$$

Since K_{a_1} is quite large, we cannot make the assumption

$$1.0 - x \cong 1.0$$

Therefore, we must solve a quadratic equation.

$$x^2 + 0.013x - 0.013 = 0$$

The above equation is a quadratic equation of the form $ax^2 + bx + c = 0$. The solution for a quadratic equation is:

$$x = \frac{-b \pm \sqrt{b^2 - 4ac}}{2a}$$

Here, we have a = 1, b = 0.013, and c = −0.013. Substituting into the above equation,

$$x = \frac{-0.013 \pm \sqrt{(0.013)^2 - 4(1)(-0.013)}}{2(1)}$$

$$x = \frac{-0.013 \pm 0.23}{2}$$

$$x = 0.11\ M \quad \text{or} \quad x = -0.12\ M$$

The second solution is physically impossible because you cannot have a negative concentration. The first solution is the correct answer.

Step 3: Having solved for x, calculate the concentrations when the equilibrium for the first stage of ionization is reached.

Because $K_{a_1} >> K_{a_2}$, we assume that essentially all the H^+ comes from the first ionization stage. Hence,

$$\mathbf{[H^+] = [HA^-] = x = 0.11\ M}$$

$$[H_2A] = 1.0 - x = 1.00 - 0.11 = \mathbf{0.89\ M}$$

Step 4: Now, consider the second stage of ionization. Set up a table showing the concentrations for the second ionization stage. Let y be the change in concentration.

	$HA^-(aq)$	$\rightleftharpoons$	$H^+(aq)$	+	$A^{2-}(aq)$
Initial (M):	0.11		0.11		0
Change (M):	$-y$		$+y$		$+y$
Equilibrium (M):	$0.11 - y$		$0.11 + y$		y

Step 5: Write the ionization constant expression for K_{a_2}. Then, solve for y.

$$K_{a_2} = \frac{[H^+][A^{2-}]}{[HA^-]}$$

$$6.3 \times 10^{-8} = \frac{(0.11 + y)(y)}{(0.11 - y)}$$

Since K_{a_2} is very small, we can make an assumption that y is very small compared to 0.11.

Hence,

$$0.11 \pm y \cong 0.11$$

$$6.3 \times 10^{-8} = \frac{(0.11)(y)}{0.11}$$

Solving for y.

$$\mathbf{y = 6.3 \times 10^{-8}\ M = [A^{2-}]}$$

Checking the validity of the assumption,

$$\frac{6.3 \times 10^{-8}}{0.11} \times 100\% = 5.7 \times 10^{-5}\% < 5\%$$

The assumption was valid.

PRACTICE EXERCISE

8. The first and second ionization constants of H_2CO_3 are 4.2×10^{-7} and 4.8×10^{-11}, respectively. Calculate the concentrations of H^+, HCO_3^-, CO_3^{2-}, and unionized H_2CO_3 in a 0.080 *M* H_2CO_3 solution.

Text Problem: 52

PROBLEM TYPE 7: CALCULATING THE pH OF A WEAK BASE

The procedure used to calculate the pH of a weak base solution is essentially the same as the one used for weak acids.

EXAMPLE 16.9

What is the pH of a 0.10 *M* C_5H_5N (pyridine) solution?

Step 1: Express the equilibrium concentrations of all species in terms of initial concentrations and a single unknown *x*, that represents the change in concentration. Let $(-x)$ be the depletion in concentration (mol/L) of C_5H_5N. From the stoichiometry of the reaction, it follows that the increase in concentration for both OH^- and $C_5H_5NH^+$ must be *x*. Complete a table that lists the initial concentrations, the change in concentrations, and the equilibrium concentrations.

	$C_5H_5N\,(aq) + H_2O\,(l) \rightleftharpoons$	$C_5H_5NH^+\,(aq)$	$+\ OH^-\,(aq)$
Initial (*M*):	0.10	0	0
Change (*M*):	$-x$	$+x$	$+x$
Equilibrium (*M*):	$0.10 - x$	x	x

Step 2: Write the ionization constant expression in terms of the equilibrium concentrations. Knowing the value of the equilibrium constant (K_b), solve for *x*.

$$K_b = \frac{[C_5H_5NH^+][OH^-]}{[C_5H_5N]}$$

You can look up the K_b value for pyridine in Table 16.5 of your text.

$$1.7 \times 10^{-9} = \frac{(x)(x)}{(0.10 - x)}$$

At this point, we can make an assumption that *x* is very small compared to 0.10. Hence,

$$0.10 - x \cong 0.10$$

Oftentimes, assumptions such as these are valid if *K* is very small. A very small value of *K* means that a very small amount of reactants go to products. Hence, *x* is small. If we did not make this assumption, we would have to solve a quadratic equation.

$$1.7 \times 10^{-9} \cong \frac{(x)(x)}{0.10}$$

Solving for *x*.

$$x = 1.3 \times 10^{-5}\,M = [OH^-]$$

Step 3: Having solved for the $[OH^-]$, calculate the pOH of the solution. Then use the relationship, pH + pOH = 14, to solve for the pH of the solution.

$$pOH = -\log[OH^-] = -\log(1.3 \times 10^{-5}) = 4.89$$

$$\mathbf{pH} = 14 - pOH = 14.00 - 4.89 = \mathbf{9.11}$$

PRACTICE EXERCISE

9. What is the pH of a 2.0 *M* NH_3 solution?

Text Problems: 56, 58

PROBLEM TYPE 8: PREDICTING THE ACID-BASE PROPERTIES OF SALT SOLUTIONS

Salts, when dissolved in water, can produce neutral, acidic, or basic solutions. We need to consider four possibilities.

1. **Salts that produce neutral solutions.** It is generally true that salts containing an alkali metal ion or an alkaline earth metal ion (except Be^{2+}) and the conjugate base of a strong acid (for example, Cl^-, NO_3^-, and ClO_4^-) do not undergo hydrolysis, and thus their solutions are neutral.

2. **Salts that produce basic solutions.** Salts that contain the conjugate base of a weak acid and an alkali metal or alkaline earth metal ion, produce basic solutions. As an example, consider potassium fluoride. The dissociation of KF in water is given by:

$$KF\ (s) \xrightarrow{H_2O} K^+\ (aq) + F^-\ (aq)$$

A hydrated alkali or alkaline earth metal ion has no acidic or basic properties. However, the conjugate base of a weak acid is a weak base (it has an affinity for H^+ ions). F^- is the conjugate base of the weak acid HF. The hydrolysis reaction is given by:

$$F^-\ (aq) + H_2O\ (l) \rightleftharpoons HF\ (aq) + OH^-\ (aq)$$

Because this reaction produces OH^- ions, the potassium fluoride solution will be basic.

3. **Salts that produce acidic solutions.** Salts that contain the conjugate acid of a weak base and the conjugate base of a strong acid, produce acidic solutions. As an example, consider ammonium iodide. The dissociation of NH_4I in water is given by

$$NH_4I\ (s) \xrightarrow{H_2O} NH_4^+\ (aq) + I^-\ (aq)$$

The conjugate base of a strong acid, such as I^-, does not undergo hydrolysis. However, the conjugate acid of a weak base is a weak acid and does undergo hydrolysis. The reaction is

$$NH_4^+\ (aq) + H_2O\ (l) \rightleftharpoons NH_3\ (aq) + H_3O^+\ (aq)$$

Since this reaction produces H_3O^+ ions, the ammonium iodide solution will be acidic.

4. **Salts in which both the cation and anion hydrolyze.** In these cases, you must compare the base strength of the anion to the acid strength of the cation. We consider three situations.

- $K_b > K_a$. If K_b for the anion is greater than K_a for the cation, the solution is basic because the anion will hydrolyze to a greater extent than the cation. At equilibrium, there will be more OH^- ions than H^+ ions.
- $K_b < K_a$. Conversely, if K_b for the anion is smaller than K_a for the cation, the solution will be acidic because cation hydrolysis will be more extensive than anion hydrolysis. At equilibrium, there will be more H^+ ions than OH^- ions.
- $K_b \cong K_a$. If K_b is approximately equal to K_a, the solution will be close to neutral.

EXAMPLE 16.10

Is a 0.10 *M* solution of Na_2CO_3 acidic, basic, or neutral?

Na_2CO_3 is a salt. Salts that contain the conjugate base of a weak acid and an alkali metal or alkaline earth metal ion, produce basic solutions. The dissociation of Na_2CO_3 in water is given by

$$Na_2CO_3\,(s) \xrightarrow{H_2O} 2\,Na^+\,(aq) + CO_3^{2-}\,(aq)$$

A hydrated alkali or alkaline earth metal ion has no acidic or basic properties. However, the conjugate base of a weak acid is a weak base (it has an affinity for H^+ ions). CO_3^{2-} is the conjugate base of the weak acid HCO_3^-.

The hydrolysis reaction is given by

$$CO_3^{2-}\,(aq) + H_2O\,(l) \rightleftharpoons HCO_3^-\,(aq) + OH^-\,(aq)$$

Because this reaction produces OH^- ions, the sodium carbonate solution will be basic.

PRACTICE EXERCISE

10. Predict whether the following aqueous solutions will be acidic, basic, or neutral.

(a) KI
(b) NH_4Cl
(c) CH_3COOK

Text Problems: 66, 68, 70

ANSWERS TO PRACTICE EXERCISES

1. (a) CH_3COO^- (b) HS^- (c) SO_3^{2-} (d) ClO^-

2. $[OH^-] = 1.5 \times 10^{-11}\,M$

3. $pOH = 3.14$
$pH = 10.86$

4. $[H^+] = 3.5 \times 10^{-5}\,M$

5. $pH = 1.60$

6. $pH = 1.82$

7. $K_a = 6.69 \times 10^{-5}$

8. $[H^+] = [HCO_3^-] = 1.8 \times 10^{-4}\,M$ $[H_2CO_3] = 0.080\,M$ $[CO_3^{2-}] = 4.8 \times 10^{-11}\,M$

9. $pH = 11.8$

10. (a) neutral (b) acidic (c) basic

SOLUTIONS TO SELECTED TEXT PROBLEMS

16.4 Recall that the conjugate base of a Brønsted acid is the species that remains when *one* proton has been removed from the acid.

(a) nitrite ion: NO_2^-

(b) hydrogen sulfate ion (also called bisulfate ion): HSO_4^-

(c) hydrogen sulfide ion (also called bisulfide ion): HS^-

(d) cyanide ion: CN^-

(e) formate ion: $HCOO^-$

16.6 The conjugate acid of any base is just the base with a proton added.

(a) H_2S	**(b)** H_2CO_3	**(c)** HCO_3^-	**(d)** H_3PO_4	**(e)** $H_2PO_4^-$
(f) HPO_4^{2-}	**(g)** H_2SO_4	**(h)** HSO_4^-	**(i)** HNO_2	**(j)** HSO_3^-

16.8 **(a)** The Lewis structures are

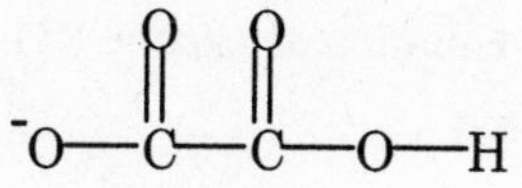

and

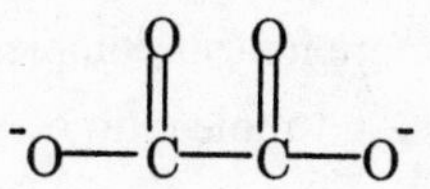

(b) H^+ and $C_2H_2O_4$ can act only as acids, $C_2HO_4^-$ can act as both an acid and a base, and $C_2O_4^{2-}$ can act only as a base.

16.16 For (a) and (b) we can calculate the H^+ concentration using the equation representing the definition of pH. Problem Type 3B.

(a) $\text{pH} = -\log[H^+]$

$$[H^+] = 10^{-5.20} = \mathbf{6.3 \times 10^{-6}}\ \boldsymbol{M}$$

(b) $\text{pH} = -\log[H^+]$

$$[H^+] = 10^{-16.00} = \mathbf{1.0 \times 10^{-16}}\ \boldsymbol{M}$$

(c) For part (c), it is probably easiest to calculate the $[H^+]$ from the ion product of water, Problem Type 2. The ion product of water is applicable to all aqueous solutions. At 25°C,

$$K_w = 1.0 \times 10^{-14} = [H^+][OH^-]$$

Rearrange the equation to solve for $[H^+]$.

$$[H^+] = \frac{1.0 \times 10^{-14}}{[OH^-]} = \frac{1.0 \times 10^{-14}}{3.7 \times 10^{-9}} = \mathbf{2.7 \times 10^{-6}}\ \boldsymbol{M}$$

16.18 The ion product for water is applicable to all aqueous solutions. However, at 40°C, K_w has a larger value.

$$K_w = 3.8 \times 10^{-14} = [H^+][OH^-]$$

The mole ratio of H^+ to OH^- is 1:1 for the autoionization of water. Let x equal the H^+ concentration.

$$3.8 \times 10^{-14} = x^2$$

$$x = [H^+] = 1.9 \times 10^{-7}\ M$$

$$\mathbf{pH} = -\log[H^+] = -\log(1.9 \times 10^{-7}) = \mathbf{6.72}$$

Is this water sample neutral?

16.20 **(a)** acidic **(b)** neutral **(c)** basic

16.22 $5.50\ \text{mL}\left(\dfrac{1\ \text{L}}{1000\ \text{mL}}\right)\left(\dfrac{0.360\ \text{mol}}{1\ \text{L}}\right) = \mathbf{1.98 \times 10^{-3}\ mol\ KOH}$

KOH is a strong base and therefore ionizes completely. The OH^- concentration equals the KOH concentration, because there is a 1:1 mole ratio between KOH and OH^-.

$$[OH^-] = 0.360\ M$$

$$\mathbf{pOH} = -\log[OH^-] = \mathbf{0.444}$$

16.24 We can calculate the OH^- concentration from the pOH.

$$\text{pOH} = 14.00 - \text{pH} = 14.00 - 10.00 = 4.00$$

$$[OH^-] = 10^{-\text{pOH}} = 1.0 \times 10^{-4}\ M$$

Since NaOH is a strong base, it ionizes completely. The OH^- concentration equals the initial concentration of NaOH.

$$[\text{NaOH}] = 1.0 \times 10^{-4}\ \text{mol/L}$$

So, we need to prepare 546 mL of $1.0 \times 10^{-4}\ M$ NaOH.

This is a factor-label problem. We need to perform the following unit conversions.

mol/L → mol NaOH → grams NaOH

546 mL = 0.546 L

$$\mathbf{?\ g\ NaOH} = \left(\frac{1.0 \times 10^{-4}\ \text{mol NaOH}}{1\ \text{L soln}}\right) \times (0.546\ \text{L soln}) \times \left(\frac{40.00\ \text{g NaOH}}{1\ \text{mol NaOH}}\right)$$

$$= \mathbf{2.2 \times 10^{-3}\ g\ NaOH}$$

16.30 **(a)** strong base **(b)** weak base **(c)** weak base **(d)** weak base **(e)** strong base

16.32 **(a)** false, they are equal **(b)** true, find the value of log(1.00) on your calculator
(c) true (d) false, if the acid is strong, [HA] = 0.00 *M*

16.34 Cl^- is the conjugate base of the strong acid, HCl. It is a negligibly weak base and has no affinity for protons. Therefore, the reaction will *not* proceed from left to right to any measurable extent.

Another way to think about this problem is to consider the possible products of the reaction.

$$CH_3COOH\ (aq) + Cl^-\ (aq) \rightarrow HCl\ (aq) + CH_3COO^-\ (aq)$$

The favored reaction is the one that proceeds from right to left. HCl is a strong acid and will ionize completely, donating all its protons to the base, CH_3COO^-.

16.36 The conjugate bases are $C_6H_5O^-$ from phenol and CH_3O^- from methanol. The $C_6H_5O^-$ is stabilized by resonance:

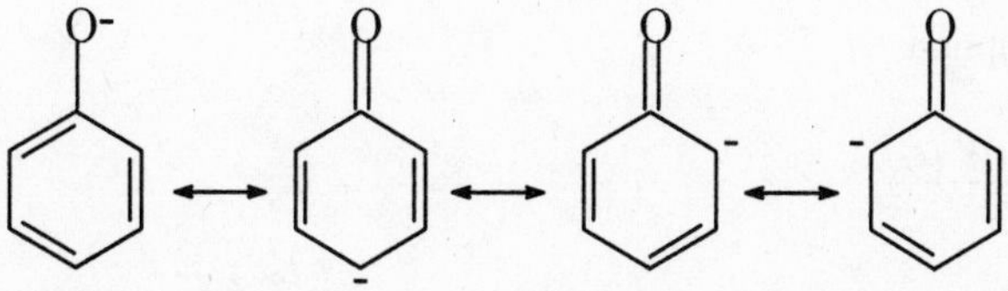

The CH_3O^- ion has no such resonance stabilization. A more stable conjugate base means an increase in the strength of the acid.

16.42 Ionization of a weak monoprotic acid, Problem Type 5A.

Step 1: Calculate the concentration of acetic acid before ionization.

$$0.0560\ \text{g acetic acid} \times \frac{1\ \text{mol acetic acid}}{60.05\ \text{g acetic acid}} = 9.33 \times 10^{-4}\ \text{mol acetic acid}$$

$$\frac{9.33 \times 10^{-4}\ \text{mol}}{0.0500\ \text{L soln}} = 0.0187\ M\ \text{acetic acid}$$

Step 2: Next, recognize that acetic acid is a weak, monoprotic acid. It is not one of the six strong acids, so it must be a weak acid. Express the equilibrium concentrations of all species in terms of initial concentrations and a single unknown *x*, that represents the change in concentration. Let (−*x*) be the depletion in concentration (mol/L) of CH_3COOH. From the stoichiometry of the reaction, it follows that the increase in concentration for both H^+ and CH_3COO^- must be *x*. Complete a table that lists the initial concentrations, the change in concentrations, and the equilibrium concentrations.

	$CH_3COOH\ (aq)$	$\rightleftharpoons$	$H^+\ (aq)$ +	$CH_3COO^-\ (aq)$
Initial (*M*):	0.0187		0	0
Change (*M*):	$-x$		$+x$	$+x$
Equilibrium (*M*):	$0.0187 - x$		x	x

Step 3: Write the ionization constant expression in terms of the equilibrium concentrations. Knowing the value of the equilibrium constant (K_a), solve for *x*.

$$K_a = \frac{[H^+][CH_3COO^-]}{[CH_3COOH]}$$

You can look up the K_a value for acetic acid in Table 16.3 of your text.

$$1.8 \times 10^{-5} = \frac{(x)(x)}{(0.0187 - x)}$$

At this point, we can make an assumption that x is very small compared to 0.0187. Hence,

$$0.0187 - x \cong 0.0187$$

$$1.8 \times 10^{-5} = \frac{(x)(x)}{0.0187}$$

Solving for x.

$$\mathbf{x = 5.8 \times 10^{-4}\ M = [H^+] = [CH_3COO^-]}$$

$$\mathbf{[CH_3COOH]} = (0.0187 - 5.8 \times 10^{-4})M = \mathbf{0.0181\ M}$$

Checking the validity of the assumption,

$$\frac{5.8 \times 10^{-4}}{0.0187} \times 100\% = 3.1\% < 5\%$$

The assumption is valid.

16.44 A pH of 3.26 corresponds to a $[H^+]$ of 5.5×10^{-4} M. Let the original concentration of formic acid be x so that:

	$HCOOH\ (aq)$	$\rightleftharpoons$	$H^+\ (aq)$ +	$HCOO^-\ (aq)$
Initial (M):	x		0	0
Change (M):	-5.5×10^{-4}		5.5×10^{-4}	5.5×10^{-4}
Equilibrium (M):	$x - 5.5 \times 10^{-4}$		5.5×10^{-4}	5.5×10^{-4}

Substitute K_a and the equilibrium concentrations into the ionization constant expression to solve for x.

$$\frac{[H^+][HCOO^-]}{[HCOOH]} = 1.7 \times 10^{-4}$$

$$\frac{(5.5 \times 10^{-4})^2}{(x - 5.5 \times 10^{-4})} = 1.7 \times 10^{-4}$$

$$\mathbf{x = [HCOOH] = 2.3 \times 10^{-3}\ M}$$

16.46 Percent ionization is defined as:

$$\text{percent ionization} = \frac{\text{ionized acid concentration at equilibrium}}{\text{initial concentration of acid}} \times 100\%$$

For a monoprotic acid, HA, the concentration of acid that undergoes ionization is equal to the concentration of H^+ ions or the concentration of A^- ions at equilibrium. Thus, we can write:

$$\text{percent ionization} = \frac{[H^+]}{[HA]_0} \times 100\%$$

(a) First, recognize that hydrofluoric acid is a weak acid. It is not one of the six strong acids, so it must be a weak acid.

Step 1: Express the equilibrium concentrations of all species in terms of initial concentrations and a single unknown x, that represents the change in concentration. Let $(-x)$ be the depletion in concentration (mol/L) of HF. From the stoichiometry of the reaction, it follows that the increase in concentration for both H^+ and F^- must be x. Complete a table that lists the initial concentrations, the change in concentrations, and the equilibrium concentrations.

	$HF\ (aq) \rightleftharpoons$	$H^+\ (aq)$ +	$F^-\ (aq)$
Initial (M):	0.60	0	0
Change (M):	$-x$	$+x$	$+x$
Equilibrium (M):	$0.60 - x$	x	x

Step 2: Write the ionization constant expression in terms of the equilibrium concentrations. Knowing the value of the equilibrium constant (K_a), solve for x.

$$K_a = \frac{[H^+][F^-]}{[HF]}$$

You can look up the K_a value for hydrofluoric acid in Table 16.3 of your text.

$$7.1 \times 10^{-4} = \frac{(x)(x)}{(0.60 - x)}$$

At this point, we can make an assumption that x is very small compared to 0.60. Hence,

$$0.60 - x \cong 0.60$$

Oftentimes, assumptions such as these are valid if K is very small. A very small value of K means that a very small amount of reactants go to products. Hence, x is small. If we did not make this assumption, we would have to solve a quadratic equation.

$$7.1 \times 10^{-4} = \frac{(x)(x)}{0.60}$$

Solving for x.

$$x = 0.021\ M = [H^+]$$

Step 3: Having solved for the $[H^+]$, calculate the percent ionization.

$$\textbf{percent ionization} = \frac{[H^+]}{[HF]_0} \times 100\%$$

$$= \frac{0.021\ M}{0.60\ M} \times 100\% = \mathbf{3.5\%}$$

(b) – (d) are worked in a similar manner to part (a). However, as the initial concentration of HF becomes smaller, the assumption that x is very small compared to this concentration will no longer be valid. You must solve a quadratic equation.

(b) $K_a = \dfrac{[H^+][F^-]}{[HF]} = \dfrac{x^2}{(0.080 - x)} = 7.1 \times 10^{-4}$

$x^2 + (7.1 \times 10^{-4})x - (5.7 \times 10^{-5}) = 0$

$x = 7.2 \times 10^{-3}\ M$

$$\textbf{Percent ionization} = \frac{7.2 \times 10^{-3}\ M}{0.080\ M} \times 100\% = \mathbf{9.0\%}$$

(c) $K_a = \dfrac{[H^+][F^-]}{[HF]} = \dfrac{x^2}{(0.0046 - x)} = 7.1 \times 10^{-4}$

$x^2 + (7.1 \times 10^{-4})x - (3.3 \times 10^{-6}) = 0$

$x = 1.5 \times 10^{-3}\ M$

$$\textbf{Percent ionization} = \frac{1.5 \times 10^{-3}\ M}{0.0046\ M} \times 100\% = \mathbf{33\%}$$

(d) $K_a = \dfrac{[H^+][F^-]}{[HF]} = \dfrac{x^2}{(0.00028 - x)} = 7.1 \times 10^{-4}$

$x^2 + (7.1 \times 10^{-4})x - (2.0 \times 10^{-7}) = 0$

$x = 2.2 \times 10^{-4}\ M$

$$\textbf{Percent ionization} = \frac{2.2 \times 10^{-4}\ M}{0.00028\ M} \times 100\% = \mathbf{79\%}$$

As the solution becomes more dilute, the percent ionization increases.

16.48 $C_9H_8O_4\ (aq) \rightleftharpoons H^+\ (aq) + C_9H_7O_4^-$

(a) $K_a = 3.0 \times 10^{-4} = \dfrac{[H^+][C_9H_7O_4^-]}{[C_9H_8O_4]} = \dfrac{x^2}{(0.20 - x)}$

Assuming $(0.20 - x) \cong 0.20$

$$x = [H^+] = 7.7 \times 10^{-3}\ M$$

$$\textbf{Percent ionization} = \frac{[C_9H_7O_4^-]}{[C_9H_8O_4]} \times 100\%$$

$$= \frac{7.7 \times 10^{-3}\ M}{0.20\ M} \times 100\% = \mathbf{3.9\%}$$

(b) At pH 1.00 the concentration of hydrogen ion is 0.10 *M* (Why only two significant figures?) This will tend to suppress the ionization of the weak acid (LeChatelier's principle, Section 15.4 of the text). The extra hydrogen ion shifts the position of equilibrium in the direction of the un-ionized acid, and to two-significant-figure accuracy, we can safely ignore the contribution of the weak acid to the total hydrogen ion concentration. The percent ionization of the acid is then

$$\text{Percent ionization} = \frac{[C_9H_7O_4^-]}{[C_9H_8O_4]} \times 100\%$$

and,

$$K_a = \frac{[H^+][C_9H_7O_4^-]}{[C_9H_8O_4]} \quad \text{or} \quad \frac{K_a}{[H^+]} = \frac{[C_9H_7O_4^-]}{[C_9H_8O_4]}$$

Substituting into the equation for percent ionization gives

$$\textbf{percent ionization} = \frac{K_a}{[H^+]} \times 100\%$$

$$= \frac{3.0 \times 10^{-4}}{0.10} \times 100\% = \mathbf{0.30\%}$$

16.52 For the first stage of ionization:

	$H_2CO_3\ (aq)$	$\rightleftharpoons$	$H^+\ (aq)$	+	$HCO_3^-\ (aq)$
Initial (*M*):	0.025		0.00		0.00
Change (*M*):	$-x$		$+x$		$+x$
Equilibrium (*M*):	$(0.025 - x)$		x		x

$$K_{a_1} = \frac{[H^+][HCO_3^-]}{[H_2CO_3]} = \frac{x^2}{(0.025 - x)} = 4.2 \times 10^{-7}$$

Assuming $(0.025 - x) \cong 0.025$,

$$x = 1.0 \times 10^{-4}\ M$$

For the second ionization,

$$HCO_3^-\ (aq) \rightleftharpoons H^+\ (aq) + CO_3^{2-}\ (aq)$$

$$K_{a_2} = \frac{[H^+][CO_3^{2-}]}{[HCO_3^-]} = 4.8 \times 10^{-11}$$

Since HCO_3^- is a very weak acid, there is little ionization at this stage. Therefore we have:

$$\mathbf{[H^+] = [HCO_3^-] = 1.0 \times 10^{-4}}\ M$$

The CO_3^{2-} concentration is:

$$K_{a_2} = \frac{[H^+][CO_3^{2-}]}{[HCO_3^-]}$$

$$4.8 \times 10^{-11} = \frac{(1.0 \times 10^{-4}\ M)[CO_3^{2-}]}{(1.0 \times 10^{-4}\ M)}$$

$$\mathbf{[CO_3^{2-}] = 4.8 \times 10^{-11}}\ M$$

16.56 Similar to Problem Type 5B, Determining K_a from a pH measurement.

Step 1: Calculate the pOH from the pH. Then, calculate the OH^- concentration from the pOH.

$$pOH = 14.00 - pH = 14.00 - 10.66 = 3.34$$

$$pOH = -\log[OH^-]$$

or,

$$-pOH = \log[OH^-]$$

Taking the antilog of both sides of the equation,

$$10^{-pOH} = [OH^-]$$

$$[OH^-] = 10^{-3.34} = 4.6 \times 10^{-4}\ M$$

Step 2: Complete a table that lists the initial concentrations, the change in concentrations, and the equilibrium concentrations. Let B represent the weak base.

	$B\ (aq) + H_2O\ (l) \rightleftharpoons$	$BH^+\ (aq) +$	$OH^-\ (aq)$
Initial (M):	0.30	0	0
Change (M):	$-x$	$+x$	$+x$
Equilibrium (M):	$0.30 - x$	x	x

In this case, x is known.

$$x = [BH^+] = [OH^-] = 4.6 \times 10^{-4}\ M$$

$$[B] = 0.30 - x = (0.30 - 4.6 \times 10^{-4})M = 0.30\ M$$

Step 3: Substitute the equilibrium concentrations into the ionization constant expression to solve for K_b.

$$K_b = \frac{[BH^+][OH^-]}{[B]}$$

$$\mathbf{K_b} = \frac{(4.6 \times 10^{-4})^2}{(0.30)} = \mathbf{7.1 \times 10^{-7}}$$

16.58 The reaction is:

$$NH_3\ (aq) + H_2O\ (l) \rightleftharpoons NH_4^+\ (aq) + OH^-\ (aq)$$

Let $[OH^-] = [NH_4^+] = x$

At equilibrium we have:

$$K_b = \frac{[NH_4^+][OH^-]}{[NH_3]}$$

$$1.8 \times 10^{-5} = \frac{x^2}{(0.080 - x)} \cong \frac{x^2}{0.080}$$

$$x = 1.2 \times 10^{-3}\ M$$

$$\textbf{Percent NH}_3\ \textbf{present as NH}_4^+ = \frac{1.2 \times 10^{-3}}{0.080} \times 100\% = \mathbf{1.5\%}$$

16.66 Predicting the Acid-Base Properties of Salt Solutions, Problem Type 8.

There is an inverse relationship between acid strength and conjugate base strength. As acid strength decreases, the proton accepting power of the conjugate base increases. In general the weaker the acid, the stronger the conjugate base. All three of the potassium salts ionize completely to form the conjugate base of the respective acid. The greater the pH, the stronger the conjugate base, and therefore, the weaker the acid.

The order of increasing acid strength is **HZ < HY < HX.**

16.68 The salt ammonium chloride completely ionizes upon dissolution, producing 0.42 *M* $[NH_4^+]$ and 0.42 *M* $[Cl^-]$ ions. NH_4^+ will undergo hydrolysis because it is a weak acid (NH_4^+ is the conjugate acid of the weak base, NH_3).

Step 1: Express the equilibrium concentrations of all species in terms of initial concentrations and a single unknown x, that represents the change in concentration. Let $(-x)$ be the depletion in concentration (mol/L) of NH_4^+. From the stoichiometry of the reaction, it follows that the increase in concentration for both H_3O^+ and NH_3 must be x. Complete a table that lists the initial concentrations, the change in concentrations, and the equilibrium concentrations.

	$NH_4^+\ (aq)$ + $H_2O\ (l)$ $\rightleftharpoons$	$NH_3\ (aq)$ +	$H_3O^+\ (aq)$
Initial (*M*):	0.42	0.00	0.00
Change (*M*):	$-x$	$+x$	$+x$
Equilibrium (*M*):	$(0.42 - x)$	x	x

Step 2: You can calculate the K_a value for NH_4^+ from the K_b value of NH_3. The relationship is

$$K_a \times K_b = K_w$$

or

$$K_a = \frac{K_w}{K_b} = \frac{1.0 \times 10^{-14}}{1.8 \times 10^{-5}} = 5.6 \times 10^{-10}$$

Step 3: Write the ionization constant expression in terms of the equilibrium concentrations. Knowing the value of the equilibrium constant (K_a), solve for x.

$$K_a = \frac{[NH_3][H_3O^+]}{[NH_4^+]} = \frac{x^2}{0.42 - x} = 5.6 \times 10^{-10}$$

Assuming $(0.42 - x) \cong 0.42$,

$$x = [H^+] = 1.5 \times 10^{-5}\ M$$

$$\textbf{pH} = -\log(1.5 \times 10^{-5}) = \textbf{4.82}$$

Since NH_4Cl is the salt of a weak base (aqueous ammonia) and a strong acid (HCl), we expect the solution to be slightly acidic, which is confirmed by the calculation.

16.70 $HCO_3^- \rightleftharpoons H^+ + CO_3^{2-}$ $\qquad K_a = 4.8 \times 10^{-11}$

$HCO_3^- + H_2O \rightleftharpoons H_2CO_3 + OH^-$ $\qquad K_b = \frac{K_w}{K_a} = \frac{1.0 \times 10^{-14}}{4.2 \times 10^{-7}} = 2.4 \times 10^{-8}$

HCO_3^- has a greater tendency to hydrolyze than to ionize ($K_b > K_a$). The solution will be basic (pH > 7).

16.74 $AlCl_3$ is a Lewis acid with an incomplete octet of electrons and Cl^- is the Lewis base donating a pair of electrons.

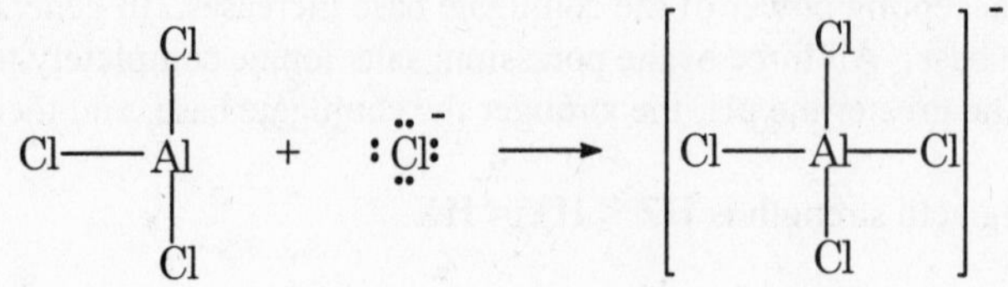

16.76 By definition Brønsted acids are proton donors, therefore such compounds must contain at least one hydrogen atom. In Problem 16.73, Lewis acids that do not contain hydrogen, and therefore are not Brønsted acids, are CO_2, SO_2, and BCl_3. Can you name others?

16.78 We first find the number of moles of CO_2 produced in the reaction:

$$0.350\text{ g NaHCO}_3\left(\frac{1\text{ mol NaHCO}_3}{84.01\text{ g NaHCO}_3}\right)\left(\frac{1\text{ mol CO}_2}{1\text{ mol NaHCO}_3}\right) = 4.17\times10^{-3}\text{ mol CO}_2$$

$$V_{CO_2} = \frac{n_{CO_2}RT}{P} = \frac{(4.17\times10^{-3}\text{ mol})(0.0821\text{ L}\cdot\text{atm/K}\cdot\text{mol})(37.0+273)\text{K}}{(1.00\text{ atm})} = \mathbf{0.106\ L}$$

16.80 If we assume that the unknown monoprotic acid is a strong acid that is 100% ionized, then the $[H^+]$ concentration will be 0.0642 *M*.

$$\text{pH} = -\log(0.0642) = 1.19$$

Since the actual pH of the solution is higher, the acid must be a weak acid.

16.82 The reaction of a weak acid with a strong base is driven to completion due to the strength of the strong base. Irrespective of whether the strong base is reacting with a strong monoprotic acid or a weak monoprotic acid, the same number of moles of acid is required to react with a constant number of moles of base. Therefore the volume of base required to react with the same concentration of acid solutions (either both weak, both strong, or one strong and one weak) will be the same.

16.84 High oxidation state leads to covalent compounds and low oxidation state leads to ionic compounds. Therefore, CrO is ionic and basic and CrO_3 is covalent and acidic.

16.86 We can write two equilibria that add up to the equilibrium in the problem.

$$CH_3COOH\,(aq) \rightleftharpoons H^+\,(aq) + CH_3COO^-\,(aq) \qquad K_a = \frac{[H^+][CH_3COO^-]}{[CH_3COOH]} = 1.8\times10^{-5}$$

$$H^+\,(aq) + NO_2^-\,(aq) \rightleftharpoons HNO_2\,(aq) \qquad K_a' = \frac{1}{K_a(HNO_2)} = \frac{1}{4.5\times10^{-4}} = 2.2\times10^3$$

$$K_a' = \frac{[HNO_2]}{[H^+][NO_2^-]}$$

$$CH_3COOH\,(aq) + NO_2^-\,(aq) \rightleftharpoons CH_3COO^-\,(aq) + HNO_2\,(aq) \qquad K = \frac{[CH_3COO^-][HNO_2]}{[CH_3COOH][NO_2^-]} = K_a \times K_a'$$

The equilibrium constant for this sum is the product of the equilibrium constants of the component reactions.

$$K = K_a \times K_a' = (1.8 \times 10^{-5})(2.2 \times 10^{3}) = \mathbf{4.0 \times 10^{-2}}$$

16.88 Let Nov represent the novocaine molecule which is a weak base in water.

$$\text{Nov} + H_2O \rightleftharpoons \text{NovH}^+ + OH^-$$

$$K_b = \frac{[\text{NovH}^+][OH^-]}{[\text{Nov}]}$$

Calculate the $[OH^-]$ from the pH.

pH = 7.40 and pOH = 6.60

$$[OH^-] = 10^{-pOH} = 10^{-6.60} = 2.5 \times 10^{-7}\ M$$

Substitute the $[OH^-]$ into the K_b expression.

$$8.91 \times 10^{-6} = \frac{[\text{NovH}^+](2.5 \times 10^{-7}\ M)}{[\text{Nov}]}$$

$$\frac{[\text{NovH}^+]}{[\text{Nov}]} = 36$$

This is the ratio of the concentration of the acid to that of its base. The ratio of the concentration of the base to that of its acid is:

$$\frac{[\text{Nov}]}{[\text{NovH}^+]} = \frac{1}{36} = \mathbf{0.028}$$

16.90 $0.100\ M\ Na_2CO_3 \rightarrow 0.200\ M\ Na^+ + 0.100\ M\ CO_3^{2-}$

See Table 16.4 of the text for the K_b values.

<u>First stage:</u>

$$\underset{0.100\ M}{CO_3^{2-}(aq)} + H_2O(l) \rightleftharpoons \underset{x\ M}{HCO_3^-(aq)} + \underset{x\ M}{OH^-(aq)}$$

$$K_{b_1} = 2.1 \times 10^{-4} = \frac{[HCO_3^-][OH^-]}{[CO_3^{2-}]} = \frac{x^2}{0..100 - x}$$

$$x = 4.6 \times 10^{-3}\ M = [HCO_3^-] = [OH^-]$$

<u>Second stage:</u>

$$\underset{(4.6 \times 10^{-3} - y)\ M}{HCO_3^-(aq)} + H_2O(l) \rightleftharpoons \underset{y\ M}{H_2CO_3(aq)} + \underset{(4.6 \times 10^{-3} + y)\ M}{OH^-(aq)}$$

$$K_{b_2} = \frac{[H_2CO_3][OH^-]}{[HCO_3^-]}$$

$$2.4 \times 10^{-8} = \frac{y(4.6 \times 10^{-3} + y)}{4.6 \times 10^{-3} - y}$$

$$y = 2.4 \times 10^{-8}\ M$$

At equilibrium:

$$\mathbf{[Na^+] = 0.200\ \mathit{M}}$$

$$[\mathbf{HCO_3^-}] = (4.6 \times 10^{-3})\ M - (2.4 \times 10^{-8})\ M \cong \mathbf{4.6 \times 10^{-3}\ \mathit{M}}$$

$$\mathbf{[H_2CO_3] = 2.4 \times 10^{-8}\ \mathit{M}}$$

$$[\mathbf{OH^-}] = (4.6 \times 10^{-3})\ M + (2.4 \times 10^{-8}]\ M \cong \mathbf{4.6 \times 10^{-3}\ \mathit{M}}$$

$$[\mathbf{H^+}] = \frac{1.0 \times 10^{-14}}{4.6 \times 10^{-3}} = \mathbf{2.2 \times 10^{-12}\ \mathit{M}}$$

16.92 When NaCN is treated with HCl, the following reaction occurs.

$$NaCN + HCl \rightarrow NaCl + HCN$$

HCN is a very weak acid, and only partially ionizes in solution.

$$HCN\ (aq) \rightleftharpoons H^+(aq) + CN^-(aq)$$

The main species in solution is HCN which has a tendency to escape into the gas phase.

$$HCN(aq) \rightleftharpoons HCN(g)$$

Since the HCN(*g*) that is produced is a highly poisonous compound, it would be dangerous to treat NaCN with acids without proper ventilation.

16.94 The equilibrium is established:

	$CH_3COOH(aq)$	$\rightleftharpoons$ $CH_3COO^-(aq)$	+ $H^+(aq)$
Initial (M):	0.150	0	0.100
Change (M):	$-x$	$+x$	$+x$
Equilibrium (M)	$(0.150 - x)$	x	$(0.100 + x)$

$$K_a = \frac{[CH_3COO^-][H^+]}{[CH_3COOH]} = \frac{x(0.100 + x)}{0.150 - x} = \frac{0.100\ x}{0.150} = 1.8 \times 10^{-5}$$

$$x = 2.7 \times 10^{-5}\ M = [H^+]$$

2.7×10^{-5} M is the $[H^+]$ contributed by CH_3COOH. HCl is a strong acid that completely ionizes. It contributes a $[H^+]$ of 0.100 M to the solution.

$$[H^+]_{total} = (0.100 + 2.7 \times 10^{-5})\ M \cong 0.100\ M$$

$$\mathbf{pH = 1.000}$$

The pH is totally determined by the HCl and is independent of the CH_3COOH.

16.96 The important equation is the hydrolysis of NO_2^-: $NO_2^- + H_2O \rightleftharpoons HNO_2 + OH^-$

(a) Addition of HCl will result in the reaction of the H^+ from the HCl with the OH^- that was present in the solution. The OH^- will effectively be removed and the equilibrium will shift to the right to compensate (more hydrolysis).

(b) Addition of NaOH is effectively addition of more OH^- which places stress on the right hand side of the equilibrium. The equilibrium will shift to the left (less hydrolysis) to compensate for the addition of OH^-.

(c) Addition of NaCl will have no effect.

(d) Recall that the percent ionization of a weak acid increases with dilution (see Figure 16.3 of the text). The same is true for weak bases. Thus dilution will cause more hydrolysis, shifting the equilibrium to the right.

16.98

HCOOH	$\rightleftharpoons$	H^+ +	$HCOO^-$
$0.400 - x$		x	x

Total concentration of particles in solution: $(0.400 - x) + x + x = 0.400 + x$

Assuming the molarity of the solution is equal to the molality, we can write:

$$\Delta T_f = K_f m$$

$$0.758 = (1.86)(0.400 + x)$$

$$0.408 = 0.400 + x$$

$$x = 0.00800 = [H^+] = [HCOO^-]$$

$$\boldsymbol{K_a} = \frac{[H^+][HCOO^-]}{[HCOOH]} = \frac{(0.00800)(0.00800)}{0.400 - 0.00800} = \mathbf{1.6 \times 10^{-4}}$$

16.100 $SO_2(g) + H_2O(l) \rightleftharpoons H^+(aq) + HSO_3^-(aq)$

Recall that 0.12 ppm SO_2 would mean 0.12 parts SO_2 per 1 million (10^6) parts of air by volume. The number of particles of SO_2 per volume will be directly related to the pressure.

$$P_{SO_2} = \frac{0.12 \text{ parts } SO_2}{10^6 \text{ parts air}} \text{ atm} = 1.2 \times 10^{-7} \text{ atm}$$

We can now calculate the $[H^+]$ from the equilibrium constant expression.

$$K = \frac{[H^+][HSO_3^-]}{P_{SO_2}}$$

$$1.3 \times 10^{-2} = \frac{x^2}{1.2 \times 10^{-7}}$$

$$x^2 = (1.3 \times 10^{-2})(1.2 \times 10^{-7})$$

$$x = 3.9 \times 10^{-5}\ M = [H^+]$$

$$\textbf{pH} = -\log(3.9 \times 10^{-5}) = \mathbf{4.40}$$

16.102 **(a)** The overall equation is

$$Fe_2O_3(s) + 6HCl(aq) \longrightarrow 2FeCl_3(aq) + 3H_2O(l)$$

and the net ionic equation is

$$Fe_2O_3(s) + 6H^+(aq) \longrightarrow 2Fe^{3+}(aq) + 3H_2O(l)$$

Since HCl donates the H^+ ion, it is the Brønsted acid. Each Fe_2O_3 unit accepts six H^+ ions; therefore, it is the Brønsted base.

(b) The first stage is

$$CaCO_3(s) + HCl(aq) \longrightarrow Ca^{2+}(aq) + HCO_3^-(aq) + Cl^-(aq)$$

and the second stage is

$$HCl(aq) + HCO_3^-(aq) \longrightarrow CO_2(g) + Cl^-(aq) + H_2O(l)$$

The overall equation is

$$CaCO_3(s) + 2HCl(aq) \longrightarrow CaCl_2(aq) + H_2O(l) + CO_2(g)$$

The $CaCl_2$ formed is soluble in water.

(c) We need to find the concentration of the HCl solution in order to determine its pH. Let's assume a volume of 1.000 L = 1000 mL. The mass of 1000 mL of solution is:

$$1000\text{ mL} \times \frac{1.073\text{ g}}{1\text{ mL}} = 1073\text{ g}$$

The number of moles of HCl in a 15 percent solution is:

$$\frac{15\%\text{ HCl}}{100\%\text{ soln}} \times 1073\text{ g soln} = 1.6 \times 10^2\text{ g HCl} \times \frac{1\text{ mol HCl}}{36.46\text{ g HCl}} = 4.4\text{ mol HCl}$$

Thus, there are 4.4 moles of HCl in one liter of solution, and the concentration is 4.4 *M*. The pH of the solution is

$$\text{pH} = -\log(4.4) = \mathbf{-0.64}$$

This is a highly acidic solution (note that the pH is negative) which is needed to dissolve large quantities of rocks in the oil recovery process.

CHAPTER 17
ACID-BASE EQUILIBRIA AND SOLUBILITY EQUILIBRIA

PROBLEM-SOLVING STRATEGIES AND TUTORIAL SOLUTIONS

TYPES OF PROBLEMS

Problem Type 1: Buffers.
- **(a)** Identifying buffer systems.
- **(b)** Calculating the pH of a buffer system.
- **(c)** Preparing a buffer solution with a specific pH.

Problem Type 2: Titrations.
- **(a)** Strong acid–strong base titrations.
- **(b)** Weak acid–strong base titrations.
- **(c)** Strong acid–weak base titrations.

Problem Type 3: Choosing Suitable Acid-Base Indicators.

Problem Type 4: Solubility Equilibria.
- **(a)** Calculating K_{sp} from molar solubility.
- **(b)** Calculating solubility from K_{sp}.
- **(c)** Predicting a precipitation reaction.
- **(d)** The effect of a common ion on solubility.

Problem Type 5: Complex Ion Equilibria and Solubility.

PROBLEM TYPE 1: BUFFERS

A **buffer solution** is a solution of (1) a weak acid or a weak base and (2) its salt; both components must be present. The solution has the ability to resist change in pH upon the addition of small amounts of either acid or base. A buffer resists change in pH, because the weak acid component reacts with small amounts of added base. The weak base component of the buffer reacts with small amounts of added acid.

$$\underset{\text{weak acid of buffer}}{HA\,(aq)} + \underset{\text{added base}}{OH^-\,(aq)} \longrightarrow A^-\,(aq) + H_2O\,(l)$$

$$\underset{\text{weak base of buffer}}{A^-\,(aq)} + \underset{\text{added acid}}{H_3O^+\,(aq)} \longrightarrow HA\,(aq) + H_2O\,(l)$$

A. Identifying buffer systems

To identify a buffer system, look for a weak acid and its conjugate base (usually the anion in a soluble salt) or a weak base and its conjugate acid (usually the cation in a soluble salt).

EXAMPLE 17.1

Which of the following solutions are buffer systems: (a) CH_3COOH/CH_3COONa, (b) KNO_3/HNO_3, and (c) NH_4Cl/NH_3?

(a) CH_3COOH (acetic acid) is a weak acid, and its conjugate base, CH_3COO^- (acetate ion, the anion of the salt CH_3COONa), is a weak base. Therefore, this is a buffer system.

(b) Because HNO_3 is a strong acid, its conjugate base, NO_3^-, is an extremely weak base. This means that NO_3^- will not combine with H^+ in solution to form HNO_3. Thus, the system cannot act as a buffer system.

(c) NH_3 (ammonia) is a weak base and its conjugate acid, NH_4^+ (ammonium ion, the cation of the salt NH_4Cl), is a weak acid. Therefore, this is a buffer system.

PRACTICE EXERCISE

1. Which of the following solutions are buffer systems: (a) KCN/HCN, (b) NaCl/HCl, and (c) HNO_2/KNO_2?

Text Problem: 6

B. Calculating the pH of a buffer system

Calculating the pH of a buffer system is similar to calculating the pH of a weak acid solution (Problem Type 5A, Chapter 16) or the pH of a weak base solution (Problem Type 7, Chapter 16). The difference is that the initial concentration of the conjugate of the weak acid or weak base is *not zero*. The initial concentration of the conjugate comes from the salt component of the buffer.

For example, consider a solution that is 1.0 *M* in CH_3COOH (acetic acid) and 1.0 *M* in CH_3COONa (sodium acetate). If this were only a 1.0 *M* acetic acid solution, the initial concentration of the conjugate base (CH_3COO^-) would be *zero*. However, CH_3COONa is a soluble salt that ionizes completely.

	$CH_3COONa\,(aq)$	$\longrightarrow$	$Na^+\,(aq)$	+	$CH_3COO^-\,(aq)$
Initial	1.0 *M*		0		0
After ionization	0		1.0 *M*		1.0 *M*

The CH_3COO^- ion is present initially and enters into the weak acid equilibrium of acetic acid. Thus, if we look at the weak acid equilibrium for acetic acid, we have:

	$CH_3COOH\,(aq)$	$\rightleftharpoons$	$H^+\,(aq)$	+	$CH_3COO^-\,(aq)$
Initial (*M*):	1.0		0		1.0
Change (*M*):	$-x$		$+x$		$+x$
Equilibrium (*M*):	$1.0 - x$		x		$1.0 + x$

You should notice that the only difference between this equilibrium, and a weak acid equilibrium (see Problem Type 5A, Chapter 16), is that the initial concentration of the conjugate base, CH_3COO^-, is *not zero*, as it would be for a weak acid equilibrium. Remember, a buffer contains both a weak acid and its conjugate base or a weak base and its conjugate acid. The above system is a buffer because it initially contains both the weak acid (CH_3COOH) and its conjugate base (CH_3COO^-). Example 17.2 illustrates how to calculate the pH of a buffer system.

EXAMPLE 17.2

(a) Calculate the pH of a buffer system containing 0.25 *M* HF and 0.50 *M* NaF.

(b) What is the pH of the buffer system after the addition of 0.060 mol of gaseous HCl to 1.00 L of the solution? Assume that the volume of the solution does not change upon addition of the HCl.

(a)

Step 1: Recognize that NaF is a soluble salt that will completely dissociate. From the dissociation of NaF, we can calculate the initial concentration of the weak base, F^-.

	$NaF\ (aq)$	$\longrightarrow$	$Na^+\ (aq)$	+	$F^-\ (aq)$
Initial	0.50 *M*		0		0
After dissociation	0		0.50 *M*		0.50 *M*

Step 2: Next, consider the weak acid equilibrium. Express the equilibrium concentrations of all species in terms of initial concentrations and a single unknown, *x*, which represents the change in concentration. Let $(-x)$ be the depletion in concentration (mol/L) of HF. From the stoichiometry of the reaction, it follows that the increase in concentration for both H^+ and F^- must be *x*. Complete a table that lists the initial concentrations, the change in concentrations, and the equilibrium concentrations.

	$HF\ (aq)$	$\rightleftharpoons$	$H^+\ (aq)$	+	$F^-\ (aq)$
Initial (*M*):	0.25		0		0.50
Change (*M*):	$-x$		$+x$		$+x$
Equilibrium (*M*):	$0.25 - x$		x		$0.50 + x$

Step 3: Write the ionization constant expression in terms of the equilibrium concentrations. Knowing the value of the equilibrium constant (K_a), solve for *x*.

$$K_a = \frac{[H^+][F^-]}{[HF]}$$

You can look up the K_a value for hydrofluoric acid in Table 16.3 of your text.

$$7.1 \times 10^{-4} = \frac{(x)(0.50 + x)}{(0.25 - x)}$$

At this point, we will make two assumptions.

$$0.25 - x \cong 0.25$$

and

$$0.50 + x \cong 0.50$$

Oftentimes, assumptions such as these are valid if *K* is very small. A very small value of *K* means that a very small amount of reactants go to products. Hence, *x* is small. If we did not make this assumption, we would have to solve a quadratic equation!

$$7.1 \times 10^{-4} \cong \frac{(x)(0.50)}{0.25}$$

Solving for *x*,

$$x = 3.6 \times 10^{-4}\ M = [H^+]$$

Checking the validity of the assumption,

$$\frac{3.6 \times 10^{-4}}{0.25} \times 100\% = 0.14\% < 5\%$$

The assumption is valid.

Step 3: Having solved for the $[H^+]$, calculate the pH of the solution.

$$\textbf{pH} = -\log[H^+] = -\log(3.6 \times 10^{-4}) = \textbf{3.44}$$

(b) We have added 0.060 mol of HCl (strong acid), which will react completely with the weak base component of the buffer. This reaction will change the equilibrium concentrations of both the acid and base components of the equilibrium. Therefore, after the reaction between HCl and the weak base of the buffer, the equilibrium must be reestablished.

Step 1: Write the reaction that occurs between the strong acid, HCl, and the weak base component of the buffer, F^-. After addition of HCl, complete ionization of HCl occurs.

	$HCl\ (aq)$ $\longrightarrow$	$H^+\ (aq)$ +	$Cl^-\ (aq)$
Initial	0.060 mol	0	0
After ionization	0	0.060 mol	0.060 mol

Next, H^+ will react with the weak base component of the buffer, F^-. Since H^+ is the strongest acid that can exist in water, this reaction will be driven to completion.

$$H^+\ (aq) + F^-\ (aq) \rightarrow HF\ (aq)$$

Since we have 1.00 L of buffer solution, the number of moles of F^- in solution is:

$$\text{mol } F^- = \left(0.50\ \frac{\text{mol}}{\text{L}}\right) \times 1.00\text{ L} = 0.50\text{ mol}$$

Similarly, the number of moles of HF is 0.25 mol.

We can now calculate the moles of F^- and HF that remain after F^- reacts with the strong acid HCl.

	$H^+\ (aq)$ +	$F^-\ (aq)$ $\longrightarrow$	$HF\ (aq)$
Initial (mol)	0.060	0.50	0.25
Change (mol)	−0.060	−0.060	+0.060
After reaction (mol)	0	0.44	0.31

Step 2: After the reaction above, HF and F^- are no longer in equilibrium. Equilibrium must be reestablished. The concentrations of HF and F^- are:

$$[HF] = \frac{0.31\text{ mol}}{1.00\text{ L}} = 0.31\ M$$

$$[F^-] = \frac{0.44\text{ mol}}{1.00\text{ L}} = 0.44\ M$$

Reestablishing equilibrium between HF and F^-, we have:

	$HF\ (aq)$ $\rightleftharpoons$	$H^+\ (aq)$ +	$F^-\ (aq)$
Initial (M):	0.31	0	0.44
Change (M):	$-x$	$+x$	$+x$
Equilibrium (M):	$0.31 - x$	x	$0.44 + x$

Step 3: Write the ionization constant expression in terms of the equilibrium concentrations. Knowing the value of the equilibrium constant (K_a), solve for x.

$$K_a = \frac{[H^+][F^-]}{[HF]}$$

$$7.1 \times 10^{-4} = \frac{(x)(0.44 + x)}{(0.31 - x)}$$

$$7.1 \times 10^{-4} \cong \frac{(x)(0.44)}{0.31}$$

$$x = [H^+] = 5.0 \times 10^{-4}\ M$$

Step 4: Having solved for the $[H^+]$, calculate the pH of the solution.

$$\mathbf{pH} = -\log[H^+] = -\log(5.0 \times 10^{-4}) = \mathbf{3.30}$$

The pH of the solution dropped only 0.15 pH units upon addition of 0.060 mol of the strong acid, HCl. Thus, this buffer solution resisted change in pH.

Does it make sense that the pH of the solution decreased upon the addition of a strong acid to the buffer?

PRACTICE EXERCISE

2. A buffer solution is prepared by mixing 500 mL of 0.600 *M* CH_3COOH with 500 mL of a 1.00 *M* CH_3COONa solution. What is the pH of the solution?

3. (a) Calculate the pH of a buffer system that is 0.0600 *M* HNO_2 and 0.160 *M* $NaNO_2$.
(b) What is the pH after 2.00 mL of 2.00 *M* NaOH are added to 1.00 L of this buffer?

Text Problems: 8, 10, 14

C. Preparing a buffer solution with a specific pH

To prepare a buffer solution with a specific pH, we need to consider the Henderson-Hasselbalch equation. For a derivation of this equation, see Section 17.2 of your text.

$$\text{pH} = \text{p}K_a + \log\frac{[\text{conjugate base}]}{[\text{acid}]}$$

where,

$$\text{p}K_a = -\log K_a$$

If the molar concentrations of the acid and its conjugate base are approximately equal, that is,

$$[\text{acid}] \cong [\text{conjugate base}]$$

then,

$$\log\frac{[\text{conjugate base}]}{[\text{acid}]} \cong 0$$

Substituting into the Henderson-Hasselbalch equation gives:

$$\text{pH} \cong \text{p}K_a$$

Thus, to prepare a buffer solution, we should choose a weak acid with a $\text{p}K_a$ value close to the desired pH. This choice not only gives the desired pH value for the buffer system, but also ensures that we have comparable amounts of the weak acid and its conjugate base present. Having pH $\cong$ $\text{p}K_a$ is a prerequisite for the buffer system to function effectively.

EXAMPLE 17.3

Which of the following mixtures is suitable for making a buffer solution with an optimum pH of about 9.2?

(a) CH_3COONa/CH_3COOH (b) NH_4Cl/NH_3 (c) HF/NaF
(d) $HNO_2/NaNO_2$ (e) NaCl/HCl

We can rule out choice (e) immediately, because it contains a strong acid and its conjugate base. This solution is not a buffer.

For the other choices we need to calculate the pK_a and compare it to the desired pH. The solutions with a pK_a values close to the desired pH will be effective buffers.

(a) $pK_a = 4.74$
(b) $pK_a = 9.26$
(c) $pK_a = 3.15$
(d) $pK_a = 3.35$

Thus, the only solution that would make an effective buffer at pH = 9.2 is choice (b), NH_4Cl/NH_3.

> **Tip:** We could have saved some time by recognizing that if the K_a value is less than 1×10^{-7}, the pK_a value will be greater than 7. The only choice that had a K_a value less than 1×10^{-7} was choice (b). Thus, it was the only choice that has a pK_a value greater than 7.

PRACTICE EXERCISE

4. How would you prepare a liter of an HF/F^- buffer at a pH of 2.85?

Text Problem: 16

PROBLEM TYPE 2: TITRATIONS

A. Strong acid–strong base titrations

The reaction between a strong acid (HNO_3) and a strong base (KOH) can be written as:

$$HNO_3\,(aq) + KOH\,(aq) \longrightarrow KNO_3\,(aq) + H_2O\,(l)$$

or in terms of the net ionic equation

$$H^+\,(aq) + OH^-\,(aq) \longrightarrow H_2O\,(l)$$

At the equivalence point of a titration, the $[H^+] = [OH^-]$. Thus, the only species present in solution other than water at the equivalence point is $KNO_3\,(aq)$. Potassium nitrate is a salt that does not undergo hydrolysis (see Problem Type 8, Chapter 16); therefore, the pH at the equivalence point of a strong acid–strong base titration is 7.

A typical strong acid–strong base titration problem involves calculating the pH at various points in the titration. Example 17.4 below illustrates this type of problem.

EXAMPLE 17.4

A 20.0 mL sample of 0.0200 *M* HNO_3 is titrated with 0.0100 *M* KOH.

(a) What is the pH at the equivalence point?
(b) How many mL of KOH are required to reach the equivalence point?
(c) What is the pH before any KOH is added?
(d) What will be the pH after 10.0 mL of KOH are added?
(e) What will be the pH after 45.0 mL of KOH are added?

(a) Since this is a strong acid–strong base titration, the pH at the equivalence point is 7.

(b) We worked this type of problem in Chapter 4 (see Problem Type 8, Titrations).

Step 1: In order to have the correct mole ratio to solve the problem, you must start with a balanced chemical equation.

$$HNO_3\,(aq) + KOH\,(aq) \longrightarrow KNO_3\,(aq) + H_2O\,(l)$$

Step 2: From the molarity and volume of the HNO_3 solution, you can calculate moles of HNO_3. Then, using the mole ratio from the balanced equation above, you can calculate moles of KOH.

20.0 mL = 0.0200 L

$$\text{mol KOH} = \frac{0.0200\ \text{mol HNO}_3}{1\ \text{L of solution}} \times (0.0200\ \text{L solution}) \times \frac{1\ \text{mol KOH}}{1\ \text{mol HNO}_3} = 4.00 \times 10^{-4}\ \text{mol KOH}$$

Step 3: Solve the molarity equation algebraically for liters of solution. Then, substitute in the moles of KOH and molarity of KOH to solve for volume of KOH.

$$M = \frac{\text{moles of solute}}{\text{liters of solution}}$$

$$\text{liters of solution} = \frac{\text{moles of solute}}{M}$$

$$\textbf{volume of KOH} = \frac{4.00 \times 10^{-4}\ \text{mol KOH}}{0.0100\ \text{mol / L}} = \mathbf{0.0400\ L = 40.0\ mL}$$

(c) We only have the strong acid, HNO_3, in solution before any KOH is added. Thus, we need to calculate the pH of a strong acid (Problem Type 4, Chapter 16).

Step 1: Strong acids ionize completely in solution. Let's write the reaction and set up a table to calculate the H^+ ion concentration in solution.

	$HNO_3\,(aq)$	$\longrightarrow$ $H^+\,(aq)$	+ $NO_3^-\,(aq)$
initial conc.	0.0200 *M*	0	0
conc. after ionization	0	0.0200 *M*	0.0200 *M*

Since there is a 1:1 mole ratio between H^+ and HNO_3, the H^+ concentration equals the HNO_3 concentration.

Step 2: Calculate the pH from the H^+ concentration.

$$\textbf{pH} = -\log[H^+] = -\log[0.0200\ M] = \mathbf{1.70}$$

(d) Any strong base added will react completely with the strong acid in solution. The reaction is:

$$HNO_3\,(aq) + KOH\,(aq) \longrightarrow KNO_3\,(aq) + H_2O\,(l)$$

Step 1: Calculate the number of moles of HNO_3 in 20.0 mL, and calculate the number of moles of KOH in 10.0 mL.

$$\text{mol } HNO_3 = 20.0\text{ mL soln} \times \frac{1.00\text{ L}}{1000\text{ mL}} \times \frac{0.0200\text{ mol } HNO_3}{1.00\text{ L soln}} = 4.00 \times 10^{-4}\text{ mol}$$

$$\text{mol KOH} = 10.0\text{ mL soln} \times \frac{1.00\text{ L}}{1000\text{ mL}} \times \frac{0.0100\text{ mol KOH}}{1.00\text{ L soln}} = 1.00 \times 10^{-4}\text{ mol}$$

Step 2: Set up a table showing the number of moles of HNO_3 and KOH before and after the reaction.

	$HNO_3\,(aq)$	+ $KOH\,(aq)$	$\longrightarrow$ $KNO_3\,(aq)$	+ $H_2O\,(l)$
initial mol	4.00×10^{-4}	1.00×10^{-4}	0	
mol after reaction	3.00×10^{-4}	0	1.00×10^{-4}	

Thus, 3.00×10^{-4} mol of the strong acid HNO_3 remain after the addition of 10.0 mL of KOH.

Step 3: The total volume of solution is now the sum of the volume of the acid solution and the volume of the base solution.

$$V_{soln} = V_{acid} + V_{base} = 20.0\text{ mL} + 10.0\text{ mL} = 30.0\text{ mL}$$

We can now calculate the concentration of H^+ in solution. Since HNO_3 ionizes completely, the number of moles of H^+ in solution is 3.00×10^{-4} mol. Calculate the molarity by dividing the moles by liters of solution.

30.0 mL = 0.0300 L

$$[H^+] = \frac{3.00 \times 10^{-4}\text{ mol}}{0.0300\text{ L}} = 0.0100\,M$$

Step 4: Calculate the pH from the H^+ concentration.

$$\mathbf{pH} = -\log[H^+] = -\log[0.0100\,M] = \mathbf{2.00}$$

(e) Start this part by following the same procedure as in part (d).

Step 1: Calculate the number of moles of KOH in 45.0 mL.

$$\text{mol KOH} = 45.0\text{ mL soln} \times \frac{1.00\text{ L}}{1000\text{ mL}} \times \frac{0.0100\text{ mol KOH}}{1.00\text{ L soln}} = 4.50 \times 10^{-4}\text{ mol}$$

Step 2: Set up a table showing the number of moles of HNO_3 and KOH before and after the reaction.

	$HNO_3\,(aq)$	+ $KOH\,(aq)$	$\longrightarrow$ $KNO_3\,(aq)$	+ $H_2O\,(l)$
initial mol	4.00×10^{-4}	4.50×10^{-4}	0	
mol after reaction	0	5.00×10^{-5}	4.00×10^{-4}	

We have passed the equivalence point of the titration. The only species of significance that remains in solution is the strong base, KOH. 5.00×10^{-5} mol of the strong base KOH remain after the addition of 45.0 mL of KOH.

Step 3: The total volume of solution is now the sum of the volume of the acid solution and the volume of the base solution.

$$V_{soln} = V_{acid} + V_{base} = 20.0\text{ mL} + 45.0\text{ mL} = 65.0\text{ mL}$$

We can now calculate the concentration of OH^- in solution. Since KOH ionizes completely, the number of moles of OH^- in solution is 5.00×10^{-5} mol. Calculate the molarity by dividing the moles by liters of solution.

65.0 mL = 0.0650 L

$$[OH^-] = \frac{5.00 \times 10^{-5}\text{ mol}}{0.0650\text{ L}} = 7.69 \times 10^{-4}\ M$$

Step 4: Calculate the pOH from the OH^- concentration.

$$\text{pOH} = -\log[OH^-] = -\log[7.69 \times 10^{-4}\ M] = 3.11$$

Step 5: Use the relationship, pH + pOH = 14.00, to calculate the pH of the solution.

$$\mathbf{pH} = 14.00 - 3.11 = \mathbf{10.89}$$

PRACTICE EXERCISE

5. Consider the titration of 25.0 mL of 0.250 *M* KOH with 0.100 *M* HCl.

(a) What is the pH at the equivalence point?
(b) How many mL of HCl are required to reach the equivalence point?
(c) What is the pH before any HCl is added?
(d) What will be the pH after 15.0 mL of HCl are added?
(e) What will be the pH after 75.0 mL of HCl are added?

Text Problems: 18, 20

B. Weak acid–strong base titrations

Consider the neutralization between nitrous acid (a weak acid) and sodium hydroxide (a strong base):

$$HNO_2\,(aq) + NaOH\,(aq) \longrightarrow NaNO_2\,(aq) + H_2O\,(l)$$

The net ionic equation is:

$$HNO_2\,(aq) + OH^-\,(aq) \longrightarrow NO_2^-\,(aq) + H_2O\,(l)$$

At the equivalence point of a titration, the $[HNO_2] = [OH^-]$. Thus, the major species present in solution other than water at the equivalence point are NO_2^- *(aq)* and Na^+ *(aq)*. NO_2^- is a weak base; it is the conjugate base of the weak acid, HNO_2 (see Problem Type 8, Chapter 16). The hydrolysis reaction is given by:

$$NO_2^-\,(aq) + H_2O\,(l) \rightleftharpoons HNO_2\,(aq) + OH^-\,(aq)$$

Therefore, at the equivalence point, the pH will be *greater than* 7 as a result of the excess OH^- formed.

C. Strong acid–weak base titrations

Consider the neutralization between ammonia (a weak base) and nitric acid (a strong acid):

$$NH_3\,(aq) + HNO_3\,(aq) \rightarrow NH_4NO_3\,(aq)$$

The net ionic equation is:

$$NH_3\,(aq) + H^+\,(aq) \rightarrow NH_4^+\,(aq)$$

At the equivalence point of a titration, the $[NH_3] = [H^+]$. Thus, the major species present in solution other than water at the equivalence point are NH_4^+ (aq) and NO_3^- (aq). NH_4^+ is a weak acid; it is the conjugate base of the weak base, NH_3 (see Problem Type 8, Chapter 16). The hydrolysis reaction is given by:

$$NH_4^+\,(aq) + H_2O\,(l) \rightleftharpoons NH_3\,(aq) + H_3O^+\,(aq)$$

Therefore, at the equivalence point, the pH will be *less than* 7 as a result of the excess H_3O^+ ions formed.

PROBLEM TYPE 3: CHOOSING SUITABLE ACID-BASE INDICATORS

The criterion for choosing an appropriate indicator for a given titration is whether the pH range over which the indicator changes color corresponds with the steep portion of the titration curve. If the indicator does not change color during this portion of the curve, it will not accurately identify the equivalence point. Table 17.2 of your text lists some common acid-base indicators and the pH ranges in which they change color.

Typically, the pH range for an indicator is equal to the pK_a of the indicator, plus or minus one pH unit.

$$\text{pH range} = pK_a \pm 1$$

EXAMPLE 17.5

Will bromophenol blue be a good choice as an indicator for a weak acid–strong base titration?

A typical titration curve for a weak acid–strong base titration is shown in Figure 17.4 of your text. The pH at the equivalence point is greater than 7. Checking Table 17.2, bromophenol blue changes color over the pH range, 3.0–4.6. This indicator would change color before the steep portion of the titration curve for a typical weak acid–strong base titration. Thus, bromophenol blue would *not* be a good choice as an indicator for a weak acid–strong base titration.

PRACTICE EXERCISE

6. Would methyl red be a suitable indicator for a titration that has a pH of 5.0 at the equivalence point?

Text Problem: 26

PROBLEM TYPE 4: SOLUBILITY EQUILIBRIA

Solubility equilibria typically involve salts of low solubility. For example, calcium phosphate, $[Ca_3(PO_4)_2]$, is practically insoluble in water. However, a very small amount of calcium phosphate will dissolve and dissociate completely into Ca^{2+} and PO_4^{3-} ions. An equilibrium will then be established between undissolved calcium phosphate and the ions in solution. Consider a saturated solution of calcium phosphate that is in contact with solid calcium phosphate. The solubility equilibrium can be represented as

$$Ca_3(PO_4)_2\,(s) \rightleftharpoons 3\,Ca^{2+}\,(aq) + 2\,PO_4^{3-}\,(aq)$$

We know from Chapter 15 that for heterogeneous equilibria, the concentration of a solid is a constant. Thus, we can write the equilibrium constant for the dissolution of $Ca_3(PO_4)_2$ as

$$K_{sp} = [Ca^{2+}]^3[PO_4^{3-}]^2$$

where K_{sp} is called the solubility product constant or simply the solubility product. In general, the **solubility product** of a compound is the product of the molar concentrations of the constituent ions, each raised to the power of its stoichiometric coefficient in the balanced equilibrium equation.

A. Calculating K_{sp} from molar solubility

Molar solubility is the number of moles of solute in 1 L of a saturated solution (moles per liter). The concentrations of the ions in the solubility product expression are also molar concentrations. From the molar solubility, we can calculate the molar concentrations of the ions in solution from the stoichiometry of the balanced equilibrium equation.

EXAMPLE 17.6

If the solubility of $Fe(OH)_2$ in water is 7.7×10^{-6} mol/L at a certain temperature, what is its K_{sp} value at that temperature?

Step 1: Write the equilibrium reaction. Then, from the equilibrium equation, write the solubility product expression.

$$Fe(OH)_2\,(s) \rightleftharpoons Fe^{2+}\,(aq) + 2\,OH^-\,(aq)$$

$$K_{sp} = [Fe^{2+}][OH^-]^2$$

Step 2: The molar solubility is the amount of $Fe(OH)_2$ that dissolves. From the stoichiometry of the equilibrium equation, you should find that:

$$[Fe^{2+}] = [Fe(OH)_2] = 7.7 \times 10^{-6}\,M$$

and

$$[OH^-] = 2[Fe(OH)_2] = 1.5 \times 10^{-5}\,M$$

Step 3: Substitute the equilibrium concentrations of Fe^{2+} and OH^- into the solubility product expression to calculate K_{sp}.

$$\mathbf{K_{sp}} = [Fe^{2+}][OH^-]^2 = (7.7 \times 10^{-6})(1.5 \times 10^{-5})^2 = \mathbf{1.7 \times 10^{-15}}$$

PRACTICE EXERCISE

7. At a certain temperature, the solubility of barium chromate ($BaCrO_4$) is 1.8×10^{-5} mol/L. What is the K_{sp} value at this temperature?

Text Problems: 34, 36, 40

B. Calculating solubility from K_{sp}

As in other equilibrium calculations, the expected concentrations at equilibrium can be calculated from a knowledge of the initial concentrations and the equilibrium constant (see Problem Type 5, Chapter 15). In these types of problems, it will be very helpful to recall that

equilibrium concentration = initial concentration ± the change due to reaction.

Example 17.7 illustrates this important type of calculation.

EXAMPLE 17.7

What is the molar solubility of silver phosphate (Ag_3PO_4) in water? $K_{sp} = 1.8 \times 10^{-18}$.

Step 1: Write the equilibrium reaction. Then, from the equilibrium equation, write the solubility product expression.

$$Ag_3PO_4\,(s) \rightleftharpoons 3\,Ag^+\,(aq) + PO_4^{3-}\,(aq)$$

$$K_{sp} = [Ag^+]^3[PO_4^{3-}]$$

Step 2: A certain amount of silver phosphate will dissociate in solution. Let's represent this amount as $-s$. Since one unit of Ag_3PO_4 yields 3 Ag^+ ions and one PO_4^{3-} ion, at equilibrium $[Ag^+]$ is $3s$ and $[PO_4^{3-}]$ is s. We summarize the changes in concentration as follows:

	$Ag_3PO_4\,(s)$	$\rightleftharpoons$ $3\,Ag^+\,(aq)$	+ $PO_4^{3-}\,(aq)$
Initial (*M*):		0	0
Change (*M*):	$-s$	$+3s$	$+s$
Equilibrium (*M*):		$3s$	s

Recall that the concentration of a pure solid does not enter into an equilibrium constant expression. Therefore, the concentration of Ag_3PO_4 is not important.

Step 3: Substitute the value of K_{sp} and the concentrations of Ag^+ and PO_4^{3-} in terms of s into the solubility product expression to solve for s, the molar solubility.

$$K_{sp} = [Ag^+]^3[PO_4^{3-}]$$

$$1.8 \times 10^{-18} = (3s)^3(s)$$

$$1.8 \times 10^{-18} = 27s^4$$

$$s = \text{molar solubility} = \mathbf{1.6 \times 10^{-5}\ mol/L}$$

The molar solubility indicates that 1.6×10^{-5} mol of Ag_3PO_4 will dissolve in 1 L of an aqueous solution.

PRACTICE EXERCISE

8. The K_{sp} value of silver carbonate (Ag_2CO_3) is 8.1×10^{-12} at 25°C. What is the solubility in g/L of silver carbonate in water at 25°C?

Text Problems: 38, 44

C. Predicting a precipitation reaction

To predict whether a precipitate will form, we must calculate the **ion product, *Q***. We will follow the same procedure outlined in Section 15.4 of the text or in Problem Type 4, Chapter 15. The ion product represents the molar concentrations of the ions raised to the power of their stoichiometric coefficients. Thus, for an aqueous solution containing Pb^{2+} and S^{2-} ions at 25°C,

$$Q = [Pb^{2+}]_0[S^{2-}]_0$$

The subscript 0 indicates that these are initial concentrations and do not necessarily correspond to equilibrium concentrations.

There are three possible relationships between Q and K_{sp}.

- $Q = K_{sp}$ $[Pb^{2+}]_0[S^{2-}]_0 = 3.4 \times 10^{-28}$ Since $Q = K_{sp}$, the reaction is already at equilibrium. The solution is saturated.

- $Q < K_{sp}$ $[Pb^{2+}]_0[S^{2-}]_0 < 3.4 \times 10^{-28}$ With $Q < K_{sp}$, the solution is not saturated. More solid could dissolve to produce more ions in solution. No precipitate will form.

- $Q > K_{sp}$ $[Pb^{2+}]_0[S^{2-}]_0 > 3.4 \times 10^{-28}$ With $Q > K_{sp}$, the solution is supersaturated. There are too many ions in solution. Some ions will combine to form PbS, which will precipitate out until the product of the ion concentrations is equal to 3.4×10^{-28}.

EXAMPLE 17.8

Predict whether a precipitate of PbI_2 will form when 200 mL of 0.015 *M* $Pb(NO_3)_2$ and 300 mL of 0.050 *M* NaI are mixed together. K_{sp} (PbI_2) = 1.4×10^{-8}.

The question asks whether the following reaction will occur:

$$Pb^{2+}(aq) + 2I^-(aq) \longrightarrow PbI_2(s)$$

To answer that question, we need to consider the following equilibrium:

$$PbI_2(s) \rightleftharpoons Pb^{2+}(aq) + I^-(aq)$$

Step 1: Calculate the concentrations of $Pb(NO_3)_2$ and NaI after mixing the two solutions together. This is a dilution problem. See Problem Type 6, Chapter 4. The total solution volume is 500 mL.

$$M_{initial}V_{initial} = M_{final}V_{final}$$

or,

$$M_{final} = \frac{M_{initial}V_{initial}}{V_{final}}$$

$$[Pb(NO_3)_2] = \frac{(0.015\ M)(200\ \text{mL})}{500\ \text{mL}} = 6.0 \times 10^{-3}\ M$$

$$[NaI] = \frac{(0.050\ M)(300\ \text{mL})}{500\ \text{mL}} = 0.030\ M$$

Both of these salts are soluble and ionize completely. We can write the concentrations of Pb^{2+} and I^- directly from the salt concentrations.

$$[Pb^{2+}] = 6.0 \times 10^{-3}\ M \quad \text{and} \quad [I^-] = 0.030\ M$$

Step 2: Substitute the concentrations of Pb^{2+} and I^- into the ion product expression to calculate Q. Then, compare Q to K_{sp} to determine if a precipitate will form.

$$PbI_2(s) \rightleftharpoons Pb^{2+}(aq) + I^-(aq)$$

$$Q = [Pb^{2+}]_0[I^-]_0^2$$

$$Q = (6.0 \times 10^{-3})(0.030)^2 = 5.4 \times 10^{-6}$$

Comparing Q to K_{sp}, we find the $Q > K_{sp}$. This means that the solution is supersaturated. There are too many ions in solution. Some ions will combine to form PbI_2, which will precipitate out until the product of the ion concentrations is equal to K_{sp}, 1.4×10^{-8}.

PRACTICE EXERCISE

9. Will a precipitate of MgF_2 form when 6.00×10^2 mL of a solution that is 2.0×10^{-4} M in $MgCl_2$ is added to 3.00×10^2 mL of a 1.1×10^{-2} M NaF solution? K_{sp} (MgF_2) = 6.6×10^{-9}.

Text Problem: 84

D. The effect of a common ion on solubility

Consider the slightly soluble salt copper (I) iodide. We can write its equilibrium reaction in water as follows:

$$\text{CuI}\ (s) \rightleftharpoons \text{Cu}^+\ (aq) + \text{I}^-\ (aq)$$

Now, suppose that instead of dissolving CuI in water, we attempted to dissolve a certain quantity of CuI in a potassium iodide (KI) solution. Would the presence of KI affect the solubility of the copper(I) iodide?

Remember, that KI is a soluble salt, so it will ionize completely into K^+ and I^- ions.

$$\text{KI}\ (aq) \rightarrow \text{K}^+\ (aq) + \text{I}^-\ (aq)$$

The I^- ions from KI *will affect* the copper(I) iodide equilibrium. Refer to Problem Type 6, Chapter 15, which discusses LeChatelier's principle. LeChatelier's principle states that when an external stress is applied to a system at equilibrium, the system adjusts in such a way that the stress is partially offset. In this system, the stress is additional I^- ions in solution (from KI). The CuI equilibrium will shift to the left to remove some of the additional I^- ions, decreasing the solubility of CuI.

In summary, the effect of adding a **common ion**, in this case I^-, is to **decrease** the solubility of the salt (CuI) in solution.

Calculating the solubility of a salt when a common ion is present is very similar to calculating the solubility from K_{sp} (Problem Type 4B). The only difference is that the initial concentration of the common ion is *not* zero. The initial concentration of the common ion comes from the soluble salt that is present in solution. Example 17.9 below illustrates this type of problem.

EXAMPLE 17.9

What is the molar solubility of $PbCl_2$ in a 0.50 M NaCl solution? K_{sp} ($PbCl_2$) = 2.4×10^{-4}.

This problem is worked in a similar manner to Example 17.8.

Step 1: Write the equilibrium reaction. Then, from the equilibrium equation, write the solubility product expression.

$$\text{PbCl}_2\ (s) \rightleftharpoons \text{Pb}^{2+}\ (aq) + 2\ \text{Cl}^-\ (aq)$$

$$K_{sp} = [\text{Pb}^{2+}][\text{Cl}^-]^2$$

Step 2: A certain amount of lead(II) chloride will dissociate in solution. Let's represent this amount as $-s$. Since one unit of $PbCl_2$ yields one Pb^{2+} ion, at equilibrium $[Pb^{2+}]$ is s. There are two sources of Cl^- ions. NaCl ionizes completely to give a Cl^- concentration of 0.50 *M*. Also, one unit of $PbCl_2$ yields two Cl^- ions, so at equilibrium this will add $2s$ to the Cl^- concentration. We summarize the changes in concentration as follows:

	$PbCl_2(s) \rightleftharpoons$	$Pb^{2+}(aq)$ +	$2\,Cl^-(aq)$
Initial (*M*):		0	0.50
Change (*M*):	$-s$	$+s$	$+2s$
Equilibrium (*M*):		s	$0.50 + s$

Recall that the concentration of a pure solid does not enter into an equilibrium constant expression. Therefore, the concentration of $PbCl_2$ is not important.

Step 3: Substitute the value of K_{sp} and the concentrations of Pb^{2+} and Cl^- into the solubility product expression to solve for s, the molar solubility.

$$K_{sp} = [Pb^{2+}][Cl^-]^2$$

$$2.4 \times 10^{-4} = (s)(0.50 + 2s)^2$$

Let's assume that $0.50 >> 2s$. This is a valid assumption because K_{sp} is small.

$$2.4 \times 10^{-4} \cong (s)(0.50)^2$$

$$s = \text{molar solubility} = \mathbf{9.6 \times 10^{-4}\ mol/L}$$

Checking the validity of the assumption,

$$\frac{2(9.6 \times 10^{-4})}{0.50} \times 100\% = 0.38\% < 5\%$$

The molar solubility of lead(II) chloride in pure water is 3.9×10^{-2} *M*. Does it make sense that the molar solubility decreased in 0.50 *M* NaCl?

PRACTICE EXERCISE

10. What is the molar solubility of Ag_3PO_4 in 0.20 *M* $AgNO_3$? $K_{sp}(Ag_3PO_4) = 1.8 \times 10^{-18}$.

Text Problems: 46, 48

PROBLEM TYPE 5: COMPLEX ION EQUILIBRIA AND SOLUBILITY

We can define a **complex ion** as an ion containing a central metal cation bonded to one or more molecules or ions. Transition metals have a particular tendency to form complex ions. For example, silver chloride (AgCl) is insoluble in water. However, when aqueous ammonia is added to AgCl in water, the silver chloride dissolves to form $Ag(NH_3)_2^+(aq)$ and $Cl^-(aq)$. The equilibrium equation is:

$$Ag^+(aq) + 2\,NH_3(aq) \rightleftharpoons Ag(NH_3)_2^+(aq)$$

A measure of the tendency of a metal ion to form a particular complex ion is given by the **formation constant K_f** (also called the stability constant). The formation constant is simply the equilibrium constant for complex ion formation. The larger the K_f value, the more stable the complex ion. Table 17.5 of the text lists the formation constants of a number of complex ions.

The formation constant expression for the formation of $Ag(NH_3)_2^+$ can be written as:

$$K_f = \frac{[Ag(NH_3)_2^+]}{[Ag^+][NH_3]^2} = 1.5 \times 10^7$$

The very large value of K_f in this case indicates the great stability of the complex ion $[Ag(NH_3)_2^+]$ in solution. Thus, the insoluble silver chloride dissolves to form the stable complex ion, $Ag(NH_3)_2^+$.

A typical complex ion equilibrium problem involves using the formation constant to calculate the equilibrium concentrations of species in solution. See Example 17.10 below.

EXAMPLE 17.10

Calculate the concentration of free Ag^+ ions in a solution formed by adding 0.20 mol of $AgNO_3$ to 1.0 L of 1.0 *M* NaCN. Assume no volume change upon addition of $AgNO_3$.

In solution, Ag^+ ions will complex with CN^- ions. The concentration of Ag^+ will be determined by the following equilibrium:

$$Ag^+(aq) + 2\,CN^-(aq) \rightleftharpoons Ag(CN)_2^- \qquad K_f = 1.0 \times 10^{21}$$

Since K_f is so large, this equilibrium lies far to the right. We can safely assume that all the Ag^+ reacts to form 0.20 mol of $Ag(CN)_2^-$. There is more than enough CN^- available.

Step 1: To find the concentration of free Ag^+ at equilibrium, use the formation constant expression.

$$K_f = \frac{[Ag(CN)_2^-]}{[Ag^+][CN^-]^2}$$

Rearranging,

$$[Ag^+] = \frac{[Ag(CN)_2^-]}{K_f[CN^-]^2}$$

We already know that the concentration of $Ag(CN)_2^-$ is 0.20 *M*. We can calculate the concentration of CN^- from the stoichiometry of the reaction.

$$[CN^-] = [CN^-]_0 - 2[Ag^+] = 1.0\,M - 2(0.20\,M) = 0.60\,M$$

Step 2: Substitute the equilibrium concentrations into the formation constant expression to calculate the equilibrium concentration of Ag^+.

$$[Ag^+] = \frac{[Ag(CN)_2^-]}{K_f[CN^-]^2} = \frac{0.20}{(1.0 \times 10^{21})(0.60)^2} = \mathbf{5.6 \times 10^{-22}}\ \boldsymbol{M}$$

This concentration corresponds to only *three* Ag^+ ions per 10 mL of solution!

PRACTICE EXERCISE

11. If 0.0100 mol of $Cu(NO_3)_2$ are dissolved in 1.00 L of 1.00 *M* NH_3, what are the concentrations of Cu^{2+}, $Cu(NH_3)_4^{2+}$, and NH_3 at equilibrium? Assume no volume change upon addition of the $Cu(NO_3)_2$ to the ammonia solution.

Text Problems: 52, 54, 56

ANSWERS TO PRACTICE EXERCISES

1. Both (a) and (c). **2.** pH = 4.96 **3.** **(a)** pH = 3.77 **(b)** pH = 3.81

4. To prepare a pH = 2.85 buffer you would need to use a ratio of [F^-] to [HF] of 0.50, or 1 to 2.

One way to prepare this buffer would be to add 0.50 mol of NaF (21 g) to 1.0 L of 1.0 *M* HF (assuming no change in volume).

5. **(a)** pH = 7 **(b)** 62.5 mL of HCl **(c)** pH = 13.40
(d) pH = 13.08 **(e)** pH = 1.90

6. Yes, methyl red would be a suitable indicator.

7. $K_{sp} = 3.2 \times 10^{-10}$ **8.** solubility = 0.035 g/L **9.** No, a precipitate will *not* form.

10. s = molar solubility = 2.3×10^{-16} *M* **11.** [NH_3] = 0.96 *M*
[$Cu(NH_3)_4^{2+}$] = 0.0100 *M*
[Cu^{2+}] = 2.4×10^{-16} *M*

SOLUTIONS TO SELECTED TEXT PROBLEMS

17.6 Identifying buffer systems, Problem Type 1A.

(a) HNO_2 (nitrous acid) is a weak acid, and its conjugate base, NO_2^- (nitrite ion, the anion of the salt KNO_2), is a weak base. Therefore, this is a buffer system.

(b) H_2SO_4 (sulfuric acid) is a strong acid. A buffer is a solution containing both a weak acid and a weak base. Therefore, this is *not* a buffer system.

(c) HCOOH (formic acid) is a weak acid, and its conjugate base, $HCOO^-$ (formate ion, the anion of the salt HCOOK), is a weak base. Therefore, this is a buffer system.

17.8 Calculating the pH of a buffer system, Problem Type 1B.

(a) This problem is greatly simplified because the concentration of the weak acid (acetic acid) is equal to the concentration of its conjugate base (acetate ion). Let's set up a table of initial concentrations, change in concentrations, and equilibrium concentrations.

	$CH_3COOH\ (aq) \rightleftharpoons$	$H^+\ (aq)$ +	$CH_3COO^-\ (aq)$
Initial	$2.0\ M$	0	$2.0\ M$
Change	$-x$	$+x$	$+x$
Equilibrium	$2.0\ M - x$	x	$2.0\ M + x$

$$K_a = \frac{[H^+][CH_3COO^-]}{[CH_3COOH]}$$

$$K_a = \frac{[H^+](2.0 + x)}{(2.0 - x)} \cong \frac{[H^+](2.0)}{2.0}$$

$$K_a = [H^+]$$

Taking the −log of both sides,

$$pK_a = pH$$

Thus, for a buffer system in which the [weak acid] = [weak base],

$$pH = pK_a$$

$$\mathbf{pH} = -\log(1.8 \times 10^{-5}) = \mathbf{4.74}$$

(b) Similar to part (a),

$$\mathbf{pH} = pK_a = \mathbf{4.74}$$

Buffer (a) will be a more effective buffer because the concentrations of acid and base components are ten times higher than those in (b). Thus, buffer (a) can neutralize 10 times more added acid or base compared to buffer (b).

17.10 *Step 1:* Write the equilibrium that occurs between $H_2PO_4^-$ and HPO_4^{2-}. Set up a table relating the initial concentrations, the change in concentration to reach equilibrium, and the equilibrium concentrations.

	$H_2PO_4^-(aq)$	$\rightleftharpoons$ $H^+(aq)$	+ $HPO_4^{2-}(aq)$
Initial (M):	0.15	0	0.10
Change (M):	$-x$	$+x$	$+x$
Equilibrium (M):	$0.15 - x$	x	$0.10 + x$

Step 2: Write the ionization constant expression in terms of the equilibrium concentrations. Knowing the value of the equilibrium constant (K_a), solve for x.

$$K_a = \frac{[H^+][HPO_4^{2-}]}{[H_2PO_4^-]}$$

You can look up the K_a value for dihydrogen phosphate in Table 16.4 of your text.

$$6.2 \times 10^{-8} = \frac{(x)(0.10 + x)}{(0.15 - x)}$$

$$6.2 \times 10^{-8} \cong \frac{(x)(0.10)}{(0.15)}$$

$$x = [H^+] = 9.3 \times 10^{-8}\ M$$

Step 3: Having solved for the $[H^+]$, calculate the pH of the solution.

$$\mathbf{pH} = -\log[H^+] = -\log(9.3 \times 10^{-8}) = \mathbf{7.03}$$

17.12 We can use the Henderson-Hasselbalch equation to calculate the ratio $[HCO_3^-]/[H_2CO_3]$. The Henderson-Hasselbalch equation is:

$$pH = pK_a + \log\frac{[\text{conjugate base}]}{[\text{acid}]}$$

For the buffer system of interest, HCO_3^- is the conjugate base of the acid, H_2CO_3. We can write:

$$pH = 7.40 = -\log(4.2 \times 10^{-7}) + \log\frac{[HCO_3^-]}{[H_2CO_3]}$$

$$7.40 = 6.38 + \log\frac{[HCO_3^-]}{[H_2CO_3]}$$

The [conjugate base]/[acid] ratio is:

$$\log\frac{[HCO_3^-]}{[H_2CO_3]} = 7.40 - 6.38 = 1.02$$

$$\frac{\mathbf{[HCO_3^-]}}{\mathbf{[H_2CO_3]}} = 10^{1.02} = \mathbf{1.0 \times 10^1}$$

The buffer should be more effective against an added acid because ten times more base is present compared to acid. Note that a pH of 7.40 is only a two significant figure number (Why?); the final result should only have two significant figures.

17.14 As calculated in Problem 17.8, the pH of this buffer system is equal to pK_a.

$$\mathbf{pH} = pK_a = -\log(1.8 \times 10^{-5}) = \mathbf{4.74}$$

(a) The added NaOH will react completely with the acid component of the buffer, CH_3COOH. NaOH ionizes completely; therefore, 0.080 mol of OH^- are added to the buffer.

Step 1: The neutralization reaction is:

	$CH_3COOH\,(aq)$	+ $OH^-\,(aq)$	$\longrightarrow$ $CH_3COO^-\,(aq)$	+ $H_2O\,(l)$
Initial (mol)	1.00	0.080	1.00	
After reaction (mol)	0.92	0	1.08	

Step 2: Now, the acetic acid equilibrium is reestablished. Since the volume of the solution is 1.00 L, we can convert directly from moles to molar concentration.

	$CH_3COOH\,(aq)$	$\rightleftharpoons$ $H^+\,(aq)$	+ $CH_3COO^-\,(aq)$
Initial (*M*)	0.92	0	1.08
Change (*M*)	$-x$	$+x$	$+x$
Equilibrium (*M*)	$0.92 - x$	x	$1.08 + x$

Write the K_a expression, then solve for x.

$$K_a = \frac{[H^+][CH_3COO^-]}{[CH_3COOH]}$$

$$1.8 \times 10^{-5} = \frac{(x)(1.08 + x)}{(0.92 - x)} \cong \frac{x(1.08)}{0.92}$$

$$x = [H^+] = 1.5 \times 10^{-5}\,M$$

Step 3: Having solved for the $[H^+]$, calculate the pH of the solution.

$$\mathbf{pH} = -\log[H^+] = -\log(1.5 \times 10^{-5}) = \mathbf{4.82}$$

The pH of the buffer increased from 4.74 to 4.82 upon addition of 0.080 mol of strong base.

(b) The added acid will react completely with the base component of the buffer, CH_3COO^-. HCl ionizes completely; therefore, 0.12 mol of H^+ ion are added to the buffer

Step 1: The neutralization reaction is:

	$CH_3COO^-\,(aq)$	+ $H^+\,(aq)$	$\longrightarrow$ $CH_3COOH\,(aq)$
Initial (mol)	1.00	0.12	1.00
After reaction (mol)	0.88	0	1.12

Step 2: Now, the acetic acid equilibrium is reestablished. Since the volume of the solution is 1.00 L, we can convert directly from moles to molar concentration.

	$CH_3COOH\,(aq)$	$\rightleftharpoons$ $H^+\,(aq)$	+ $CH_3COO^-\,(aq)$
Initial (*M*)	1.12	0	0.88
Change (*M*)	$-x$	$+x$	$+x$
Equilibrium (*M*)	$1.12 - x$	x	$0.88 + x$

Write the K_a expression, then solve for x.

$$K_a = \frac{[H^+][CH_3COO^-]}{[CH_3COOH]}$$

$$1.8 \times 10^{-5} = \frac{(x)(0.88 + x)}{(1.12 - x)} \cong \frac{x(0.88)}{1.12}$$

$$x = [H^+] = 2.3 \times 10^{-5}\ M$$

Step 3: Having solved for the $[H^+]$, calculate the pH of the solution.

$$\mathbf{pH} = -\log[H^+] = -\log(2.3 \times 10^{-5}) = \mathbf{4.64}$$

The pH of the buffer decreased from 4.74 to 4.64 upon addition of 0.12 mol of strong acid.

17.16 Preparing a buffer solution with a specific pH, Problem Type 1C.

Recall that to prepare a solution of a desired pH, we should choose a weak acid with a pK_a value close to the desired pH.

Calculating the pK_a for each acid:

For HA, $\quad \mathbf{p}\boldsymbol{K}_\mathbf{a} = -\log(2.7 \times 10^{-3}) = \mathbf{2.57}$

For HB, $\quad \mathbf{p}\boldsymbol{K}_\mathbf{a} = -\log(4.4 \times 10^{-6}) = \mathbf{5.36}$

For HC, $\quad \mathbf{p}\boldsymbol{K}_\mathbf{a} = -\log(2.6 \times 10^{-9}) = \mathbf{8.59}$

The buffer solution with a pK_a closest to the desired pH is HC. Thus, **HC** is the best choice to prepare a buffer solution with pH = 8.60.

17.18 The neutralization reaction is:

$$2\ KOH\ (aq) + H_2A\ (aq) \longrightarrow K_2A\ (aq) + 2\ H_2O\ (l)$$

The number of moles of H_2A reacted is:

$$11.1\ \text{mL KOH} \times \left(\frac{1.00\ \text{mol KOH}}{1000\ \text{mL}}\right)\left(\frac{1\ \text{mol } H_2A}{2\ \text{mol KOH}}\right) = 5.55 \times 10^{-3}\ \text{mol } H_2A$$

We know that 0.500 g of the diprotic acid were reacted (1/10 of the 250 mL was tested). Divide the number of grams by the number of moles to calculate the molar mass.

$$\mathcal{M}(\mathbf{H_2A}) = \frac{0.500\ \text{g } H_2A}{5.55 \times 10^{-3}\ \text{mol } H_2A} = \mathbf{90.1\ g/mol}$$

17.20 **(a)** Since the acid is monoprotic, the moles of acid equals the moles of base added.

$$HA\ (aq) + NaOH\ (aq) \longrightarrow NaA\ (aq) + H_2O\ (l)$$

$$\text{Moles acid} = 18.4\ \text{mL} \times \left(\frac{1\ \text{L}}{1000\ \text{mL}}\right) \times \left(\frac{0.0633\ \text{mol}}{1\ \text{L}}\right) = 0.00116\ \text{mol}$$

We know the mass of the unknown acid in grams and the number of moles of the unknown acid.

$$\textbf{Molar mass} = \frac{0.1276\text{ g}}{0.00116\text{ mol}} = \mathbf{1.10 \times 10^2\ g/mol}$$

(b) The number of moles of NaOH in 10.0 mL of solution is

$$10.0\ \text{mL} \times \left(\frac{1\ \text{L}}{1000\ \text{mL}}\right) \times \left(\frac{0.0633\ \text{mol}}{1\ \text{L}}\right) = 6.33 \times 10^{-4}\ \text{mol}$$

The neutralization reaction is:

	HA (*aq*) +	NaOH (*aq*) ⟶	NaA (*aq*) +	H_2O (*l*)
Initial (mol)	0.00116	6.33×10^{-4}	0	
After reaction (mol)	5.3×10^{-4}	0	6.33×10^{-4}	

Now, the weak acid equilibrium will be reestablished. The total volume of solution is 35.0 mL.

$$[\text{HA}] = \frac{5.3 \times 10^{-4}\ \text{mol}}{0.035\ \text{L}} = 0.015\ M$$

$$[\text{A}^-] = \frac{6.33 \times 10^{-4}\ \text{mol}}{0.035\ \text{L}} = 0.0181\ M$$

We can calculate the $[H^+]$ from the pH.

$$[\text{H}^+] = 10^{-\text{pH}} = 10^{-5.87} = 1.35 \times 10^{-6}\ M$$

	HA (*aq*) ⇌	H^+ (*aq*) +	A^- (*aq*)
Initial (*M*)	0.015	0	0.0181
Change (*M*)	-1.35×10^{-6}	$+1.35 \times 10^{-6}$	$+1.35 \times 10^{-6}$
Equilibrium (*M*)	0.015	1.35×10^{-6}	0.0181

Substitute the equilibrium concentrations into the equilibrium constant expression to solve for K_a.

$$\boldsymbol{K_a} = \frac{[\text{H}^+][\text{A}^-]}{[\text{HA}]} = \frac{(1.35 \times 10^{-6})(0.0181)}{0.015} = \mathbf{1.6 \times 10^{-6}}$$

17.26 The weak acid equilibrium is

$$\text{HIn}\ (aq) \rightleftharpoons \text{H}^+\ (aq) + \text{In}^-\ (aq)$$

We can write a K_a expression for this equilibrium.

$$K_a = \frac{[\text{H}^+][\text{In}^-]}{[\text{HIn}]}$$

Rearranging,

$$\frac{[\text{HIn}]}{[\text{In}^-]} = \frac{[\text{H}^+]}{K_a}$$

From the pH, we can calculate the H^+ concentration.

$$[H^+] = 10^{-pH} = 10^{-4} = 1.0 \times 10^{-4}\ M$$

$$\frac{\mathbf{[HIn]}}{\mathbf{[In^-]}} = \frac{[H^+]}{K_a} = \frac{1.0 \times 10^{-4}}{1.0 \times 10^{-6}} = \mathbf{100}$$

Since the concentration of HIn is 100 times greater than the concentration of In^-, the color of the solution will be that of HIn, the nonionized formed. The color of the solution will be **red**.

17.34 Calculating K_{sp} from molar solubility, Problem Type 4A.

In each case, we first calculate the number of moles of compound dissolved in one liter of solution (the molar solubility).

(a) $$\left(\frac{7.3 \times 10^{-2}\ \text{g SrF}_2}{1\ \text{L soln}}\right) \times \left(\frac{1\ \text{mol SrF}_2}{125.6\ \text{g SrF}_2}\right) = 5.8 \times 10^{-4}\ \text{mol / L}$$

Step 1: Write the equilibrium reaction. Then, from the equilibrium equation, write the solubility product expression.

$$SrF_2\,(s) \rightleftharpoons Sr^{2+}\,(aq) + 2\,F^-\,(aq)$$

$$K_{sp} = [Sr^{2+}][F^-]^2$$

Step 2: The molar solubility is the amount of SrF_2 that dissolves. From the stoichiometry of the equilibrium equation, you should find that

$$[Sr^{2+}] = [SrF_2] = 5.8 \times 10^{-4}\ M$$

and

$$[F^-] = 2[SrF_2] = 1.16 \times 10^{-3}\ M$$

Step 3: Substitute the equilibrium concentrations of Sr^{2+} and F^- into the solubility product expression to calculate K_{sp}.

$$\mathbf{K_{sp}} = [Sr^{2+}][F^-]^2 = (5.8 \times 10^{-4})(1.16 \times 10^{-3})^2 = \mathbf{7.8 \times 10^{-10}}$$

(b) $$\left(\frac{6.7 \times 10^{-3}\ \text{g Ag}_3\text{PO}_4}{1\ \text{L soln}}\right) \times \left(\frac{1\ \text{mol Ag}_3\text{PO}_4}{418.7\ \text{g Ag}_3\text{PO}_4}\right) = 1.6 \times 10^{-5}\ \text{mol / L}$$

(b) is solved in a similar manner to (a)

The equilibrium equation is:

$$Ag_3PO_4\,(s) \rightleftharpoons 3\,Ag^+\,(aq) + PO_4^{3-}\,(aq)$$

$$\mathbf{K_{sp}} = [Ag^+]^3[PO_4^{3-}] = [3 \times (1.6 \times 10^{-5})]^3(1.6 \times 10^{-5}) = \mathbf{1.8 \times 10^{-18}}$$

17.36 First, we can convert the solubility of MX in g/L to mol/L.

$$\left(\frac{4.63 \times 10^{-3}\ \text{g MX}}{1\ \text{L soln}}\right) \times \left(\frac{1\ \text{mol MX}}{346\ \text{g MX}}\right) = 1.34 \times 10^{-5}\ \text{mol/L} = s\ \text{(molar solubility)}$$

The equilibrium reaction is:

$$MX\ (s) \rightleftharpoons M^{n+}\ (aq) + X^{n-}\ (aq)$$

Since the mole ratio of MX to each of the ions is 1:1, the equilibrium concentrations of each of the ions can also be represented by s. Solving for K_{sp},

$$\mathbf{K_{sp}} = [M^{n+}][X^{n-}] = s^2 = (1.34 \times 10^{-5})^2 = \mathbf{1.80 \times 10^{-10}}$$

17.38 Calculating solubility from K_{sp}, Problem Type 4B.

Step 1: Write the equilibrium reaction. Then, from the equilibrium equation, write the solubility product expression.

$$CaF_2\ (s) \rightleftharpoons Ca^{2+}\ (aq) + 2\ F^-\ (aq)$$

$$K_{sp} = [Ca^{2+}][F^-]^2$$

Step 2: A certain amount of calcium fluoride will dissociate in solution. Let's represent this amount as $-s$. Since one unit of CaF_2 yields one Ca^{2+} ion and two F^- ions, at equilibrium $[Ca^{2+}]$ is s and $[F^-]$ is $2s$. We summarize the changes in concentration as follows:

	$CaF_2\ (s)$	$\rightleftharpoons$ $Ca^{2+}\ (aq)$	+ $2\ F^-\ (aq)$
Initial (M):		0	0
Change (M):	$-s$	$+s$	$+2s$
Equilibrium (M):		s	$2s$

Recall, that the concentration of a pure solid does not enter into an equilibrium constant expression. Therefore, the concentration of CaF_2 is not important.

Step 3: Substitute the value of K_{sp} and the concentrations of Ca^{2+} and F^- in terms of s into the solubility product expression to solve for s, the molar solubility.

$$K_{sp} = [Ca^{2+}][F^-]^2$$

$$4.0 \times 10^{-11} = (s)(2s)^2$$

$$4.0 \times 10^{-11} = 4s^3$$

$$s = \text{molar solubility} = \mathbf{2.2 \times 10^{-4}\ mol/L}$$

The molar solubility indicates that 2.2×10^{-4} mol of CaF_2 will dissolve in 1 L of an aqueous solution.

17.40 First we can calculate the OH^- concentration from the pH.

$$pOH = 14.00 - pH$$

$$pOH = 14.00 - 9.68 = 4.32$$

$$[OH^-] = 10^{-pOH} = 10^{-4.32} = 4.8 \times 10^{-5}\ M$$

The equilibrium equation is:

$$MOH\ (s) \rightleftharpoons M^+\ (aq) + OH^-\ (aq)$$

From the balanced equation we know that $[M^+] = [OH^-]$

$$K_{sp} = [M^+][OH^-] = (4.8 \times 10^{-5})^2 = \mathbf{2.3 \times 10^{-9}}$$

17.42 The net ionic equation is:

$$Sr^{2+}(aq) + 2F^-(aq) \longrightarrow SrF_2(s)$$

Let's find the limiting reagent in the precipitation reaction.

$$\text{Moles } F^- = 75\ \text{mL} \times \left(\frac{1\ \text{L}}{1000\ \text{mL}}\right)\left(\frac{0.060\ \text{mol}}{1\ \text{L}}\right) = 0.0045\ \text{mol}$$

$$\text{Moles } Sr^{2+} = 25\ \text{mL} \times \left(\frac{1\ \text{L}}{1000\ \text{mL}}\right)\left(\frac{0.15\ \text{mol}}{1\ \text{L}}\right) = 0.0038\ \text{mol}$$

From the stoichiometry of the balanced equation, twice as many moles of F^- are required to react with Sr^{2+}. This would require 0.0076 mol of F^-, but we only have 0.0045 mol. Thus, F^- is the limiting reagent.

Let's assume that the above reaction goes to completion. Then, we will consider the equilibrium that is established when SrF_2 partially dissociates into ions.

	$Sr^{2+}(aq)$	$+\ 2F^-(aq)$	$\longrightarrow SrF_2(s)$
Initial (mol)	0.0038	0.0045	0
Change (mol)	−0.0023	−0.0045	+0.0023
After reaction (mol)	0.0015	0	0.0023

Now, let's establish the equilibrium reaction. The total volume of the solution is 100 mL = 0.100 L. Divide the above moles by 0.100 L to convert to molar concentration.

	$SrF_2(s)$	$\rightleftharpoons Sr^{2+}(aq)$	$+\ 2F^-(aq)$
Initial (M)	0.023	0.015	0
Change (M)	$-x$	$+x$	$+2x$
Equilibrium (M)	$0.023 - x$	$0.015 + x$	$2x$

Write the solubility product expression, then solve for x.

$$K_{sp} = [Sr^{2+}][F^-]^2$$

$$2.0 \times 10^{-10} = (0.015 + x)(2x)^2 \cong (0.015)(2x)^2$$

$$x = 5.8 \times 10^{-5}\ M$$

$$[\mathbf{F^-}] = 2x = \mathbf{1.2 \times 10^{-4}\ M}$$

$$[\mathbf{Sr^{2+}}] = 0.015 + x = \mathbf{0.015\ M}$$

Both sodium ions and nitrate ions are spectator ions and therefore do not enter into the precipitation reaction.

$$[\mathbf{NO_3^-}] = \frac{2(0.0038)\ \text{mol}}{0.10\ \text{L}} = \mathbf{0.076\ M}$$

$$[\mathbf{Na^+}] = \frac{0.0045\ \text{mol}}{0.10\ \text{L}} = \mathbf{0.045\ M}$$

17.46 The effect of a common ion on solubility, Problem Type 4D.

(a) Set up a table to find the equilibrium concentrations in pure water.

	$PbBr_2(s) \rightleftharpoons$	$Pb^{2+}(aq)$	+ $2\,Br^-(aq)$
Initial (M)		0	0
Change (M)	$-s$	$+s$	$+2s$
Equilibrium (M)		s	$2s$

$$K_{sp} = [Pb^{2+}][Br^-]^2$$

$$8.9 \times 10^{-6} = (s)(2s)^2$$

$$s = \text{molar solubility} = \mathbf{0.013\ M}$$

(b) Set up a table to find the equilibrium concentrations in 0.20 M KBr. KBr is a soluble salt that ionizes completely giving a initial concentration of Br^- = 0.20 M.

	$PbBr_2(s) \rightleftharpoons$	$Pb^{2+}(aq)$	+ $2\,Br^-(aq)$
Initial (M)		0	0.20
Change (M)	$-s$	$+s$	$+2s$
Equilibrium (M)		s	$0.20 + 2s$

$$K_{sp} = [Pb^{2+}][Br^-]^2$$

$$8.9 \times 10^{-6} = (s)(0.20 + 2s)^2$$

$$8.9 \times 10^{-6} \cong (s)(0.20)^2$$

$$s = \text{molar solubility} = \mathbf{2.2 \times 10^{-4}\ M}$$

Thus, the molar solubility of $PbBr_2$ is reduced from 0.013 M to 2.2×10^{-4} M as a result of the common ion (Br^-) effect.

(c) Set up a table to find the equilibrium concentrations in 0.20 M $Pb(NO_3)_2$. $Pb(NO_3)_2$ is a soluble salt that dissociates completely giving a initial concentration of Pb^{2+} = 0.20 M.

	$PbBr_2(s) \rightleftharpoons$	$Pb^{2+}(aq)$	+ $2\,Br^-(aq)$
Initial (M):		0.20	0
Change (M):	$-s$	$+s$	$+2s$
Equilibrium (M):		$0.20 + s$	$2s$

$$K_{sp} = [Pb^{2+}][Br^-]^2$$

$$8.9 \times 10^{-6} = (0.20 + s)(2s)^2$$

$$8.9 \times 10^{-6} \cong (0.20)(2s)^2$$

$$s = \text{molar solubility} = \mathbf{3.3 \times 10^{-3}\ M}$$

Thus, the molar solubility of $PbBr_2$ is reduced from 0.013 M to 3.3×10^{-3} M as a result of the common ion (Pb^{2+}) effect.

17.48 The equilibrium reaction is:

$$BaSO_4\,(s) \rightleftharpoons Ba^{2+}\,(aq) + SO_4^{2-}\,(aq)$$

For both parts of the problem:

$$K_{sp} = [Ba^{2+}][SO_4^{2-}] = 1.1 \times 10^{-10}$$

(a) In pure water, let $[Ba^{2+}] = [SO_4^{2-}] = s$

$$K_{sp} = [Ba^{2+}][SO_4^{2-}]$$

$$1.1 \times 10^{-10} = s^2$$

$$\mathbf{s = 1.0 \times 10^{-5}\,\mathit{M}}$$

The molar solubility of $BaSO_4$ in pure water is 1.0×10^{-5} mol/L.

(b) Assuming the molar solubility of $BaSO_4$ to be s, then

$$[Ba^{2+}] = s\,M \quad \text{and} \quad [SO_4^{2-}] = (1.0 + s)\,M \cong 1.0\,M$$

$$K_{sp} = [Ba^{2+}][SO_4^{2-}]$$

$$1.1 \times 10^{-10} = (s)(1.0)$$

$$\mathbf{s = 1.1 \times 10^{-10}\,\mathit{M}}$$

Due to the common ion effect, the molar solubility of $BaSO_4$ decreases to 1.1×10^{-10} mol/L in $1.0\,M\ SO_4^{2-}(aq)$ compared to 1.0×10^{-5} mol/L in pure water.

17.52 See Table 17.5.

(a) The equations are as follows:

$$CuI_2\,(s) \rightleftharpoons Cu^{2+}\,(aq) + 2\,I^-\,(aq)$$

$$Cu^{2+}\,(aq) + 4\,NH_3\,(aq) \rightleftharpoons [Cu(NH_3)_4]^{2+}\,(aq)$$

The ammonia combines with the Cu^{2+} ions formed in the first step to form the complex ion $[Cu(NH_3)_4]^{2+}$, effectively removing the Cu^{2+} ions, causing the first equilibrium to shift to the right (resulting in more CuI_2 dissolving).

(b) Similar to part (a):

$$AgBr\,(s) \rightleftharpoons Ag^+\,(aq) + Br^-\,(aq)$$

$$Ag^+\,(aq) + 2\,CN^-\,(aq) \rightleftharpoons [Ag(CN)_2]^-\,(aq)$$

(c) Similar to parts (a) and (b).

$$HgCl_2\,(s) \rightleftharpoons Hg^{2+}\,(aq) + 2Cl^-\,(aq)$$

$$Hg^{2+}\,(aq) + 4Cl^-\,(aq) \rightleftharpoons [HgCl_4]^{2-}\,(aq)$$

17.54 Complex Ion Equilibria and Solubility, Problem Type 5.

In solution, Cd^{2+} ions will complex with CN^- ions. The concentration of Cd^{2+} will be determined by the following equilibrium

$$Cd^{2+}(aq) + 4\,CN^-(aq) \rightleftharpoons Cd(CN)_4^{2-} \qquad K_f = 7.1 \times 10^{16}$$

Since K_f is so large, this equilibrium lies far to the right. We can safely assume that all the Cd^{2+} reacts.

Step 1: Calculate the initial concentration of Cd^{2+} ions.

$$[Cd^{2+}]_0 = 0.50\,\text{g} \times \frac{1\text{ mol }Cd(NO_3)_2}{236.42\text{ g }Cd(NO_3)_2} \times \frac{1\text{ mol Cd}}{1\text{ mol }Cd(NO_3)_2} \times \frac{1}{0.50\text{ L}} = 4.2 \times 10^{-3}\,M$$

Step 2: If we assume that the above equilibrium goes to completion, we can write

	$Cd^{2+}(aq)$ +	$4\,CN^-(aq)$ ⟶	$Cd(CN)_4^{2-}$
Initial (M)	4.2×10^{-3}	0.50	0
After reaction (M)	0	0.48	4.2×10^{-3}

Step 3: To find the concentration of free Cd^{2+} at equilibrium, use the formation constant expression.

$$K_f = \frac{[Cd(CN)_4^{2-}]}{[Cd^{2+}][CN^-]^4}$$

Rearranging,

$$[Cd^{2+}] = \frac{[Cd(CN)_4^{2-}]}{K_f[CN^-]^4}$$

Substitute the equilibrium concentrations calculate in *Step 2* into the formation constant expression to calculate the equilibrium concentration of Cd^{2+}.

$$\mathbf{[Cd^{2+}]} = \frac{[Cd(CN)_4^{2-}]}{K_f[CN^-]^4} = \frac{4.2 \times 10^{-3}}{(7.1 \times 10^{16})(0.48)^4} = \mathbf{1.1 \times 10^{-18}\,M}$$

$$\mathbf{[CN^-]} = 0.48\,M + (1.1 \times 10^{-18}\,M) = \mathbf{0.48\,M}$$

$$\mathbf{[Cd(CN)_4^{2-}]} = (4.2 \times 10^{-3}\,M) - (1.1 \times 10^{-18}) = \mathbf{4.2 \times 10^{-3}\,M}$$

17.56 Silver iodide is only slightly soluble. It dissociates to form a small amount of Ag^+ and I^- ions. The Ag^+ ions then complex with NH_3 in solution to form the complex ion $Ag(NH_3)_2^+$. The balanced equations are:

$$AgI(s) \rightleftharpoons Ag^+(aq) + I^-(aq) \qquad K_{sp} = [Ag^+][I^-] = 8.3 \times 10^{-17}$$

$$Ag^+(aq) + 2\,NH_3(aq) \rightleftharpoons Ag(NH_3)_2^+(aq) \qquad K_f = \frac{[Ag(NH_3)_2^+]}{[Ag^+][NH_3]^2} = 1.5 \times 10^7$$

Overall: $$AgI(s) + 2\,NH_3(aq) \rightleftharpoons Ag(NH_3)_2^+(aq) + I^-(aq) \qquad K = K_{sp} \times K_f = 1.2 \times 10^{-9}$$

If s is the molar solubility of AgI then,

	$AgI(s)$ +	$2\,NH_3(aq)$ ⇌	$Ag(NH_3)_2^+(aq)$ +	$I^-(aq)$
initial (M):		1.0	0.0	0.0
change (M):	$-s$	$-2s$	$+s$	$+s$
equilibrium (M)		$(1.0 - 2s)$	s	s

Because K_f is large, we can assume all of the silver ions exist as $Ag(NH_3)_2^+$. Thus,

$$[Ag(NH_3)_2^+] = [I^-] = s$$

We can write the equilibrium constant expression for the above reaction, then solve for s.

$$K = 1.2 \times 10^{-9} = \frac{(s)(s)}{(1.0 - 2s)^2} \cong \frac{(s)(s)}{(1.0)^2}$$

$$\mathbf{s = 3.5 \times 10^{-5}\ M}$$

At equilibrium, 3.5×10^{-5} moles of AgI dissolves in 1 L of 1.0 M NH_3 solution.

17.60 Since some $PbCl_2$ precipitates, the solution is saturated. From Table 17.3, the value of K_{sp} for lead(II) chloride is 2.4×10^{-4}. The equilibrium is:

$$PbCl_2\ (aq) \rightleftharpoons Pb^{2+}\ (aq) + 2\ Cl^-\ (aq)$$

We can write the solubility product expression for the equilibrium.

$$K_{sp} = [Pb^{2+}][Cl^-]^2$$

K_{sp} and $[Cl^-]$ are known. Solving for the Pb^{2+} concentration,

$$[\mathbf{Pb^{2+}}] = \frac{K_{sp}}{[Cl^-]^2} = \frac{2.4 \times 10^{-4}}{(0.15)^2} = \mathbf{0.011\ M}$$

17.62 Chloride ion will precipitate Ag^+ but not Cu^{2+}. So, dissolve some solid in H_2O and add HCl. If a precipitate forms, the salt was $AgNO_3$. A flame test will also work. Cu^{2+} gives a green flame test.

17.64 The acid-base neutralization reaction is

$$H^+\ (aq) + NH_3\ (aq) \rightarrow NH_4^+\ (aq)$$

(a)

Step 1: Calculate the initial number of moles of both H^+ and NH_3.

$$\text{mol } H^+ = 0.025\ \cancel{L} \times \frac{0.100\text{ mol}}{1\ \cancel{L}} = 0.0025\text{ mol}$$

$$\text{mol } NH_3 = 0.010\ \cancel{L} \times \frac{0.100\text{ mol}}{1\ \cancel{L}} = 0.0010\text{ mol}$$

Step 2: Set up a table showing the initial moles and the moles after reaction.

	$H^+\ (aq)$ +	$NH_3\ (aq)$ ⟶	$NH_4^+\ (aq)$
Initial (mol)	0.0025	0.0010	0
After reaction (mol)	0.0015	0	0.0010

Step 3: Calculate the molar concentration of H^+, then calculate the pH. The total volume of solution is 35.0 mL = 0.0350 L. You don't need to worry about the weak acid NH_4^+ at this point because H^+ is a much stronger acid.

$$[H^+] = \frac{0.0015 \text{ mol}}{0.0350 \text{ L}} = 0.043\ M$$

$$\textbf{pH} = -\log[H^+] = -\log(0.043) = \textbf{1.37}$$

(b)

Step 1: Calculate the number of moles of NH_3 added.

$$\text{mol } NH_3 = 0.025 \not{L} \times \frac{0.100 \text{ mol}}{1 \not{L}} = 0.0025 \text{ mol}$$

Step 2: We are at the equivalence point of the titration. The moles of base added equal the moles of acid in solution. Set up a table showing the initial moles and the moles after reaction.

	$H^+(aq)$ +	$NH_3(aq)$ ⟶	$NH_4^+(aq)$
Initial (mol)	0.0025	0.0025	0
After reaction (mol)	0	0	0.0025

The only species of significance left in solution is the weak acid NH_4^+. We need to set up the equilibrium reaction for this weak acid.

Step 3: Calculate the molar concentration of NH_4^+ ions, then set up the equilibrium reaction. The total volume of solution is 50.0 mL = 0.0500 L.

$$[NH_4^+] = \frac{0.0025 \text{ mol}}{0.0500 \text{ L}} = 0.050\ M$$

	$NH_4^+(aq)$ ⇌	$NH_3(aq)$ +	$H^+(aq)$
Initial (*M*):	0.050	0	0
Change (*M*):	$-x$	$+x$	$+x$
Equilibrium (*M*):	$0.050 - x$	x	x

$$K_a = \frac{[NH_3][H^+]}{[NH_4^+]}$$

$$K_a = \frac{K_w}{K_b(NH_3)} = \frac{1.0 \times 10^{-14}}{1.8 \times 10^{-5}} = 5.6 \times 10^{-10}$$

$$5.6 \times 10^{-10} = \frac{(x)(x)}{(0.050 - x)} \cong \frac{(x)(x)}{0.050}$$

$$x = [H^+] = 5.3 \times 10^{-6}\ M$$

Step 4: Having solved for the $[H^+]$, calculate the pH of the solution.

$$\textbf{pH} = -\log[H^+] = -\log(5.3 \times 10^{-6}) = \textbf{5.28}$$

(c)

Step 1: Calculate the number of moles of NH_3 added.

$$\text{mol } NH_3 = 0.035 \cancel{L} \times \frac{0.100 \text{ mol}}{1 \cancel{L}} = 0.0035 \text{ mol}$$

Step 2: We are past the equivalence point of the titration. Set up a table showing the initial moles and the moles after reaction.

	$H^+ (aq)$ +	$NH_3 (aq)$ ⟶	$NH_4^+ (aq)$
Initial (mol)	0.0025	0.0035	0
After reaction (mol)	0	0.0010	0.0025

At this point, we have both the weak base (NH_3) and its conjugate acid (NH_4^+) in solution. We have a buffer.

Step 3: Calculate the molar concentrations of NH_3 and NH_4^+ ions, then set up the equilibrium reaction. The total volume of solution is 60.0 mL = 0.0600 L.

$$[NH_3] = \frac{0.0010 \text{ mol}}{0.0600 \text{ L}} = 0.017\, M$$

$$[NH_4^+] = \frac{0.0025 \text{ mol}}{0.0600 \text{ L}} = 0.042\, M$$

	$NH_4^+ (aq)$ ⇌	$NH_3 (aq)$ +	$H^+ (aq)$
Initial (M):	0.042	0.017	0
Change (M):	$-x$	$+x$	$+x$
Equilibrium (M):	$0.042 - x$	$0.017 + x$	x

$$K_a = \frac{[NH_3][H^+]}{[NH_4^+]}$$

$$5.6 \times 10^{-10} = \frac{(x)(0.017 + x)}{(0.042 - x)} \cong \frac{(x)(0.017)}{0.042}$$

$$x = [H^+] = 1.4 \times 10^{-9}\, M$$

Step 4: Having solved for the $[H^+]$, calculate the pH of the solution.

$$\mathbf{pH} = -\log[H^+] = -\log(1.4 \times 10^{-9}) = \mathbf{8.85}$$

17.66 We can use the Henderson-Hasselbalch equation to solve for the pH when the indicator is 90% acid/10% conjugate base and when the indicator is 10% acid/90% conjugate base.

$$pH = pK_a + \log\frac{[\text{conjugate base}]}{[\text{acid}]}$$

Solving for the pH with 90% of the indicator in the HIn form:

$$pH = 3.46 + \log\frac{[10]}{[90]} = 3.46 - 0.95 = 2.51$$

Next, solving for the pH with 90% of the indicator in the In^- form:

$$pH = 3.46 + \log\frac{[90]}{[10]} = 3.46 + 0.95 = 4.41$$

Thus the pH range varies from **2.51 to 4.41** as the [HIn] varies from 90% to 10%.

17.68 First, calculate the pH of the 2.00 M weak acid (HNO_2) solution before any NaOH is added.

$$K_a = \frac{[H^+][NO_2^-]}{[HNO_2]}$$

$$4.5 \times 10^{-4} = \frac{x^2}{2.00 - x}$$

$$x = [H^+] = 0.030\ M$$

$$pH = -\log(0.030) = 1.52$$

Since the pH after the addition is 1.5 pH units greater, the new pH = 1.52 + 1.50 = 3.02.

From this new pH, we can calculate the $[H^+]$ in solution.

$$[H^+] = 10^{-pH} = 10^{-3.02} = 9.55 \times 10^{-4}\ M$$

When the NaOH is added, we dilute our original 2.00 M HNO_2 solution to:

$$M_I V_I = M_F V_F$$

$$(2.00\ M)(400\ mL) = M_F(600\ mL)$$

$$M_F = 1.33\ M$$

Since we have not reached the equivalence point, we have a buffer solution. The reaction between HNO_2 and NaOH is:

$$HNO_2\ (aq) + NaOH\ (aq) \longrightarrow NaNO_2\ (aq) + H_2O\ (l)$$

Since the mole ratio between HNO_2 and NaOH is 1:1, the decrease in $[HNO_2]$ is the same as the decrease in [NaOH].

We can calculate the decrease in $[HNO_2]$ by setting up the weak acid equilibrium. From the pH of the solution, we know that the $[H^+]$ at equilibrium is $9.55 \times 10^{-4}\ M$.

	$HNO_2\ (aq)$ $\rightleftharpoons$	$H^+\ (aq)$ +	$NO_2^-\ (aq)$
Initial (M)	1.33	0	0
Change (M)	$-x$		$+x$
Equilibrium (M)	$1.33 - x$	9.55×10^{-4}	x

We can calculate x from the equilibrium constant expression.

$$K_a = \frac{[H^+][NO_2^-]}{[HNO_2]}$$

$$4.5 \times 10^{-4} = \frac{(9.55 \times 10^{-4})(x)}{1.33 - x}$$

$$x = 0.426\ M$$

Thus, x is the decrease in $[HNO_2]$ which equals the concentration of added OH^-. However, this is the concentration of NaOH after it has been diluted to 600 mL. We need to correct for the dilution from 200 mL to 600 mL to calculate the concentration of the original NaOH solution.

$$M_I V_I = M_F V_F$$
$$M_I(200\text{ mL}) = (0.426\ M)(600\text{ mL})$$
$$\mathbf{[NaOH]} = M_I = \mathbf{1.28\ M}$$

17.70 The resulting solution is not a buffer system. There is excess NaOH and the neutralization is well past the equivalence point. We have a solution of sodium acetate and sodium hydroxide.

$$\text{Moles NaOH} = 0.500\cancel{L}\left(\frac{0.167\text{ mol}}{1\cancel{L}}\right) = 0.0835\text{ mol}$$

$$\text{Moles } CH_3COOH = 0.500\cancel{L}\left(\frac{0.100\text{ mol}}{1\cancel{L}}\right) = 0.0500\text{ mol}$$

The reaction between sodium hydroxide and acetic acid is:

	NaOH (*aq*)	+	CH_3COOH (*aq*)	$\longrightarrow$	$NaCH_3COO$ (*aq*)	+	H_2O (*l*)
Initial (mol)	0.0835		0.0500		0		
After Reaction (mol)	0.0335		0		0.0500		

Since the total volume of the solution is 1.00 L, we can convert the number of moles directly to molarity.

$$\mathbf{[OH^-] = 0.0335\ M}$$
$$\mathbf{[Na^+] = 0.0335\ M + 0.0500\ M = 0.0835\ M}$$
$$\mathbf{[CH_3COO^-] = 0.0500\ M}$$

We can calculate the $[H^+]$ from the $[OH^-]$.

$$\mathbf{[H^+]} = \frac{K_w}{[OH^-]} = \frac{1.00 \times 10^{-14}}{0.0335} = \mathbf{2.99 \times 10^{-13}\ M}$$

Finally, we can calculate the acetic acid concentration from the K_a expression.

$$K_a = \frac{[H^+][CH_3COO^-]}{[CH_3COOH]}$$

or,

$$[CH_3COOH] = \frac{[H^+][CH_3COO^-]}{K_a}$$

$$\mathbf{[CH_3COOH]} = \frac{(2.99 \times 10^{-13})(0.0500)}{1.8 \times 10^{-5}} = \mathbf{8.3 \times 10^{-10}\ M}$$

17.72 For the first part we use K_a for ammonium ion. (Why?) The Henderson–Hasselbalch equation is

$$\mathbf{pH} = -\log(5.6\times10^{-10}) + \log\frac{(0.20\text{ M})}{(0.20\text{ M})} = \mathbf{9.25}$$

For the second part, the acid–base reaction is

$$NH_3(g) + H^+(aq) \rightarrow NH_4^+(aq)$$

We find the number of moles of HCl added

$$10.0\text{ mL}\left(\frac{1\text{ L}}{1000\text{ mL}}\right)\left(\frac{0.10\text{ mol HCl}}{1\text{ L}}\right) = 0.0010\text{ mol HCl}$$

The number of moles of NH_3 and NH_4^+ originally present are

$$65.0\text{ mL}\left(\frac{1\text{ L}}{1000\text{ mL}}\right)\left(\frac{0.20\text{ mol}}{1\text{ L}}\right) = 0.013\text{ mol}$$

Using the acid–base reaction, we find the number of moles of NH_3 and NH_4^+ after addition of the HCl. Setting up a table:

	$H^+(aq)$	+ $NH_3(aq)$	$\longrightarrow$	$NH_4^+(aq)$
Initial (mol)	0.0010	0.013		0.013
Change (mol)	−0.0010	−0.0010		+0.0010
After reaction (mol)	0	0.012		0.014

We find the new pH:

$$\mathbf{pH} = 9.25 + \log\frac{(0.012)}{(0.014)} = \mathbf{9.18}$$

17.74 The number of moles of Ba^{2+} present in the original 50.0 mL of solution is:

$$50.0\text{ mL} \times \frac{1.00\text{ mol Ba}^{2+}}{1\text{ L soln}} \times \frac{1\text{ L}}{1000\text{ mL}} = 0.0500\text{ mol Ba}^{2+}$$

The number of moles of SO_4^{2-} present in the original 86.4 mL of solution, assuming complete dissociation, is:

$$86.4\text{ mL} \times \frac{0.494\text{ mol SO}_4^{2-}}{1\text{ L soln}} \times \frac{1\text{ L}}{1000\text{ mL}} = 0.0427\text{ mol SO}_4^{2-}$$

The complete ionic equation is:

	$Ba^{2+}(aq)$	+ $2\,OH^-(aq)$	+ $2\,H^+(aq)$	+ $SO_4^{2-}(aq)$	$\rightarrow$ $BaSO_4(s)$	+ $H_2O(l)$
Initial (mol):	0.0500	0.100	0.0854	0.0427	0	
Change (mol)	−0.0427	−2(0.0427)	−2(0.0427)	−0.0427	+0.0427	
After rxn (mol)	0.0073	0.015	0	0	0.0427	

Thus the mass of $BaSO_4$ formed is:

$$(0.0427\text{ mol BaSO}_4) \times \frac{233.4\text{ g BaSO}_4}{1\text{ mol BaSO}_4} = \mathbf{9.97\text{ g BaSO}_4}$$

The pH can be calculated from the excess OH^- in solution. First, calculate the molar concentration of OH^-. The total volume of solution is 136.4 mL = 0.1364 L.

$$[OH^-] = \frac{0.015 \text{ mol}}{0.1364 \text{ L}} = 0.11\ M$$

$$pOH = -\log(0.11) = 0.96$$

$$\mathbf{pH} = 14.00 - pOH = 14.00 - 0.96 = \mathbf{13.04}$$

17.76 Since equal volumes of the two solutions were used, the initial molar concentrations will be halved.

$$[Ag^+] = \frac{0.12\ M}{2} = 0.060\ M$$

$$[Cl^-] = \frac{2(0.14\ M)}{2} = 0.14\ M$$

Let's assume that the Ag^+ ions and Cl^- ions react completely to form AgCl (*s*). Then, we will reestablish the equilibrium between AgCl, Ag^+, and Cl^-.

	$Ag^+\ (aq)$ +	$Cl^-\ (aq)$ ⟶	$AgCl\ (s)$
Initial (*M*)	0.060	0.14	0
Change (*M*)	−0.060	−0.060	+0.060
After reaction (*M*)	0	0.080	0.060

Now, setting up the equilibrium,

	$AgCl(s)$ ⇌	$Ag^+(aq)$ +	$Cl^-(aq)$
Initial (*M*):	0.060	0	0.080
Change (*M*):	$-s$	$+s$	$+s$
Equilibrium (*M*):	$0.060 - s$	s	$0.080 + s$

Set up the K_{sp} expression to solve for *s*.

$$K_{sp} = [Ag^+][Cl^-]$$

$$1.6 \times 10^{-10} = (s)(0.080 + s)$$

$$s = 2.0 \times 10^{-9}\ M$$

$$\mathbf{[Ag^+]} = s = \mathbf{2.0 \times 10^{-9}\ M}$$

$$\mathbf{[Cl^-]} = 0.080\ M - s = \mathbf{0.080\ M}$$

$$\mathbf{[Zn^{2+}]} = \frac{0.14\ M}{2} = \mathbf{0.070\ M}$$

$$\mathbf{[NO_3^-]} = \frac{0.12\ M}{2} = \mathbf{0.060\ M}$$

17.78 For $Fe(OH)_3$, $K_{sp} = 1.1 \times 10^{-36}$. When $[Fe^{3+}] = 0.010\ M$, the $[OH^-]$ value is

$$K_{sp} = [Fe^{3+}][OH^-]^3$$

or

$$[OH^-] = \left(\frac{K_{sp}}{[Fe^{3+}]}\right)^{\frac{1}{3}}$$

$$[OH^-] = \left(\frac{1.1 \times 10^{-36}}{0.010}\right)^{\frac{1}{3}} = 4.8 \times 10^{-12}\ M$$

This $[OH^-]$ corresponds to a pH of 2.68. In other words, $Fe(OH)_3$ will begin to precipitate from this solution at pH of 2.68.

For $Zn(OH)_2$, $K_{sp} = 1.8 \times 10^{-14}$. When $[Zn^{2+}] = 0.010\ M$, the $[OH^-]$ value is

$$[OH^-] = \left(\frac{K_{sp}}{[Zn^{2+}]}\right)^{\frac{1}{2}}$$

$$[OH^-] = \left(\frac{1.8 \times 10^{-14}}{0.010}\right)^{\frac{1}{2}} = 1.3 \times 10^{-6}\ M$$

This corresponds to a pH of 8.11. In other words $Zn(OH)_2$ will begin to precipitate from the solution at pH = 8.11. These results show that $Fe(OH)_3$ will precipitate when the pH just exceeds 2.68 and that $Zn(OH)_2$ will precipitate when the pH just exceeds 8.11. Therefore, to selectively remove iron as $Fe(OH)_3$, the pH must be greater than 2.68 but less than 8.11.

17.80 **(b)** SO_4^{2-} *(aq)* is a weak base
(c) OH^- *(aq)* is a strong base
(d) $C_2O_4^{2-}$ *(aq)* is a weak base
(e) PO_4^{3-} *(aq)* is a weak base.

The solubilities of the above will increase in acidic solution. Only (a), which contains an extremely weak base (I^- is the conjugate base of the strong acid HI) is unaffected by the acid solution.

17.82 The equilibrium reaction is:

	$Pb(IO_3)_2\ (aq)$ $\rightleftharpoons$	$Pb^{2+}\ (aq)$ +	$2\ IO_3^-\ (aq)$
Initial (*M*)		0	0.10
Change (*M*)	-2.4×10^{-11}	$+2.4 \times 10^{-11}$	$+2(2.4 \times 10^{-11})$
Equilibrium (*M*)		2.4×10^{-11}	$\cong 0.10$

Substitute the equilibrium concentrations into the solubility product expression to calculate K_{sp}.

$$K_{sp} = [Pb^{2+}][IO_3^-]^2$$

$$\mathbf{K_{sp}} = (2.4 \times 10^{-11})(0.10)^2 = \mathbf{2.4 \times 10^{-13}}$$

17.84 We first determine the effect of the added ammonia.

$$M_I V_I = M_F V_F$$
$$(0.60\ M)(2.00\ mL) = M_F(1002\ mL)$$
$$M_F = 0.0012\ M\ NH_3$$

Ammonia is a weak base (K_b = 1.8×10^{-5}), which will add OH^- to the solution. Iron(II) hydroxide, $Fe(OH)_2$, is an insoluble salt. We need to determine if there is enough Fe^{2+} and OH^- in solution to precipitate $Fe(OH)_2$. First, calculate the amount of OH^- in solution.

	$NH_3 + H_2O \rightleftharpoons$	NH_4^+	$+ OH^-$
initial (*M*):	0.0012	0	0
change (*M*):	$-x$	$+x$	$+x$
equil. (*M*):	$0.0012 - x$	x	x

$$K_b = \frac{[NH_4^+][OH^-]}{[NH_3]} = \frac{x^2}{(0.0012 - x)} = 1.8 \times 10^{-5}$$

Solving the resulting quadratic equation gives x = 0.00014, or $[OH^-]$ = 0.00014 *M*

Next, we compute the value of Q_c for this solution and compare it to the K_{sp} for iron(II) hydroxide.

$$Q_c = [Fe^{2+}]_0[OH^-]_0^2 = (1.0 \times 10^{-3})(0.00014)^2 = 2.0 \times 10^{-11}$$

Q_c is larger than $K_{sp}(1.6 \times 10^{-14})$; therefore a precipitate of $Fe(OH)_2$ will form.

17.86 The precipitate is HgI_2.

$$Hg^{2+}(aq) + 2I^-(aq) \longrightarrow HgI_2(s)$$

With further addition of I^-, a soluble complex ion is formed and the precipitate redissolves.

$$HgI_2(s) + 2I^-(aq) \longrightarrow HgI_4^{2-}(aq)$$

17.88 We can use the Henderson-Hasselbalch equation to solve for the pH when the indicator is 95% acid/ 5% conjugate base and when the indicator is 5% acid/95% conjugate base.

$$pH = pK_a + \log\frac{[\text{conjugate base}]}{[\text{acid}]}$$

Solving for the pH with 95% of the indicator in the HIn form:

$$pH = 9.10 + \log\frac{[5]}{[95]} = 9.10 - 1.28 = 7.82$$

Next, solving for the pH with 95% of the indicator in the In^- form:

$$pH = 9.10 + \log\frac{[95]}{[5]} = 9.10 + 1.28 = 10.38$$

Thus the pH range varies from **7.82 to 10.38** as the [HIn] varies from 95% to 5%.

17.90 **(a)** The solubility product expressions for both substances have exactly the same mathematical form and are therefore directly comparable. The substance having the smaller K_{sp} (AgI) will precipitate first. (Why?)

(b) When CuI just begins to precipitate the solubility product expression will just equal K_{sp} (saturated solution). The concentration of Cu^+ at this point is 0.010 M (given in the problem), so the concentration of iodide ion must be:

$$K_{sp} = [Cu^+][I^-] = (0.010)[I^-] = 5.1 \times 10^{-12}$$

$$[I^-] = \frac{5.1 \times 10^{-12}}{0.010} = 5.1 \times 10^{-10}\ M$$

Using this value of $[I^-]$, we find the silver ion concentration

$$[Ag^+] = \frac{K_{sp}}{[I^-]} = \frac{8.3 \times 10^{-17}}{5.1 \times 10^{-10}} = 1.6 \times 10^{-7}\ M$$

(c) The percent of silver ion remaining in solution is:

$$\textbf{\% Ag}^+(aq) = \frac{1.6 \times 10^{-7}\ M}{0.010\ M} \times 100\% = \mathbf{0.0016\%\ or\ 1.6 \times 10^{-3}\%}$$

Is this an effective way to separate silver from copper?

17.92 **(a)** $MCO_3 + 2HCl \rightarrow MCl_2 + H_2O + CO_2$
$HCl + NaOH \rightarrow NaCl + H_2O$

(b) Moles of HCl reacted with MCO_3 = Total moles of HCl − Moles of excess HCl

$$\text{Total moles of HCl} = 20.00\ \text{mL} \times \frac{1\ \text{L}}{1000\ \text{mL}} \times \frac{0.0800\ \text{mol}}{1\ \text{L}} = 1.60 \times 10^{-3}\ \text{mol HCl}$$

$$\text{Moles of excess HCl} = 5.64\ \text{mL} \times \frac{1\ \text{L}}{1000\ \text{mL}} \times \frac{0.1000\ \text{mol}}{1\ \text{L}} = 5.64 \times 10^{-4}\ \text{mol HCl}$$

$$\text{Moles of HCl reacted with } MCO_3 = 1.60 \times 10^{-3}\ \text{mol} - 5.64 \times 10^{-4}\ \text{mol} = 1.04 \times 10^{-3}\ \text{mol HCl}$$

$$\text{Moles of } MCO_3 \text{ reacted} = 1.04 \times 10^{-3}\ \text{mol HCl} \times \frac{1\ \text{mol } MCO_3}{2\ \text{mol HCl}} = 5.20 \times 10^{-4}\ \text{mol } MCO_3$$

$$\textbf{Molar mass of } \mathbf{MCO_3} = \frac{0.1022\ \text{g}}{5.20 \times 10^{-4}\ \text{mol}} = \mathbf{197\ g/mol}$$

Molar mass of CO_3 = 60.01 g/mol

Molar mass of M = 197 g/mol − 60.01 g/mol = 137 g/mol

The metal, M, is **Ba**!

17.94 The balanced equation is: $H_2C_2O_4 \cdot xH_2O + 2\,NaOH \rightarrow Na_2C_2O_4 + (2 + x)\,H_2O$

The number of moles of NaOH reacted is:

$$15.9\ \text{mL NaOH}\left(\frac{0.500\ \text{mol NaOH}}{1000\ \text{mL}}\right) = 7.95 \times 10^{-3}\ \text{mol NaOH}$$

Since two moles of NaOH combine with one mole of oxalic acid, the number of moles of oxalic acid reacted is 3.98×10^{-3} mol. This is the number of moles of oxalic acid hydrate in 25.0 mL of solution. In 250 mL, the number of moles present is 3.98×10^{-2} mol.

Thus the molar mass is:

$$\frac{5.00\text{ g}}{3.98 \times 10^{-2}\text{ mol}} = 126\text{ g/mol}$$

From the molecular formula we can write:

$$2(1.008)\text{ g} + 2(12.01)\text{ g} + 4(16.00)\text{ g} + x(18.02)\text{g} = 126\text{ g}$$

Solving for x:

$$\boldsymbol{x = 2}$$

The formula is $H_2C_2O_4 \cdot 2H_2O$

17.96 **(a)** Add sulfate. Na_2SO_4 is soluble, $BaSO_4$ is not.

(b) Add sulfide. K_2S is soluble, PbS is not

(c) Add iodide. ZnI_2 is soluble, HgI_2 is not.

17.98 $H_2PO_4^- \rightleftharpoons H^+ + HPO_4^{2-}$

$$K_a = 6.2 \times 10^{-8}$$

$$pK_a = 7.20$$

$$7.50 = 7.20 + \log\frac{[HPO_4^{2-}]}{[H_2PO_4^-]}$$

$$\frac{[HPO_4^{2-}]}{[H_2PO_4^-]} = 2.0$$

We need to add enough NaOH so that

$$[HPO_4^{2-}] = 2[H_2PO_4^-]$$

Initially there was

$$0.200\text{ L} \times 0.10\text{ mol/L} = 0.020\text{ mol } NaH_2PO_4 \text{ present.}$$

For $[HPO_4^{2-}] = 2[H_2PO_4^-]$, we must add enough NaOH to react with 2/3 of the $H_2PO_4^-$. After reaction with NaOH, we have

$$\frac{0.020}{3}\text{ mol } H_2PO_4^- = 0.0067\text{ mol } H_2PO_4^-$$

$$\text{mol } HPO_4^{2-} = 2 \times 0.0067\text{ mol} = 0.013\text{ mol } HPO_4^{2-}$$

The moles of NaOH reacted is equal to the moles of HPO_4^{2-} produced because the mole ratio between OH^- and HPO_4^{2-} is 1:1.

$$OH^- + H_2PO_4^- \rightarrow HPO_4^{2-} + H_2O$$

$$V_{NaOH} = \frac{\text{mol}_{NaOH}}{M_{NaOH}} = \frac{0.013\text{ mol}}{1.0\text{ mol/L}} = 0.013\text{ L} = \mathbf{13\ mL}$$

17.100 **(c)** has the highest $[H^+]$

$F^- + SbF_5 \rightarrow SbF_6^-$

Removal of F^- promotes further ionization of HF, producing more H^+.

17.102 **(a)** This is a common ion (CO_3^{2-}) problem.

The dissociation of Na_2CO_3 is:

$$Na_2CO_3\,(s) \xrightarrow{H_2O} \underset{2(0.050\,M)}{2\,Na^+\,(aq)} + \underset{0.050\,M}{CO_3^{2-}\,(aq)}$$

Let s be the molar solubility of $CaCO_3$ in Na_2CO_3 solution. We summarize the changes as:

	$CaCO_3\,(s) \rightleftharpoons$	$Ca^{2+}\,(aq)$ +	CO_3^{2-}
Initial (M)		0.00	0.050
Change (M)		$+s$	$+s$
Equil. (M)		$+s$	$0.050 + s$

$$K_{sp} = [Ca^{2+}][CO_3^{2-}]$$

$$8.7 \times 10^{-9} = s(0.050 + s)$$

Since s is small, we can assume that $0.050 + s \cong 0.050$

$$8.7 \times 10^{-9} = 0.050s$$

$$s = 1.7 \times 10^{-7}\,M$$

Thus, the addition of washing soda to permanent hard water removes most of the Ca^{2+} ions as a result of the common ion effect.

(b) Mg^{2+} is not removed by this procedure, because $MgCO_3$ is fairly soluble ($K_{sp} = 4.0 \times 10^{-5}$).

(c) The K_{sp} for $Ca(OH)_2$ is 8.0×10^{-6}.

$$Ca(OH)_2 \rightleftharpoons Ca^{2+} + 2OH^-$$

At equil: $\quad s \quad 2s$

$$K_{sp} = 8.0 \times 10^{-6} = [Ca^{2+}][OH^-]^2$$

$$4s^3 = 8.0 \times 10^{-6}$$

$$s = 0.0126\,M$$

$$[OH^-] = 2s = 0.0252\,M$$

$$pOH = -\log(0.0252) = 1.60$$

$$\mathbf{pH = 12.40}$$

(d) The [OH^-] calculated above is 0.0252 *M*. At this rather high concentration of OH^-, most of the Mg^{2+} will be removed as $Mg(OH)_2$. The small amount of Mg^{2+} remaining in solution is due to the following equilibrium:

$$Mg(OH)_2\,(s) \rightleftharpoons Mg^{2+}\,(aq) + 2\,OH^-\,(aq)$$

$$K_{sp} = [Mg^{2+}][OH^-]^2$$

$$1.2 \times 10^{-11} = [Mg^{2+}](0.0252)^2$$

$$\mathbf{[Mg^{2+}] = 1.9 \times 10^{-8}\ \mathit{M}}$$

(e) Remove Ca^{2+} first because it is present in larger amounts.

CHAPTER 18
THE CHEMISTRY OF COORDINATION COMPOUNDS

PROBLEM-SOLVING STRATEGIES AND TUTORIAL SOLUTIONS

TYPES OF PROBLEMS

Problem Type 1: Assigning Oxidation Numbers to the Metal Atom in Coordination Compounds.

Problem Type 2: Naming Coordination Compounds.

Problem Type 3: Writing Formulas for Coordination Compounds.

Problem Type 4: Predicting the Number of Unpaired Spins in a Coordination Compound.

PROBLEM TYPE 1: ASSIGNING OXIDATION NUMBERS TO THE METAL ATOM IN COORDINATION COMPOUNDS

Transition metals exhibit variable oxidation states in their compounds. The charge on the central metal atom and its surrounding ligands sum to zero in a neutral coordination compound. In a complex ion, the charges on the central metal atom and the surrounding ligands sum to the net charge of the ion.

EXAMPLE 18.1

Specify the oxidation number of the central metal atom in each of the following compounds:
(a) $[Co(NH_3)_6]Cl_3$ and (b) $Co(CN)_6^{3-}$

(a) NH_3 is a neutral species. Since each chloride ion carries a −1 charge, and there are three Cl^- ions, the oxidation number of Co must be +3.

(b) Each cyanide ion has a charge of −1. The sum of the oxidation number of Co and the −6 charge for the six cyanide ions is −3. Therefore, the oxidation number of Co must be +3.

PRACTICE EXERCISE

1. Specify the oxidation number of the central metal atom in each of the following compounds:
(a) $[Pt(NH_3)_3Cl_3]NO_3$ and (b) $Ni(CO)_4$.

Text Problems: 12, 14

PROBLEM TYPE 2: NAMING COORDINATION COMPOUNDS

The complete set of rules for naming coordination compounds is given in Section 18.2 of your text. Presented below is an abridged version of those rules.

1. The cation is named before the anion, as is the case for other ionic compounds.
2. Within a complex ion the ligands are named first, in alphabetical order, and the metal ion is named last.

3. The names of anionic ligands end with the letter *o*, whereas a neutral ligand is usually called by the name of the molecule. Exceptions are listed in Table 18.3 of the text.

4. Greek prefixes are used to indicate the number of ligands of a particular kind present. If the ligand itself contains a Greek prefix, the prefixes *bis* (2), *tris* (3), and *tetrakis* (4) are used to indicate the number of ligands present.

5. The oxidation number of the metal is written in Roman numerals following the name of the metal.

6. If the complex is an anion, its name ends in "–ate".

EXAMPLE 18.2
Name the following coordination compounds and complex ions: (a) $[Ni(NH_3)_6]^{2+}$, (b) $K_2[Cu(CN)_4]$, (c) $[Pt(NH_3)_4Cl_2]Cl_2$.

(a) NH_3 is a neutral species; therefore, the Ni must have a +2 charge. Ammonia as a ligand in a coordination compound is called ammine. The complex ion is called **hexaamminenickel(II) ion**.

(b) Potassium is an alkali metal; it always has a +1 charge in ionic compounds. Since the two potassium ions have a total charge of +2, the charge on the complex ion $[Cu(CN)_4]$ must be −2. Cyanide ion has a −1 charge; therefore, Cu must have a +2 charge. Potassium is the cation and is named first. The compound is called **potassium tetracyanocuprate(II)**. We use an "–ate" ending because the complex is an anion.

(c) The complex ion has a +2 charge balanced by the total charge of −2 for the two chloride ions. Focusing on the complex ion, NH_3 is a neutral species and the two chloride ligands each have a −1 charge. Since the charge of the complex ion is +2, platinum must have a +4 charge. This compound is called **tetraamminedichloroplatinum(IV) chloride**.

PRACTICE EXERCISE

2. Name the following coordination compounds and complex ions:

(a) $[Cr(OH)_4]^-$ (b) $[Pt(NH_3)_3Cl_3]NO_3$.

Text Problem: 16

PROBLEM TYPE 3: WRITING FORMULAS FOR COORDINATION COMPOUNDS

Follow the nomenclature rules given in Problem Type 2 above. Remember that the cation is named before the anion.

EXAMPLE 18.3
Write the formulas for the following compounds or complex ions:
(a) tetrahydroxoaluminate(III) ion, (b) potassium hexachloropalladate(IV), (c) diaquodicyanocopper(II)

(a) Tetrahydroxo refers to four hydroxide ligands. The "–ate" of aluminate indicates that the complex is an anion. The Roman numeral (III) indicates a +3 charge on aluminum. Since each hydroxide ligand has a −1 charge, the charge on the ion is −1. The formula for the complex ion is $\mathbf{[Al(OH_4)]^-}$.

(b) The complex anion, hexachloropalladate(IV), has a −2 charge. Each of the six chloride ligands has a −1 charge (total of −6) and palladium has a +4 charge. To balance the −2 charge of the anion, there must be two potassium +1 ions. The formula for the compound is $\mathbf{K_2[PdCl_6]}$.

(c) Diaquo refers to two water ligands, and dicyano refers to two cyanide ligands. Each cyanide ligand has a −1 charge (total of −2), which balances the +2 charge on copper. The compound is electrically neutral. The formula for the compound is $\mathbf{[Cu(H_2O)_2(CN)_2]}$.

PRACTICE EXERCISE

3. Write the formulas for the following compounds:

   (a) tris(ethylenediamine)nickel(II) sulfate

   (b) tetraamminediaquocobalt(III) chloride

   (c) potassium hexacyanoferrate(II)

Text Problem: 18

PROBLEM TYPE 4: PREDICTING THE NUMBER OF UNPAIRED SPINS IN A COORDINATION COMPOUND

First, we must write the electron configuration for the transition metal of the coordination compound. Remember that the *ns* shell fills before the $(n - 1)d$ shell. When the number of *d* electrons in the transition metal is known, we must decide how to place them in the *d* orbitals.

Crystal Field Theory tells us that in an octahedral complex, the five *d* orbitals are *not* equivalent in energy. The d_{xy}, d_{xz}, and d_{yz} are degenerate and at a lower energy than the degenerate set, d_{z^2} and $d_{x^2-y^2}$. The energy difference between these two sets of d orbitals is called the **crystal field splitting (Δ)**.

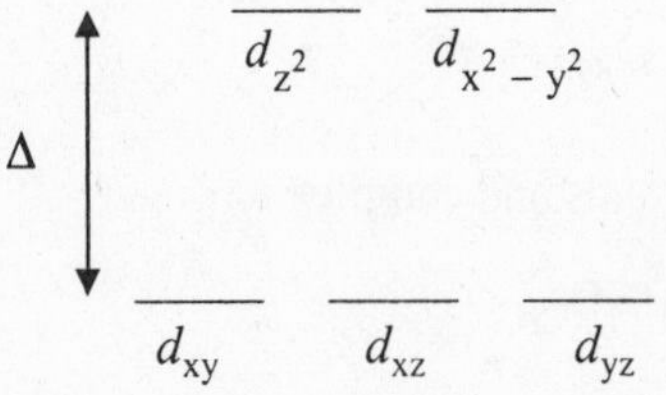

Due to the crystal field splitting, for metals with electron configurations of d^4, d^5, d^6, or d^7, there are two ways to place the electrons in the five *d* orbitals. Let's consider a d^4 case. According to Hund's rule (see Section 7.9 of the text), maximum stability is reached if the electrons are placed in four separate orbitals with parallel spin. But, this arrangement can be achieved only at a cost; one of the four electrons must be energetically promoted to the higher energy d_{z^2} or $d_{x^2-y^2}$ orbital. This energy investment is not needed if all four electrons enter the d_{xy}, d_{xz}, and d_{yz} orbitals. However, in this electron arrangement, we must pair up two of the electrons. This pairing also takes energy. The two possible electron configurations are shown below.

high spin complex

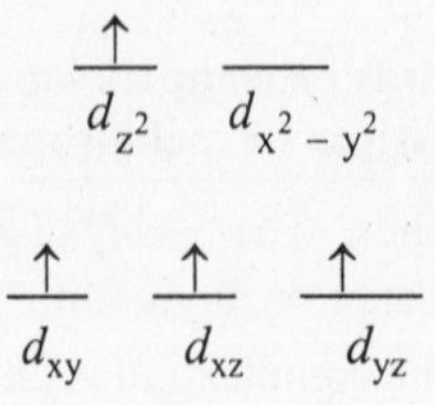

d_{z^2} $d_{x^2-y^2}$

low spin complex

↑↓ ↑ ↑

d_{xy} d_{xz} d_{yz}

Whether a complex is high spin or low spin depends on the ligands that are bonded to the metal center. If a ligand is a weak-field ligand, the crystal field splitting (Δ) is small. Therefore, only a small energy expenditure is needed to keep all spins parallel, resulting in *high spin complexes*. On the other hand, the crystal field splitting is larger when a strong-field ligand is bound to the metal center. In these cases, it is energetically more favorable to pair up the electrons, resulting in *low spin complexes*.

In summary, the actual arrangement of the *d*-electrons is determined by the amount of stability gained by having maximum parallel spins versus the investment in energy required to promote electrons to higher *d* orbitals.

EXAMPLE 18.4

Predict the number of unpaired spins in the $[Co(CN)_6]^{3-}$ ion.

The electron configuration of Co^{3+} is $[Ar]3d^6$. Since CN^- is a strong-field ligand, we expect $[Co(CN)_6]^{3-}$ to be a low spin complex. All six electrons will be placed in the lower energy *d* orbitals (d_{xy}, d_{xz}, and d_{yz}), and there will be no unpaired spins.

d_{z^2} $d_{x^2-y^2}$

↑↓ ↑↓ ↑↓

d_{xy} d_{xz} d_{yz}

PRACTICE EXERCISE

4. Predict the number of unpaired spins in the $[Fe(H_2O)_6]^{3+}$ ion. Water is a weak-field ligand.

Text Problem: 30

ANSWERS TO PRACTICE EXERCISES

1. **(a)** Pt, +4 **(b)** Ni, 0

2. **(a)** tetrahydroxochromate(III) ion **(b)** triamminetrichloroplatinum(IV) nitrate

3. **(a)** $[Ni(NH_2CH_2CH_2NH_2)_3]SO_4$ **(b)** $[Co(NH_3)_4(H_2O)_2]Cl_3$ **(c)** $K_4[Fe(CN)_6]$

4. five unpaired spins

SOLUTIONS TO SELECTED TEXT PROBLEMS

18.12 (a) The oxidation number of Cr is +3.

(b) The coordination number of Cr is 6.

(c) Oxalate ion ($C_2O_4^{2-}$) is a bidentate ligand.

18.14 Assigning Oxidation Numbers to the Metal Atom in Coordination Compounds, Problem Type 1.

(a) Since **sodium** is always **+1** and the oxygens are −2, **Mo** must have an oxidation number of **+6**.

(b) **Magnesium** is **+2** and oxygen −2; therefore **W** is **+6**.

(c) **Potassium** ion is always **+1**, so the charge on the complex ion is −4. Each cyanide ion has a −1 charge; therefore, **iron** must have a **+2** charge.

18.16 Naming Coordination Compounds, Problem Type 2.

(a) Ethylenediamine is a neutral ligand, and each chloride has a −1 charge. Therefore, cobalt has a oxidation number of +3. The correct name for the ion is ***cis*–dichlorobis(ethylenediamine)cobalt(III)**. The prefix *bis* means two; we use this instead of *di* because *di* already appears in the name ethylenediamine.

(b) There are four chlorides each with a −1 charge; therefore, Pt has a +4 charge. The correct name for the compound is **pentaamminechloroplatinum(IV) chloride**.

(c) NH_3 is a neutral ligand, and there are three chlorides each with a −1 charge; therefore, Co has a +3 charge. The correct name for the compound is **hexaaminecobalt(III) chloride**.

(d) There are three chlorides each with a −1 charge; therefore, Co has a +3 charge. The correct name for the compound is **pentaamminechlorocobalt(III) chloride**.

(e) There are two chlorides each with a −1 charge; therefore, Pt has a +2 charge. The correct name for the compound is ***trans*–diamminedichloroplatinum(II)**.

18.18 Writing Formulas for Coordination Compounds, Problem Type 3.

(a) There are two ethylenediamine ligands and two chloride ligands. The correct formula is $\mathbf{[Cr(en)_2Cl_2]^+}$.

(b) There are five carbonyl (CO) ligands. The correct formula is $\mathbf{Fe(CO)_5}$.

(c) There are four cyanide ligands each with a −1 charge. Therefore, the complex ion has a −2 charge, and two K^+ ions are needed to balance the −2 charge of the anion. The correct formula is $\mathbf{K_2[Cu(CN)_4]}$.

(d) There are four NH_3 ligands and two H_2O ligands. Two chloride ions are needed to balance the +2 charge of the complex ion. The correct formula is $\mathbf{[Co(NH_3)_4(H_2O)Cl]Cl_2}$.

18.20 (a) In general for any MA_2B_4 octahedral molecule, only two geometric isomers are possible. The only real distinction is whether the two A–ligands are cis or trans. In Figure 18.8 of the text, (a) and (c) are the same compound (Cl atoms cis in both), and (b) and (d) are identical (Cl atoms trans in both).

(b) A model or a careful drawing is very helpful to understand the MA_3B_3 octahedral structure. There are only two possible geometric isomers. The first has all A's (and all B's) cis; this is called the facial isomer. The second has two A's (and two B's) at opposite ends of the molecule (trans). Try to make or draw other possibilities. What happens?

18.22 The isomers are:

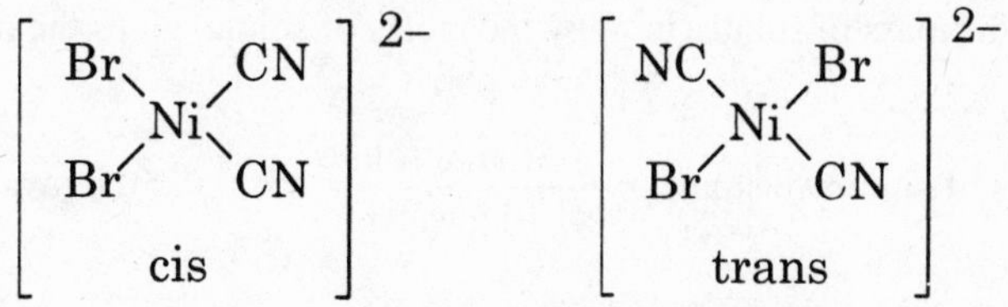

18.28 When a substance appears to be yellow, it is absorbing light from the blue–violet, high energy end of the visible spectrum. Often this absorption is just the tail of a strong absorption in the ultraviolet. Substances that appear green or blue to the eye are absorbing light from the lower energy red or orange part of the spectrum.

Cyanide ion is a very strong field ligand. It causes a larger crystal field splitting than water, resulting in the absorption of higher energy (shorter wavelength) radiation when a *d* electron is excited to a higher energy *d* orbital.

18.30 **(a)** Each cyanide ligand has a −1 charge, so the oxidation state of Cr is +2. The electron configuration of Cr^{2+} is $[Ar]3d^4$. Cyanide is a strong field ligand (see Problem 18.28). The four 3*d* electrons should occupy the three lower orbitals as shown; there should be two unpaired electrons.

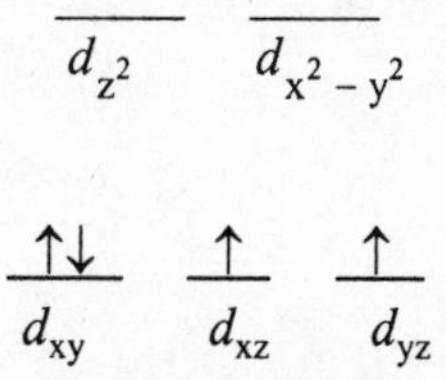

$[Cr(CN)_6]^{4-}$

(b) Water is a weak field ligand. The four 3*d* electrons should occupy the five orbitals as shown. There should be four unpaired electrons.

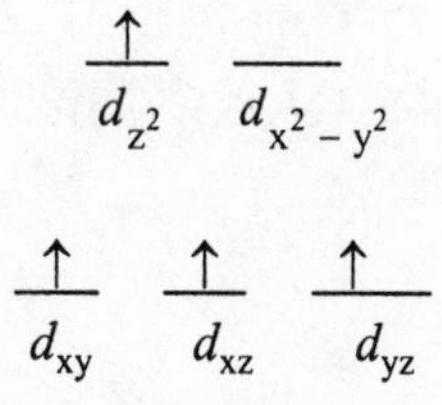

$[Cr(H_2O)_6]^{2+}$

18.32 Calculating molar mass from freezing-point depression, Problem Type 6A, Chapter 12.

Step 1: The equation for freezing-point depression is

$$\Delta T_f = K_f m$$

Solve this equation algebraically for molality (m), then substitute ΔT_f and K_f into the equation to calculate the molality.

$$m = \frac{\Delta T_f}{K_f} = \frac{0.56\ \text{°C}}{1.86\ \text{°C}/m} = 0.30\ m$$

Step 2: Multiplying the molality by the mass of solvent (in kg) gives moles of unknown solute. Then, dividing the mass of solute (in g) by the moles of solute, gives the molar mass of the unknown solute.

$$\text{? mol of unknown solute} = \frac{0.30 \text{ mol solute}}{1 \text{ kg water}} \times 0.0250 \text{ kg water} = 0.0075 \text{ mol solute}$$

$$\text{molar mass of unknown} = \frac{0.875 \text{ g}}{0.0075 \text{ mol}} = 117 \text{ g/mol}$$

The molar mass of $Co(NH_3)_4Cl_3$ is 233.4 g/mol, which is twice the computed molar mass. This implies dissociation into **two ions** in solution. The formula must be:

$\mathbf{[Co(NH_3)_4Cl_2]Cl}$

which contains the complex ion $[Co(NH_3)_4Cl_2]^+$ and a chloride ion, Cl^-. Possible structures of the complex ion are:

18.34 Reversing the first equation:

$$Ag(NH_3)_2^+(aq) \rightleftharpoons Ag^+(aq) + 2NH_3(aq)$$

$$K_1 = \frac{1}{1.5 \times 10^7} = 6.7 \times 10^{-8}$$

and adding it to the second equation,

$$Ag^+(aq) + 2CN^-(aq) \rightleftharpoons Ag(CN)_2^-(aq)$$

$$K_2 = 1.0 \times 10^{21}$$

Gives the equilibrium,

$$Ag(NH_3)_2^+(aq) + 2CN^-(aq) \rightleftharpoons Ag(CN)_2^-(aq) + 2NH_3\ (aq)$$

$$\boldsymbol{K} = K_1K_2 = (6.7 \times 10^{-8})(1.0 \times 10^{21}) = \mathbf{6.7 \times 10^{13}}$$

18.36 The orbital splitting diagram below shows five unpaired electrons. The Pauli exclusion principle requires that an electron jumping to a higher energy 3*d* orbital would have to change its spin. A transition that involves a change in spin state is forbidden, and thus does not occur to any appreciable extent.

The colors of transition metal complexes are due to the emission of energy in the form of visible light when an excited *d* electron relaxes to the ground state. In Mn^{2+} complexes, the promotion of a *d* electron to an excited state is spin forbidden; therefore, there is little in the way of an emission spectrum and the complexes are practically colorless.

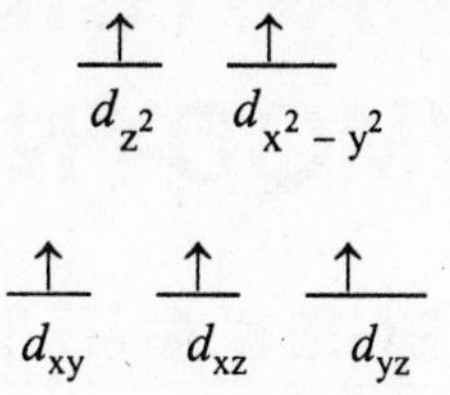

$Mn(H_2O)_6^{2+}$

18.38 Recall the wavelength and energy are inversely proportional. Thus, absorption of longer wavelength radiation corresponds to a lower energy transition. Lower energy corresponds to a smaller crystal field splitting.

(a) H_2O is a weaker field ligand than NH_3. Therefore, the crystal field splitting will be smaller for the aquo complex, and it will absorb at longer wavelengths.

(b) Fluoride is a weaker field ligand than cyanide. The fluoro complex will absorb at longer wavelengths.

(c) Chloride is a weaker field ligand than NH_3. The chloro complex will absorb at longer wavelengths.

18.40 **(a)** $4Au\ (s) + 8\ CN^-\ (aq) + O_2\ (g) + 2\ H_2O\ (l) \rightarrow 4[Au(CN)_2]^-\ (aq) + 4\ OH^-\ (aq)$

(b) $Zn\ (s) + 2[Au(CN)_2]^-\ (aq) \rightarrow [Zn(CN_4)]^{2-}\ (aq) + 2\ Au\ (s)$

(c) There are two cyanide ligands, so the coordination number is two. The geometry is linear.

18.42 Mn^{3+} is $3d^4$ and Cr^{3+} is $3d^5$. Therefore, Mn^{3+} has a greater tendency to accept an electron and is a stronger oxidizing agent. The $3d^5$ electron configuration of Cr^{3+} is a stable configuration.

18.44 Fe^{2+} is $3d^6$; Fe^{3+} is $3d^5$.

Therefore, Fe^{3+} (like Mn^{2+}, see problem 18.36) is nearly colorless, so it must be light yellow in color.

18.46 **(a)** Since carbon is less electronegative than oxygen, it is more likely that carbon will share electrons with Fe forming a metal-to-ligand sigma bond. The *sp* orbital on carbon containing the lone pair overlaps with the empty sp^3d^2 orbital on Fe.

(b)

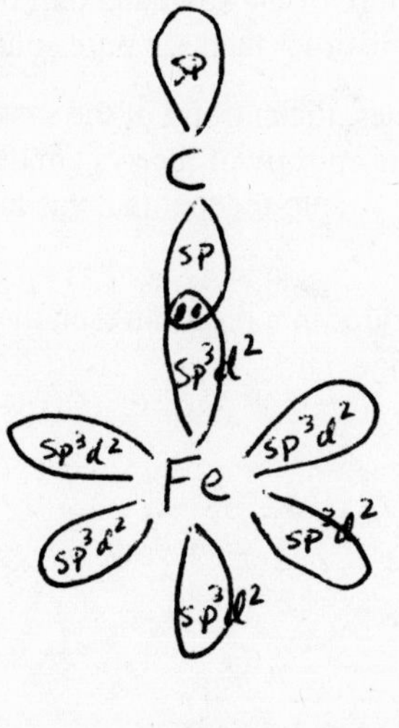

CHAPTER 19
THERMODYNAMICS

PROBLEM-SOLVING STRATEGIES AND TUTORIAL SOLUTIONS

TYPES OF PROBLEMS

Problem Type 1: Entropy.
(a) Predicting entropy changes.
(b) Calculating entropy changes of a system.

Problem Type 2: Free Energy.
(a) Calculating standard free energy changes.
(b) Calculating standard free energy change from enthalpy and entropy changes.
(c) Entropy changes due to phase transitions.

Problem Type 3: Free Energy and Chemical Equilibrium.
(a) Using the standard free energy change to calculate the equilibrium constant.
(b) Using the free energy change to predict the direction of a reaction.

PROBLEM TYPE 1: ENTROPY

Entropy (S) is a direct measure of the randomness or disorder of a system. The greater the disorder of a system, the greater its entropy. Conversely, the more ordered a system, the lower its entropy.

A. Predicting entropy changes

For any substance, the particles in the solid are more ordered than those in the liquid state, which in turn are more ordered than those in the gaseous state. See Figure 19.2 of your text. Thus, for any substance, the entropy for the same molar amount always increases in the following order.

$$S_{\text{solid}} < S_{\text{liquid}} < S_{\text{gas}}$$

There are a number of other factors that you need to consider to predict entropy changes.

- Heating increases the entropy of a system because it increases the random motion of atoms and molecules.
- When a solid dissolves in water, the highly ordered structure of the solid and part of the order of the water are destroyed. Consequently, the solution possesses greater disorder than the pure solute and pure solvent.
- If a reaction produces more gas molecules than it consumes, the entropy of the system increases. If the total number of gas molecules diminishes during a reaction, the entropy of the system decreases. The dependence of the entropy of reaction on the number of gas molecules is due to the fact that gas molecules possess much greater entropy than either solid or liquid molecules.
- If there is no net change in the total number of gas molecules in a reaction, then the entropy of the system may increase or decrease, but it will be a relatively small change.

EXAMPLE 19.1

Predict whether the entropy increases, decreases, or remains essentially unchanged for the following reactions.

(a) $H_2O_2\,(l) \longrightarrow H_2O\,(l) + 1/2\,O_2\,(g)$

(b) $H^+\,(aq) + OH^-\,(aq) \longrightarrow H_2O\,(l)$

(c) $Ca(OH)_2\,(s) + CO_2\,(g) \longrightarrow CaCO_3\,(s) + H_2O\,(g)$

(a) The number of moles of gaseous compounds in the products is greater than in the reactant. Entropy increases during this reaction.

(b) Two reactants combine into one product in this reaction. Order is increased and so entropy decreases.

(c) The number of moles of gas phase products is the same as the reactants. The entropy remains essentially unchanged.

PRACTICE EXERCISE

1. Predict whether the entropy increases, decreases, or remains essentially unchanged for the following reactions:

   (a) $CaO\,(s) + CO_2\,(g) \longrightarrow CaCO_3\,(s)$

   (b) $CuSO_4\,(s) \longrightarrow Cu^{2+}\,(aq) + SO_4^{2-}\,(aq)$

   (c) $2\,HCl\,(g) + Br_2\,(l) \longrightarrow 2\,HBr\,(g) + Cl_2\,(g)$

Text Problems: 8, 12

B. Calculating entropy changes of a system

The universe is made up of the system and the surroundings. The entropy change in the universe (ΔS_{univ}) for any process is the *sum* of the entropy changes in the system (ΔS_{sys}) and in the surroundings (ΔS_{surr}).

$$\Delta S_{univ} = \Delta S_{sys} + \Delta S_{surr}$$

For a spontaneous process, the entropy of the universe increases.

$$\Delta S_{univ} = \Delta S_{sys} + \Delta S_{surr} > 0$$

In chemistry, we typically focus on the entropy of the system. Let's suppose that the system is represented by the following reaction:

$$a\text{A} + b\text{B} \longrightarrow c\text{C} + d\text{D}$$

As is the case for the enthalpy of a reaction [see Equation (6.7) in the text], the standard entropy change $\Delta S°$ is given by:

$$\Delta S^\circ_{rxn} = [cS°(\text{C}) + dS°(\text{D})] - [aS°(\text{A}) + bS°(\text{B})]$$

or, using Σ to represent summation and m and n for the stoichiometric coefficients in the reaction,

$$\Delta S^\circ_{rxn} = \Sigma nS°(\text{products}) - \Sigma mS°(\text{reactants})$$

Using the standard entropy values ($S°$) listed in Appendix 2, we can calculate ΔS°_{rxn}, which corresponds to ΔS_{sys}.

EXAMPLE 19.2

Use standard entropy values to calculate the standard entropy change (ΔS°_{rxn}) for the reaction

$$H_2(g) + 1/2\ O_2(g) \longrightarrow H_2O(l)$$

The standard entropy change is given by:

$$\Delta S^\circ_{rxn} = S^\circ[H_2O(l)] - \{S^\circ[H_2(g)] + (1/2)S^\circ[O_2(g)]\}$$

You can look up the standard entropy values in Appendix 2 of the text.

$$\Delta S^\circ_{rxn} = (1\ \text{mol})\left(\frac{69.9\ \text{J}}{\text{K}\cdot\text{mol}}\right) - \left[(1\ \text{mol})\left(\frac{131.0\ \text{J}}{\text{mol}\cdot\text{K}}\right) + \left(\frac{1}{2}\ \text{mol}\right)\left(\frac{205.0\ \text{J}}{\text{mol}\cdot\text{K}}\right)\right]$$

$$\Delta S^\circ_{rxn} = 69.9\ \text{J/K} - 233.5\ \text{J/K} = \mathbf{-163.6\ J/K}$$

> **Tip:** This reaction is known to be spontaneous, so you would think that ΔS should be positive. Remember, that the value of ΔS°_{rxn} applies only to the system; ΔS_{univ} will have a positive value.

PRACTICE EXERCISE

2. Hydrate lime, $Ca(OH)_2$, can be reformed into quicklime, CaO, by heating:

$$Ca(OH)_2(s) \xrightarrow{\text{heat}} CaO(s) + H_2O(g)$$

Use standard entropy values to calculate the standard entropy change (ΔS°_{rxn}) for this reaction.

Text Problem: 10

PROBLEM TYPE 2: FREE ENERGY

The second law of thermodynamics tells us that a spontaneous reaction increases the entropy of the universe; that is, $\Delta S_{univ} > 0$. To calculate ΔS_{univ}, we must calculate both ΔS_{surr} and ΔS_{sys}. However, it can be difficult to calculate ΔS_{surr}, and typically we are only concerned with what happens in a particular system.

Therefore, considering only the system, we use another thermodynamic function to determine if a reaction will occur spontaneously. This function, called **Gibbs free energy (*G*)**, or simply free energy is given by the following equation:

$$G = H - TS$$

where,

H is the enthalpy
S is the entropy
T is the temperature (in K)

The change in free energy (ΔG) of a system for a reaction at constant temperature is:

$$\Delta G = \Delta H - T\Delta S$$

The sign of ΔG will allow us to predict whether a reaction is spontaneous at equilibrium. At constant temperature and pressure, we can summarize the following conditions in terms of ΔG.

- $\Delta G < 0$ — A spontaneous process in the forward direction.
- $\Delta G > 0$ — A nonspontaneous reaction as written. However, the reaction is spontaneous in the reverse direction.
- $\Delta G = 0$ — The system is at equilibrium. There is no net change in the system.

A. Calculating standard free energy changes

Let's again suppose that the system is represented by the following reaction:

$$a\text{A} + b\text{B} \longrightarrow c\text{C} + d\text{D}$$

As is the case for the entropy of a reaction the standard free energy change $\Delta G°$ is given by:

$$\Delta G^\circ_{\text{rxn}} = [c\,\Delta G^\circ_f(\text{C}) + d\,\Delta G^\circ_f(\text{D})] - [a\,\Delta G^\circ_f(\text{A}) + b\,\Delta G^\circ_f(\text{B})]$$

or,

$$\Delta G^\circ_{\text{rxn}} = \Sigma n\,\Delta G^\circ_f\,(\text{products}) - \Sigma m\,\Delta G^\circ_f\,(\text{reactants})$$

where m and n are stoichiometric coefficients. The term ΔG°_f is the **standard free energy of formation** of a compound. It is the free energy change that occurs when 1 mole of the compound is synthesized from its elements in their standard states (see Table 19.2 of your text for conventions used for standard states). By definition, the standard free energy of formation of any element in its stable form is *zero*.

For example,

$$\Delta G^\circ_f\,[\text{O}_2(\text{g})] = 0 \quad \text{and} \quad \Delta G^\circ_f\,[\text{Na(s)}] = 0$$

Other standard free energies of formation can be found in Appendix 2 of the text.

EXAMPLE 19.3

Calculate $\Delta G^\circ_{\text{rxn}}$ at 25°C for the following reaction given that $\Delta G^\circ_f\,(Fe_2O_3) = -741.0$ kJ/mol.

$$\mathbf{2\,Al\,(\mathit{s}) + Fe_2O_3\,(\mathit{s}) \longrightarrow Al_2O_3\,(\mathit{s}) + 2\,Fe\,(\mathit{s})}$$

The standard free energy change is given by:

$$\Delta G^\circ_{\text{rxn}} = \{\Delta G^\circ_f\,[\text{Al}_2\text{O}_3\,(s)] + 2\,\Delta G^\circ_f\,[\text{Fe}\,(s)]\} - \{2\,\Delta G^\circ_f\,[\text{Al}\,(s)] + \Delta G^\circ_f\,[\text{Fe}_2\text{O}_3\,(s)]\}$$

You can look up the standard free energy of formation values in Appendix 2 of the text.

$$\Delta G^\circ_{\text{rxn}} = [(1\,\cancel{\text{mol}})(-1576.41\text{ kJ}/\cancel{\text{mol}}) + 0] - [0 + (1\,\cancel{\text{mol}})(-741.0\text{ kJ}/\cancel{\text{mol}})]$$

$$\mathbf{\Delta G^\circ_{rxn}} = -1576.41\text{ kJ} + 741.0\text{ kJ} = \mathbf{-835.4\ kJ}$$

> **Tip:** Remember that the standard free energy of formation of any element in its stable form is *zero*. Therefore, the standard free energy of formation values for both Al (*s*) and Fe (*s*) are *zero*.

PRACTICE EXERCISE

3. Using Appendix 2 of the text, calculate $\Delta G^\circ_{\text{rxn}}$ values for the following reactions:

(a) $3\,\text{CaO}\,(s) + 2\,\text{Al}\,(s) \longrightarrow 3\,\text{Ca}\,(s) + \text{Al}_2\text{O}_3\,(s)$

(b) $\text{ZnO}\,(s) \longrightarrow \text{Zn}\,(s) + 1/2\,\text{O}_2\,(g)$

Text Problems: 16, 30

B. Calculating standard free energy change from enthalpy and entropy changes

We can determine the sign of ΔG if we know the signs of both ΔH and ΔS. A negative ΔH (an exothermic reaction) and a positive ΔS (increase in disorder), give a negative ΔG. In addition, temperature may influence the direction of a spontaneous reaction. There are four possible outcomes for the relationship

$$\Delta G = \Delta H - T\Delta S$$

- If both ΔH and ΔS are positive, ΔG will be negative only when $T\Delta S$ is greater in magnitude than ΔH. This condition is met when T is large (high temperature).
- If ΔH is positive and ΔS is negative, ΔG will always be positive, regardless of temperature. The reaction is nonspontaneous.
- If ΔH is negative and ΔS is positive, ΔG will always be negative, regardless of temperature. The reaction is spontaneous.
- If ΔH is negative and ΔS is negative, ΔG will be negative only when ΔH is greater in magnitude than $T\Delta S$. This condition is met when T is small (low temperature).

We can also calculate the value of ΔG if we know the values of ΔH, ΔS, and the temperature. Typically, $\Delta H°$ is calculated from standard enthalpies of formation.

$$\Delta H° = \Sigma n\,\Delta H_f^\circ\,(\text{products}) - \Sigma m\,\Delta H_f^\circ\,(\text{reactants})$$

Also, $\Delta S°$ is calculated from standard entropy values (see Problem Type 1B).

$$\Delta S° = \Sigma nS°(\text{products}) - \Sigma mS°(\text{reactants})$$

Then, for reactions carried out under standard-state conditions, we can substitute $\Delta H°$, $\Delta S°$, and the temperature into the following equation to calculate $\Delta G°$.

$$\Delta G° = \Delta H° - T\Delta S°$$

EXAMPLE 19.4

Calculate $\Delta G°$ for the following reaction at 298 K:

$$\mathbf{2\,H_2\,(g) + CO\,(g) \rightleftharpoons CH_3OH\,(g)}$$

given that $\Delta H° = -90.7$ kJ and $\Delta S° = -221.5$ J/K for this process.

The standard free energy change is given by:

$$\Delta G° = \Delta H° - T\Delta S°$$

Step 1: Let's convert $\Delta S°$ to units of kJ/K, so that we have consistent units.

$$-221.5\,\frac{\cancel{\text{J}}}{\text{K}} \times \frac{1\text{ kJ}}{1000\,\cancel{\text{J}}} = -0.2215\text{ kJ/K}$$

Step 2: Substitute $\Delta H°$, $\Delta S°$, and the temperature into the above equation to calculate $\Delta G°$.

$$\Delta G° = \Delta H° - T\Delta S°$$

$$\Delta G° = -90.7\text{ kJ} - [(298\,\cancel{\text{K}})(-0.2215\text{ kJ}/\cancel{\text{K}})]$$

$$\mathbf{\Delta G° = -90.7\text{ kJ} + 66.0\text{ kJ} = -24.7\text{ kJ}}$$

PRACTICE EXERCISE

4. The synthesis of O_2 (*g*) is often carried out in general chemistry laboratories by the decomposition of $KClO_3$,

$$KClO_3\,(s) \longrightarrow KCl\,(s) + 3/2\,O_2\,(g)$$

for which $\Delta H° = -44.7$ kJ and $\Delta S° = +59.1$ J/K. Is this reaction spontaneous at 25°C under standard conditions?

Text Problem: 18

C. Entropy changes due to phase transitions

At the transition temperature, the melting or boiling point, a system is at equilibrium and $\Delta G = 0$. Thus, we can write,

$$\Delta G = \Delta H - T\Delta S$$

or

$$0 = \Delta H - T\Delta S$$

and

$$\Delta S = \frac{\Delta H}{T} \tag{19.1}$$

ΔS is the entropy change due to the phase transition.

EXAMPLE 19.5

The heat of fusion of water, ΔH_{fus}, at 0°C is 6.02 kJ/mol. What is ΔS_{fus} for 1 mol of H_2O at the melting point?

The entropy change due to the phase transition (the melting of water), can be calculated using Equation (19.1). Recall that the temperature must be in units of Kelvin (0°C = 273 K).

$$\Delta S = \frac{\Delta H}{T}$$

$$\Delta S = \frac{6.02\ \frac{\text{kJ}}{\text{mol}}}{273\ \text{K}} = \mathbf{+0.0221\ kJ/mol{\cdot}K} = \mathbf{+22.1\ J/mol{\cdot}K}$$

The increase in entropy upon melting the solid corresponds to the increase in molecular disorder in the liquid state compared to the solid state.

PRACTICE EXERCISE

5. The enthalpy of vaporization of mercury is 58.5 kJ/mol and the normal boiling point is 630 K. What is the entropy of vaporization of mercury?

Text Problem: 54

PROBLEM TYPE 3: FREE ENERGY AND CHEMICAL EQUILIBRIUM

A. Using the standard free energy change to calculate the equilibrium constant

There is a relationship between the free energy change (ΔG) and the standard free energy change.

$$\Delta G = \Delta G^\circ + RT\ln Q \qquad (19.2)$$

where,

R is the gas constant (8.314 J/mol·K)
T is the absolute temperature of the reaction (in K)
Q is the reaction quotient (see Problem Type 4, Chapter 15)

At equilibrium, by definition, $\Delta G = 0$ and $Q = K$, where K is the equilibrium constant. Thus,

$$0 = \Delta G^\circ + RT\ln K$$

or

$$\Delta G^\circ = -RT\ln K \qquad (19.3)$$

K can be either K_p, used for gases, or K_c used for reactions in solution. Note that the larger the value of K, the more negative the value of ΔG°. This should make sense because a large value of K means that the equilibrium lies far to the right (toward products), and a negative value for ΔG° means that the reaction is spontaneous in the forward direction. See Table 19.4 of the text for a summary of the relation between K and ΔG°.

EXAMPLE 19.6
The standard free energy change for the reaction

$$1/2\ N_2\,(g) + 3/2\ H_2\,(g) \rightleftharpoons NH_3\,(g)$$

is $\Delta G^\circ = 26.9$ kJ/mol at 700 K. Calculate the equilibrium constant at this temperature.

The equilibrium constant is related to the standard free energy change by the Equation (19.3).

$$\Delta G^\circ = -RT\ln K$$

Step 1: R has units of J/mol·K, so we must convert ΔG° from units of kJ/mol to J/mol.

$$26.9\ \frac{\text{kJ}}{\text{mol}} \times \frac{1000\text{ J}}{1\text{ kJ}} = 2.69 \times 10^4\text{ J/mol}$$

Step 2: Substitute ΔG°, R, and T into Equation (19.3) to calculate the equilibrium constant, K_p. We are calculating K_p in this problem since this a gas-phase reaction.

$$\Delta G^\circ = -RT\ln K_p$$

$$2.69 \times 10^4\text{ J/mol} = -(8.314\text{ J/mol·K})(700\text{ K})\ln K_p$$

$$-4.62 = \ln K_p$$

Taking the antilog of both sides gives:

$$e^{-4.62} = K_p$$

$$\mathbf{K_p = 9.85 \times 10^{-3}}$$

PRACTICE EXERCISE

6. In Chapter 15, we saw that for the reaction

$$H_2\,(g) + I_2\,(g) \longrightarrow 2\,HI\,(g)$$

the equilibrium constant at 400°C is $K_p = 64$. Calculate the value of ΔG° at this temperature.

Text Problems: 22, 24, 26a, 28

B. Using the free energy change to predict the direction of a reaction

To predict the direction of a reaction, you can calculate ΔG from Equation (19.2).

$$\Delta G = \Delta G° + RT\ln Q \qquad (19.2)$$

To calculate the free energy change (ΔG), you must be given or you must calculate $\Delta G°$ and you must calculate the reaction quotient Q. Substitute the values of $\Delta G°$ and Q into Equation (19.2) to solve for ΔG. Recall the meaning of the sign of ΔG:

- $\Delta G < 0$ A spontaneous process in the forward direction.
- $\Delta G > 0$ A nonspontaneous reaction as written. However, the reaction is spontaneous in the reverse direction.
- $\Delta G = 0$ The system is at equilibrium. There is no net change in the system.

EXAMPLE 19.7

Using the reaction and data given in Example 19.6, calculate ΔG at 700 K if the reaction mixture consists of 30.0 atm of H_2, 20.0 atm of N_2, and 0.500 atm of NH_3.

Under nonstandard conditions, ΔG is related to the reaction quotient Q by Equation (19.2).

$$\Delta G = \Delta G° + RT\ln Q_p$$

We are using Q_p in the equation because this is a gas-phase reaction.

Step 1: $\Delta G°$ is given in Example 19.6. We must calculate Q_p.

$$Q_p = \frac{P_{NH_3}}{P_{N_2}^{1/2} \cdot P_{H_2}^{3/2}} = \frac{0.50}{(20.0)^{1/2}(30.0)^{3/2}} = 6.80 \times 10^{-4}$$

Step 2: Substitute $\Delta G° = 2.69 \times 10^4$ J/mol and Q_p into Equation (19.2) to calculate ΔG.

$$\Delta G = \Delta G° + RT\ln Q_p$$

$$\Delta G = 2.69 \times 10^4 \text{ J/mol} + (8.314 \text{ J/mol}\cdot\text{K})(700 \text{ K})\ln(6.80 \times 10^{-4})$$

$$\Delta G = 2.69 \times 10^4 \text{ J/mol} - 4.24 \times 10^4 \text{ J/mol} = \mathbf{-1.55 \times 10^4 \text{ J/mol}} = \mathbf{-15.5 \text{ kJ/mol}}$$

By making the partial pressures of N_2 and H_2 high and that of NH_3 low, the reaction is spontaneous in the forward direction. This condition corresponds to $Q_p < K_p$, and so the reaction proceeds in the forward direction until $Q_p = K_p$ (see Problem Type 4, Chapter 15).

PRACTICE EXERCISE

7. Calculate ΔG for the following reaction at 25°C when the pressure of CO_2 is 0.0010 atm, given that $\Delta H° = 177.8$ kJ and $\Delta S° = 160.5$ J/K.

$$CaCO_3(s) \longrightarrow CaO(s) + CO_2(g)$$

Text Problem: 26b

ANSWERS TO PRACTICE EXERCISES

1. **(a)** entropy decreases **(b)** entropy increases **(c)** entropy increases

2. $\Delta S^\circ_{rxn} = +145.1$ J/K

3. **(a)** $\Delta G^\circ_{rxn} = 236$ kJ **(b)** $\Delta G^\circ_{rxn} = 318.2$ kJ

4. $\Delta G^\circ = -62.3$ kJ. Since $\Delta G^\circ < 0$, the reaction is spontaneous.

5. $\Delta S_{vap} = 92.9$ J/mol·K

6. $\Delta G^\circ = -23.2$ kJ

7. $\Delta G = 113$ kJ

SOLUTIONS TO SELECTED TEXT PROBLEMS

19.4 The probability (P) of finding all the molecules in the same flask becomes progressively smaller as the number of molecules increases. An equation that relates the probability to the number of molecules is given in the text.

$$P = \left(\frac{1}{2}\right)^N$$

where,

N is the total number of molecules present.

Using the above equation, we find:

(a) $P = 0.02$ **(b)** $P = 9 \times 10^{-19}$ **(c)** $P = 2 \times 10^{-181}$

19.8 In order of increasing entropy per mole at 25°C:

(c) < (d) < (a) < (e) < (b)

(c) Na (*s*): ordered, crystalline material.
(d) NaCl (*s*): ordered crystalline material, but with more particles per mole than Na(s).
(a) Ne (*g*): a monatomic gas, hence of higher entropy than a solid.
(e) NH_3 (*g*): a polyatomic gas.
(b) SO_2 (*g*): a polyatomic gas of higher molar mass than ammonia.

19.10 Calculating entropy changes of a system, Problem Type 1B. The standard entropy change for a reaction can be calculated using the following equation.

$$\Delta S^\circ_{\text{rxn}} = \Sigma n S^\circ(\text{products}) - \Sigma m S^\circ(\text{reactants})$$

(a) $\Delta S^\circ_{\text{rxn}} = S^\circ(\text{Cu}) + S^\circ(\text{H}_2\text{O}) - [S^\circ(\text{H}_2) + S^\circ(\text{CuO})]$

$= (1\text{ mol})(33.3\text{ J/K·mol}) + (1\text{ mol})(188.7\text{ J/K·mol}) - [(1\text{ mol})(131.0\text{ J/K·mol}) + (1\text{ mol})(43.5\text{ J/K·mol})]$

$=$ **47.5 J/K**

(b) $\Delta S^\circ_{\text{rxn}} = S^\circ(\text{Al}_2\text{O}_3) + 3S^\circ(\text{Zn}) - [2S^\circ(\text{Al}) + 3S^\circ(\text{ZnO})]$

$= (1\text{ mol})(50.99\text{ J/K·mol}) + (3\text{ mol})(41.6\text{ J/K·mol}) - [(2\text{ mol})(28.3\text{ J/K·mol}) + (3\text{ mol})(43.9\text{ J/K·mol})]$

$=$ **−12.5 J/K**

(c) $\Delta S^\circ_{\text{rxn}} = S^\circ(\text{CO}_2) + 2S^\circ(\text{H}_2\text{O}) - [S^\circ(\text{CH}_4) + 2S^\circ(\text{O}_2)]$

$= (1\text{ mol})(213.6\text{ J/K·mol}) + (2\text{ mol})(69.9\text{ J/K·mol}) - [(1\text{ mol})(186.2\text{ J/K·mol}) + (2\text{ mol})(205.0\text{ J/K·mol})]$

$=$ **−242.8 J/K**

Why was the entropy value for water different in parts (a) and (c)?

19.12 **(a)** $\Delta S < 0$; gas reacting with a liquid to form a solid (decrease in number of moles of gas).

(b) $\Delta S > 0$; solid decomposing to give a liquid and a gas.

(c) $\Delta S > 0$; increase in number of moles of gas.

(d) $\Delta S < 0$; gas reacting with a solid to form a solid (decrease in number of moles of gas).

19.16 Calculating standard free energy changes, Problem Type 2A. The standard free energy change for a reaction can be calculated using the following equation.

$$\Delta G^\circ_{\text{rxn}} = \Sigma n\,\Delta G^\circ_{\text{f}}\,(\text{products}) - \Sigma m\,\Delta G^\circ_{\text{f}}\,(\text{reactants})$$

(a) $\Delta G^\circ_{\text{rxn}} = 2\,\Delta G^\circ_{\text{f}}(\text{MgO}) - [2\,\Delta G^\circ_{\text{f}}(\text{Mg}) + \Delta G^\circ_{\text{f}}(\text{O}_2)]$

$$\boldsymbol{\Delta G^\circ_{\text{rxn}}} = (2\text{ mol})(-569.6\text{ kJ/mol}) - [(2\text{ mol})(0) + (1\text{ mol})(0)] = \mathbf{-1139\ kJ}$$

(b) $\Delta G^\circ_{\text{rxn}} = 2\,\Delta G^\circ_{\text{f}}(\text{SO}_3) - [2\,\Delta G^\circ_{\text{f}}(\text{SO}_2) + \Delta G^\circ_{\text{f}}(\text{O}_2)]$

$$\boldsymbol{\Delta G^\circ_{\text{rxn}}} = (2\text{ mol})(-370.4\text{ kJ/mol}) - [(2\text{ mol})(-300.4\text{ kJ/mol}) + (1\text{ mol})(0)] = \mathbf{-140.0\ kJ}$$

(c) $\Delta G^\circ_{\text{rxn}} = 4\,\Delta G^\circ_{\text{f}}[\text{CO}_2(g)] + 6\,\Delta G^\circ_{\text{f}}[\text{H}_2\text{O}(l)] - \{2\,\Delta G^\circ_{\text{f}}[\text{C}_2\text{H}_6(g)] + 7\,\Delta G^\circ_{\text{f}}[\text{O}_2(g)]\}$

$$\Delta G^\circ_{\text{rxn}} = (4\text{ mol})(-394.4\text{ kJ/mol}) + (6\text{ mol})(-237.2\text{ kJ/mol}) - [(2\text{ mol})(-32.89\text{ kJ/mol}) + (7\text{ mol})(0)]$$

$$\boldsymbol{\Delta G^\circ_{\text{rxn}}} = \mathbf{-2935.0\ kJ}$$

19.18 **Reaction A:** Calculate ΔG from ΔH and ΔS.

$$\boldsymbol{\Delta G} = \Delta H - T\Delta S = -126{,}000\text{ J} - (298\text{ K} \times 84\text{ J/K}) = \mathbf{-151{,}000\ J}$$

The free energy change is negative so the reaction is spontaneous at 298 K. Since ΔH is negative and ΔS is positive, the reaction is spontaneous at all temperatures.

Reaction B: Calculate ΔG.

$$\boldsymbol{\Delta G} = \Delta H - T\Delta S = -11{,}700\text{ J} - [(298\text{ K}) \times (-105\text{ J/K})] = \mathbf{+19{,}600\ J}$$

The free energy change is positive at 298 K which means the reaction is not spontaneous at that temperature. The positive sign of ΔG results from the large negative value of ΔS. At lower temperatures, the $-T\Delta S$ term will be smaller thus allowing the free energy change to be negative.

ΔG will equal zero when $\Delta H = T\Delta S$.

Rearranging,

$$\boldsymbol{T} = \frac{\Delta H}{\Delta S} = \frac{-11700\text{ J}}{-105\text{ J/K}} = \mathbf{111\ K}$$

At temperatures below 111 K, ΔG will be negative and the reaction will be spontaneous.

19.22 Similar to using the standard free energy change to calculate the equilibrium constant, Problem Type 3A.

The equilibrium constant is related to the standard free energy change by the following equation.

$$\Delta G^\circ = -RT\ln K$$

Substitute K_w, R, and T into the above equation to calculate the standard free energy change, ΔG°. The temperature at which $K_w = 1.0 \times 10^{-14}$ is 25°C = 298 K.

$$\Delta G^\circ = -RT\ln K_w$$

$$\mathbf{\Delta G^\circ} = -(1\ \cancel{\text{mol}})(8.314\ \text{J}/\cancel{\text{mol}}\cdot\cancel{\text{K}})(298\ \cancel{\text{K}})\ln(1.0 \times 10^{-14}) = \mathbf{8.0 \times 10^4\ J} = \mathbf{8.0 \times 10^1\ kJ}$$

19.24 Use standard free energies of formation from Appendix 2 to find the standard free energy difference.

$$\Delta G^\circ_{rxn} = 2\,\Delta G^\circ_f[H_2(g)] + \Delta G^\circ_f[O_2(g)] - 2\,\Delta G^\circ_f[H_2O(g)]$$

$$\Delta G^\circ_{rxn} = (2\ \text{mol})(0) + (1\ \text{mol})(0) - (2\ \cancel{\text{mol}})(-228.6\ \text{kJ}/\cancel{\text{mol}})$$

$$\Delta G^\circ_{rxn} = 457.2\ \text{kJ} = 4.572 \times 10^5\ \text{J}$$

We can calculate K_p using the following equation.

$$\Delta G^\circ = -RT\ln K_p$$

$$4.572 \times 10^5\ \cancel{\text{J}} = -(1\ \cancel{\text{mol}})(8.314\ \cancel{\text{J}}/\cancel{\text{mol}}\cdot\cancel{\text{K}})(298\ \cancel{\text{K}})\ln K_p$$

$$-185 = \ln K_p$$

Taking the antilog of both sides,

$$e^{-185} = K_p$$

$$\mathbf{K_p = 5 \times 10^{-81}}$$

19.26 **(a)** Similar to using the standard free energy change to calculate the equilibrium constant, Problem Type 3A.

The equilibrium constant is related to the standard free energy change by the following equation.

$$\Delta G^\circ = -RT\ln K$$

Substitute K_p, R, and T into the above equation to the standard free energy change, ΔG°.

$$\Delta G^\circ = -RT\ln K_p$$

$$\mathbf{\Delta G^\circ} = -(1\ \cancel{\text{mol}})(8.314\ \text{J}/\cancel{\text{mol}}\cdot\cancel{\text{K}})(2000\ \cancel{\text{K}})\ln(4.40) = \mathbf{-2.46 \times 10^4\ J} = \mathbf{-24.6\ kJ}$$

(b) Similar to using the free energy change to predict the direction of a reaction, Problem Type 3B.

Under nonstandard conditions, ΔG is related to the reaction quotient Q by the following equation.

$$\Delta G = \Delta G^\circ + RT\ln Q_p$$

We are using Q_p in the equation because this is a gas-phase reaction.

Step 1: $\Delta G°$ was calculated in part (a). We must calculate Q_p.

$$Q_p = \frac{P_{H_2O} \cdot P_{CO}}{P_{H_2} \cdot P_{CO_2}} = \frac{(0.66)(1.20)}{(0.25)(0.78)} = 4.1$$

Step 2: Substitute $\Delta G° = -2.46 \times 10^4$ J and Q_p into the following equation to calculate ΔG.

$$\Delta G = \Delta G° + RT\ln Q_p$$

$$\Delta G = -2.46 \times 10^4 \text{ J} + (1 \text{ mol})(8.314 \text{ J/mol·K})(2000 \text{ K})\ln(4.1)$$

$$\Delta G = -2.46 \times 10^4 \text{ J} + 2.35 \times 10^4 \text{ J}$$

$$\mathbf{\Delta G = -1.10 \times 10^3 \text{ J} = -1.10 \text{ kJ}}$$

19.28 We use the given K_p to find the standard free energy change.

$$\Delta G° = -RT\ln K$$

$$\Delta G° = -(1 \text{ mol})(8.314 \text{ J/K·mol})(298 \text{ K})\ln(5.62 \times 10^{35}) = 2.04 \times 10^5 \text{ J} = -204 \text{ kJ}$$

The standard free energy of formation of one mole of $COCl_2$ can now be found using the standard free energy of reaction calculated above and the standard free energies of formation of CO (*g*) and Cl_2 (*g*).

$$\Delta G°_{rxn} = \Sigma n\, \Delta G°_f(\text{products}) - \Sigma m\, \Delta G°_f(\text{reactants})$$

$$\Delta G°_{rxn} = \Delta G°_f[COCl_2(g)] - \{\Delta G°_f[CO(g)] + \Delta G°_f[Cl_2(g)]\}$$

$$-204 \text{ kJ} = (1\text{mol})\,\Delta G°_f[COCl_2(g)] - [(1 \text{ mol})(-137.3 \text{ kJ/mol}) + (1 \text{ mol})(0)]$$

$$\mathbf{\Delta G°_f[COCl_2(g)] = -341 \text{ kJ/mol}}$$

19.30 The standard free energy change is given by:

$$\Delta G°_{rxn} = \Delta G°_f(\text{graphite}) - \Delta G°_f(\text{diamond})$$

You can look up the standard free energy of formation values in Appendix 2 of the text.

$$\mathbf{\Delta G°_{rxn}} = (1 \text{ mol})(0) - (1 \text{ mol})(2.87 \text{ kJ/mol}) = \mathbf{-2.87 \text{ kJ}}$$

Thus, the reaction is spontaneous under standard state conditions at 25°C. However, the rate of the diamond to graphite conversion is very slow so that it will take millions of years before the process is complete.

19.32 In each part of this problem we can use the following equation to calculate ΔG.

$$\Delta G = \Delta G° + RT\ln Q$$

or,

$$\Delta G = \Delta G° + RT\ln [H^+][OH^-]$$

(a) In this case, the given concentrations are equilibrium concentrations at 25°C. Since the reaction is at equilibrium, $\mathbf{\Delta G = 0}$. This is advantageous, because it allows us to calculate $\Delta G°$. Also recall that at equilibrium, $Q = K$. We can write:

$$\Delta G° = -RT\ln K_w$$

$$\mathbf{\Delta G°} = -(1 \text{ mol})(8.314 \text{ J/K·mol})(298 \text{ K})\ln(1.0 \times 10^{-14}) = \mathbf{8.0 \times 10^4 \text{ J}}$$

(b) $\Delta G = \Delta G° + RT\ln Q = \Delta G° + RT\ln [H^+][OH^-]$

$$\mathbf{\Delta G} = 8.0 \times 10^4\ \text{J} + (1\ \text{mol})(8.314\ \text{J/K·mol})(298\ \text{K})\ln (1.0 \times 10^{-3})(1.0 \times 10^{-4}) = \mathbf{4.0 \times 10^4\ J}$$

(c) $\Delta G = \Delta G° + RT\ln Q = \Delta G° + RT\ln [H^+][OH^-]$

$$\mathbf{\Delta G} = 8.0 \times 10^4\ \text{J} + (1\ \text{mol})(8.314\ \text{J/K·mol})(298\ \text{K})\ln (1.0 \times 10^{-12})(2.0 \times 10^{-8}) = \mathbf{-3.2 \times 10^4\ J}$$

(d) $\Delta G = \Delta G° + RT\ln Q = \Delta G° + RT\ln [H^+][OH^-]$

$$\mathbf{\Delta G} = 8.0 \times 10^4\ \text{J} + (1\ \text{mol})(8.314\ \text{J/K·mol})(298\ \text{K})\ln (3.5)(4.8 \times 10^{-4}) = \mathbf{6.4 \times 10^4\ J}$$

19.34 One possible explanation is simply that no reaction is possible, namely that there is an unfavorable free energy difference between products and reactants ($\Delta G > 0$).

A second possibility is that the potential for spontaneous change is there ($\Delta G < 0$), but that the reaction is extremely slow (very large activation energy).

A remote third choice is that the student accidentally prepared a mixture in which the components were already at their equilibrium concentrations.

Which of the above situations would be altered by the addition of a catalyst?

19.36 For a solid to liquid phase transition (melting) the entropy always increases ($\Delta S > 0$) and the reaction is always endothermic ($\Delta H > 0$).

(a) Melting is always spontaneous above the melting point, so $\Delta G < 0$.

(b) At the melting point (−77.7°C), solid and liquid are in equilibrium, so $\Delta G = 0$.

(c) Melting is not spontaneous below the melting point, so $\Delta G > 0$.

19.38 If the process is *spontaneous* as well as *endothermic*, the signs of ΔG and ΔH must be negative and positive, respectively. Since $\Delta G = \Delta H - T\Delta S$, the sign of ΔS must be positive for ΔG to be negative.

19.40 **(a)** Using the relationship:

$$\frac{\Delta H_{vap}}{T_{b.p.}} = \Delta S_{vap} \cong 90\ \text{J/K·mol}$$

benzene	ΔS_{vap} = 87.8 J/K·mol
hexane	ΔS_{vap} = 90.1 J/K·mol
mercury	ΔS_{vap} = 93.7 J/K·mol
toluene	ΔS_{vap} = 91.8 J/K·mol

Most liquids have ΔS_{vap} approximately equal to a constant value because the order of the molecules in the liquid state is similar. The order of most gases is totally random; thus, ΔS for liquid → vapor should be similar for most liquids.

(b) Using the data in Table 11.5, we find:

ethanol ΔS_{vap} = 111.9 J/K·mol

water ΔS_{vap} = 109.4 J/K·mol

Both water and ethanol have a larger ΔS_{vap} because the liquid molecules are more ordered due to hydrogen bonding.

19.42 **(a)** At two different temperatures T_1 and T_2,

$$\Delta G_1^\circ = \Delta H^\circ - T_1\Delta S^\circ = -RT\ln K_1 \qquad (1)$$

$$\Delta G_2^\circ = \Delta H^\circ - T_2\Delta S^\circ = -RT\ln K_2 \qquad (2)$$

Rearranging Equations (1) and (2),

$$\ln K_1 = \frac{-\Delta H^\circ}{RT_1} + \frac{\Delta S^\circ}{R} \qquad (3)$$

$$\ln K_2 = \frac{-\Delta H^\circ}{RT_2} + \frac{\Delta S^\circ}{R} \qquad (4)$$

Subtracting equation (3) from equation (4) gives,

$$\ln K_2 - \ln K_1 = \left(\frac{-\Delta H^\circ}{RT_2} + \frac{\Delta S^\circ}{R}\right) - \left(\frac{-\Delta H^\circ}{RT_1} + \frac{\Delta S^\circ}{R}\right)$$

$$\ln\frac{K_2}{K_1} = \frac{\Delta H^\circ}{R}\left(\frac{1}{T_1} - \frac{1}{T_2}\right)$$

$$\ln\frac{K_2}{K_1} = \frac{\Delta H^\circ}{R}\left(\frac{T_2 - T_1}{T_1T_2}\right)$$

(b) Using the equation that we just derived, we can calculate the equilibrium constant at 65°C.

$K_1 = 4.63 \times 10^{-3}$ $T_1 = 298\text{ K}$

$K_2 = ?$ $T_2 = 338\text{ K}$

$$\ln\frac{K_2}{4.63\times10^{-3}} = \frac{58.0\times10^3\text{ J}}{8.314\text{ J/K}\cdot\text{mol}}\left(\frac{338\text{ K} - 298\text{ K}}{338\text{ K}\times298\text{ K}}\right)$$

$$\ln\frac{K_2}{4.63\times10^{-3}} = 2.77$$

Taking the antilog of both sides of the equation,

$$\frac{K_2}{4.63\times10^{-3}} = e^{2.77}$$

$$\mathbf{K_2 = 0.074}$$

$K_2 > K_1$, as we would predict for a positive ΔH°. Recall that an increase in temperature will shift the equilibrium for an endothermic reaction to the right (toward products).

19.44 The equilibrium reaction is:

$$AgCl\,(s) \rightleftharpoons Ag^+\,(aq) + Cl^-\,(aq)$$

$$K_{sp} = [Ag^+][Cl^-] = 1.6 \times 10^{-10}$$

We can calculate the standard enthalpy of reaction from the standard enthalpies of formation in Appendix 2.

$$\Delta H° = \Delta H_f°(Ag^+) + \Delta H_f°(Cl^-) - \Delta H_f°(AgCl)$$

$$\Delta H° = (1\ \cancel{mol})(105.9\ kJ/\cancel{mol}) + (1\ \cancel{mol})(-167.2\ kJ/\cancel{mol}) - (1\ \cancel{mol})(-127.0\ kJ/\cancel{mol}) = 65.7\ kJ$$

From Problem 19.42(a):

$$\ln\frac{K_2}{K_1} = \frac{\Delta H°}{R}\left(\frac{T_2 - T_1}{T_1 T_2}\right)$$

$K_1 = 1.6 \times 10^{-10}$ $\quad T_1 = 298\ K$

$K_2 = ?$ $\quad T_2 = 333\ K$

$$\ln\frac{K_2}{1.6\times 10^{-10}} = \frac{6.57\times 10^4\ J}{8.314\ J/K\cdot mol}\left(\frac{333\ K - 298\ K}{333\ K \times 298\ K}\right)$$

$$\ln\frac{K_2}{1.6\times 10^{-10}} = 2.79$$

$$\frac{K_2}{1.6\times 10^{-10}} = e^{2.79}$$

$$\mathbf{K_2 = 2.6 \times 10^{-9}}$$

The increase in K indicates that the solubility increases with temperature.

19.46 Assuming that both $\Delta H°$ and $\Delta S°$ are temperature independent, we can calculate both $\Delta H°$ and $\Delta S°$.

$$\Delta H° = [\Delta H_f°(CO) + \Delta H_f°(H_2)] - [\Delta H_f°(H_2O) + \Delta H_f°(C)]$$

$$\Delta H° = [(1\ mol)(-110.5\ kJ/mol) + (1\ mol)(0\ kJ/mol)] - [(1\ mol)(-241.8\ kJ/mol) + (1\ mol)(0\ kJ/mol)]$$

$$\Delta H° = 131.3\ kJ$$

$$\Delta S° = [S°(CO) + S°(H_2)] - [S°(H_2O) + S°(C)]$$

$$\Delta S° = [(1\ mol)(197.9\ J/K\cdot mol) + (1\ mol)(131.0\ J/K\cdot mol)] - [(1\ mol)(188.7\ J/K\cdot mol) + (1\ mol)(5.69\ J/K\cdot mol)]$$

$$\Delta S° = 134.5\ J/K$$

It is obvious from the given conditions that the reaction must take place at a fairly high temperature (in order to have red–hot coke). Setting $\Delta G° = 0$

$$0 = \Delta H° - T\Delta S°$$

$$T = \frac{\Delta H°}{\Delta S°} = \frac{(131.3\ \cancel{kJ})\left(\frac{1000\ \cancel{J}}{1\ \cancel{kJ}}\right)}{134.5\ \cancel{J}/K} = \mathbf{976\ K = 703°C}$$

The temperature must be greater than 703°C for the reaction to be spontaneous.

19.48 For a reaction to be spontaneous at constant temperature and pressure, $\Delta G < 0$. The process of crystallization proceeds with more order (less disorder), so $\mathbf{\Delta S < 0}$. We also know that

$$\Delta G = \Delta H - T\Delta S$$

Since ΔG must be negative, and since the entropy term will be positive ($-T\Delta S$, where ΔS is negative), then ΔH must be negative **($\Delta H < 0$)**. The reaction will be exothermic.

19.50 For the reaction to be spontaneous, ΔG must be negative.

$$\Delta G = \Delta H - T\Delta S$$

Given that $\Delta H = 19 \text{ kJ} = 19{,}000 \text{ J}$, then

$$\Delta G = 19{,}000 \text{ J} - (273 \text{ K} + 72 \text{ K})(\Delta S)$$

Solving the equation with the value of $\Delta G = 0$

$$0 = 19{,}000 \text{ J} - (273 \text{ K} + 72 \text{ K})(\Delta S)$$

$$\mathbf{\Delta S = 55 \text{ J/K}}$$

This value of ΔS which we solved for is the value needed to produce a ΔG value of zero. The *minimum* value of ΔS that will produce a spontaneous reaction will be any value of entropy *greater than* 55 J/K.

19.52 The second law states that the entropy of the universe must increase in a spontaneous process. But the entropy of the universe is the sum of two terms: the entropy of the system plus the entropy of the surroundings. One of the entropies can decrease, but not both. In this case, the decrease in system entropy is offset by an increase in the entropy of the surroundings. The reaction in question is exothermic, and the heat released raises the temperature (and the entropy) of the surroundings.

Could this process be spontaneous if the reaction were endothermic?

19.54 Entropy changes due to phase transitions, Problem Type 2C.

Step 1: The entropy change due to the phase transition (the vaporization of ethanol), can be calculated using the following equation. Recall that the temperature must be in units of Kelvin (78.3°C = 351 K).

$$\Delta S_{\text{vap}} = \frac{\Delta H_{\text{vap}}}{T_{\text{b.p.}}}$$

$$\Delta S_{\text{vap}} = \frac{39.3 \dfrac{\text{kJ}}{\text{mol}}}{351 \text{ K}} = +0.112 \text{ kJ/mol·K} = +112 \text{ J/mol·K}$$

Step 2: The problem asks for the change in entropy for the vaporization of 0.50 moles of ethanol. The ΔS calculated above is for 1 mole of ethanol.

$$\mathbf{\Delta S \textbf{ for 0.50 mol}} = (112 \text{ J/}\cancel{\text{mol}}\text{·K})(0.50 \ \cancel{\text{mol}}) = \mathbf{56 \text{ J/K}}$$

19.56 Combination of Problem Type 2A and 3A.

Step 1: Calculate $\Delta G°$ from the standard enthalpies of formation.

$$\Delta G° = 2\,\Delta G_f^\circ(\text{NO}) - [\,\Delta G_f^\circ(\text{N}_2) + \Delta G_f^\circ(\text{O}_2)]$$

$$= (2 \ \cancel{\text{mol}})(86.7 \text{ kJ/}\cancel{\text{mol}}) - [(1 \text{ mol})(0) + (1 \text{ mol})(0)] = 173 \text{ kJ}$$

Step 2: The equilibrium constant is related to the standard free-energy change by the following equation.

$$\Delta G° = -RT\ln K$$

R has units of J/mol·K, so we must convert $\Delta G°$ from units of kJ/mol to J/mol.

$$173\ \frac{\text{kJ}}{\text{mol}} \times \frac{1000\ \text{J}}{1\ \text{kJ}} = 1.73 \times 10^5\ \text{J/mol}$$

Substitute $\Delta G°$, R, and T into the above equation to calculate the equilibrium constant, K_p.

$$\Delta G° = -RT\ln K_p$$

$$1.73 \times 10^5\ \text{J/mol} = -(8.314\ \text{J/mol·K})(298\ \text{K})\ln K_p$$

$$-69.8 = \ln K_p$$

Taking the antilog of both sides gives

$$e^{-69.8} = K_p$$

$$\mathbf{K_p = 4.9 \times 10^{-31}}$$

Since the magnitude of K_p is so small, under atmospheric conditions oxygen does not react with nitrogen to produce nitrogen monoxide (nitric oxide).

19.58 The equilibrium constant is related to the standard free energy change by the following equation:

$$\Delta G° = -RT\ln K_p$$

$$2.12 \times 10^5\ \text{J} = -(1\ \text{mol})(8.314\ \text{J/mol·K})(298\ \text{K})\ln K_p$$

$$-85.6 = \ln K_p$$

$$K_p = 6.7 \times 10^{-38}$$

We can write the equilibrium constant expression for the reaction.

$$K_p = \sqrt{P_{O_2}}$$

$$P_{O_2} = (K_p)^2$$

$$\mathbf{P_{O_2}} = (6.7 \times 10^{-38})^2 = \mathbf{4.5 \times 10^{-75}\ atm}$$

This pressure is far too small to measure.

19.60 Both **(a)** and **(b)** apply to a reaction with a negative $\Delta G°$ value. Statement (c) is not always true. An endothermic reaction that has a positive $\Delta S°$ (increase in entropy) will have a negative $\Delta G°$ value at high temperatures.

19.62 For the given reaction we can calculate the standard free energy change from the standard free energies of formation (see Appendix 2). Then, we can calculate the equilibrium constant, K_p, from the standard free energy change.

$$\Delta G° = \Delta G_f°[Ni(CO)_4] - [4\Delta G_f°(CO) + \Delta G_f°(Ni)]$$

$$\Delta G° = (1 \text{ mol})(-587.4 \text{ kJ/mol}) - [(4 \text{ mol})(-137.3 \text{ kJ/mol}) + (1 \text{ mol})(0)] = -38.2 \text{ kJ} = -3.82 \times 10^4 \text{ J}$$

Substitute $\Delta G°$, R, and T (in K) into the following equation to solve for K_p.

$$\Delta G° = -RT\ln K_p$$

$$\ln K_p = \frac{-\Delta G°}{RT} = \frac{-(-3.82\times 10^4 \text{ J})}{(8.314 \text{ J/K}\cdot\text{mol})(353 \text{ K})}$$

$$\mathbf{K_p = 4.5 \times 10^5}$$

19.64 We write the two equations as follows. The standard free-energy change for the overall reaction will be the sum of the two steps.

$$Cu_2O\,(s) \rightleftharpoons 2\,Cu\,(s) + \frac{1}{2}O_2\,(g) \qquad \Delta G° = 125 \text{ kJ}$$

$$C\,(\text{graphite}) + \frac{1}{2}O_2\,(g) \rightleftharpoons CO\,(g) \qquad \Delta G° = -137.3 \text{ kJ}$$

$$Cu_2O + C\,(\text{graphite}) \rightleftharpoons 2\,Cu\,(s) + CO\,(g) \qquad \Delta G° = -12 \text{ kJ}$$

We can now calculate the equilibrium constant from the standard free-energy change, $\Delta G°$.

$$\ln K = \frac{-\Delta G°}{RT} = \frac{-(-12 \times 10^3 \text{ J})}{(8.314 \text{ J/K}\cdot\text{mol})(673 \text{ K})}$$

$$\ln K = 2.1$$

$$\mathbf{K = 8}$$

19.66 **(a)** $\Delta G° = \Delta G_f°(H_2) + \Delta G_f°(Fe^{2+}) - \Delta G_f°(Fe) - 2\Delta G_f°(H^+)$

$$\Delta G° = (1 \text{ mol})(0) + (1 \text{ mol})(-84.9 \text{ kJ/mol}) - (1 \text{ mol})(0) - (2 \text{ mol})(0)$$

$$\Delta G° = -84.9 \text{ kJ}$$

$$\Delta G° = -RT\ln K$$

$$-84.9 \times 10^3 \text{ J} = -(1 \text{ mol})(8.314 \text{ J/mol}\cdot\text{K})(298 \text{ K})\ln K$$

$$\mathbf{K = 7.6 \times 10^{14}}$$

(b) $\Delta G° = (1 \text{ mol})(\Delta G_f°(H_2) + (1 \text{ mol})(\Delta G_f°(Cu^{2+}) - \Delta G_f°(Cu) - (2 \text{ mol})(\Delta G_f°(H^+)$

$$\Delta G° = 64.98 \text{ kJ}$$

$$\Delta G° = -RT\ln K$$

$$64.98 \times 10^3 \text{ J} = -(1 \text{ mol})(8.314 \text{ J/mol}\cdot\text{K})(298 \text{ K})\ln K$$

$$\mathbf{K = 4.1 \times 10^{-12}}$$

The activity series is correct. The very large value of K for reaction (a) indicates that *products* are highly favored; whereas, the very small value of K for reaction (b) indicates that *reactants* are highly favored.

19.68 $2NO + O_2 \rightleftharpoons 2NO_2$

$$\Delta G° = (2\text{ mol})(51.8\text{ kJ/mol}) - (2\text{ mol})(86.7\text{ kJ/mol}) - 0 = -69.8\text{ kJ}$$

$$\Delta G° = -RT\ln K$$

$$-69.8 \times 10^3 = -(1\text{ mol})(8.314\text{ J/mol·K})(298\text{ K})\ln K$$

$$K = 1.7 \times 10^{12}$$

$$K = \frac{k_f}{k_r}$$

$$1.7 \times 10^{12}\ M^{-1} = \frac{7.1 \times 10^9\ M^{-2}s^{-1}}{k_r}$$

$$\mathbf{k_r = 4.2 \times 10^{-3}\ M^{-1}s^{-1}}$$

19.70 First convert to moles of ice.

$$74.6\text{ g }H_2O(s) \times \frac{1\text{ mol }H_2O(s)}{18.02\text{ g }H_2O(s)} = 4.14\text{ mol }H_2O(s)$$

For a phase transition:

$$\Delta S_{sys} = \frac{\Delta H_{sys}}{T}$$

$$\Delta S_{sys} = \frac{4.14\text{ mol} \times 6010\ \frac{\text{J}}{\text{mol}}}{273\text{ K}} = \mathbf{91.1\ J/K}$$

$$\Delta S_{surr} = \frac{-\Delta H_{sys}}{T}$$

$$\Delta S_{surr} = \frac{-4.14\text{ mol} \times 6010\ \frac{\text{J}}{\text{mol}}}{273\text{ K}} = \mathbf{-91.1\ J/K}$$

$$\Delta S_{univ} = \Delta S_{sys} + \Delta S_{surr} = \mathbf{0}$$

This is an equilibrium process. There is no net change.

19.72 Since we are dealing with the same ion (K^+), Equation 19.8 of the text can be written as:

$$\Delta G = \Delta G° + RT\ln Q$$

$$\Delta G = 0 + (1\text{ mol})(8.314\text{ J/mol·K})(310\text{ K})\ln\frac{400\text{ mM}}{15\text{ mM}}$$

$$\mathbf{\Delta G = 8.5 \times 10^3\ J = 8.5\ kJ}$$

19.74 **(a)** $\Delta G°$ for CH_3COOH:

$$\Delta G° = -(1 \text{ mol})(8.314 \text{ J/mol·K})(298 \text{ K}) \ln (1.8 \times 10^{-5})$$

$$\mathbf{\Delta G°} = 2.7 \times 10^4 \text{ J} = \mathbf{27 \text{ kJ}}$$

$\Delta G°$ for $CH_2ClCOOH$:

$$\Delta G° = -(1 \text{ mol})(8.314 \text{ J/mol·K})(298 \text{ K}) \ln (1.4 \times 10^{-3})$$

$$\mathbf{\Delta G°} = 1.6 \times 10^4 \text{ J} = \mathbf{16 \text{ kJ}}$$

(b) The $T\Delta S°$ is the dominant term.

(c) The breaking of the O–H bond in ionization of the acid and the forming of the O–H bond in H_3O^+.

(d) The CH_3COO^- ion is smaller than CH_2ClCOO^- and can participate in hydration to a greater extent, leading to a more ordered solution.

19.76 **(a)** Let us label the molecules as follows:

Pentane	2–methylbutane	2,2–dimethylpropane
A	B	C

For the conversion of pentane to 2–methylbutane, we can calculate the standard free-energy change from the standard free-energies of formation.

$$A \longrightarrow B$$

$$\Delta G° = (1 \text{ mol})\Delta G_f°(B) - (1 \text{ mol})\ \Delta G_f°(A)$$

$$= (1 \text{ mol})(-14.8 \text{ kJ/mol}) - (1 \text{ mol})(-8.37 \text{ kJ/mol})$$

$$= -6.4 \text{ kJ} = -6.4 \times 10^3 \text{ J}$$

Next, we can calculate the equilibrium constant from the standard free-energy change, $\Delta G°$.

$$\ln K = \frac{-\Delta G°}{RT} = \frac{-(-6.4 \times 10^3 \text{ J})}{(8.314 \text{ J/K·mol})(298 \text{ K})}$$

$$\ln K = 2.6$$

$$K = e^{2.6}$$

$$K = 13 = \frac{[B]}{[A]}$$

For the conversion of 2–methylbutane to 2,2–dimethylpropane, we can follow the same strategy as above.

$$B \longrightarrow C$$

$$\Delta G° = (1 \text{ mol})(-15.2 \text{ kJ/mol}) - (1 \text{ mol})(-14.8 \text{ kJ/mol})$$

$$= -0.4 \text{ kJ} = -4 \times 10^2 \text{ J}$$

$$\ln K = \frac{-\Delta G^\circ}{RT} = \frac{-(-4 \times 10^2\ \text{J})}{(8.314\ \text{J/K}\cdot\text{mol})(298\ \text{K})}$$

$$\ln K = 0.2$$

$$K = 1.2 = \frac{[\text{C}]}{[\text{B}]}$$

The equilibrium constant for the first conversion tells us that in the mixture we have 13 times more moles of B compared to A. Let's assume that at equilibrium we have 1 mole of A. Then, we have 13 moles of B in the mixture.

The equilibrium constant for the second conversion tells us that we have 1.2 times more moles of C compared to B. Therefore, the number of moles of C in the mixture is

$$\text{mol C} = 1.2 \times \text{mol B} = 1.2 \times 13\ \text{mol} = 16\ \text{mol}$$

We can now calculate the mole percents of each component. We can calculate the mole percent by multiplying the mole fraction by 100%. Recall that mole fraction is defined as:

$$X_\text{a} = \frac{\text{moles of A}}{\text{total number of moles}}$$

$$\textbf{mol\% A} = \frac{1\ \text{mol}}{(1 + 13 + 16)\ \text{mol}} \times 100\% = \textbf{3\%}$$

$$\textbf{mol\% B} = \frac{13\ \text{mol}}{(1 + 13 + 16)\ \text{mol}} \times 100\% = \textbf{43\%}$$

We can calculate the mol % of component C by difference.

$$\textbf{mol\% C} = 100\% - (3 + 43)\% = \textbf{54\%}$$

(b) Based on the mole percents of the mixture at equilibrium, the stability increases (from A to C) as branching increases.

CHAPTER 20
REDOX REACTIONS AND ELECTROCHEMISTRY

PROBLEM-SOLVING STRATEGIES AND TUTORIAL SOLUTIONS

TYPES OF PROBLEMS

Problem Type 1: Balancing Redox Equations.

Problem Type 2: Standard Electrode Potentials.
- **(a)** Comparing strengths of oxidizing agents.
- **(b)** Calculating the standard emf ($\mathcal{E}°$) of a galvanic cell.

Problem Type 3: Spontaneity of Redox Reactions.
- **(a)** Predicting whether a redox reaction is spontaneous.
- **(b)** Calculating $\Delta G°$ and K from $\mathcal{E}°$.

Problem Type 4: The Nernst Equation.
- **(a)** Using the Nernst equation to predict the spontaneity of a redox reaction.
- **(b)** Using the Nernst equation to calculate concentration.

Problem Type 5: Electrolysis.
- **(a)** Predicting the products of electrolysis.
- **(b)** Calculating the quantity of products in electrolysis.

PROBLEM TYPE 1: BALANCING REDOX EQUATIONS

We will use the **ion-electron method** to balance redox reactions. In this approach, the overall reaction is divided into two half-reactions, one for oxidation and one for reduction. The two equations are balanced separately and then added together to give the overall balanced equation. Example 20.1 demonstrates how to balance a redox reaction by the ion-electron method.

EXAMPLE 20.1
Balance the following redox reaction in an acidic solution.

$$\mathbf{Sn + NO_3^- \longrightarrow Sn^{2+} + NO}$$

Step 1: Write the unbalanced equation for the reaction in ionic form.

The equation given in the problem is already in ionic form.

Step 2: Separate the equation into two half-reactions.

$$\overset{0}{\mathrm{Sn}} \xrightarrow{\text{oxidation}} \mathrm{Sn}^{2+}$$

$$\overset{+5}{\mathrm{N}}\mathrm{O_3^-} \xrightarrow{\text{reduction}} \overset{+2}{\mathrm{N}}\mathrm{O}$$

Step 3: Balance the atoms other than O and H in each half-reaction separately.

The oxidation half-reaction is already balanced for Sn atoms, and the reduction half-reaction is balanced for nitrogen atoms.

Step 4: For reactions in an acidic medium, add H_2O to balance the O atoms and H^+ to balance the H atoms.

Since the reaction takes place in an acidic medium, we add *two* H_2O molecules to the right-hand side of the reduction half-reaction to balance the O atoms.

$$NO_3^- \longrightarrow NO + 2\,H_2O$$

To balance the H atoms, we add 4 H^+ to the left-hand side.

$$4\,H^+ + NO_3^- \longrightarrow NO + 2\,H_2O$$

Step 5: Add electrons to one side of each half-reaction to balance the charge.

For the oxidation half-reaction we write:

$$Sn \longrightarrow Sn^{2+} + 2e^-$$

We added two electrons to the right-hand side so that the charge is *zero* on each side of the half-reaction.

For the reduction half-reaction, there are three net positive charges on the left-hand side and zero charge on the right-hand side. Therefore, we add three electrons on the left-hand side to balance the charge.

$$4\,H^+ + NO_3^- + 3e^- \longrightarrow NO + 2\,H_2O$$

These are the balanced half-reactions.

Step 6: In a redox reaction, the number of electrons gained must equal the number of electrons lost. If necessary, we must equalize the number of electrons in the two half-reactions by multiplying one or both half-reactions by appropriate coefficients.

We see that two electrons are lost in the oxidation half-reaction and three electrons are gained in the reduction half-reaction. We need to multiply the oxidation half-reaction by 3 and the reduction half-reaction by 2. Thus, six electrons are lost and six electrons are gained.

$$3\,Sn \longrightarrow 3\,Sn^{2+} + 6e^-$$

$$8\,H^+ + 2\,NO_3^- + 6e^- \longrightarrow 2\,NO + 4\,H_2O$$

Step 7: Add the two half-reactions and balance the final equation by inspection. The electrons on both sides of the equation must cancel.

$$8\,H^+ + 2\,NO_3^- + 3\,Sn + 6e^- \longrightarrow 2\,NO + 3\,Sn^{2+} + 4\,H_2O + 6e^-$$

The electrons on both sides cancel, and we are left with the balanced net ionic equation:

$$8\,H^+ + 2\,NO_3^- + 3\,Sn \longrightarrow 2\,NO + 3\,Sn^{2+} + 4\,H_2O$$

Step 8: Verify that the equation contains the same types and numbers of atoms and the same charges on both sides of the equation.

The equation is "atomically" balanced. There are 8 H, 2 N, 6 O, and 3 Sn atoms on each side of the equation. The equation is also "electrically" balanced. The net charge is +6 on each side of the equation.

PRACTICE EXERCISE

1. The redox reaction between permanganate ion and iron(II) ions in acidic solution can be used to analyze iron ore for its iron content. Balance this redox reaction.

$$MnO_4^- (aq) + Fe^{2+} (aq) \longrightarrow Fe^{3+} (aq) + Mn^{2+} (aq)$$

Text Problem: 2

PROBLEM TYPE 2: STANDARD ELECTRODE POTENTIALS

A. Comparing strengths of oxidizing agents

For a reduction reaction at an electrode when all solutes are 1 *M* and all gases are at 1 atm, the voltage is called the **standard reduction potential**. Table 20.1 lists the standard reduction potentials for a number of half-reactions. The more positive the value of $\mathcal{E}°$, the greater the tendency for the substance to be reduced, and therefore the stronger its tendency to act as an oxidizing agent. Looking at Table 20.1, we see that F_2 is the strongest oxidizing agent and Li^+ the weakest.

EXAMPLE 20.2

Arrange the following species in order of increasing strength as oxidizing agents under standard-state conditions: Zn^{2+}, MnO_4^- (in acid solution), and Ag^+.

Consulting Table 20.1 of the text, we write the half-reactions in the order of increasing standard emf value, $\mathcal{E}°$.

$$Zn^{2+} (aq) + 2e^- \longrightarrow Zn (s) \qquad \mathcal{E}° = -0.76 \text{ V}$$

$$Ag^+ (aq) + e^- \longrightarrow Ag (s) \qquad \mathcal{E}° = +0.80 \text{ V}$$

$$MnO_4^- (aq) + 8 H^+ (aq) + 5e^- \longrightarrow Mn^{2+} (aq) + 4 H_2O (l) \qquad \mathcal{E}° = +1.51 \text{ V}$$

Since the reduction of MnO_4^- has the highest potential and Zn^{2+} has the lowest reduction potential, the order of increasing strength of oxidizing agents is:

$$Zn^{2+} < Ag^+ < MnO_4^-$$

The large positive reduction potential for MnO_4^- indicates the strong tendency for permanganate ion to be reduced, and therefore the stronger its tendency to act as an oxidizing agent.

PRACTICE EXERCISE

2. Arrange the following species in order of increasing strength as oxidizing agents: Ce^{4+}, O_2, H_2O_2, and SO_4^{2-}.

Text Problems: 10, 14

B. Calculating the standard emf ($\mathcal{E}°$) of an electrochemical cell

In an electrochemical cell, electrons flow from one electrode to the other. This indicates that there is a voltage difference between the two electrodes. This voltage difference is called the **electromotive force**, or **emf** ($\mathcal{E}$), and it can be measured by connecting a voltmeter to both electrodes. The electromotive force is also called the cell voltage or cell potential. It is usually measured in volts.

If all solutes have a concentration of 1 *M* and all gases have a pressure of 1 atm (standard conditions), the voltage difference between the two electrodes of the cell is called the **standard emf** ($\mathcal{E}^\circ_{cell}$). The standard emf is the sum of the standard oxidation potential and the standard reduction potential for each half-reaction.

$$\mathcal{E}^\circ_{cell} = \mathcal{E}^\circ_{ox} + \mathcal{E}^\circ_{red}$$

Under standard-state conditions for reactants and products, the redox reaction is *spontaneous* in the forward direction if the standard emf of the cell is *positive*. If the standard cell emf is negative, the reaction is spontaneous in the opposite direction.

We can calculate the standard emf of the cell using a table of reduction potentials (Table 20.1). As an example, consider the galvanic cell represented by the following reaction:

$$Cu\,(s) + 2\,Ag^{+}\,(aq) \longrightarrow Cu^{2+}\,(aq) + Ag\,(s)$$

Let's break this reaction down into its two half-reactions.

$$Cu\,(s) \xrightarrow{\text{oxidation}} Cu^{2+}\,(aq) + 2e^{-}$$

$$2\,Ag^{+}\,(aq) + 2e^{-} \xrightarrow{\text{reduction}} 2\,Ag\,(s)$$

We can calculate the standard cell emf by adding the oxidation and reduction potentials.

$$\mathcal{E}^\circ_{cell} = \mathcal{E}^\circ_{ox} + \mathcal{E}^\circ_{red} = \mathcal{E}^\circ_{Cu/Cu^{2+}} + \mathcal{E}^\circ_{Ag^{+}/Ag}$$

Note that the symbols for the standard oxidation and reduction potentials represent the direction of the reaction.

$$\mathcal{E}^\circ_{reactant/product}$$

Table 20.1 of the text lists the standard reduction potentials for a number of half-reactions. The standard oxidation potential for an anode half-reaction is equal in magnitude, but of *opposite sign* to that of the reduction potential for the reverse reaction. For example, the reverse reaction of the copper oxidation is:

$$Cu^{2+}\,(aq) + 2e^{-} \xrightarrow{\text{reduction}} Cu\,(s)$$

You can look up the standard reduction potential for this half-reaction in Table 20.1.

$$\mathcal{E}^\circ_{red} = \mathcal{E}^\circ_{Cu^{2+}/Cu} = 0.34\text{ V}$$

Therefore, the standard oxidation potential for the oxidation of Cu to Cu^{2+} has the opposite of the standard reduction potential above.

$$Cu\,(s) \xrightarrow{\text{oxidation}} Cu^{2+}\,(aq) + 2e^{-}$$

$$\mathcal{E}^\circ_{ox} = \mathcal{E}^\circ_{Cu/Cu^{2+}} = -0.34\text{ V}$$

We can also look up the standard reduction potential for silver in Table 20.1.

$$\mathcal{E}^\circ_{red} = \mathcal{E}^\circ_{Ag^+/Ag} = 0.80\text{ V}$$

Finally, we can calculate the standard cell emf.

$$\mathcal{E}^\circ_{cell} = \mathcal{E}^\circ_{Cu/Cu^{2+}} + \mathcal{E}^\circ_{Ag^+/Ag} = -0.34\text{ V} + 0.80\text{ V} = 0.46\text{ V}$$

EXAMPLE 20.3

Calculate the standard cell potential for an electrochemical cell in which the following reaction takes place:

$$\mathbf{Cl_2\,(g) + 2\,Br^-\,(aq) \longrightarrow Br_2\,(l) + 2\,Cl^-\,(aq)}$$

Step 1: Let's separate the reaction into half-reactions.

$$Cl_2\,(g) + 2e^- \xrightarrow{\text{reduction}} 2\,Cl^-\,(aq)$$

$$2\,Br^-\,(aq) \xrightarrow{\text{oxidation}} Br_2\,(l) + 2e^-$$

Step 2: We can look up standard reduction potentials in Table 20.1 of the text. Recall that a standard oxidation potential is equal in magnitude, but of opposite sign to that of the reduction potential for the reverse reaction.

$$\mathcal{E}^\circ_{red} = \mathcal{E}^\circ_{Cl_2/Cl^-} = +1.36\text{ V}$$

$$\mathcal{E}^\circ_{ox} = \mathcal{E}^\circ_{Br^-/Br_2} = -1.07\text{ V}$$

Step 3: Finally, the standard cell emf is the sum of the standard potentials for the two half-reactions.

$$\mathcal{E}^\circ_{cell} = \mathcal{E}^\circ_{Cl_2/Cl^-} + \mathcal{E}^\circ_{Br^-/Br_2}$$

$$\mathcal{E}^\circ_{\mathbf{cell}} = +1.36\text{ V} + -1.07\text{ V} = \mathbf{+0.29\text{ V}}$$

PRACTICE EXERCISE

3. Consider a uranium-bromine galvanic cell in which U is oxidized and Br_2 is reduced. The reduction half-reactions are:

$$U^{3+}\,(aq) + 3e^- \longrightarrow U\,(s) \qquad \mathcal{E}^\circ_{U^{3+}/U} = ?$$

$$Br_2\,(l) + 2e^- \longrightarrow 2\,Br^-\,(aq) \qquad \mathcal{E}^\circ_{Br_2/Br^-} = 1.07\text{ V}$$

If the standard cell emf is 2.91 V, what is the standard reduction potential for uranium?

Text Problem: 8

PROBLEM TYPE 3: SPONTANEITY OF REDOX REACTIONS

A. Predicting whether a redox reaction is spontaneous

There is a relationship between free energy change and cell emf.

$$\Delta G = -nF\mathcal{E}_{cell} \qquad (20.1)$$

where,

n is the number of moles of electrons transferred during the redox reaction
F is the Faraday constant, which is the electrical charge contained in 1 mole of electrons

$$1\ F = 96{,}500\ \text{C/mol} = 96{,}500\ \text{J/V}\cdot\text{mol}$$

For a derivation of this relationship, see Section 20.4 of the text.

Both n and F are positive quantities, and we know from Chapter 19 that ΔG is *negative* for a spontaneous process. Therefore, $\mathcal{E}_{\text{cell}}$ is *positive* for a spontaneous process. Table 20.2 of the text summarizes the relationship between ΔG and $\mathcal{E}_{\text{cell}}$.

For reactions in which reactants and products are in their standard states, Equation (20.1) becomes:

$$\Delta G^\circ = -nF\mathcal{E}^\circ_{\text{cell}} \qquad (20.2)$$

EXAMPLE 20.4

Predict whether a spontaneous reaction will occur when the following reactants and products are in their standard states:

$$\mathbf{Fe^{2+} + Cr_2O_7^{2-} \longrightarrow Fe^{3+} + Cr^{3+}}$$

Let's calculate the standard cell emf for this reaction. From the sign of $\mathcal{E}^\circ_{\text{cell}}$, we can determine if the reaction is spontaneous.

Separate the reaction into half-reactions to calculate the standard cell emf.

$$Fe^{2+}\,(aq) \xrightarrow{\text{oxidation}} Fe^{3+}\,(aq) + e^- \qquad \mathcal{E}^\circ_{\text{ox}} = -0.77\ \text{V}$$

At this point, we could balance the reduction half-reaction and then come up with the overall balanced equation. But, we are not asked to do that. We can save time by looking at Table 20.1 and finding a reaction that contains both $Cr_2O_7^{2-}$ and Cr^{3+}.

$$Cr_2O_7^{2-}\,(aq) + 14\,H^+\,(aq) + 6e^- \xrightarrow{\text{red}} 2\,Cr^{3+}\,(aq) + 7\,H_2O\,(l) \qquad \mathcal{E}^\circ_{\text{red}} = +1.33\ \text{V}$$

We have not come up with a balanced equation, but we do not need a balanced equation to calculate $\mathcal{E}^\circ_{\text{cell}}$.

$$\mathcal{E}^\circ_{\text{cell}} = \mathcal{E}^\circ_{\text{ox}} + \mathcal{E}^\circ_{\text{red}}$$

$$\mathcal{E}^\circ_{\textbf{cell}} = -0.77\ \text{V} + 1.33\ \text{V} = \mathbf{+0.56\ V}$$

Since $\mathcal{E}^\circ_{\text{cell}}$ is positive, the reaction is spontaneous.

PRACTICE EXERCISE

4. Predict whether a spontaneous reaction will occur when the following reactants and products are in their standard states.

(a) $2\,Fe^{3+}\,(aq) + 2\,I^-\,(aq) \longrightarrow 2\,Fe^{2+}\,(aq) + I_2\,(s)$

(b) $Cu\,(s) + 2\,H^+\,(aq) \longrightarrow Cu^{2+}\,(aq) + H_2\,(g)$

Text Problem: 12

B. Calculating $\Delta G°$ and K from $\mathcal{E}°$

Equation (20.2) shows that the relationship between standard free energy change and standard emf is:

$$\Delta G° = -nF\mathcal{E}°_{\text{cell}} \qquad (20.2)$$

Also recall that in Section 19.6 of the text, we saw that the standard free energy change for a reaction is related to its equilibrium constant K by the following equation:

$$\Delta G° = -RT \ln K$$

Therefore, from the two equations we obtain:

$$-nF\mathcal{E}°_{\text{cell}} = -RT \ln K$$

Solving for $\mathcal{E}°_{\text{cell}}$, we obtain:

$$\mathcal{E}°_{\text{cell}} = \frac{RT}{nF} \ln K \qquad (20.3)$$

At 298 K, we can simplify Equation (20.3) by substituting for R and F and converting the natural logarithm to the common (base-10) logarithm.

$$\mathcal{E}°_{\text{cell}} = \frac{(2.303)(8.314 \text{ J/mol}\cdot\text{K})(298 \text{ K})}{n(96500 \text{ J/V}\cdot\text{mol})} \log K$$

$$\mathcal{E}°_{\text{cell}} = \frac{0.0591 \text{ V}}{n} \log K \qquad (20.4)$$

We now have relationships between $\Delta G°$ and $\mathcal{E}°_{\text{cell}}$, between $\Delta G°$ and K, and between $\mathcal{E}°_{\text{cell}}$ and K.

EXAMPLE 20.5
Calculate $\Delta G°$ and the equilibrium constant at 25°C for the reaction

$$2\,Br^-\,(aq) + I_2\,(s) \longrightarrow Br_2\,(l) + 2\,I^-\,(aq)$$

First, we can calculate the standard cell emf from half-reaction potentials. Then, the standard Gibbs free energy change is:

$$\Delta G° = -nF\mathcal{E}°_{\text{cell}} \qquad (20.2)$$

The equilibrium constant at 25°C is related to $\mathcal{E}°_{\text{cell}}$ by:

$$\mathcal{E}°_{\text{cell}} = \frac{0.0591 \text{ V}}{n} \log K \qquad (20.4)$$

Step 1: Separate the reaction into half-reactions to calculate the standard cell emf.

$$I_2\,(s) + 2e^- \xrightarrow{\text{reduction}} 2\,I^-\,(aq) \qquad \mathcal{E}°_{\text{red}} = +0.53 \text{ V}$$

$$2\,Br^-\,(aq) \xrightarrow{\text{oxidation}} Br_2\,(l) + 2e^- \qquad \mathcal{E}°_{\text{ox}} = -1.07 \text{ V}$$

$$2\,Br^-\,(aq) + I_2\,(s) \longrightarrow Br_2\,(l) + 2\,I^-\,(aq) \qquad \mathcal{E}°_{\text{cell}} = -0.54 \text{ V}$$

Step 2: Substitute $\mathcal{E}^\circ_{cell}$ into Equation (20.2) to calculate the standard free energy change. In the balanced equation above, 2 moles of electrons are transferred. Therefore, $n = 2$ mol e^-.

$$\Delta G^\circ = -nF\mathcal{E}^\circ_{cell}$$

$$\Delta G^\circ = -(2 \text{ mol e}^-)(96{,}500 \text{ J/V·mol e}^-)(-0.54 \text{ V})$$

$$\mathbf{\Delta G^\circ = 1.04 \times 10^5 \text{ J} = 104 \text{ kJ}}$$

Both the positive value of ΔG° and the negative value for $\mathcal{E}^\circ_{cell}$ indicate that the reaction is not spontaneous under standard conditions.

Step 3: Rearrange Equation (20.4) to solve for the equilibrium constant, *K*.

$$\log K = \frac{n\mathcal{E}^\circ}{0.0591 \text{ V}}$$

$$\log K = \frac{(2)(-0.54 \text{ V})}{0.0591 \text{ V}} = -18.3$$

Taking the antilog of both sides of the equation,

$$\mathbf{K} = 10^{-18.3} = \mathbf{5 \times 10^{-19}}$$

You should notice that the magnitude of *K* relates to what we have learned from ΔG° and $\mathcal{E}^\circ_{cell}$. A very small value for the equilibrium constant indicates that reactants are highly favored at equilibrium. This agrees with the signs of ΔG° and $\mathcal{E}^\circ_{cell}$, which indicate that the reaction is not spontaneous in the forward direction under standard conditions.

PRACTICE EXERCISE

5. Calculate the equilibrium constant for the following redox reaction at 25°C.

$$2\,Fe^{2+}(aq) + Ni^{2+}(aq) \longrightarrow 2\,Fe^{3+}(aq) + Ni(s)$$

Text Problems: 18, 20, 22

PROBLEM TYPE 4: THE NERNST EQUATION

Thus far, we have only focused on redox reactions in which the reactants and products are in their standard states. However, standard-state conditions are often difficult and sometimes impossible to maintain. Therefore, we need a relationship between cell emf and the concentrations of reactants and products under *nonstandard-state* conditions.

In Chapter 19, we encountered a relationship between the free energy change (ΔG) and the standard free energy change (ΔG°). See Problem Type 3, Chapter 19.

$$\Delta G = \Delta G^\circ + RT\ln Q$$

where,

Q is the reaction quotient.

We also know that:

$$\Delta G = -nF\mathcal{E}_{cell}$$

and

$$\Delta G° = -nF\mathcal{E}°_{cell}$$

Substituting for ΔG and $\Delta G°$ in the first equation, we find:

$$-nF\mathcal{E}_{cell} = -nF\mathcal{E}°_{cell} + RT \ln Q$$

Dividing both sides of the equation by $-nF$ and converting to $\log_{10}$ gives:

$$\mathcal{E} = \mathcal{E}° - \frac{2.303RT}{nF}\log Q \qquad (20.5)$$

Equation (20.5) is known as the **Nernst equation**. At 298 K, Equation (20.5) can be simplified by substituting R, T, and F into the equation.

$$\mathcal{E} = \mathcal{E}° - \frac{0.0591\text{ V}}{n}\log Q \qquad (20.6)$$

At equilibrium, there is no net transfer of electrons, so $\mathcal{E} = 0$ and $Q = K$, where K is the equilibrium constant of the redox reaction. Substituting into Equation (20.6) gives

$$\mathcal{E}° = \frac{0.0591\text{ V}}{n}\log K$$

This is Equation (20.4) derived in Problem Type 3B above.

A. Using the Nernst equation to predict the spontaneity of a redox reaction

To predict the spontaneity of a redox reaction, we can use the Nernst equation to calculate the cell emf, $\mathcal{E}$. If the cell emf is *positive*, the redox reaction is *spontaneous*. Conversely, if $\mathcal{E}$ is *negative*, the reaction is *not spontaneous* in the direction written.

EXAMPLE 20.6

Calculate the cell emf and predict whether the following reaction is spontaneous at 25°C.

$$\mathbf{Zn\ (s) + 2\ H^+\ (1 \times 10^{-4}\ M) \longrightarrow Zn^{2+}\ (1.5\ M) + H_2\ (1\ atm)}$$

The cell emf can be calculated using the Nernst equation.

$$\mathcal{E} = \mathcal{E}° - \frac{0.0591\text{ V}}{n}\log Q$$

or

$$\mathcal{E} = \mathcal{E}° - \frac{0.0591\text{ V}}{n}\log\frac{[\text{Zn}^{2+}]P_{\text{H}_2}}{[\text{H}^+]^2}$$

Step 1: Calculate the standard cell emf, $\mathcal{E}°$, from standard reduction potentials (Table 20.1). Separate the reaction into its half-reactions.

$$2\ \text{H}^+\ (aq) + 2\text{e}^- \xrightarrow{\text{reduction}} \text{H}_2\ (g) \qquad \mathcal{E}°_{\text{red}} = +0.00\text{ V}$$

$$\text{Zn}\ (s) \xrightarrow{\text{oxidation}} \text{Zn}^{2+}\ (aq) + 2\text{e}^- \qquad \mathcal{E}°_{\text{ox}} = +0.76\text{ V}$$

$$\text{Zn}\ (s) + 2\ \text{H}^+\ (aq) \longrightarrow \text{Zn}^{2+}\ (aq) + \text{H}_2\ (g) \qquad \mathcal{E}°_{\text{cell}} = +0.76\text{ V}$$

Step 2: Calculate the cell emf, $\mathcal{E}$, from the Nernst equation. Two moles of electrons were transferred during the reaction, so $n = 2$.

$$\mathcal{E} = \mathcal{E}^\circ - \frac{0.0591\text{ V}}{n}\log\frac{[Zn^{2+}]P_{H_2}}{[H^+]^2}$$

$$\mathcal{E} = (0.76\text{ V}) - \frac{0.0591\text{ V}}{2}\log\frac{(1.5)(1)}{(1\times10^{-4})^2}$$

$$= 0.76\text{ V} - 0.24\text{ V}$$

$$= \mathbf{0.52\text{ V}}$$

The reaction is spontaneous because the cell emf, $\mathcal{E}$, is positive. The low concentration of H^+ (1×10^{-4} *M*) compared to its standard-state value (1 *M*) means that the driving force for the reaction will be less than in the standard state. This is confirmed because $\mathcal{E} < \mathcal{E}^\circ$. We can also think about this in terms of Le Chatelier's principle. When H^+ ions are removed, the equilibrium will shift to the left to replace some of the H^+ ions that are removed. This will decrease the driving force for the forward reaction.

PRACTICE EXERCISE

6. Calculate the cell emf for the following reaction:

$$2\,Ag^+\,(0.10\,M) + H_2\,(1\text{ atm}) \longrightarrow 2\,Ag\,(s) + 2\,H^+\,(\text{pH} = 8)$$

Text Problems: 26, 28, 30

B. Using the Nernst equation to calculate concentration

Recall that the reaction quotient Q equals the concentrations of the products raised to the power of their stoichiometric coefficients divided by the concentrations of the reactants raised to the power of their stoichiometric coefficients.

$$Q = \frac{[\text{products}]^x}{[\text{reactants}]^y}$$

Since Q is in the Nernst equation, if the cell emf is measured and the standard cell emf is calculated, we can determine the concentration of one of the components if the concentrations of the other components are known. See Example 20.7 below.

EXAMPLE 20.7

An electrochemical cell is constructed from a silver half-cell and a copper half-cell. The copper half-cell contains 0.10 *M* $Cu(NO_3)_2$ and the concentration of silver ions in the other half-cell is unknown. If the Ag electrode is the cathode and the cell emf is measured and found to be 0.10 V at 25°C, what is the Ag^+ ion concentration?

We are given the cell emf, $\mathcal{E} = 0.10$ V, in the problem. If we can write the reaction occurring in the cell, we can calculate the standard cell emf, $\mathcal{E}^\circ$. We are also given the Cu^{2+} concentration in the problem. Given all this information, we can calculate the Ag^+ ion concentration using the Nernst equation.

Step 1: Write the half-reactions to calculate the standard cell emf, $\mathcal{E}^\circ$. We are told that the Ag electrode is the cathode (reduction occurs at the cathode). We can write:

$$Ag^+\,(aq) + e^- \xrightarrow{\text{reduction}} Ag\,(s) \qquad \mathcal{E}^\circ_{red} = +0.80\text{ V}$$

If Ag^+ is reduced, Cu must be oxidized.

$$Cu(s) \xrightarrow{\text{oxidation}} Cu^{2+}(aq) + 2e^- \qquad \mathcal{E}^\circ_{ox} = -0.34\ V$$

$$\mathcal{E}^\circ_{cell} = \mathcal{E}^\circ_{ox} + \mathcal{E}^\circ_{red}$$

$$\mathcal{E}^\circ_{cell} = -0.34\ V + 0.80\ V = \mathbf{+0.46\ V}$$

We must multiply the Ag half-reaction by two so that the number of electrons gained equals the number of electrons lost. The balanced reaction is:

$$2\ Ag^+(aq) + Cu(s) \longrightarrow 2\ Ag(s) + Cu^{2+}(aq)$$

Step 2: Substitute $\mathcal{E}$, $\mathcal{E}^\circ$, and $[Cu^{2+}]$ into the Nernst equation to calculate the $[Ag^+]$. Also, $n = 2$ since two moles of electrons are transferred during the reaction.

$$\mathcal{E} = \mathcal{E}^\circ - \frac{0.0591\ V}{n}\log Q$$

or

$$\mathcal{E} = \mathcal{E}^\circ - \frac{0.0591\ V}{n}\log\frac{[Cu^{2+}]}{[Ag^+]^2}$$

$$0.10\ V = 0.46\ V - \frac{0.0591\ V}{2}\log\frac{[0.10]}{[Ag^+]^2}$$

$$\frac{2(0.10\ V - 0.46\ V)}{-0.0591\ V} = \log\frac{0.10}{[Ag^+]^2}$$

$$12.2 = \log\frac{0.10}{[Ag^+]^2}$$

Taking the antilog of both sides of the equation,

$$10^{12.2} = \frac{0.10}{[Ag^+]^2}$$

$$2 \times 10^{12} = \frac{0.10}{[Ag^+]^2}$$

$$[Ag^+] = \sqrt{\frac{0.10}{2 \times 10^{12}}}$$

$$\mathbf{[Ag^+] = 2 \times 10^{-7}\ \mathit{M}}$$

PRACTICE EXERCISE

7. When the concentration of Zn^{2+} is 0.15 *M*, the measured voltage of a Zn-Cu galvanic cell is 0.40 V. What is the Cu^{2+} ion concentration?

Text Problem: 90

PROBLEM TYPE 5: ELECTROLYSIS

Electrolysis is the process in which electrical energy is used to cause a nonspontaneous chemical reaction to occur. The same principles underlie electrolysis and the processes that take place in galvanic cells.

A. Predicting the products of electrolysis

EXAMPLE 20.8

Predict the products of the electrolysis of an aqueous $MgCl_2$ solution.

This aqueous solution contains several species that could be oxidized or reduced. Two species that could be reduced are the metal ion (Mg^{2+}) and H_2O. The reduction half-reactions that might occur at the cathode are

(1) $Mg^{2+}(aq) + 2e^- \longrightarrow Mg(s)$ $\qquad \mathcal{E}^\circ_{red} = -2.37\ V$

(2) $2\ H_2O(l) + 2e^- \longrightarrow H_2(g) + 2\ OH^-(aq)$ $\qquad \mathcal{E}^\circ_{red} = -0.83\ V$

(3) $2\ H^+(aq) + 2e^- \longrightarrow H_2(g)$ $\qquad \mathcal{E}^\circ_{red} = 0.00\ V$

We can rule out Reaction (1) immediately. The very negative standard reduction potential indicates that Mg^{2+} has essentially no tendency to undergo reduction. Reaction (3) is preferred over Reaction (2) under standard-state conditions. However, at a pH of 7 (as is the case for a $MgCl_2$ solution), they are equally probable. We generally use Reaction (2) to describe the cathode reaction because the concentration of H^+ ions is too low (about 1×10^{-7} *M*) to make (3) a reasonable choice.

The oxidation reactions that might occur at the anode are:

(4) $2\ Cl^-(aq) \longrightarrow Cl_2(g) + 2e^-$ $\qquad \mathcal{E}^\circ_{ox} = -1.36\ V$

(5) $2\ H_2O(l) \longrightarrow O_2(g) + 4\ H^+(aq) + 2e^-$ $\qquad \mathcal{E}^\circ_{ox} = -1.23\ V$

The standard oxidation potentials of (4) and (5) are not very different, but the values do suggest that H_2O should be preferentially oxidized at the anode. However, we find by experiment that the gas liberated at the anode is Cl_2, not O_2. The large overvoltage for O_2 formation prevents its production when the Cl^- ion is there to compete (see Section 20.8 of the text).

The overall reaction is:

$$2\ H_2O(l) + 2e^- \longrightarrow H_2(g) + 2\ OH^-(aq)$$
$$2\ Cl^-(aq) \longrightarrow Cl_2(g) + 2e^-$$
$$\overline{2\ H_2O(l) + 2\ Cl^-(aq) \longrightarrow H_2(g) + 2\ OH^-(aq) + Cl_2(g)}$$

B. Calculating the quantity of products in electrolysis

During electrolysis, the mass of product formed (or reactant consumed) at an electrode is proportional to both the amount of electricity transferred at the electrode and the molar mass of the substance in question.

The steps involved in calculating the quantities of substances produced in electrolysis are shown below.

Current (amperes) and time	→	Charge in coulombs	→	Number of Faradays	→	Moles of substance reduced or oxidized	→	Grams of substance reduced or oxidized

To convert to coulombs, recall that

$$1\ \text{C} = 1\ \text{A} \times 1\ \text{s}$$

To convert from coulombs to faradays, use the conversion factor

$$1\ F = 96{,}500\ \text{C/mol e}^-$$

Finally, we need to consider the number of moles of electrons transferred during the reaction. For example, to reduce Al^{3+} ions to Al metal, 3 moles of electrons are needed to reduce 1 mole of Al^{3+} ions.

$$Al^{3+} + 3e^- \longrightarrow Al$$

Since 1 *F* is the electrical charge contained in 1 mole of electrons, we will need 3 *F* of electrical charge to reduce 1 mole of Al^{3+} ions to Al metal.

EXAMPLE 20.9

How many faradays are transferred in an electrolytic cell when a current of 12 amps flows for 16 hours?

First, we can calculate the number of coulombs passing through the cell in 16 hours. Recall that

$$1\ \text{C} = 1\ \text{A·s}$$

Then, we can convert from coulombs to faradays using the following conversion factor.

$$1\ F = 96{,}500\ \text{C}$$

Step 1: Convert from amps to coulombs.

$$12\ \cancel{\text{A}} \times \frac{1\ \frac{\text{C}}{\cancel{\text{s}}}}{1\ \cancel{\text{A}}} \times \frac{3600\ \cancel{\text{s}}}{1\ \cancel{\text{h}}} \times 16\ \cancel{\text{h}} = 6.9 \times 10^5\ \text{C}$$

Step 2: Convert from coulombs to faradays.

$$\textbf{?}\ \boldsymbol{F} = 6.9 \times 10^5\ \cancel{\text{C}} \times \frac{1\ F}{96{,}500\ \cancel{\text{C}}} = \mathbf{7.2}\ \boldsymbol{F}$$

EXAMPLE 20.10

How many grams of copper metal would be deposited from a solution of $CuSO_4$ by the passage of 3.0 A of electrical current through an electrolytic cell for 2.0 h?

Step 1: Start by writing the half-reaction for the reduction of Cu^{2+} to Cu metal.

$$Cu^{2+}\ (aq) + 2e^- \longrightarrow Cu\ (s)$$

This half-reaction tells us that 2 mol of e^- are required to produce 1 mol of Cu (*s*). In other words, we need 2 *F* of electrical charge to reduce 1 mol of Cu^{2+} ions to Cu metal.

Step 2: We can calculate the mass of Cu produced using the following strategy.

current × time → coulombs → [faradays] → mol Cu → g Cu

This is a large number of steps, so let's break it down into two parts. The faraday is the link between the current measurements and the molar amount of Cu formed.

The number of faradays is:

$$3.0\,\cancel{A} \times \frac{1\,\frac{C}{\cancel{s}}}{1\,\cancel{A}} \times \frac{3600\,\cancel{s}}{1\,\cancel{h}} \times 2.0\,\cancel{h} = 2.2 \times 10^4\,\cancel{C} \times \frac{1\,F}{96{,}500\,\cancel{C}} = 0.23\,F$$

Step 3: From the number of faradays, we can calculate the moles of Cu produced. We can then convert to grams of Cu.

$$\textbf{? g Cu} = 0.23\,\cancel{F} \times \frac{1\,\cancel{mol}\text{ Cu}}{2\,\cancel{F}} \times \frac{63.55\text{ g Cu}}{1\,\cancel{mol}\text{ Cu}} = \textbf{7.3 g Cu}$$

> **Note:** This predicted amount of Cu is based on a process that is 100 percent efficient. Any side reactions or any oxidation or reduction of impurities will cause the actual yield to be less than the theoretical yield.

EXAMPLE 20.11

How long will it take to electrodeposit (plate out) 1.0 g of Ni from a $NiSO_4$ solution using a current of 2.5 A?

Step 1: First, find the number of faradays required to electrodeposit 1.0 g of Ni from solution. The half-reaction for the reduction of Ni^{2+} is:

$$Ni^{2+}\,(aq) + 2e^- \longrightarrow Ni\,(s)$$

2 *F* of electrical charge are required to reduce 1 mol of Ni^{2+} ions to Ni metal. But, we are electrodepositing less than 1 mol of Ni (*s*). We need to complete the following conversions:

g Ni → mol Ni → faradays

$$\text{? faradays} = 1.0\,\cancel{g}\text{ Ni} \times \frac{1\,\cancel{mol}\text{ Ni}}{58.7\,\cancel{g}\text{ Ni}} \times \frac{2\,F}{1\,\cancel{mol}\text{ Ni}} = 0.034\,F$$

Step 2: Determine how long it will take for 0.034 *F* of electrical charge to flow through the cell when the current is 2.5 C/s. We need to complete the following conversions:

faradays → coulombs → seconds

$$\textbf{? seconds} = 0.034\,\cancel{F} \times \frac{96{,}500\,\cancel{C}}{1\,\cancel{F}} \times \frac{1\text{ s}}{2.5\,\cancel{C}} = \mathbf{1.3 \times 10^3\text{ s}}\ (22\text{ min})$$

PRACTICE EXERCISES

8. How many faradays are transferred in an electrolytic cell when a current of 2.0 amps flows for 6.0 hours?

9. How many grams of cobalt can be electroplated by passing a constant current of 5.2 A through a solution of $CoCl_3$ for 60.0 min?

10. How long will it take to produce 54 kg of Al metal by the reduction of Al^{3+} in an electrolytic cell using a current of 5.0×10^2 amps?

Text Problems: 42, 46, **48**, 50, **52**, 54, 56

ANSWERS TO PRACTICE EXERCISES

1. $5\ Fe^{2+}\ (aq) + MnO_4^-\ (aq) + 8\ H^+\ (aq) \longrightarrow 5\ Fe^{3+}\ (aq) + Mn^{2+}\ (aq) + 4\ H_2O\ (l)$

2. $SO_4^{2-} > O_2 > Ce^{4+} > H_2O_2$

3. $\mathcal{E}^\circ_{U^{3+}/U} = -1.84$ V

4. (a) $\mathcal{E}^\circ_{cell} = +0.24$ V. The positive value of the standard cell emf indicates that the reaction is spontaneous under standard-state conditions.

   (b) $\mathcal{E}^\circ_{cell} = -0.34$ V. A negative standard cell emf indicates that this reaction is not spontaneous under standard-state conditions. Copper will not dissolve in 1 *M* HCl.

5. $K = 3.0 \times 10^{-35}$

6. $\mathcal{E}° = 1.21$ V

7. $[Cu^{2+}] = 3.1 \times 10^{-25}\ M$

8. Number of faradays transferred = 0.45 *F*

9. Grams of Co electroplated = 3.81 g Co

10. Time = 322 h

SOLUTIONS TO SELECTED TEXT PROBLEMS

20.2 Balancing Redox Equations, Problem Type 1.

(a)

Step 1: Separate the equation into two half-reactions.

$$Mn^{2+} \xrightarrow{\text{oxidation}} MnO_2$$

$$H_2O_2 \xrightarrow{\text{reduction}} H_2O$$

Step 2: Balance the atoms other than O and H in each half-reaction separately.

The oxidation half-reaction is already balanced for Mn atoms.

Step 3: Assume that the reaction is taking place in an acidic medium. Add H_2O to balance the O atoms and H^+ to balance the H atoms.

For the oxidation half-reaction, add 2 H_2O to the left-hand side of the equation to balance the O atoms.

$$Mn^{2+} + 2\,H_2O \longrightarrow MnO_2$$

To balance the H atoms, we add 4 H^+ to the right-hand side.

$$Mn^{2+} + 2\,H_2O \longrightarrow MnO_2 + 4\,H^+$$

For the reduction half-reaction, we add one H_2O to the right-hand side of the equation to balance the O atoms.

$$H_2O_2 \longrightarrow 2\,H_2O$$

To balance the H atoms, we add 2 H^+ to the left-hand side.

$$H_2O_2 + 2\,H^+ \longrightarrow 2\,H_2O$$

Step 4: Since this reaction is in basic solution, we add one OH^- to both sides for each H^+ and combine pairs of H^+ and OH^- on the same side of the arrow to form H_2O.

$$Mn^{2+} + 2\,H_2O + 4\,OH^- \longrightarrow MnO_2 + 4\,H_2O$$

$$H_2O_2 + 2\,H_2O \longrightarrow 2\,H_2O + 2\,OH^-$$

We remove extra H_2O.

$$Mn^{2+} + 4\,OH^- \longrightarrow MnO_2 + 2\,H_2O$$

$$H_2O_2 \longrightarrow 2\,OH^-$$

Step 5: Add electrons to one side of each half-reaction to balance the charge.

$$Mn^{2+} + 4\,OH^- \longrightarrow MnO_2 + 2\,H_2O + 2e^-$$

$$H_2O_2 + 2e^- \longrightarrow 2\,OH^-$$

Step 6: We can now add the reactions because the number of electrons gained (2 e^-) equals the number of electrons lost (2 e^-). Canceling extra OH^- and electrons gives the completed balanced equation.

$$Mn^{2+} + H_2O_2 + 2\,OH^- \longrightarrow MnO_2 + 2\,H_2O$$

Check to see that the equation is balanced by verifying that the equation has the same types and numbers of atoms and the same charges on both sides of the equation.

(b) This problem can be solved by the same methods used in part (a).

$$2\,Bi(OH)_3 + 3\,SnO_2^{2-} \longrightarrow 2\,Bi + 3\,H_2O + 3\,SnO_3^{2-}$$

(c)

Step 1: Separate the equation into two half-reactions.

$$C_2O_4^{2-} \xrightarrow{\text{oxidation}} CO_2$$
$$Cr_2O_7^{2-} \xrightarrow{\text{reduction}} Cr^{3+}$$

Step 2: Balance the atoms other than O and H in each half-reaction separately.

In the oxidation half-reaction, we need to balance the C atoms.

$$C_2O_4^{2-} \longrightarrow 2\,CO_2$$

In the reduction half-reaction, we need to balance the Cr atoms.

$$Cr_2O_7^{2-} \longrightarrow 2\,Cr^{3+}$$

Step 3: Add H_2O to balance the O atoms and H^+ to balance the H atoms.

In the oxidation half-reaction, the O atoms are already balanced.

$$Cr_2O_7^{2-} + 14\,H^+ \longrightarrow 2\,Cr^{3+} + 7\,H_2O$$

Step 4: Add electrons to one side of each half-reaction to balance the charge.

$$C_2O_4^{2-} \longrightarrow 2\,CO_2 + 2e^-$$
$$Cr_2O_7^{2-} + 14\,H^+ + 6e^- \longrightarrow 2\,Cr^{3+} + 7\,H_2O$$

Step 5: The number of electrons gained must equal the number of electrons lost. Multiply the oxidation half-reaction by 3, so that 6 e^- are transferred.

$$3\,C_2O_4^{2-} \longrightarrow 6\,CO_2 + 6e^-$$
$$Cr_2O_7^{2-} + 14\,H^+ + 6e^- \longrightarrow 2\,Cr^{3+} + 7\,H_2O$$

Step 6: Add the two half-reactions and balance the final equation by inspection. The electrons on both sides of the equation must cancel.

$$Cr_2O_7^{2-} + 14\,H^+ + 3\,C_2O_4^{2-} \longrightarrow 2\,Cr^{3+} + 6\,CO_2 + 7\,H_2O$$

(d) This problem can be solved by the same methods used in part (c).

$$2\,Cl^- + 2\,ClO_3^- + 4\,H^+ \longrightarrow Cl_2 + 2\,ClO_2 + 2\,H_2O$$

20.8 Calculating the standard emf of a galvanic cell, Problem Type 2B.

Step 1: Let's separate the reaction into half-reactions.

$$Al\,(s) \xrightarrow{\text{oxidation}} Al^{3+}\,(aq) + 3e^-$$

$$Ag^+\,(aq) + e^- \xrightarrow{\text{reduction}} Ag(s)$$

Step 2: We can look up standard reduction potentials in Table 20.1 of the text. Recall that a standard oxidation potential is equal in magnitude, but of opposite sign to that of the reduction potential for the reverse reaction.

$$\mathcal{E}^\circ_{ox} = \mathcal{E}^\circ_{Al/Al^{3+}} = +1.66\text{ V}$$

$$\mathcal{E}^\circ_{red} = \mathcal{E}^\circ_{Ag^+/Ag} = +0.80\text{ V}$$

Step 3: Finally, the standard cell emf is the sum of the standard potentials for the two half-reactions.

$$\mathcal{E}^\circ_{cell} = \mathcal{E}^\circ_{Al/Al^{3+}} + \mathcal{E}^\circ_{Ag^+/Ag}$$

$$\mathcal{E}^\circ_{\mathbf{cell}} = +1.66\text{ V} + 0.80\text{ V} = \mathbf{+2.46\text{ V}}$$

The cell reaction under standard conditions is:

$$Al\,(s) + 3\,Ag^+\,(aq) \longrightarrow 3\,Ag\,(s) + Al^{3+}\,(aq)$$

20.10 The half–reaction for oxidation is:

$$2\,H_2O\,(l) \longrightarrow O_2\,(g) + 4\,H^+\,(aq) + 4e^- \qquad \mathcal{E}^\circ_{ox} = -1.23\text{ V}$$

The species that can oxidize water to molecular oxygen must have an $\mathcal{E}^\circ_{red}$ more positive than +1.23 V. From Table 20.1 we see that only Cl_2 (*g*) and MnO_4^- (*aq*) in acid solution can oxidize water to oxygen.

20.12 Predicting whether a redox reaction is spontaneous, Problem Type 3A.

$\mathcal{E}^\circ_{cell}$ is *positive* for a spontaneous reaction. In each case, we can calculate the standard cell emf from the potentials for the two half-reactions.

$$\mathcal{E}^\circ_{cell} = \mathcal{E}^\circ_{ox} + \mathcal{E}^\circ_{red}$$

(a) $\mathcal{E}^\circ = 2.87\text{ V} - 0.40\text{ V} = \mathbf{2.47\text{ V}}$. The reaction is spontaneous.

(b) $\mathcal{E}^\circ = -1.07\text{ V} - 0.14\text{ V} = \mathbf{-1.21\text{ V}}$. The reaction is not spontaneous.

(c) $\mathcal{E}^\circ = -0.80\text{ V} - 0.25\text{ V} = \mathbf{-1.05\text{ V}}$. The reaction is not spontaneous.

(d) $\mathcal{E}^\circ = -0.15\text{ V} + 0.77\text{ V} = \mathbf{0.62\text{ V}}$. The reaction is spontaneous.

20.14 Similar to Problem Type 2A, Comparing the strengths of oxidizing agents.

The greater the tendency for the substance to be oxidized, the stronger its tendency to act as a reducing agent.

The species that has a stronger tendency to be oxidized will have a smaller reduction potential. Recall that a standard oxidation potential would be equal in magnitude, but of opposite sign to that of the reduction potential for the reverse reaction.

(a) Li **(b)** H_2 **(c)** Fe^{2+} **(d)** Br^-

20.18 Calculating $\mathcal{E}°$ from K. Similar to Problem Type 3B.

The equation that relates K and the standard cell emf is:

$$\mathcal{E}^\circ_{\text{cell}} = \frac{0.0591\text{ V}}{n}\log K$$

Substitute the equilibrium constant and the mol of e^- transferred ($n = 2$) during the redox reaction into the above equation to calculate $\mathcal{E}°$.

$$\mathcal{E}^\circ = \frac{0.0591\text{ V}}{n}\log K = \frac{0.0591\text{ V}}{2}\log(2.69 \times 10^{12}) = \mathbf{0.367\ V}$$

20.20 **(a)** We break the equation into two half–reactions:

$$\text{Mg}(s) \xrightarrow{\text{ox}} \text{Mg}^{2+}(aq) + 2e^- \qquad \mathcal{E}^\circ_{\text{ox}} = 2.37\text{ V}$$

$$\text{Pb}^{2+}(aq) + 2e^- \xrightarrow{\text{red}} \text{Pb}(s) \qquad \mathcal{E}^\circ_{\text{red}} = -0.13\text{ V}$$

The standard emf is the sum of the standard potentials for the two half-reactions.

$$\mathcal{E}^\circ_{\text{cell}} = \mathcal{E}^\circ_{\text{ox}} + \mathcal{E}^\circ_{\text{red}} = 2.37\text{ V} - 0.13\text{ V} = 2.24\text{ V}$$

We can calculate $\Delta G°$ from the standard emf.

$$\Delta G^\circ = -nF\mathcal{E}^\circ_{\text{cell}}$$

$$= -(2\text{ mol})(96500\text{ J/V·mol})(2.24\text{ V}) = \mathbf{-432\ kJ}$$

Next, we can calculate K using Equation 20.5 of the text.

$$\mathcal{E}^\circ_{\text{cell}} = \frac{0.0591\text{ V}}{n}\log K$$

or

$$\log K = \frac{n\mathcal{E}^\circ_{\text{cell}}}{0.0591\text{ V}}$$

and

$$K_c = 10^{\frac{n\mathcal{E}^\circ_{\text{cell}}}{0.0591}}$$

$$= 10^{\frac{(2)(2.24)}{(0.0591)}} = 10^{75.8} = \mathbf{6 \times 10^{75}}$$

Tip: You could also calculate K_c from the standard free energy change, $\Delta G°$, using the equation: $\Delta G° = -RT\ln K_c$.

(b) We break the equation into two half–reactions:

$$Br_2\,(l) + 2e^- \xrightarrow{\text{red}} 2\,Br^-\,(aq) \qquad \mathcal{E}^\circ_{red} = 1.07\text{ V}$$

$$2\,I^-\,(aq) \xrightarrow{\text{ox}} I_2\,(s) + 2e^- \qquad \mathcal{E}^\circ_{ox} = -0.53\text{ V}$$

The standard emf is the sum of the standard potentials for the two half-reactions.

$$\mathcal{E}^\circ_{cell} = \mathcal{E}^\circ_{ox} + \mathcal{E}^\circ_{red} = -0.53\text{ V} + 1.07\text{ V} = 0.54\text{ V}$$

We can calculate ΔG° from the standard emf.

$$\Delta G^\circ = -nF\mathcal{E}^\circ_{cell}$$

$$= -(2\text{ mol})(96500\text{ J/V}\cdot\text{mol})(0.54\text{ V}) = \mathbf{-104\text{ kJ}}$$

Next, we can calculate K using Equation 20.5 of the text.

$$K_c = 10^{\frac{n\mathcal{E}^\circ_{cell}}{0.0591}}$$

$$= 10^{\frac{(2)(0.54)}{(0.0591)}} = 10^{18.3} = \mathbf{2 \times 10^{18}}$$

(c) This is worked in an analogous manner to parts (a) and (b).

$$\mathcal{E}^\circ_{cell} = \mathcal{E}^\circ_{ox} + \mathcal{E}^\circ_{red}$$

$$= -0.77\text{ V} + 1.23\text{ V} = +0.46\text{ V}$$

$$\Delta G^\circ = -nF\mathcal{E}^\circ_{cell}$$

$$= -(4\text{ mol})(96500\text{ J/V}\cdot\text{mol})(0.46\text{ V}) = \mathbf{-178\text{ kJ}}$$

$$K_c = 10^{\frac{n\mathcal{E}^\circ_{cell}}{0.0591}}$$

$$= 10^{\frac{(4)(0.46)}{(0.0591)}} = 10^{31.1} = \mathbf{1 \times 10^{31}}$$

(d) This is worked in an analogous manner to parts (a), (b), and (c).

$$\mathcal{E}^\circ_{cell} = \mathcal{E}^\circ_{ox} + \mathcal{E}^\circ_{red}$$

$$= 1.66\text{ V} + 0.53\text{ V} = 2.19\text{ V}$$

$$\Delta G^\circ = -nF\mathcal{E}^\circ_{cell}$$

$$= -(6\text{ mol})\,(96500\text{ J/V}\cdot\text{mol})(2.19\text{ V}) = \mathbf{-1.27 \times 10^3\text{ kJ}}$$

$$K_c = 10^{\frac{n\mathcal{E}^\circ_{cell}}{0.0591}}$$

$$= 10^{\frac{(6)(2.19)}{(0.0591)}} = 10^{222} = \mathbf{1 \times 10^{222}}$$

20.22 Calculating $\Delta G°$ and K from $\mathcal{E}°$, Problem Type 3B.

Step 1: Separate the reaction into half-reactions to calculate the standard cell emf.

$$Cu^{+}(aq) \xrightarrow{\text{ox}} Cu^{2+}(aq) + e^{-} \qquad \mathcal{E}^{\circ}_{ox} = -0.15\text{ V}$$

$$Cu^{+}(aq) + e^{-} \xrightarrow{\text{red}} Cu(s) \qquad \mathcal{E}^{\circ}_{red} = 0.52\text{ V}$$

$$2\,Cu^{+}(aq) \longrightarrow Cu^{2+}(aq) + Cu(s)$$

$$\mathcal{E}^{\circ}_{\textbf{cell}} = \mathcal{E}^{\circ}_{ox} + \mathcal{E}^{\circ}_{red} = -0.15\text{ V} + 0.52\text{ V} = \mathbf{0.37\ V}$$

Step 2: The equation that relates $\mathcal{E}^{\circ}_{cell}$ and the standard free energy change is shown below. Substitute $\mathcal{E}^{\circ}_{cell}$ into the equation to calculate $\Delta G°$. For this redox reaction, $n = 1$.

$$\Delta G° = -nF\mathcal{E}^{\circ}_{cell}$$

$$= -(1\ \cancel{\text{mol}})(96500\ \text{J}/\cancel{\text{V}}\cdot\cancel{\text{mol}})(0.37\ \cancel{\text{V}}) = \mathbf{-36\ kJ}$$

Step 3: Next, we can calculate K using Equation 20.5 of the text.

$$\mathcal{E}^{\circ}_{cell} = \frac{0.0591\text{ V}}{n}\log K \qquad (20.4)$$

or

$$\log K = \frac{n\mathcal{E}^{\circ}_{cell}}{0.0591\text{ V}}$$

and

$$K_c = 10^{\frac{n\mathcal{E}^{\circ}_{cell}}{0.0591}}$$

$$= 10^{\frac{(1)(0.37)}{(0.0591)}} = 10^{6.26} = \mathbf{2 \times 10^{6}}$$

Is this a spontaneous reaction? Can copper(I) salts exist in aqueous solution?

20.26 Using the Nernst equation, Problem Type 4.

(a)

Step 1: Use standard potentials for the two half-reactions to calculate the standard cell emf.

$$\mathcal{E}^{\circ}_{\textbf{cell}} = \mathcal{E}^{\circ}_{ox} + \mathcal{E}^{\circ}_{red} = 2.37\text{ V} - 0.14\text{ V} = \mathbf{2.23\ V}$$

Step 2: Use the Nernst equation to compute the emf at the given nonstandard concentrations.

$$\mathcal{E} = \mathcal{E}° - \frac{0.0591\text{ V}}{n}\log Q$$

$$= 2.23\text{ V} - \frac{0.0591}{2}\log\frac{0.045}{0.035} = \mathbf{2.23\ V}$$

Step 3: We can find the free energy difference at the given concentrations using the following equation:

$$\Delta G = -nF\mathcal{E}_{cell}$$

$$= -(2\text{ mol})(96500\text{ J/V}\cdot\text{mol})(2.23\text{ V}) = \mathbf{-4.30 \times 10^{5}\ J} = \mathbf{-4.30 \times 10^{2}\ kJ}$$

(b)
Step 1: Use standard potentials for the two half-reactions to calculate the standard cell emf.

$$\mathcal{E}^\circ_{cell} = \mathcal{E}^\circ_{ox} + \mathcal{E}^\circ_{red} = 0.76\text{ V} - 0.74\text{ V} = \mathbf{0.02\text{ V}}$$

Step 2: Use the Nernst equation to compute the emf at the given nonstandard concentrations.

$$\mathcal{E} = \mathcal{E}^\circ - \frac{0.0591\text{ V}}{n}\log Q$$

$$= 0.02\text{ V} - \frac{0.0591}{6}\log\frac{(0.0085)^3}{(0.010)^2} = \mathbf{0.04\text{ V}}$$

Step 3: We can find the free energy difference at the given concentrations using the following equation:

$$\Delta G = -nF\mathcal{E}_{cell}$$

$$= -(6\text{ mol})(96500\text{ J/V}\cdot\text{mol})(0.04\text{ V}) = \mathbf{-2\times10^4\text{ J}} = \mathbf{-2\times10^1\text{ kJ}}$$

20.28 Let's write the two half-reactions to calculate the standard cell emf. (Oxidation occurs at the Pb electrode.)

$$Pb(s) \xrightarrow{ox} Pb^{2+}(aq) + 2e^- \qquad \mathcal{E}^\circ_{ox} = +0.13\text{ V}$$

$$2H^+(aq) + 2e^- \xrightarrow{red} H_2(g) \qquad \mathcal{E}^\circ_{red} = 0.00\text{ V}$$

$$2H^+(aq) + Pb(s) \longrightarrow H_2(g) + Pb^{2+}(aq)$$

$$\mathcal{E}^\circ_{cell} = \mathcal{E}^\circ_{ox} + \mathcal{E}^\circ_{red} = 0.13\text{ V} + 0.00\text{ V} = 0.13\text{ V}$$

Using the Nernst equation, we can calculate the cell emf, $\mathcal{E}$.

$$\mathcal{E} = \mathcal{E}^\circ - \frac{0.0591\text{ V}}{n}\log\frac{[Pb^{2+}]P_{H_2}}{[H^+]^2}$$

$$= 0.13\text{ V} - \frac{0.0591\text{ V}}{2}\log\frac{(0.10)(1.0)}{(0.050)^2} = \mathbf{0.083\text{ V}}$$

20.30 All concentration cells have the same standard emf: zero volts.

$$Mg^{2+}(aq) + 2e^- \xrightarrow{red} Mg(s) \qquad \mathcal{E}^\circ_{red} = -2.37\text{ V}$$

$$Mg(s) \xrightarrow{ox} Mg^{2+}(aq) + 2e^- \qquad \mathcal{E}^\circ_{ox} = +2.37\text{ V}$$

$$\mathcal{E}^\circ_{cell} = \mathcal{E}^\circ_{ox} + \mathcal{E}^\circ_{red} = 2.37\text{ V} - 2.37\text{ V} = 0.00\text{ V}$$

We use the Nernst equation to compute the emf. There are two moles of electrons transferred from the reducing agent to the oxidizing agent in this reaction, so $n = 2$.

$$\mathcal{E} = \mathcal{E}^\circ - \frac{0.0591\text{ V}}{n}\log Q = \mathcal{E}^\circ - \frac{0.0591\text{ V}}{n}\log\frac{[Mg^{2+}]_{ox}}{[Mg^{2+}]_{red}}$$

$$\mathcal{E} = 0\text{ V} - \frac{0.0591\text{ V}}{2}\log\frac{0.24}{0.53} = \mathbf{0.010\text{ V}}$$

What is the direction of spontaneous change in all concentration cells?

20.34 We can calculate the standard free energy change, $\Delta G°$, from the standard free energies of formation, ΔG_f° (see Problem Type 2A , Chapter 19). Then, we can calculate the standard cell emf, $\mathcal{E}_{cell}^\circ$, from $\Delta G°$.

The overall reaction is:

$$C_3H_8\,(g) + 5\,O_2\,(g) \longrightarrow 3\,CO_2\,(g) + 4\,H_2O\,(l)$$

$$\Delta G_{rxn}^\circ = 3\,\Delta G_f^\circ[CO_2(g)] + 4\,\Delta G_f^\circ[H_2O(l)] - \{\Delta G_f^\circ[C_3H_8(g)] + 5\,\Delta G_f^\circ[O_2(g)]\}$$

$$\Delta G_{rxn}^\circ = (3\text{ mol})(-394.4\text{ kJ/mol}) + (4\text{ mol})(-237.2\text{ kJ/mol}) - [(1\text{ mol})(-23.5\text{ kJ/mol}) + (5\text{ mol})(0)]$$

$$= -2108.5\text{ kJ}$$

We can now calculate the standard emf using the following equation:

$$\Delta G° = -nF\mathcal{E}_{cell}^\circ$$

or

$$\mathcal{E}_{cell}^\circ = \frac{-\Delta G°}{nF}$$

Check the half-reactions on p. 650 of the text to determine that 20 moles of electrons are transferred during this redox reaction.

$$\mathcal{E}_{cell}^\circ = \frac{-(-2108.5 \times 10^3\text{ J})}{(20\text{ mol})(96500\text{ J/V}\cdot\text{mol})} = \mathbf{1.09\ V}$$

Does this suggest that, in theory, it should be possible to construct a galvanic cell (battery) based on any conceivable spontaneous reaction?

20.42 Calculating the quantity of products in electrolysis, Problem Type 5B.

(a) The only ions present in molten $CaCl_2$ are Ca^{2+} and Cl^-. The electrode reactions are:

anode: $2\,Cl^-\,(aq) \longrightarrow Cl_2\,(g) + 2e^-$

cathode: $Ca^{2+}\,(aq) + 2e^- \longrightarrow Ca\,(s)$

This cathode half-reaction tells us that 2 mol of e^- are required to produce 1 mol of Ca (*s*). In other words, we need 2 *F* of electrical charge to reduce 1 mol of Ca^{2+} ions to Ca metal.

(b)
Step 1: We can calculate the mass of Ca produced using the following strategy.

current × time → coulombs → [faradays] → mol Ca → g Ca

This is a large number of steps, so let's break it down into two parts. The faraday is the link between the current measurements and the molar amount of Ca formed.

The number of faradays is:

$$0.50\,\text{A} \times \frac{1\,\frac{\text{C}}{\text{s}}}{1\,\text{A}} \times \frac{60\,\text{s}}{1\text{ min}} \times 30\text{ min} = 9.0 \times 10^2\,\text{C} \times \frac{1\,F}{96{,}500\,\text{C}} = 9.3 \times 10^{-3}\,F$$

Step 3: From the number of faradays, we can calculate the moles of Ca produced. We can then convert to grams of Ca.

$$? \textbf{ g Ca} = 9.3 \times 10^{-3} \cancel{F} \times \frac{1 \cancel{\text{mol}} \text{ Ca}}{2 \cancel{F}} \times \frac{40.08 \text{ g Ca}}{1 \cancel{\text{mol}} \text{ Ca}} = \textbf{0.19 g Ca}$$

20.44 The costs for producing various metals can be calculated by setting up the ratios of electrons needed to reduce each type of ion. You also need to take into account the atomic mass of each metal (i.e., What mass of metal ions is reduced by the moles of electrons in the balanced equation?). The reductions are:

$$Mg^{2+} + 2e^- \longrightarrow Mg$$

$$Al^{3+} + 3e^- \longrightarrow Al$$

$$Na^+ + e^- \longrightarrow Na$$

$$Ca^{2+} + 2e^- \longrightarrow Ca$$

(a) For aluminum :

$$\left(\frac{\$155}{1 \cancel{\text{ton}}}\right)\left(\frac{3 \cancel{\text{mol}} \text{ e}^- / \cancel{\text{mol}} \text{ Al}}{2 \cancel{\text{mol}} \text{ e}^- / \cancel{\text{mol}} \text{ Mg}}\right)\left(\frac{24.31 \cancel{\text{g}} \text{ Mg} / \cancel{\text{mol}} \text{ Mg}}{26.98 \cancel{\text{g}} \text{ Al} / \cancel{\text{mol}} \text{ Al}}\right) \times 10.0 \cancel{\text{tons}} = \mathbf{\$2.09 \times 10^3}$$

(b) For sodium:

$$\left(\frac{\$155}{1 \text{ ton}}\right)\left(\frac{1 \text{ mol e}^- / \text{mol Na}}{2 \text{ mol e}^- / \text{mol Mg}}\right)\left(\frac{24.31 \text{ g Mg} / \text{mol Mg}}{22.99 \text{ g Na} / \text{mol Na}}\right) \times 30.0 \text{ tons} = \mathbf{\$2.46 \times 10^3}$$

(c) For calcium:

$$\left(\frac{\$155}{1 \text{ ton}}\right)\left(\frac{2 \text{ mol e}^- / \text{mol Ca}}{2 \text{ mol e}^- / \text{mol Mg}}\right)\left(\frac{24.31 \text{ g Mg} / \text{mol Mg}}{40.08 \text{ g Ca} / \text{mol Ca}}\right) \times 50.0 \text{ tons} = \mathbf{\$4.70 \times 10^3}$$

20.46 **(a)** The half–reaction is:

$$2\ H_2O\ (l) \longrightarrow O_2\ (g) + 4\ H^+\ (aq) + 4e^-$$

First, we can calculate the number of moles of oxygen produced using the ideal gas equation (see Problem Type 2A, Chapter 5).

$$n_{O_2} = \frac{PV}{RT}$$

$$n_{O_2} = \frac{(1.0 \cancel{\text{atm}})(0.84 \cancel{\text{L}})}{(0.0821 \cancel{\text{L}} \cdot \cancel{\text{atm}} / \text{mol} \cdot \cancel{\text{K}})(298 \cancel{\text{K}})} = 0.034 \text{ mol } O_2$$

Since 4 moles of electrons are needed for every 1 mole of oxygen, we will need 4 *F* of electrical charge to produce 1 mole of oxygen.

$$? \boldsymbol{F} = 0.034 \cancel{\text{mol}} \text{ O}_2 \times \frac{4 F}{1 \cancel{\text{mol}} \text{ O}_2} = \mathbf{0.14}\ \boldsymbol{F}$$

(b) The half–reaction is:

$$2\,Cl^- (aq) \longrightarrow Cl_2 (g) + 2e^-$$

The number of moles of chlorine produced is:

$$n_{Cl_2} = \frac{PV}{RT}$$

$$n_{Cl_2} = \frac{\left(\frac{750\ \text{mmHg}}{760\ \text{mmHg/atm}}\right)(1.50\ \text{L})}{(0.0821\ \text{L}\cdot\text{atm/mol}\cdot\text{K})(293\ \text{K})} = 0.0615\ \text{mol}\ Cl_2$$

Since 2 moles of electrons are needed for every 1 mole of chlorine gas, we will need 2 F of electrical charge to produce 1 mole of chlorine gas.

$$?\ \boldsymbol{F} = 0.0615\ \text{mol}\ Cl_2 \times \frac{2\ F}{1\ \text{mol}\ Cl_2} = \mathbf{0.123}\ \boldsymbol{F}$$

(c) The half–reaction is:

$$Sn^{2+} (aq) + 2e^- \longrightarrow Sn (s)$$

The number of moles of Sn (*s*) produced is

$$?\ \text{mol Sn} = 6.0\ \text{g Sn} \times \frac{1\ \text{mol Sn}}{118.7\ \text{g Sn}} = 0.051\ \text{mol Sn}$$

Since 2 moles of electrons are needed for every 1 mole of Sn, we will need 2 F of electrical charge to reduce 1 mole of Sn^{2+} ions to Sn metal.

$$?\ \boldsymbol{F} = 0.051\ \text{mol Sn} \times \frac{2\ F}{1\ \text{mol Sn}} = \mathbf{0.10}\ \boldsymbol{F}$$

20.48 Electrolysis, Problem Type 5.

(a) The half–reaction is:

$$Ag^+ (aq) + e^- \longrightarrow Ag (s)$$

(b) Since this reaction is taking place in an aqueous solution, the probable oxidation is the oxidation of water. (Neither Ag^+ nor NO_3^- can be further oxidized.)

$$2\,H_2O (l) \longrightarrow O_2 (g) + 4\,H^+ (aq) + 4e^-$$

(c) The half-reaction tells us that 1 F of electrical charge is needed to reduce 1 mol of Ag^+ to Ag metal. We can set up the following strategy to calculate the quantity of electricity (in C) needed to deposit 0.67 g of Ag.

grams Ag → mol Ag → faradays → coulombs

$$0.67\ \text{g Ag}\left(\frac{1\ \text{mol Ag}}{107.9\ \text{g Ag}}\right)\left(\frac{1\ F}{1\ \text{mol Ag}}\right)\left(\frac{96500\ \text{C}}{1\ F}\right) = \mathbf{6.0\times10^2\ C}$$

20.50 **(a)** First find the amount of charge needed to produce 2.00 g of silver according to the half–reaction:

$$Ag^{+}(aq) + e^{-} \longrightarrow Ag(s)$$

$$2.00\,\text{g Ag}\left(\frac{1\,\text{mol Ag}}{107.9\,\text{g Ag}}\right)\left(\frac{1\,F}{1\,\text{mol Ag}}\right)\left(\frac{96500\,\text{C}}{1\,F}\right) = 1.79\times10^{3}\,\text{C}$$

The half–reaction for the reduction of copper(II) is:

$$Cu^{2+}(aq) + 2e^{-} \longrightarrow Cu(s)$$

From the amount of charge calculated above, we can calculate the mass of copper deposited in the second cell.

$$1.79\times10^{3}\,\text{C}\left(\frac{1\,F}{96500\,\text{C}}\right)\left(\frac{1\,\text{mol Cu}}{2\,F}\right)\left(\frac{63.55\,\text{g Cu}}{1\,\text{mol Cu}}\right) = \mathbf{0.589\,g\,Cu}$$

(b) We can calculate the current flowing through the cells using the following strategy.

Coulombs → Coulombs/hr → Coulombs/second = A

Recall that 1 C = 1 A·s

The current flowing through the cells is:

$$1.79\times10^{3}\,\text{A}\cdot\text{s}\left(\frac{1}{3.75\,\text{h}}\right)\left(\frac{1\,\text{h}}{3600\,\text{s}}\right) = \mathbf{0.133\,A}$$

20.52 Electrolysis, Problem Type 5.

Step 1: Balance the half–reaction.

$$Cr_2O_7^{2-}(aq) + 14\,H^{+}(aq) + 12e^{-} \longrightarrow 2\,Cr(s) + 7\,H_2O(l)$$

Step 2: Calculate the quantity of chromium metal by calculating the volume and converting this to mass using the given density.

Volume Cr = thickness × surface area

$$= 1.0\times10^{-2}\,\text{mm}\left(\frac{1\,\text{m}}{1000\,\text{mm}}\right)\times 0.25\,\text{m}^2 = 2.5\times10^{-6}\,\text{m}^3$$

Converting to cm^3,

$$2.5\times10^{-6}\,\text{m}^3\times\left(\frac{100\,\text{cm}}{1\,\text{m}}\right)^3 = 2.5\,\text{cm}^3$$

Next, calculate the mass of Cr.

Mass = density × volume

$$\text{Mass Cr} = 2.5\,\text{cm}^3\left(\frac{7.19\,\text{g}}{1\,\text{cm}^3}\right) = 18\,\text{g Cr}$$

Step 3: Find the number of faradays required to electrodeposit 18 g of Cr from solution. The half-reaction is:

$$Cr_2O_7^{2-}\ (aq) + 14\ H^+\ (aq) + 12e^- \longrightarrow 2\ Cr\ (s) + 7\ H_2O\ (l)$$

Six faradays of electrical charge are required to reduce 1 mol of Cr metal. But, we are electrodepositing less than 1 mol of Cr (*s*). We need to complete the following conversions:

g Cr → mol Cr → faradays

$$?\ \text{faradays} = 18\ \text{g Cr} \times \frac{1\ \text{mol Cr}}{52.00\ \text{g Cr}} \times \frac{6\ F}{1\ \text{mol Cr}} = 2.1\ F$$

Step 4: Determine how long it will take for 2.1 *F* of electrical charge to flow through the cell when the current is 25.0 C/s. We need to complete the following conversions:

faradays → coulombs → seconds → hours

$$\mathbf{?\ h} = 2.1\ F \times \frac{96{,}500\ \text{C}}{1\ F} \times \frac{1\ \text{s}}{25.0\ \text{C}} \times \frac{1\ \text{h}}{3600\ \text{s}} = \mathbf{2.3\ h}$$

Would any time be saved by connecting several bumpers together in a series?

20.54 Based on the half-reaction, we know that one faraday will produce half a mole of copper.

$$Cu^{2+}\ (aq) + 2e^- \longrightarrow Cu\ (s)$$

First, let's calculate the charge (in C) needed to deposit 0.300 g of Cu.

$$(3.00\ \text{A} \times 304\ \text{s})\left(\frac{1\ \text{C}}{1\ \text{A}\cdot\text{s}}\right) = 912\ \text{C}$$

We know that one faraday will produce half a mole of copper, but we don't have a half a mole of copper. We have:

$$0.300\ \text{g Cu} \times \frac{1\ \text{mol Cu}}{63.55\ \text{g Cu}} = 4.72 \times 10^{-3}\ \text{mol}$$

We calculated the number of coulombs (912 C) needed to produce 4.72×10^{-3} mol of Cu. How many coulombs will it take to produce 0.500 moles of Cu? This will be Faraday's constant.

$$\frac{912\ \text{C}}{4.72 \times 10^{-3}\ \text{mol Cu}} \times 0.500\ \text{mol Cu} = \mathbf{9.66 \times 10^4\ C = 1\ \mathit{F}}$$

20.56 First we can calculate the number of moles of hydrogen produced using the ideal gas equation.

$$n_{H_2} = \frac{PV}{RT}$$

$$n_{H_2} = \frac{(782/760)\ \text{atm}\ (0.854\ \text{L})}{(0.0821\ \text{L}\cdot\text{atm}/\text{K}\cdot\text{mol})(298\ \text{K})} = 0.0359\ \text{mol}$$

The number of faradays passed through the solution is:

$$0.0359\ \text{mol}\ H_2\left(\frac{2\ F}{1\ \text{mol}\ H_2}\right) = \mathbf{0.0718\ \mathit{F}}$$

20.58 If you have difficulty balancing redox equations, see Problem Type 1. The balanced equation is:

$$Cr_2O_7^{2-} + 6\ Fe^{2+} + 14\ H^+ \longrightarrow 2\ Cr^{3+} + 6\ Fe^{3+} + 7\ H_2O$$

The remainder of this problem is a solution stoichiometry problem. See Problem Types 5 and 8, Chapter 4.

The number of moles of potassium dichromate in 26.0 mL of the solution is:

$$26.0\ \text{mL}\left(\frac{1\ \text{L}}{1000\ \text{mL}}\right)\left(\frac{0.0250\ \text{mol}}{1\ \text{L}}\right) = 6.50 \times 10^{-4}\ \text{mol}\ K_2Cr_2O_7$$

From the balanced equation it can be seen that 1 mole of dichromate is stoichiometrically equivalent to 6 moles of iron(II). The number of moles of iron(II) oxidized is therefore

$$6.50 \times 10^{-4}\ \text{mol}\ Cr_2O_7^{2-}\left(\frac{6\ \text{mol}\ Fe^{2+}}{1\ \text{mol}\ Cr_2O_7^{2-}}\right) = 3.90 \times 10^{-3}\ \text{mol}\ Fe^{2+}$$

Finally, the molar concentration of Fe^{2+} is:

$$\frac{3.90 \times 10^{-3}\ \text{mol}}{25.0 \times 10^{-3}\ \text{L}} = 0.156\ \text{mol/L} = \mathbf{0.156}\ \boldsymbol{M}\ \mathbf{Fe^{2+}}$$

20.60 The balanced equation is:

$$MnO_4^- + 5\ Fe^{2+} + 8\ H^+ \longrightarrow Mn^{2+} + 5\ Fe^{3+} + 4\ H_2O$$

First, let's calculate the number of moles of potassium permanganate in 23.30 mL of solution.

$$23.30\ \text{mL}\left(\frac{1\ \text{L}}{1000\ \text{mL}}\right)\left(\frac{0.0194\ \text{mol}}{1\ \text{L}}\right) = 4.52 \times 10^{-4}\ \text{mol}\ KMnO_4$$

From the balanced equation it can be seen that 1 mole of permanganate is stoichiometrically equivalent to 5 moles of iron(II). The number of moles of iron(II) oxidized is therefore

$$4.52 \times 10^{-4}\ \text{mol}\ MnO_4^-\left(\frac{5\ \text{mol}\ Fe^{2+}}{1\ \text{mol}\ MnO_4^-}\right) = 2.26 \times 10^{-3}\ \text{mol}\ Fe^{2+}$$

The mass of Fe^{2+} oxidized is:

$$\text{mass}\ Fe^{2+} = 2.26 \times 10^{-3}\ \text{mol}\ Fe^{2+} \times \frac{55.85\ \text{g}\ Fe^{2+}}{1\ \text{mol}\ Fe^{2+}} = 0.126\ \text{g}\ Fe^{2+}$$

Finally, the mass percent of iron in the ore can be calculated.

$$\text{mass \% Fe} = \frac{\text{mass of iron}}{\text{total mass of sample}} \times 100\%$$

$$\mathbf{\%Fe} = \frac{0.126\ \text{g}}{0.2792\ \text{g}} \times 100\% = \mathbf{45.1\%}$$

20.62 **(a)** The half–reactions are:

(i) $MnO_4^-\ (aq) + 8\ H^+\ (aq) + 5e^- \longrightarrow Mn^{2+}\ (aq) + 4\ H_2O\ (l)$

(ii) $C_2O_4^{2-}\ (aq) \longrightarrow 2\ CO_2\ (g) + 2e^-$

We combine the half–reactions to cancel electrons, that is, [2 × equation (i)] + [5 × equation (ii)]

$$2\ MnO_4^-\ (aq) + 16\ H^+\ (aq) + 5\ C_2O_4^{2-}\ (aq) \longrightarrow 2\ Mn^{2+}\ (aq) + 10\ CO_2\ (g) + 8\ H_2O\ (l)$$

(b) We can calculate the moles of $KMnO_4$ from the molarity and volume of solution.

$$24.0\ \text{mL}\ KMnO_4\left(\frac{1\ \text{L}}{1000\ \text{mL}}\right)\left(\frac{0.0100\ \text{mol}\ KMnO_4}{1\ \text{L}}\right) = 2.40\times10^{-4}\ \text{mol}\ KMnO_4$$

We can calculate the mass of oxalic acid from the stoichiometry of the balanced equation. The mole ratio between oxalate ion and permanganate ion is 5:2.

$$2.40\times10^{-4}\ \text{mol}\ KMnO_4\left(\frac{5\ \text{mol}\ H_2C_2O_4}{2\ \text{mol}\ KMnO_4}\right)\left(\frac{90.04\ \text{g}\ H_2C_2O_4}{1\ \text{mol}\ H_2C_2O_4}\right) = 0.0540\ \text{g}\ H_2C_2O_4$$

Finally, the percent by mass of oxalic acid in the sample is:

$$\textbf{\% oxalic acid} = \frac{0.0540\ \text{g}}{1.00\ \text{g}}\times 100\% = \mathbf{5.40\ \%}$$

20.64 The balanced equation is:

$$2\ MnO_4^- + 5\ C_2O_4^{2-} + 16\ H^+ \longrightarrow 2\ Mn^{2+} + 10\ CO_2 + 8\ H_2O$$

Therefore, 2 mol MnO_4^- reacts with 5 mol $C_2O_4^{2-}$

$$\text{Moles of } MnO_4^- \text{ reacted} = (24.2\ \text{mL})\left(\frac{1\ \text{L}}{1000\ \text{mL}}\right)\left(\frac{9.56\times10^{-4}\ \text{mol}\ MnO_4^-}{1\ \text{L}}\right)$$

$$= 2.31\times10^{-5}\ \text{mol}\ MnO_4^-$$

Recognize that the mole ratio of Ca^{2+} to $C_2O_4^{2-}$ is 1:1 in CaC_2O_4. The mass of Ca^{2+} in 10.0 mL is:

$$(2.31\times10^{-5}\ \text{mol}\ MnO_4^-)\times\left(\frac{5\ \text{mol}\ Ca^{2+}}{2\ \text{mol}\ MnO_4^-}\right)\left(\frac{40.08\ \text{g}\ Ca^{2+}}{1\ \text{mol}\ Ca^{2+}}\right) = 2.31\times10^{-3}\ \text{g}\ Ca^{2+}$$

Finally, converting to mg/mL, we have:

$$\left(\frac{2.31\times10^{-3}\ \text{g}\ Ca^{2+}}{10.0\ \text{mL}}\right)\left(\frac{1000\ \text{mg}}{1\ \text{g}}\right) = \mathbf{0.231\ mg\ Ca^{2+}/\ mL\ blood}$$

20.66 **(a)** The half–reactions are:

$2\ H^+\ (aq) + 2e^- \longrightarrow H_2\ (g)$ $\qquad \mathcal{E}° = 0.00\ V$

$Ag^+\ (aq) + e^- \longrightarrow Ag\ (s)$ $\qquad \mathcal{E}° = 0.80\ V$

$$\mathcal{E}^\circ_{\textbf{cell}} = \mathcal{E}^\circ_{ox} + \mathcal{E}^\circ_{red} = 0.00\ V + 0.80\ V = \mathbf{0.80\ V}$$

(b) The spontaneous cell reaction under standard-state conditions is:

$$2\,Ag^+\,(aq) + H_2\,(g) \longrightarrow 2\,Ag\,(s) + 2\,H^+\,(aq)$$

(c) Using the Nernst equation we can calculate the cell potential under nonstandard-state conditions.

$$\mathcal{E} = \mathcal{E}^\circ - \frac{0.0591\text{ V}}{n}\log\frac{[H^+]^2}{[Ag^+]^2 P_{H_2}}$$

(i) The potential is:

$$\mathcal{E} = 0.80\text{ V} - \frac{0.0591\text{ V}}{2}\log\frac{(1.0\times10^{-2})^2}{(1.0)^2(1.0)} = \mathbf{0.92\ V}$$

(ii) The potential is:

$$\mathcal{E} = 0.80\text{ V} - \frac{0.0591\text{ V}}{2}\log\frac{(1.0\times10^{-5})^2}{(1.0)^2(1.0)} = \mathbf{1.10\ V}$$

(d) From the results in part (c), we deduce that this cell is a pH meter; its potential is a sensitive function of the hydrogen ion concentration. Each 1 unit increase in pH causes a voltage increase of 0.060 V.

20.68 The overvoltage of oxygen is not large enough to prevent its formation at the anode. Water is oxidized before fluoride ion.

$$2\,H_2O\,(l) \longrightarrow O_2\,(g) + 4\,H^+\,(aq) + 4e^- \qquad \mathcal{E}^\circ_{ox} = -1.23\text{ V}$$

$$2\,F^-\,(aq) \longrightarrow F_2\,(g) + 2e^- \qquad \mathcal{E}^\circ_{ox} = -2.87\text{ V}$$

The very negative standard oxidation potential indicates that F^- has essentially no tendency to undergo oxidation.

This fact was one of the major obstacles preventing the discovery of fluorine for many years. HF was usually chosen as the substance for electrolysis, but two problems interfered with the experiment. First, any water in the HF was oxidized before the fluoride ion. Second, pure HF without any water in it is a nonconductor of electricity (HF is a weak acid!). The problem was finally solved by dissolving KF in liquid HF to give a conducting solution.

20.70 We can calculate the amount of charge that 4.0 g of MnO_2 can produce.

$$4.0\text{ g MnO}_2\left(\frac{1\text{ mol}}{86.94\text{ g}}\right)\left(\frac{2\,F}{2\text{ mol MnO}_2}\right)\left(\frac{96500\text{ C}}{1\,F}\right) = 4.44\times10^3\text{ C}$$

Since a current of one ampere represents a flow of one coulomb per second, we can find the time it takes for this amount of charge to pass.

$0.0050\text{ A} = 0.0050\text{ C/s}$

$$4.44\times10^3\text{ C}\left(\frac{1\text{ s}}{0.0050\text{ C}}\right)\left(\frac{1\text{ h}}{3600\text{ s}}\right) = \mathbf{2.5\times10^2\ h}$$

20.72 Since this is a concentration cell, the standard emf is zero. (Why?) Using the Nernst equation, we can write equations to calculate the cell voltage for the two cells.

$$(1) \quad \mathcal{E}_{\text{cell}} = -\frac{2.303RT}{nF}\log Q = -\frac{2.303RT}{2F}\log\frac{[Hg_2^{2+}]\text{soln A}}{[Hg_2^{2+}]\text{soln B}}$$

$$(2) \quad \mathcal{E}_{\text{cell}} = -\frac{2.303RT}{nF}\log Q = -\frac{2.303RT}{1F}\log\frac{[Hg^{+}]\text{soln A}}{[Hg^{+}]\text{soln B}}$$

In the first case, two electrons are transferred per mercury ion ($n = 2$), while in the second only one is transferred ($n = 1$). Note that the concentration ratio will be 1:10 in both cases. The voltages calculated at 18°C are:

$$(1) \quad \mathcal{E}_{\text{cell}} = -\frac{(2.303)(8.314 \text{ J/K}\cdot\text{mol})(291 \text{ K})}{2(96500 \text{ J/V}\cdot\text{mol})}\log 10^{-1} = 0.0289 \text{ V}$$

$$(2) \quad \mathcal{E}_{\text{cell}} = -\frac{(2.303)(8.314 \text{ J/K}\cdot\text{mol})(291 \text{ K})}{1(96500 \text{ J/V}\cdot\text{mol})}\log 10^{-1} = 0.0577 \text{ V}$$

Since the calculated cell potential for cell (1) agrees with the measured cell emf, we conclude that the mercury(I) ion exists as Hg_2^{2+} in solution.

20.74 We begin by treating this like an ordinary stoichiometry problem (see Problem Type 3B, Chapter 3).

Step 1: Calculate the number of moles of Mg and Ag^+.

The number of moles of magnesium is:

$$1.56 \text{ g Mg}\left(\frac{1 \text{ mol Mg}}{24.31 \text{ g Mg}}\right) = 0.0642 \text{ mol Mg}$$

The number of moles of silver ion in the solution is:

$$\left(\frac{0.100 \text{ mol Ag}^+}{1 \text{ L}}\right)0.1000 \text{ L} = 0.0100 \text{ mol Ag}^+$$

Step 2: Calculate the mass of Mg remaining by determining how much Mg reacts with Ag^+.

The balanced equation for the reaction is:

$$2\,Ag^+(aq) + Mg(s) \longrightarrow 2\,Ag(s) + Mg^{2+}(aq)$$

Since you need twice as much Ag^+ compared to Mg for complete reaction, Ag^+ is the limiting reagent. The amount of Mg consumed is:

$$0.0100 \text{ mol Ag}^+\left(\frac{1 \text{ mol Mg}}{2 \text{ mol Ag}^+}\right) = 0.00500 \text{ mol Mg}$$

The amount of magnesium remaining is:

$$(0.0642 - 0.00500) \text{ mol Mg} \times 24.31 \text{ g/mol} = \mathbf{1.44 \text{ g Mg}}$$

Step 3: Assuming complete reaction, calculate the concentration of Mg^{2+} ions produced.

Since the mole ratio between Mg and Mg^{2+} is 1:1, the mol of Mg^{2+} formed will equal the mol of Mg reacted. The concentration of Mg^{2+} is:

$$[Mg^{2+}]_0 = \frac{0.00500 \text{ mol}}{0.100 \text{ L}} = 0.0500\ M$$

Step 4: We can calculate the equilibrium constant for the reaction from the standard cell emf.

$$\mathcal{E}^\circ_{cell} = \mathcal{E}^\circ_{ox} + \mathcal{E}^\circ_{red} = 2.37 \text{ V} + 0.80 \text{ V} = 3.17 \text{ V}$$

We can then compute the equilibrium constant.

$$K = 10^{\frac{n\mathcal{E}^\circ_{cell}}{0.0591}}$$

$$= 10^{\frac{(2)(3.17)}{(0.0591)}} = 10^{107} = 1 \times 10^{107}$$

Step 5: To find equilibrium concentrations of Mg^{2+} and Ag^+, we have to solve an equilibrium problem (see Problem Type 5, Chapter 15).

Let x be the small amount of Mg^{2+} that reacts to achieve equilibrium. The concentration of Ag^+ will be $2x$ at equilibrium. Assume that essentially all Ag^+ has been reduced so that the initial concentration of Ag^+ is zero.

	$2\,Ag^+(aq) + Mg(s) \rightleftharpoons$	$2\,Ag(s) + Mg^{2+}(aq)$
Initial (M):	0.0000	0.0500
Change (M):	$+2x$	$-x$
Equilibrium (M):	$2x$	$(0.0500 - x)$

$$K = \frac{[Mg^{2+}]}{[Ag^+]^2}$$

$$1 \times 10^{107} = \frac{(0.0500 - x)}{(2x)^2}$$

We can assume $0.0500 - x \cong 0.0500$.

$$1 \times 10^{107} \cong \frac{0.0500}{(2x)^2}$$

$$(2x)^2 = \frac{0.0500}{1 \times 10^{107}} = 0.0500 \times 10^{-107}$$

$$(2x)^2 = 5.00 \times 10^{-109} = 50.0 \times 10^{-110}$$

$$2x = 7 \times 10^{-55}\ M$$

$$\mathbf{[Ag^+] = 2x = 7 \times 10^{-55}\ M}$$

$$[Mg^{2+}] = 0.0500 - x = \mathbf{0.0500\ M}$$

20.76 **(a)** Since this is an acidic solution, the gas must be hydrogen gas from the reduction of hydrogen ion. The two electrode reactions and the overall cell reaction are:

$$\begin{array}{ll} \text{anode:} & Cu\ (s) \longrightarrow Cu^{2+}\ (aq) + 2e^- \\ \text{cathode:} & 2\ H^+\ (aq) + 2e^- \longrightarrow H_2\ (g) \\ \hline & Cu\ (s) + 2\ H^+\ (aq) \longrightarrow Cu^{2+}\ (aq) + H_2\ (g) \end{array}$$

Since 0.584 g of copper was consumed, the amount of hydrogen gas produced is:

$$0.584\ \text{g Cu}\left(\frac{1\ \text{mol Cu}}{63.55\ \text{g Cu}}\right)\left(\frac{1\ \text{mol H}_2}{1\ \text{mol Cu}}\right) = 9.20 \times 10^{-3}\ \text{mol H}_2$$

At STP, 1 mole of an ideal gas occupies a volume of 22.41 L. Thus, the volume of H_2 at STP is:

$$V_{H_2} = (9.20 \times 10^{-3}\ \text{mol H}_2)(22.41\ \text{L/mol}) = \mathbf{0.206\ L}$$

(b) From the current and the time, we can calculate the amount of charge:

$$(1.18\ \text{A})\left(\frac{1\ \text{C/s}}{1\ \text{A}}\right)(1.52 \times 10^3\ \text{s}) = 1.79 \times 10^3\ \text{C}$$

Since we know the charge of an electron, we can compute the number of electrons.

$$(1.79 \times 10^3\ \text{C})\left(\frac{1\ e^-}{1.6022 \times 10^{-19}\ \text{C}}\right) = 1.12 \times 10^{22}\ e^-$$

Using the amount of copper consumed in the reaction and the fact that 2 mol of e^- are produced for every 1 mole of copper consumed, we can calculate Avogadro's number.

$$\frac{1.12 \times 10^{22}\ e^-}{9.20 \times 10^{-3}\ \text{mol Cu}} \times \left(\frac{1\ \text{mol Cu}}{2\ \text{mol}\ e^-}\right) = \mathbf{6.09 \times 10^{23}\ e^-/mol}$$

In practice, Avogadro's number can be determined by electrochemical experiments like this. The charge of the electron can be found independently by Millikan's experiment.

20.78 **(a)** We can calculate $\Delta G°$ from standard free energies of formation.

$$\begin{aligned} \Delta G° &= 2\,\Delta G_f°(N_2) + 6\,\Delta G_f°(H_2O) - [4\,\Delta G_f°(NH_3) + 3\,\Delta G_f°(O_2)] \\ &= 0 + (6\ \text{mol})(-237.2\ \text{kJ/mol}) - [(4\ \text{mol})(-16.6\ \text{kJ/mol}) + 0] \\ &= \mathbf{-1356.8\ kJ} \end{aligned}$$

(b) The half–reactions are:

$$4\ NH_3\ (g) \longrightarrow 2\ N_2\ (g) + 12\ H^+\ (aq) + 12e^-$$

$$3\ O_2\ (g) + 12\ H^+\ (aq) + 12e^- \longrightarrow 6\ H_2O\ (l)$$

The overall reaction is a 12–electron process. We can calculate the standard cell emf from the standard free energy change, $\Delta G°$.

$$\Delta G° = -nF\mathcal{E}^\circ_{\text{cell}}$$

$$\mathcal{E}^\circ_{\text{cell}} = \frac{-\Delta G°}{nF} = \frac{-\left(-1356.8\ \cancel{\text{kJ}} \times \frac{1000\ \cancel{\text{J}}}{1\ \cancel{\text{kJ}}}\right)}{(12\ \cancel{\text{mol}})(96500\ \cancel{\text{J}}/\text{V}\cdot\cancel{\text{mol}})} = \mathbf{1.17\ V}$$

20.80 Electrolysis, Problem Type 5.

The reduction of Ag^+ to Ag metal is:

$$Ag^+(aq) + e^- \longrightarrow Ag$$

We can calculate both the moles of Ag deposited and the moles of Au deposited.

$$?\ \text{mol Ag} = \frac{2.64\ \cancel{\text{g}}\ \text{Ag}}{107.9\ \cancel{\text{g}}\ \text{Ag}/\text{mol Ag}} = 2.45 \times 10^{-2}\ \text{mol Ag}$$

$$?\ \text{mol Au} = \frac{1.61\ \cancel{\text{g}}\ \text{Au}}{197.0\ \cancel{\text{g}}\ \text{Au}/\text{mol Au}} = 8.17 \times 10^{-3}\ \text{mol Au}$$

We do not know the oxidation state of Au ions, so we will represent the ions as Au^{n+}. If we divide the mol of Ag by the mol of Au, we can determine the ratio of Ag^+ reduced compared to Au^{n+} reduced.

$$\frac{2.45 \times 10^{-2}\ \text{mol Ag}}{8.17 \times 10^{-3}\ \text{mol Au}} = 3$$

That is, the same number of electrons that reduced the Ag^+ ions to Ag reduced only one–third the number of moles of the Au^{n+} ions to Au. Thus, each Au^{n+} required three electrons per ion for every one electron for Ag^+. The oxidation state for the gold ion is **+3**; the ion is Au^{3+}.

$$Au^{3+}(aq) + 3e^- \longrightarrow Au$$

20.82 We reverse the first half–reaction and add it to the second to come up with the overall balanced equation

$$Hg_2^{2+} \longrightarrow 2\,Hg^{2+} + 2e^- \qquad \mathcal{E}° = -0.92\ \text{V}$$

$$Hg_2^{2+} + 2e^- \longrightarrow 2\,Hg \qquad \mathcal{E}° = 0.85\ \text{V}$$

$$2\,Hg_2^{2+} \longrightarrow 2\,Hg^{2+} + 2\,Hg \qquad \mathcal{E}° = -0.07\ \text{V}$$

Let's divide the above equation by two to arrive at the equation given in the problem. Since we divided the equation by two, only 1 mol of electrons is transferred during this redox reaction. Also, since the cell potential is an intensive property,

$$Hg_2^{2+}(aq) \longrightarrow Hg^{2+}(aq) + Hg(l) \qquad \mathcal{E}° = -0.07\ \text{V}$$

We calculate $\Delta G°$ from $\mathcal{E}°$.

$$\Delta G° = -nF\mathcal{E}° = -(1\ \text{mol})(96500\ \text{J/V}\cdot\text{mol})(-0.07\ \text{V}) = \mathbf{7 \times 10^4\ J = 7\ kJ}$$

The corresponding equilibrium constant is:

$$K = \frac{[Hg^{2+}]}{[Hg_2^{2+}]}$$

We calculate K from $\Delta G°$.

$$\Delta G° = -RT \ln K$$

$$\ln K = \frac{-7 \times 10^3\ J}{(8.314\ J/K \cdot mol)(298\ K)}$$

$$\mathbf{K = 6 \times 10^{-2}}$$

20.84 The reactions for the electrolysis of NaCl (aq) are:

Anode:	$2\ Cl^-\ (aq) \longrightarrow Cl_2\ (g) + 2e^-$
Cathode:	$2\ H_2O\ (l) + 2e^- \longrightarrow H_2\ (g) + 2\ OH^-(aq)$
Overall:	$2H_2O(l) + 2Cl^-(aq) \rightarrow H_2(g) + Cl_2(g) + 2OH^-(aq)$

From the pH of the solution, we can calculate the OH^- concentration. From the $[OH^-]$, we can calculate the moles of OH^- produced. Then, from the moles of OH^- we can calculate the average current used.

$$pH = 12.24$$

$$pOH = 14.00 - 12.24 = 1.76$$

$$[OH^-] = 1.74 \times 10^{-2}\ M$$

The moles of OH^- produced are:

$$(1.74 \times 10^{-2}\ mol/\not{L}) \times (0.300\ \not{L}) = 5.22 \times 10^{-3}\ mol\ OH^-$$

From the balanced equation, it takes 1 mole of e^- to produce 1 mole of OH^- ions.

$$5.22 \times 10^{-3}\ \not{mol}\ OH^- \times \frac{1\ \not{F}}{1\ \not{mol}\ OH^-} \times \frac{96500\ C}{1\ \not{F}} = 504\ C$$

Recall that 1 C = 1 A·s

$$504\ A \cdot \not{s} \times \frac{1\ \not{min}}{60\ \not{s}} \times \frac{1}{6.00\ \not{min}} = \mathbf{1.4\ A}$$

20.86 The reaction is:

$$Pt^{n+} + ne^- \longrightarrow Pt$$

Thus, we can calculate the charge of the platinum ions by realizing that n mol of e^- are required per mol of Pt formed.

The moles of Pt formed are:

$$9.09\ \not{g}\ Pt \times \frac{1\ mol\ Pt}{195.1\ \not{g}\ Pt} = 0.0466\ mol\ Pt$$

Next, calculate the charge passed in C.

$$C = 2.00\,\text{h} \times \frac{3600\,\text{s}}{1\,\text{h}} \times \frac{2.50\,\text{C}}{1\,\text{s}} = 1.80 \times 10^4\ \text{C}$$

Convert to faradays.

$$?\,F = (1.80 \times 10^4\ \text{C}) \times \frac{1\,F}{96500\ \text{C}} = 0.187\,F$$

We now know the number of faradays (0.187 *F*) needed to produce 0.0466 mol of Pt metal. We can calculate the number of faradays needed to produce 1 mole of Pt metal.

$$\frac{0.187\,F}{0.0466\ \text{mol Pt}} = 4.01\ F/\text{mol Pt}$$

Since we need 4 *F* of electrical charge to reduce 1 mole of Pt ions, the charge on the Pt ions must be **+4**.

20.88 The half–reaction for the oxidation of water to oxygen is:

$$2\,H_2O\,(l) \longrightarrow O_2\,(g) + 4\,H^+\,(aq) + 4e^- \qquad \mathcal{E}^\circ_{ox} = -1.23\ \text{V}$$

Knowing that one mole of any gas at STP occupies a volume of 22.4 L, we find the number of moles of oxygen.

$$(4.26\ \text{L}\ O_2)\left(\frac{1\ \text{mol}}{22.4\ \text{L}}\right) = 0.190\ \text{mol}\ O_2$$

Since four electrons are required to form one oxygen molecule, the number of electrons must be:

$$(0.190\ \text{mol}\ O_2)\left(\frac{4\ \text{mol}\ e^-}{1\ \text{mol}\ O_2}\right)\left(\frac{6.022 \times 10^{23}\ e^-}{1\ \text{mol}}\right) = 4.58 \times 10^{23}\ e^-$$

The amount of charge passing through the solution is:

$$(6.00\ \text{A})\left(\frac{1\ \text{C/s}}{1\ \text{A}}\right)\left(\frac{3600\ \text{s}}{1\ \text{h}}\right)(3.40\ \text{h}) = 7.34 \times 10^4\ \text{C}$$

We find the electron charge by dividing the amount of charge by the number of electrons.

$$\frac{7.34 \times 10^4\ \text{C}}{4.58 \times 10^{23}\ e^-} = \mathbf{1.60 \times 10^{-19}\ C/e^-}$$

In actual fact, this sort of calculation can be used to find Avogadro's number, not the electron charge. The latter can be measured independently, and one can use this charge together with electrolytic data like the above to calculate the number of objects in one mole. See also Problem 20.76.

20.90 The overall cell reaction is:

$$2Ag^+(aq) + H_2(g) \rightarrow 2Ag(s) + 2H^+(aq)$$

We write the Nernst equation for this system.

$$\mathcal{E} = \mathcal{E}^\circ - \frac{0.0591\text{ V}}{n}\log Q = 0.080\text{ V} - \frac{0.0591\text{ V}}{2}\log\frac{[H^+]^2}{[Ag^+]^2 P_{H_2}}$$

The measured voltage is 0.589 V, and we can find the silver ion concentration as follows. Recall that the pressure of H_2 (*g*) is 1 atm and that the $[H^+] = 1\ M$ in an SHE.

$$0.589\text{ V} = 0.80\text{ V} - \frac{0.0591\text{ V}}{2}\log\frac{1}{[Ag^+]^2}$$

$$\log\frac{1}{[Ag^+]^2} = 7.14$$

$$\frac{1}{[Ag^+]^2} = 1.4 \times 10^7$$

$$[Ag^+] = 2.7 \times 10^{-4}\ M$$

Knowing the silver ion concentration, we can calculate the oxalate ion concentration and the solubility product constant.

$$[C_2O_4^{2-}] = (2.7 \times 10^{-4}\ M\ Ag^+) \times \frac{1\text{ mol } C_2O_4^{2-}}{2\text{ mol } Ag^+} = 1.35 \times 10^{-4}\ M$$

$$K_{sp} = [Ag^+]^2[C_2O_4^{2-}] = (2.7 \times 10^{-4})^2(1.35 \times 10^{-4}) = \mathbf{9.8 \times 10^{-12}}$$

20.92 **(a)** Gold does not tarnish in air because the reduction potential for oxygen is insufficient to result in the oxidation of gold.

$$O_2 + 4H^+ + 4e^- \rightarrow 2H_2O \qquad \mathcal{E}^\circ = 1.23\text{ V}$$

That is, $\mathcal{E}^\circ_{ox} + \mathcal{E}^\circ_{red} < 0$, for either oxidation by O_2 to Au^+ or Au^{3+}.

(b)

$$3(Au^+ + e^- \rightarrow Au) \qquad \mathcal{E}^\circ = 1.69\text{ V}$$
$$Au \rightarrow Au^{3+} + 3e^- \qquad \mathcal{E}^\circ = -1.50\text{ V}$$
$$3Au^+ \rightarrow 2Au + Au^{3+} \qquad \mathcal{E}^\circ = 0.19\text{ V}$$

Calculating ΔG,

$$\Delta G^\circ = -nF\mathcal{E}^\circ = -3 \times F \times \mathcal{E}^\circ = -3 \times 96{,}500\text{ J/V}\cdot\text{mol} \times 0.19\text{ V} = \mathbf{-55.0\text{ kJ}}$$

For spontaneous electrochemical equations, ΔG° must be negative. Thus the disproportionation occurs spontaneously.

(c) Since the most stable oxidation state for gold is Au^{3+}, the predicted reaction is:

$$2Au + 3F_2 \rightarrow 2AuF_3$$

20.94 The balanced equation is: $5Fe^{2+} + MnO_4^- + 8H^+ \longrightarrow Mn^{2+} + 5Fe^{3+} + 4\,H_2O$

Calculate the amount of iron(II) in the original solution using the mole ratio from the balanced equation.

$$23.0\ \text{mL}\left(\frac{1\ \text{L}}{1000\ \text{mL}}\right)\left(\frac{0.0200\ \text{mol KMnO}_4}{1\ \text{L}}\right)\left(\frac{5\ \text{mol Fe}^{2+}}{1\ \text{mol KMnO}_4}\right) = 0.00230\ \text{mol Fe}^{2+}$$

The concentration of iron(II) must be:

$$[\mathbf{Fe^{2+}}] = \left(\frac{0.00230\ \text{mol}}{0.0250\ \text{L}}\right) = \mathbf{0.0920}\ \boldsymbol{M}$$

The total iron concentration can be found by simple proportion because the same sample volume (25.0 mL) and the same $KMnO_4$ solution were used.

$$[\text{Fe}]_{\text{total}} = \frac{40.0\ \text{mL KMnO}_4}{23.0\ \text{mL KMnO}_4} \times 0.0920\ M = 0.160\ M$$

$$[\mathbf{Fe^{3+}}] = [\text{Fe}]_{\text{total}} - [\text{Fe}^{2+}] = 0.160\ M - 0.0920\ M = \mathbf{0.068}\ \boldsymbol{M}$$

Why are the two titrations with permanganate necessary in this problem?

20.96 **(a)** $1\text{A·h} = 1\text{A} \times 3600\text{s} = 3600\ \text{A·s} = 3600\ \text{C}$

(b) Anode: $Pb + SO_4^{2-} \rightarrow PbSO_4 + 2e^-$ $\quad \mathcal{E}° = 0.31\ \text{V}$

The charge produced by 1 mole of Pb is nF. Therefore, if we multiply nF by the moles of Pb in the battery, we will obtain the capacity of the battery.

$Q = nF \times$ number of moles of Pb

$$Q = (2\ \text{mol})(96500\ \text{C/mol})\frac{406\,\text{g}}{207.2\ \text{g/mol}} = 3.74 \times 10^5\ \text{C} = 3.74 \times 10^5\ \text{A·s} \times \frac{1\ \text{h}}{3600\ \text{s}} = \mathbf{104\ A{\cdot}h}$$

This ampere·hour cannot be fully realized because the concentration of H_2SO_4 keeps decreasing.

(c) $\mathcal{E}°_{\textbf{cell}} = 0.31\ \text{V} + 1.70\ \text{V} = \mathbf{2.01\ V}$ (From Table 20.1)

$\Delta G° = -nF\mathcal{E}°$

$\boldsymbol{\Delta G°} = -(2\ \text{mol})(96500\ \text{J/V·mol})(2.01\ \text{V}) = \mathbf{-3.88 \times 10^5\ J}$

Spontaneous as expected.

20.98 **(a)** unchanged **(b)** unchanged **(c)** squared **(d)** doubled **(e)** doubled

20.100

$Pb \rightarrow Pb^{2+} + 2e^-$	$\mathcal{E}° = 0.13\ \text{V}$
$2H^+ + 2e^- \rightarrow H_2$	$\mathcal{E}° = 0$
$Pb + 2H^+ \rightarrow Pb^{2+} + H_2$	$\mathcal{E}° = 0.13\ \text{V}$

$$\text{pH} = 1.60$$

$$[H^+] = 10^{-1.60} = 0.025\ M$$

$$\mathscr{E} = \mathscr{E}° - \frac{0.0591\ \text{V}}{n}\log Q = \mathscr{E}° - \frac{0.0591\ \text{V}}{n}\log\frac{[\text{Pb}^{2+}]P_{\text{H}_2}}{[\text{H}^+]^2}$$

$$0 = 0.13 - \frac{0.0591\ \text{V}}{2}\log\frac{(0.035)P_{\text{H}_2}}{(0.025)^2}$$

$$4.4 = \log\frac{(0.035)P_{\text{H}_2}}{(0.025)^2}$$

$$2.5 \times 10^4 = \frac{(0.035)P_{\text{H}_2}}{(0.025)^2}$$

$$\mathbf{P_{H_2} = 4.5 \times 10^2\ atm}$$

20.102 **(a)** At the anode (Mg): $Mg \rightarrow Mg^{2+} + 2e^-$

Also: $Mg + 2HCl \rightarrow MgCl_2 + H_2\,(g)$

At the cathode (Cu): $2H^+ + 2e^- \rightarrow H_2\,(g)$

(b) The solution does not turn blue.

(c) After all the HCl has been neutralized, the white precipitate is:

$$Mg^{2+} + 2OH^- \rightarrow Mg(OH)_2(s)$$

CHAPTER 21
NUCLEAR CHEMISTRY

PROBLEM-SOLVING STRATEGIES AND TUTORIAL SOLUTIONS

TYPES OF PROBLEMS

Problem Type 1: Balancing Nuclear Equations.

Problem Type 2: Nuclear Stability

Problem Type 3: Calculating Nuclear Binding Energy.

Problem Type 4: Kinetics of Radioactive Decay.

Problem Type 5: Balancing Nuclear Transmutation Equations.

PROBLEM TYPE 1: BALANCING NUCLEAR EQUATIONS

Writing a nuclear equation differs slightly from writing equations for chemical reactions. In addition to writing the symbols for various chemical elements, we must also explicitly indicate protons, neutrons, and electrons. In fact, we must show the numbers of protons and neutrons present in *every* species in the reaction.

You must know the symbols for elementary particles in order to balance a nuclear equation.

$^{1}_{1}p$ or $^{1}_{1}H$	$^{1}_{0}n$	$^{0}_{-1}e$ or $^{0}_{-1}\beta$	$^{0}_{+1}e$ or $^{0}_{+1}\beta$	$^{4}_{2}He$ or $^{4}_{2}\alpha$
proton	neutron	electron	positron	α particle

In balancing any nuclear equation, we observe the following rules:

- The total number of protons plus neutrons in the products and in the reactants must be the same (conservation of mass number).
- The total number of nuclear charges in the products and in the reactants must be the same (conservation of atomic number).

If the atomic numbers and mass numbers of all the species but one in a nuclear equation are known, the unknown species can be identified by applying the above rules.

EXAMPLE 21.1
Identify X in the following nuclear reactions:

(a) $^{14}_{7}N + ^{1}_{0}n \longrightarrow ^{14}_{6}C + X$

(b) $^{226}Ra \longrightarrow ^{4}_{2}\alpha + X$

(a) According to rule 1, the sum of the mass numbers must be conserved. Thus, the unknown product will have a mass number of 1. According to rule 2, the sum of the atomic numbers must be conserved. Thus, the nuclear charge of the unknown product must be 1. The particle is a proton.

$$^{14}_{7}N + ^{1}_{0}n \longrightarrow ^{14}_{6}C + \mathbf{^{1}_{1}p}$$

(b) Note that the atomic number of radium is missing. Look at a periodic table to find that Ra is element number 88. Balancing the mass numbers first, we find that the unknown product must have a mass of 222. Balancing the nuclear charges, we find that the nuclear charge of the unknown must be 86. Element number 86 is radon (Rn).

$$^{226}_{88}\text{Ra} \longrightarrow {}^{4}_{2}\alpha + {}^{222}_{86}\textbf{Rn}$$

PRACTICE EXERCISE

1. Identify X in the following nuclear reactions:

(a) $^{239}_{94}\text{Pu} \longrightarrow {}^{235}_{92}\text{U} + \text{X}$

(b) $^{90}_{38}\text{Sr} \longrightarrow {}^{90}_{37}\text{Rb} + \text{X}$

Text Problems: 6, 30

PROBLEM TYPE 2: NUCLEAR STABILITY

The principal factor for determining the stability of a nucleus is the *neutron-to-proton ratio* (n:p). For stable elements of low atomic number, the n:p ratio is close to 1. As the atomic number increases, the n:p ratios of stable nuclei become greater than 1. The deviation in the n:p ratio at higher atomic numbers arises because a larger number of neutrons is needed to stabilize the nucleus by counteracting the strong repulsion among the large number of protons. Figure 21.1 of the text shows a plot of the number of neutrons versus the number of protons in various isotopes. The stable nuclei are located in an area of the graph called the *belt of stability*. Most radioactive nuclei lie outside the belt.

The following rules are useful in predicting nuclear stability.

- Nuclei that contain 2, 8, 20, 50, 82, or 126 protons or neutrons are generally more stable than nuclei that do not possess these numbers. These numbers are called *magic numbers*.
- Nuclei with even numbers of both protons and neutrons are generally more stable than those with odd numbers of these particles (see Table 21.2 of the text).
- All isotopes of the elements starting with polonium (Po, Z = 84) are radioactive. All isotopes of technetium (Tc, Z = 43) and promethium (Pm, Z = 61) are also radioactive.

EXAMPLE 21.2
Rank the following nuclei in order of increasing nuclear stability:
$^{40}_{20}\textbf{Ca}$ $^{39}_{20}\textbf{Ca}$ $^{11}_{5}\textbf{B}$

Boron-11 has both an odd number of protons and an odd number of neutrons; therefore, it should be the least stable of the three nuclei. Calcium-39 has a magic number of protons (20), but an odd number of neutrons. Calcium-40 has both a magic number of protons and neutrons (20), so it should be more stable than calcium-39. The order of increasing nuclear stability is

$$^{11}_{5}\text{B} < {}^{39}_{20}\text{Ca} < {}^{40}_{20}\text{Ca}$$

PRACTICE EXERCISE

2. Rank the following isotopes in order of increasing nuclear stability:
$^{39}_{20}\text{Ca}$ $^{222}_{86}\text{Rn}$ $^{98}_{43}\text{Tc}$

Text Problems: 18, 20

PROBLEM TYPE 3: CALCULATING NUCLEAR BINDING ENERGY

A quantitative measure of nuclear stability is the **nuclear binding energy**, which is the energy required to break up a nucleus into its component protons and neutrons. The concept of nuclear binding energy evolved from studies showing that the masses of nuclei are always less than the sum of the masses of the **nucleons** (the protons and neutrons in a nucleus). The difference between the mass of an atom and the sum of the masses of its protons, neutrons, and electrons is called the **mass defect**. According to Einstein's mass-energy equivalence relationship

$$E = mc^2$$

where,

E is energy
m is mass
c is the velocity of light

the loss in mass shows up as energy (heat) given off to the surroundings. We can calculate the amount of energy released by writing:

$$\Delta E = (\Delta m)c^2$$

where,

ΔE = energy of products – energy of reactants

Δm = mass of products – mass of reactants

See Example 21.3 below for a detailed calculation.

EXAMPLE 21.3

Calculate the nuclear binding energy of the light isotope of helium, ^{3}He. The atomic mass of $^3_2$He is 3.01603 amu.

The binding energy is the energy required for the process

$$^3_2\text{He} \longrightarrow 2\,^1_1\text{p} + \,^1_0\text{n}$$

Step 1: Calculate the mass difference (Δm) between the products and reactants.

$$\Delta m = [2(\text{proton mass}) + (\text{neutron mass})] - \,^3_2\text{He atomic mass}$$

$$= [2(1.007825\text{ amu}) + 1.008665\text{ amu}] - 3.01603\text{ amu}$$

$$= 8.29 \times 10^{-3}\text{ amu}$$

Step 2: Use Einstein's equation to calculate the energy change for the process, ΔE.

$$\Delta E = (\Delta m)c^2$$

$$= 8.29 \times 10^{-3}\text{ amu} \times (3.00 \times 10^8\text{ m/s})^2$$

$$= 7.46 \times 10^{14}\ \frac{\text{amu}\cdot\text{m}^2}{\text{s}^2}$$

Let's convert to useful units (J/He atom).

$$7.46 \times 10^{14}\ \frac{\text{amu}\cdot\text{m}^2}{\text{s}^2} \times \frac{1.00\text{ g}}{6.022 \times 10^{23}\text{ amu}} \times \frac{1\text{ kg}}{1000\text{ g}} \times \frac{1\text{ J}}{\frac{1\text{ kg}\cdot\text{m}^2}{\text{s}^2}} = \mathbf{1.24 \times 10^{-12}\ J/atom}$$

This is the nuclear binding energy. It's the energy required to break up one helium-3 nucleus into 2 protons and 1 neutron.

Step 3: When comparing the stability of any two nuclei we must account for the fact that they have different numbers of nucleons. For this reason, it is more meaningful to use the *nuclear binding energy per nucleon*, defined as

$$\text{nuclear binding energy per nucleon} = \frac{\text{nuclear binding energy}}{\text{number of nucleons}}$$

For the helium-3 nucleus,

$$\text{nuclear binding energy per nucleon} = \frac{1.24 \times 10^{-12}\ \text{J / He }\cancel{\text{atom}}}{3\ \text{nucleons / He }\cancel{\text{atom}}}$$

$$= \mathbf{4.13 \times 10^{-13}\ J/nucleon}$$

PRACTICE EXERCISE

3. (a) Calculate the binding energy and the binding energy per nucleon of ${}^{27}_{13}\text{Al}$. The atomic mass of ${}^{27}_{13}\text{Al}$ is 26.98154 amu.

(b) Compare the result from part (a) to the binding energy of ${}^{28}_{14}\text{Si}$, which has an even number of protons and neutrons. The atomic mass of ${}^{28}_{14}\text{Si}$ is 27.976928 amu.

Text Problem: 24

PROBLEM TYPE 4: KINETICS OF RADIOACTIVE DECAY

All radioactive decays obey first-order kinetics. To review first-order reactions, see Sections 14.2 and 14.3 of your text. The decay rate at any time t is given by

$$\text{rate of decay at time } t = kN$$

where,

k is the first-order rate constant
N is the number of radioactive nuclei present at time t

The relationship between the number of radioactive nuclei present at time zero (N_0) and the number of nuclei remaining at a later time t (N_t) is given by:

$$\ln\frac{N_0}{N_t} = kt \qquad (21.1)$$

The corresponding half-life for a first-order reaction is given by:

$$t_{\frac{1}{2}} = \frac{0.693}{k} \qquad (21.2)$$

Unlike ordinary chemical reactions, the rate constants for nuclear reactions are unaffected by changes in environmental conditions such as temperature and pressure.

EXAMPLE 21.4
Cobalt-60 is used in radiation therapy. It has a half-life of 5.26 years.
(a) Calculate the rate constant for radioactive decay.
(b) What fraction of a certain sample will remain after 12 years?

(a) The rate constant for the decay can be calculated using Equation (21.2) for the half-life of a first-order reaction.

$$t_{\frac{1}{2}} = \frac{0.693}{k}$$

Rearrange the equation to solve for the rate constant (k).

$$\boldsymbol{k} = \frac{0.693}{t_{\frac{1}{2}}} = \frac{0.693}{5.26\text{ yr}} = \mathbf{0.132\ yr^{-1}}$$

(b) The fraction of a certain sample that will remain after 12 years is

$$\frac{N_t}{N_0}$$

where $t = 12$ yr.

$\frac{N_t}{N_0}$ can be calculated from Equation (21.1) above. We can rearrange the equation to solve for $\frac{N_t}{N_0}$.

$$\ln\frac{N_0}{N_t} = kt$$

$$\frac{N_0}{N_t} = e^{kt}$$

$$\frac{N_t}{N_0} = \frac{1}{e^{kt}}$$

$$\boldsymbol{\frac{N_t}{N_0}} = \frac{1}{e^{(0.132\ \cancel{\text{yr}}^{-1})(12\ \cancel{\text{yr}})}} = \mathbf{0.205}$$

In other words, 20.5 percent of the original sample will remain after 12 years.

PRACTICE EXERCISE

4. Estimate the age of a bottle of wine that has a tritium $\left({}^{3}_{1}\text{H}\right)$ content 75.0 percent of the tritium content of environmental water obtained from the area where the grapes were grown. $t_{\frac{1}{2}} = 12.3$ yr.

Text Problems: 26, 28, 32

PROBLEM TYPE 5: BALANCING NUCLEAR TRANSMUTATION EQUATIONS

Nuclei can undergo change as a result of bombardment by neutrons, protons, or other nuclei. This process is called **nuclear transmutation**. Unlike radioactive decay, nuclear transmutation is *not* a spontaneous process; consequently, nuclear transmutation reactions have more than one reactant. As an example, consider the synthesis of neptunium (Np), which was the first transuranium element to be synthesized by scientists.

First, uranium-238 is bombarded with neutrons to produce uranium-239.

$${}^{238}_{92}\text{U} + {}^{1}_{0}\text{n} \longrightarrow {}^{239}_{92}\text{U}$$

This is a nuclear transmutation reaction. Uranium-239 is unstable and decays spontaneously to neptunium-239 by emitting a β particle.

$$^{239}_{92}\text{U} \longrightarrow {}^{239}_{93}\text{Np} + {}^{0}_{-1}\beta$$

To balance a nuclear transmutation reaction, follow the same rules used to balance nuclear equations, Problem Type 1.

- The total number of protons plus neutrons in the products and in the reactants must be the same (conservation of mass number).
- The total number of nuclear charges in the products and in the reactants must be the same (conservation of atomic number).

EXAMPLE 21.5

Write and balance the following reactions. When aluminum-27 is bombarded with α particles, phosphorus-30 and one other particle are produced. Phosphorus-30 has a low n:p ratio and decays spontaneously by positron emission.

The first reaction is a nuclear transmutation. You are given both reactants and one of the two products in the problem. To balance the equation, remember that both mass number and atomic number must be conserved.

$$^{27}_{13}\text{Al} + {}^{4}_{2}\alpha \longrightarrow {}^{30}_{15}\text{P} + \text{X}$$

To balance the mass number, the missing particle (X) must have a mass number of 1. To balance the atomic number, X must have an atomic number of 0. X must be a neutron.

$$^{27}_{13}\text{Al} + {}^{4}_{2}\alpha \longrightarrow {}^{30}_{15}\text{P} + {}^{1}_{0}\text{n}$$

The second reaction is spontaneous so there is only one reactant, phosphorus-30. The problem indicates that phosphorus-30 decays by positron emission. Let's write down what we know so far.

$$^{30}_{15}\text{P} \longrightarrow \text{X} + {}^{0}_{+1}\beta$$

To balance the mass number, the missing element (X) must have a mass number of 30. To balance the atomic number, X must have an atomic number of 14. X must be silicon-30.

$$^{30}_{15}\text{P} \longrightarrow {}^{30}_{14}\text{Si} + {}^{0}_{+1}\beta$$

PRACTICE EXERCISE

5. Write and balance the following reactions. When chlorine-37 is bombarded with neutrons, only one product is produced. The product is unstable and spontaneously decays by beta emission.

Text Problems: 8, 10

ANSWERS TO PRACTICE EXERCISES

1. (a) $X = {}^{4}_{2}\text{He}$

   (b) $X = {}^{0}_{+1}\beta$

2. Technetium-98 should be unstable because it has an odd number of both protons and neutrons. In fact, all isotopes of technetium are radioactive. Radon-222 will perhaps be slightly more stable than technetium-98, because it has an even number of both protons and neutrons. Remember, however, that all isotopes of the elements starting with polonium (Po, Z = 84) are radioactive. Calcium-39 has 20 protons (a "magic number") and should be the most stable of the three isotopes. The correct order of increasing nuclear stability is

$$ {}^{98}_{43}\text{Tc} \quad < \quad {}^{222}_{86}\text{Rn} \quad < \quad {}^{39}_{20}\text{Ca} $$

3. (a) Binding energy $= 3.6 \times 10^{-11}$ J/Al atom $= 1.3 \times 10^{-12}$ J/nucleon

   (b) Binding energy $= 3.8 \times 10^{-11}$ J/Si atom $= 1.4 \times 10^{-12}$ J/nucleon

   When comparing the stability of any two nuclei, it is best to compare the binding energy in units of J/nucleon. Silicon has the greater binding energy per nucleon and hence is more stable than Al. You should have expected this result, because Si has an even number of protons and neutrons; whereas, Al has an odd number of protons and an even number of neutrons. Nuclei with even numbers of both protons and neutrons are generally more stable than those with odd numbers of these particles.

4. You can calculate the rate constant (k) from the half-life.

$$ k = 0.0563\ \text{yr}^{-1} $$

   Next, $\dfrac{N_t}{N_0} = 0.750$ or $\dfrac{N_0}{N_t} = 1.33$. The age (t) of the bottle of wine can then be calculated from the equation,

$$ \ln\frac{N_0}{N_t} = kt $$

$$ \mathbf{t = 5.07\ yr} $$

5. $$ {}^{1}_{0}\text{n} + {}^{37}_{17}\text{Cl} \longrightarrow {}^{38}_{17}\text{Cl} $$

$$ {}^{38}_{17}\text{Cl} \longrightarrow {}^{38}_{18}\text{Ar} + {}^{0}_{-1}\beta $$

SOLUTIONS TO SELECTED TEXT PROBLEMS

21.6 Balancing Nuclear Equations, Problem Type 1.

(a) According to rule 1, the sum of the mass numbers must be conserved. Thus, the unknown product will have a mass number of 0. According to rule 2, the atomic number must be conserved. Thus, the nuclear charge of the unknown product must be −1. The particle is a β particle.

$$^{135}_{53}\text{I} \longrightarrow {}^{135}_{54}\text{Xe} + {}^{0}_{-1}\beta$$

(b) Balancing the mass numbers first, we find that the unknown product must have a mass of 40. Balancing the nuclear charges, we find that the atomic number of the unknown must be 20. Element number 20 is calcium (Ca).

$$^{40}_{19}\text{K} \longrightarrow {}^{0}_{-1}\beta + {}^{40}_{20}\textbf{Ca}$$

(c) Balancing the mass numbers, we find that the unknown product must have a mass of 4. Balancing the nuclear charges, we find that the nuclear charge of the unknown must be 2. The unknown particle is an alpha (α) particle.

$$^{59}_{27}\text{Co} + {}^{1}_{0}\text{n} \longrightarrow {}^{56}_{25}\text{Mn} + {}^{4}_{2}\alpha$$

(d) Balancing the mass numbers, we find that the unknown products must have a combined mass of 2. Balancing the nuclear charges, we find that the combined nuclear charge of the two unknown particles must be 0. The unknown particles are neutrons.

$$^{235}_{92}\text{U} + {}^{1}_{0}\text{n} \longrightarrow {}^{99}_{40}\text{Sr} + {}^{135}_{52}\text{Te} + 2\,{}^{1}_{0}\textbf{n}$$

21.8 **(a)** $^{80}_{34}\text{Se} + {}^{2}_{1}\text{d} \longrightarrow {}^{81}_{34}\textbf{Se} + {}^{1}_{1}\text{p}$

(b) $^{9}_{4}\textbf{Be} + {}^{2}_{1}\text{d} \longrightarrow {}^{9}_{3}\text{Li} + 2\,{}^{1}_{1}\text{p}$

(c) $^{10}_{5}\text{B} + {}^{1}_{0}\text{n} \longrightarrow {}^{7}_{3}\textbf{Li} + {}^{4}_{2}\alpha$

21.10 Upon bombardment with neutrons, mercury−198 is first converted to mercury−199, which then emits a proton. The reaction is:

$$^{198}_{80}\text{Hg} + {}^{1}_{0}\text{n} \longrightarrow {}^{199}_{80}\text{Hg} \longrightarrow {}^{198}_{79}\text{Au} + {}^{1}_{1}\text{p}$$

21.18 Nuclear Stability, Problem Type 2.

(a) Lithium-9 should be less stable. The neutron-to-proton ratio is too high. For small atoms, the n:p ratio will be close to 1:1.

(b) Sodium-25 is less stable. Its neutron-to-proton ratio is probably too high.

(c) Scandium-48 is less stable because of odd numbers of protons and neutrons. We would not expect calcium-48 to be stable even though it has a magic number of protons. Its n:p ratio is too high.

21.20 **(a)** Neon-17 should be radioactive. It falls below the belt of stability (low n:p ratio).

(b) Calcium-45 should be radioactive. It falls above the belt of stability (high n:p ratio).

(c) All technetium isotopes are radioactive.

(d) Mercury-195 should be radioactive. Mercury-196 has an even number of both neutrons and protons.

(e) All curium isotopes are unstable.

21.22 We can use the equation, $\Delta E = \Delta mc^2$, to solve the problem. Recall the following conversion factor:

$$1\text{ J} = \frac{1\text{ kg}\cdot\text{m}^2}{\text{s}^2}$$

The energy loss in one second is:

$$\Delta m = \frac{\Delta E}{c^2} = \frac{5\times 10^{26}\ \frac{\text{kg}\cdot\cancel{\text{m}^2}}{\cancel{\text{s}^2}}}{(3.00\times 10^8\ \cancel{\frac{\text{m}}{\text{s}}})^{\cancel{2}}} = 6\times 10^9\text{ kg}$$

Therefore the rate of mass loss is $\mathbf{6\times 10^9}$ **kg/s.**

21.24 Calculating Nuclear Binding Energy, Problem Type 3.

(a) The binding energy is the energy required for the process

$${}^4_2\text{He} \rightarrow 2\,{}^1_1\text{p} + 2\,{}^1_0\text{n}$$

Step 1: Calculate the mass difference (Δm) between the products and reactants.

$$\begin{aligned}\Delta m &= [2(\text{proton mass}) + 2(\text{neutron mass})] - {}^4_2\text{He atomic mass}\\ &= [2(1.007825\text{ amu}) + 2(1.008665\text{ amu})] - 4.0026\text{ amu}\\ &= 0.0304\text{ amu}\end{aligned}$$

Step 2: Use Einstein's equation to calculate the energy change for the process, ΔE.

$$\begin{aligned}\Delta E &= (\Delta m)c^2\\ &= 0.0304\text{ amu} \times (3.00\times 10^8\text{ m/s})^2\\ &= 2.74\times 10^{15}\ \frac{\text{amu}\cdot\text{m}^2}{\text{s}^2}\end{aligned}$$

Let's convert to useful units (J/He atom).

$$2.74\times 10^{15}\ \frac{\cancel{\text{amu}}\cdot\cancel{\text{m}^2}}{\cancel{\text{s}^2}} \times \frac{1.00\ \cancel{\text{g}}}{6.022\times 10^{23}\ \cancel{\text{amu}}} \times \frac{1\ \cancel{\text{kg}}}{1000\ \cancel{\text{g}}} \times \frac{1\text{ J}}{\frac{1\ \cancel{\text{kg}}\cdot\cancel{\text{m}^2}}{\cancel{\text{s}^2}}}$$

$$= \mathbf{4.55\times 10^{-12}\ J}$$

This is the nuclear binding energy. It's the energy required to break up one helium-4 nucleus into 2 protons and 2 neutrons.

Step 3: When comparing the stability of any two nuclei we must account for the fact that they have different numbers of nucleons. For this reason, it is more meaningful to use the *nuclear binding energy per nucleon*, defined as

$$\text{nuclear binding energy per nucleon} = \frac{\text{nuclear binding energy}}{\text{number of nucleons}}$$

For the helium-4 nucleus,

$$\text{nuclear binding energy per nucleon} = \frac{4.55 \times 10^{-12}\ \text{J / He}\ \cancel{\text{atom}}}{4\ \text{nucleons / He}\ \cancel{\text{atom}}}$$

$$= \mathbf{1.14 \times 10^{-12}\ J/nucleon}$$

(b) The binding energy is the energy required for the process

$$^{184}_{74}\text{W} \rightarrow 74\ ^{1}_{1}\text{p} + 110\ ^{1}_{0}\text{n}$$

Step 1: Calculate the mass difference (Δm) between the products and reactants.

$$\Delta m = [74(\text{proton mass}) + 110(\text{neutron mass})] - {}^{184}_{74}\text{W atomic mass}$$

$$= [74(1.007825\ \text{amu}) + 110(1.008665\ \text{amu})] - 183.9510\ \text{amu}$$

$$= 1.5812\ \text{amu}$$

Step 2: Use Einstein's equation to calculate the energy change for the process, ΔE.

$$\Delta E = (\Delta m)c^2$$

$$= 1.5812\ \text{amu} \times (3.00 \times 10^8\ \text{m/s})^2$$

$$= 1.42 \times 10^{17}\ \frac{\text{amu} \cdot \text{m}^2}{\text{s}^2}$$

Let's convert to useful units (J/He atom).

$$1.42 \times 10^{17}\ \frac{\cancel{\text{amu}} \cdot \cancel{\text{m}^2}}{\cancel{\text{s}^2}} \times \frac{1.00\ \cancel{\text{g}}}{6.022 \times 10^{23}\ \cancel{\text{amu}}} \times \frac{1\ \cancel{\text{kg}}}{1000\ \cancel{\text{g}}} \times \frac{1\ \text{J}}{\frac{1\ \cancel{\text{kg}} \cdot \cancel{\text{m}^2}}{\cancel{\text{s}^2}}}$$

$$= \mathbf{2.36 \times 10^{-10}\ J}$$

This is the nuclear binding energy. It's the energy required to break up one tungsten-184 nucleus into 74 protons and 110 neutrons.

Step 3: When comparing the stability of any two nuclei we must account for the fact that they have different numbers of nucleons. For this reason, it is more meaningful to use the *nuclear binding energy per nucleon*, defined as

$$\text{nuclear binding energy per nucleon} = \frac{\text{nuclear binding energy}}{\text{number of nucleons}}$$

For the tungsten-184 nucleus,

$$\text{nuclear binding energy per nucleon} = \frac{2.36 \times 10^{-10}\ \text{J / W}\ \cancel{\text{atom}}}{184\ \text{nucleons / W}\ \cancel{\text{atom}}}$$

$$= \mathbf{1.28 \times 10^{-12}\ J/nucleon}$$

21.26 Kinetics of Radioactive Decay, Problem Type 4.

Step 1: We can use the following equation to calculate the rate constant k for each point.

$$\ln\frac{N_0}{N_t} = kt$$

From day 0 to day 1, we have

$$\ln\frac{500}{389} = k(1\text{ day})$$

$$k = 0.251\text{ day}^{-1}$$

Following the same procedure for the other days,

t (day)	mass (g)	*k* (day^{-1})
0	500	
1	389	0.251
2	303	0.250
3	236	0.250
4	184	0.250
5	143	0.250
6	112	0.249

The average value of k is **0.250 day^{-1}**.

Step 2: Using the average value of k, calculate the half-life.

$$t_{\frac{1}{2}} = \frac{0.693}{k} = \frac{0.693}{0.250\text{ day}^{-1}} = \mathbf{2.77\ days}$$

21.28 Since all radioactive decay processes have first–order rate laws, the decay rate is proportional to the amount of radioisotope at any time. The half-life is given by the following equation:

$$t_{\frac{1}{2}} = \frac{0.693}{k} \qquad (1)$$

There is also an equation that relates the number of nuclei at time zero (N_0) and time t (N_t).

$$\ln\frac{N_0}{N_t} = kt$$

We can use this equation to solve for the rate constant, k. Then, we can substitute k into Equation (1) to calculate the half-life.

The time interval is:

(2:15 p.m., 12/17/99) – (1:00 p.m., 12/3/99) = 14 days + 1 hour + 15 min = 20,235 min

$$\ln\left(\frac{9.8\times10^5\text{ dis/min}}{2.6\times10^4\text{ dis/min}}\right) = k(20{,}235\text{ min})$$

$$k = 1.8\times10^{-4}\text{ min}^{-1}$$

Substitute k into equation (1) to calculate the half-life.

$$t_{\frac{1}{2}} = \frac{0.693}{k} = \frac{0.693}{1.8 \times 10^{-4}\ \text{min}^{-1}} = \mathbf{3.9 \times 10^{3}\ min\ or\ 2.7\ days}$$

21.30 The equation for the overall process is:

$$^{232}_{90}\text{Th} \longrightarrow 6\,^{4}_{2}\alpha + 4\,^{0}_{-1}\beta + \text{X}$$

The final product isotope must be $\mathbf{^{208}_{82}Pb}$.

21.32 Let's consider the decay of A first.

$$k = \frac{0.693}{t_{\frac{1}{2}}} = \frac{0.693}{4.50\ \text{s}} = 0.154\ \text{s}^{-1}$$

Let's convert k to units of day^{-1}.

$$0.154\ \frac{1}{\text{s}} \times \frac{3600\ \text{s}}{1\ \text{h}} \times \frac{24\ \text{h}}{1\ \text{day}} = 1.33 \times 10^{4}\ \text{day}^{-1}$$

Next, use the first-order rate equation to calculate the amount of A left after 30 days.

$$\ln\frac{[\text{A}]_0}{[\text{A}]} = kt$$

Let x be the amount of A left after 30 days.

$$\ln\frac{100}{x} = (1.33 \times 10^{4}\ \text{day}^{-1})(30\ \text{days}) = 3.99 \times 10^{5}$$

$$x \cong 0$$

Thus, **no A remains.**

For B: As calculated above, all of A is converted to B in less than 30 days. In fact, essentially all of A is gone in less than 1 day! This means that at the beginning of the 30 day period, there is 1.00 mol of B present. The half life of B is 15 days, so that after two half-lives (30 days), there should be **0.25 mol** of B left.

For C: As in the case of A, the half-life of C is also very short. Therefore, at the end of the 30–day period, **no C is left.**

For D: D is not radioactive. 0.75 mol of B reacted in 30 days; therefore, due to a 1:1 mole ratio between B and D, there should be **0.75 mole of D** present after 30 days.

21.44 The fact that the radioisotope appears only in the I_2 shows that the IO_3^- is formed only from the IO_4^-. Does this result rule out the possibility that I_2 could be formed from IO_4^- as well? Can you suggest an experiment to answer the question?

21.46 Add iron-59 to the person's diet, and allow a few days for the iron–59 isotope to be incorporated into the person's body. Isolate red blood cells from a blood sample and monitor radioactivity from the hemoglobin molecules present in the red blood cells.

21.48 Apparently there is a sort of Pauli exclusion principle for nucleons as well as for electrons. When neutrons pair with neutrons and when protons pair with protons, their spins cancel. Even–even nuclei are the only ones with no net spin.

21.50 **(a)** One millicurie represents 3.70×10^7 disintegrations/s. The rate of decay of the isotope is given by the rate law: rate = kN, where N is the number of atoms in the sample. We find the value of k in units of s^{-1}:

$$k = \frac{0.693}{t_{\frac{1}{2}}} = \left(\frac{0.693}{2.20 \times 10^6\ \cancel{yr}}\right)\left(\frac{1\ \cancel{yr}}{365\ \cancel{days}}\right)\left(\frac{1\ \cancel{day}}{24\ \cancel{h}}\right)\left(\frac{1\ \cancel{h}}{3600\ s}\right) = 9.99 \times 10^{-15}\ s^{-1}$$

The number of atoms (N) in a 0.500 g sample of neptunium–237 is:

$$0.500\ \cancel{g} \times \left(\frac{1\ \cancel{mol}}{237.0\ \cancel{g}}\right)\left(\frac{6.022 \times 10^{23}\ \text{atoms}}{1\ \cancel{mol}}\right) = 1.27 \times 10^{21}\ \text{atoms}$$

$$\text{rate of decay} = kN$$

$$= (9.99 \times 10^{-15}\ s^{-1})(1.27 \times 10^{21}\ \text{atoms}) = 1.27 \times 10^7\ \text{atoms/s}$$

We can also say that:

$$\text{rate of decay} = 1.27 \times 10^7\ \text{disintegrations/s}$$

The activity in millicuries is:

$$1.27 \times 10^7\ \cancel{\text{disintegrations/s}} \times \left(\frac{1\ \text{millicurie}}{3.70 \times 10^7\ \cancel{\text{disintegrations/s}}}\right) = \mathbf{0.343\ millicurie}$$

(b) The decay equation is:

$${}^{237}_{93}Np \longrightarrow {}^{4}_{2}\alpha + {}^{233}_{91}Pa$$

21.52 We use the same procedure as in Problem 21.24.

	Isotope	Atomic Mass (amu)	Nuclear Binding Energy (J/nucleon)
(a)	^{10}B	10.0129	1.04×10^{-12}
(b)	^{11}B	11.00931	1.11×10^{-12}
(c)	^{14}N	14.00307	1.20×10^{-12}
(d)	^{56}Fe	55.9349	1.41×10^{-12}

21.54 When an isotope is above the belt of stability, the neutron/proton ratio is too high. The only mechanism to correct this situation is beta emission; the process turns a neutron into a proton. Direct neutron emission does not occur.

$${}^{18}_{7}N \longrightarrow {}^{18}_{8}O + {}^{0}_{-1}\beta$$

Oxygen–18 is a stable isotope.

21.56 The age of the fossil can be determined by radioactively dating the age of the deposit that contains the fossil.

21.58 **(a)** $^{209}_{83}\text{Bi} + ^{4}_{2}\alpha \longrightarrow ^{211}_{85}\text{At} + 2^{1}_{0}\text{n}$

(b) $^{209}_{83}\text{Bi}(\alpha, 2\text{n})^{211}_{85}\text{At}$

21.60 Because of the relative masses, the force of gravity on the sun is much greater than it is on Earth. Thus the nuclear particles on the sun are already held much closer together than the equivalent nuclear particles on the earth. Less energy (lower temperature) is required on the sun to force fusion collisions between the nuclear particles.

21.62 Kinetics of Radioactive Decay, Problem Type 4.

Step 1: The half-life of carbon-14 is 5730 years. From the half-life, we can calculate the rate constant, *k*.

$$k = \frac{0.693}{t_{\frac{1}{2}}} = \frac{0.693}{5730\text{ yr}} = 1.21 \times 10^{-4}\text{ yr}^{-1}$$

Step 2: The age of the object can now be calculated using the following equation.

$$\ln\frac{N_0}{N_t} = kt$$

N = the number of radioactive nuclei. In the problem, we are given disintegrations per second per gram. The number of disintegrations is directly proportional to the number of radioactive nuclei. We can write,

$$t = \frac{1}{k}\ln\frac{\text{decay rate of fresh sample}}{\text{decay rate of old sample}}$$

$$t = \frac{1}{1.21\times10^{-4}\text{ yr}^{-1}}\ln\frac{0.260\text{ cps / g C}}{0.186\text{ cps / g C}}$$

$$\mathbf{t = 2.77 \times 10^3\ yr}$$

21.64 **(a)** The balanced equation is:

$$^{40}_{19}\text{K} \longrightarrow ^{40}_{18}\text{Ar} + ^{0}_{+1}\beta$$

(b) First, calculate the rate constant *k*.

$$k = \frac{0.693}{t_{\frac{1}{2}}} = \frac{0.693}{1.2 \times 10^9\text{ yr}} = 5.8 \times 10^{-10}\text{ yr}^{-1}$$

Then, calculate the age of the rock by substituting *k* into the following equation. ($N_t = 0.18N_0$)

$$t = \frac{1}{k}\ln\frac{N_0}{N_t} = \left(\frac{1}{5.8 \times 10^{-10}\text{ yr}^{-1}}\right)\ln\left(\frac{1.00}{0.18}\right) = \mathbf{3.0 \times 10^9\ yr}$$

21.66 **(a)** In the ^{90}Sr decay, the mass defect is:

$$\Delta m = (\text{mass } ^{90}\text{Y} + \text{mass e}^-) - \text{mass } ^{90}\text{Sr}$$

$$= [(89.907152 \text{ amu} + 5.4857 \times 10^{-4} \text{ amu}) - 89.907738 \text{ amu}] \times \frac{1 \text{ g}}{6.022 \times 10^{23} \text{ amu}}$$

$$= -6.216 \times 10^{-29} \text{ g} = -6.216 \times 10^{-32} \text{ kg}$$

The energy change is given by:

$$\Delta E = (\Delta m)c^2$$

$$= (-6.126 \times 10^{-32} \text{ kg})(3.00 \times 10^8 \text{ m/s})^2$$

$$= -5.59 \times 10^{-15} \text{ kg m}^2/\text{s}^2 = \mathbf{-5.59 \times 10^{-15} \text{ J}}$$

Similarly, for the ^{90}Y decay, we have

$$\Delta m = (\text{mass } ^{90}\text{Zr} + \text{mass e}^-) - \text{mass } ^{90}\text{Y}$$

$$= [(89.904703 \text{ amu} + 5.4857 \times 10^{-4} \text{ amu}) - 89.907152 \text{ amu}] \times \frac{1 \text{ kg}}{6.022 \times 10^{26} \text{ amu}}$$

$$= -3.156 \times 10^{-30} \text{ kg}$$

and the energy change is:

$$\Delta E = (-3.156 \times 10^{-30} \text{ kg})(3.00 \times 10^8 \text{ m/s})^2$$

$$= \mathbf{-2.84 \times 10^{-13} \text{ J}}$$

Thus the total amount of energy released is:

$$-5.59 \times 10^{-15} \text{ J} - 2.84 \times 10^{-13} \text{ J} = -2.90 \times 10^{-13} \text{ J.}$$

(b) This calculation requires that we know the rate constant for the decay. From the half–life, we can calculate k.

$$k = \frac{0.693}{t_{\frac{1}{2}}} = \frac{0.693}{28.1 \text{ yr}} = 0.0247 \text{ yr}^{-1}$$

To calculate the number of moles of ^{90}Sr decaying in a year, we apply the following equation:

$$\ln\frac{N_0}{N_t} = kt$$

$$\ln\frac{1.00}{x} = (0.0247 \text{ yr}^{-1})(1.00 \text{ yr})$$

where x is the number of moles of ^{90}Sr nuclei left over. Solving, we obtain:

$$x = 0.976 \text{ mol } ^{90}\text{Sr}$$

Thus the number of moles of nuclei which decay in a year is

$$(1.00 - 0.976) \text{ mol} = \mathbf{0.024 \text{ mol}}$$

This is a reasonable number since it takes 28.1 years for 0.5 mole of ^{90}Sr to decay.

(c) Since the half–life of ^{90}Y is much shorter than that of ^{90}Sr, we can safely assume that *all* the ^{90}Y formed from ^{90}Sr will be converted to ^{90}Zr. The energy changes calculated in part (a) refer to the decay of individual nuclei. In 0.024 mole, the number of nuclei that have decayed is:

$$0.024\ \text{mol} \times 6.022 \times 10^{23}\ \text{nuclei/mol} = 1.4 \times 10^{22}\ \text{nuclei}$$

Realize that there are two decay processes occurring, so we need to add the energy released for each process calculated in part (a). Thus, the heat released from 1 mole of ^{90}Sr waste in a year is given by:

$$\textbf{heat released} = (-2.90 \times 10^{-13}\ \text{J/nucleus})(1.4 \times 10^{22}\ \text{nuclei}) = \mathbf{-4.1 \times 10^{9}\ J = -4.1 \times 10^{6}\ kJ}$$

This amount is roughly equivalent to the heat generated by burning 50 tons of coal! Although the heat is released slowly during the course of a year, effective ways must be devised to prevent heat damage to the storage containers and subsequent leakage of radioactive material to the surroundings.

21.68 Kinetics of radioactive decay, Problem Type 4.

In problem 62, we calculated the rate constant (k) for carbon-14 decay.

$$k = 1.21 \times 10^{-4}\ \text{yr}^{-1}$$

We can now calculate the age of the Shroud of Turin using the following equation:

$$\ln \frac{N_0\ (\text{decay rate of fresh sample})}{N_t\ (\text{decay rate of old sample})} = kt$$

$$t = \frac{1}{k} \ln \frac{N_0\ (\text{decay rate of fresh sample})}{N_t\ (\text{decay rate of old sample})}$$

$$t = \frac{1}{1.21 \times 10^{-4}\ \text{yr}^{-1}} \ln \frac{15.3\ \text{cpm / g C}}{13.0\ \text{cpm / g C}}$$

$$= \mathbf{1.35 \times 10^{3}\ yr}$$

The shroud of Turin cannot be the burial cloth of Christ; it is only about 1350 years old.

21.70 One curie represents 3.70×10^{10} disintegrations/s. The rate of decay of the isotope is given by the rate law: rate = kN, where N is the number of atoms in the sample and k is the first-order rate constant. We find the value of k in units of s^{-1}:

$$k = \frac{0.693}{t_{\frac{1}{2}}} = \frac{0.693}{1.6 \times 10^{3}\ \text{yr}} = 4.3 \times 10^{-4}\ \text{yr}^{-1}$$

$$\frac{4.3 \times 10^{-4}}{1\ \text{yr}} \times \frac{1\ \text{yr}}{365\ \text{d}} \times \frac{1\ \text{d}}{24\ \text{h}} \times \frac{1\ \text{h}}{3600\ \text{s}} = 1.4 \times 10^{-11}\ \text{s}^{-1}$$

Now, we can calculate N, the number of Ra atoms in the sample.

rate = kN

$$3.7 \times 10^{10}\ \text{disintegrations/s} = (1.4 \times 10^{-11}\ \text{s}^{-1})N$$

$$N = 2.6 \times 10^{21}\ \text{Ra atoms}$$

By definition, 1 curie corresponds to exactly 3.7×10^{10} nuclear disintegrations per second which is the decay rate equivalent to that of *1 g of radium*. Thus, the mass of 2.6×10^{21} Ra atoms is 1 g.

$$\frac{2.6 \times 10^{21} \text{ Ra atoms}}{1.0 \text{ g Ra}} \times \frac{226.03 \text{ g Ra}}{1 \text{ mol Ra}} = \mathbf{5.9 \times 10^{23}\ atoms/mol} = \boldsymbol{N_A}$$

21.72 U–238, $t_{\frac{1}{2}} = 4.5 \times 10^9$ yr and Th–232, $t_{\frac{1}{2}} = 1.4 \times 10^{10}$ yr.

They are still present because of their long half lives.

21.74 $E = \dfrac{hc}{\lambda}$

$$\lambda = \frac{hc}{E} = \frac{(3.00 \times 10^8 \text{ m/s})(6.63 \times 10^{-34} \text{ J·s})}{2.4 \times 10^{-13} \text{ J}} = 8.3 \times 10^{-13} \text{ m} = \mathbf{8.3 \times 10^{-4}\ nm}$$

This wavelength is clearly in the γ-ray region of the electromagnetic spectrum.

21.76 All except gravitational have a nuclear origin.

21.78 **(1)** The nuclear submarine can be submerged for a long period without refueling.

(2) Conventional diesel engines receive an input of oxygen. A nuclear reactor does not.

CHAPTER 22
ORGANIC POLYMERS—SYNTHETIC AND NATURAL

PROBLEM-SOLVING STRATEGIES AND TUTORIAL SOLUTIONS

TYPES OF PROBLEMS

Problem Type 1: Synthetic Organic Polymers.
- **(a)** Addition reactions.
- **(b)** Condensation reactions.

Problem Type 2: Proteins.

Problem Type 3: Nucleic Acids.

PROBLEM TYPE 1: SYNTHETIC ORGANIC POLYMERS

The word **polymer** means "many parts". A **polymer** is a compound with an unusually high molecular mass, consisting of a large number of molecular units linked together. The small unit that is repeated many times is called a **monomer**. A typical polymer contains chains of monomers several thousand units long.

A. Addition reactions

Addition polymers are made by adding monomer to monomer until a long chain is produced. Ethylene (ethene) and its derivatives are excellent monomers for forming addition polymers. In Chapter 13, we saw that addition reactions occur with unsaturated compounds containing C=C and C C bonds. In an addition reaction, the polymerization process is initiated by a radical or an ion. When ethylene is heated to 250°C under high pressure (1000–3000 atm) in the presence of a little oxygen or benzoyl peroxide (the initiator), addition polymers with masses of about 30,000 amu are obtained. This reaction is represented by

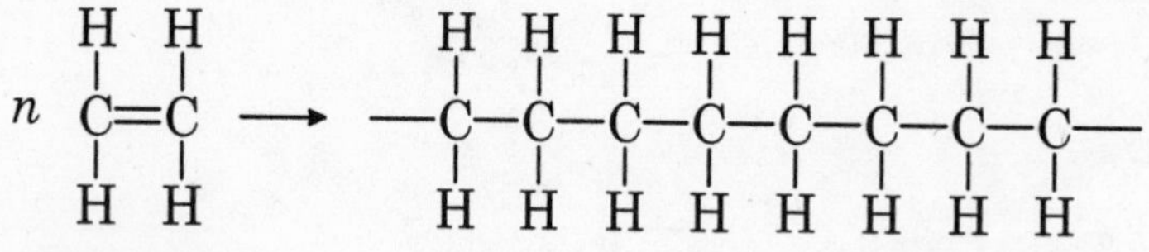

ethylene a segment of polyethylene

The general equation for addition polymerization is:

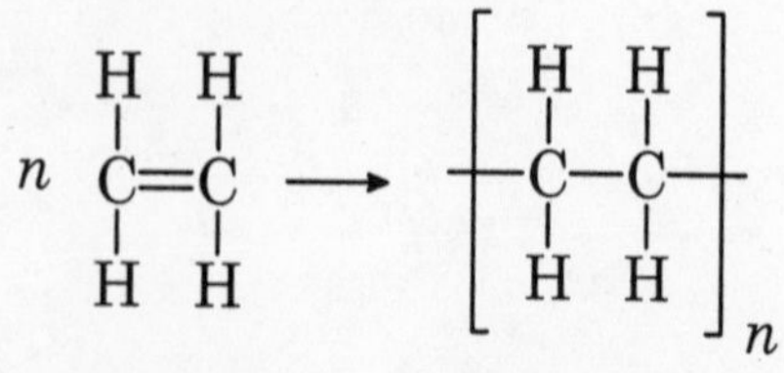

monomer repeating unit

Substitution for one or more hydrogen atoms in ethylene with Cl atoms, acetate, CN, or F provides a wide selection of monomers from which to make addition polymers with various properties. For instance, substitution of a Cl atom for a H atom in ethylene gives the monomer called vinyl chloride, CH_2=CHCl. Polymerization of vinyl chloride yields the polymer polyvinyl chloride (PVC).

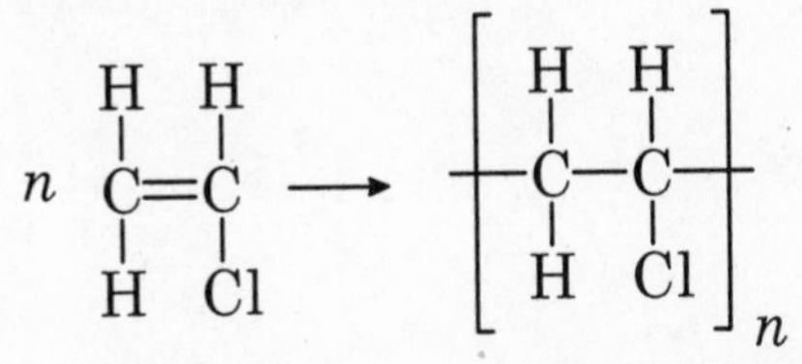

vinyl chloride monomer — polyvinyl chloride repeating unit

Polymers that are made from one type of monomer such as polyvinyl chloride are called **homopolymers.** Table 22.1 of your text gives the names, structures, and uses of a number of monomers and addition polymers.

For more detailed information on the reaction mechanism of addition polymerization, see Section 22.2 of your text.

EXAMPLE 22.1
Write the formulas of the monomers used to prepare the following polymers:
(a) Teflon, (b) Polystyrene, and (c) PVC.

Refer to Table 22.1 of your text.

(a) Teflon is an addition polymer with the formula

$$\left(CF_2-CF_2 \right)_n$$

It is prepared from the monomer tetrafluoroethylene (CF_2=CF_2).

(b) The monomer used to prepare polystyrene is styrene.

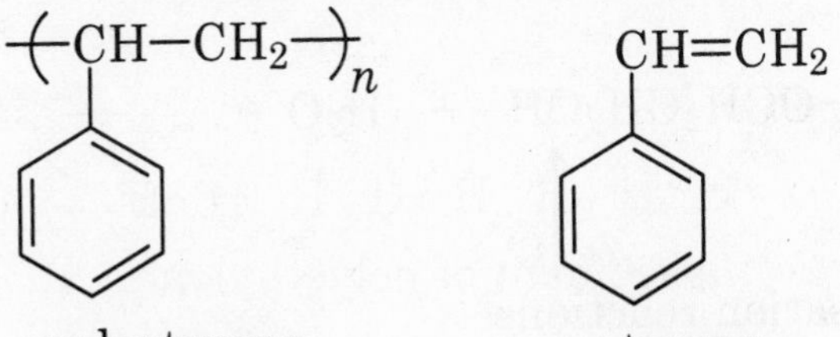

polystyrene — styrene

(c) PVC (polyvinylchloride) is prepared by the successive addition of vinyl chloride molecules (CH_2=CHCl).

PRACTICE EXERCISE

1. Draw the structures of the monomers from which the following polymers are formed:

(a)

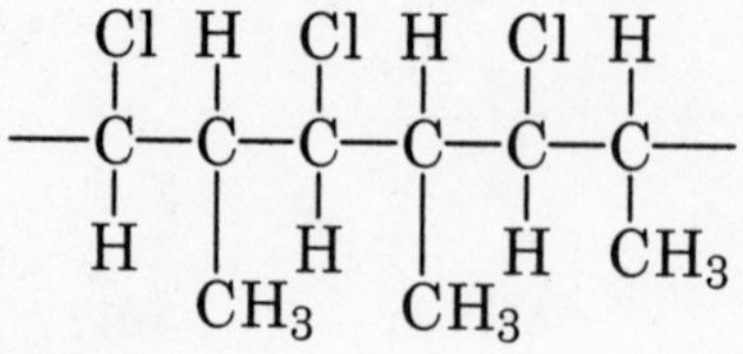

(b) $-CH_2-CCl_2-CH_2-CCl_2-CH_2-CCl_2-$

Text Problems: 8, 10, 12a

B. Condensation reactions

Copolymers are polymers that contain two or more different types of monomers. Polyesters, such as the well-known Dacron, are copolymers. In Dacron, one monomer is an alcohol and the other is a carboxylic acid, which can be joined by an *esterification* reaction. The alcohol and the acid both must contain two functional groups. The monomers of Dacron are the dicarboxylic acid called terephthalic acid and the dialcohol called ethylene glycol.

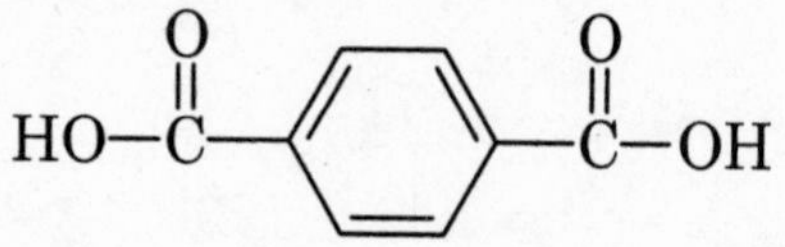

terephthalic acid

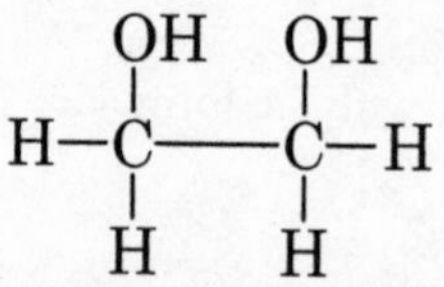

ethylene glycol

Condensation reactions differ from addition reactions in that the former always result in the formation of a small molecule such as water. When terephthalic acid and ethylene glycol react to form an ester, the first products are:

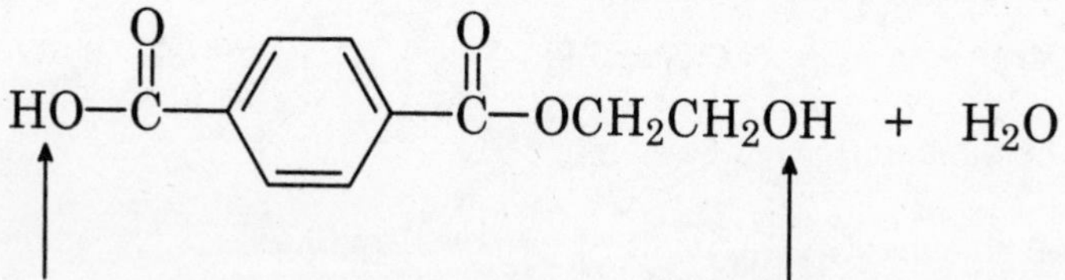

sites for further condensation reactions

When this product reacts with another molecule of the diacid, the polymer chain grows longer.

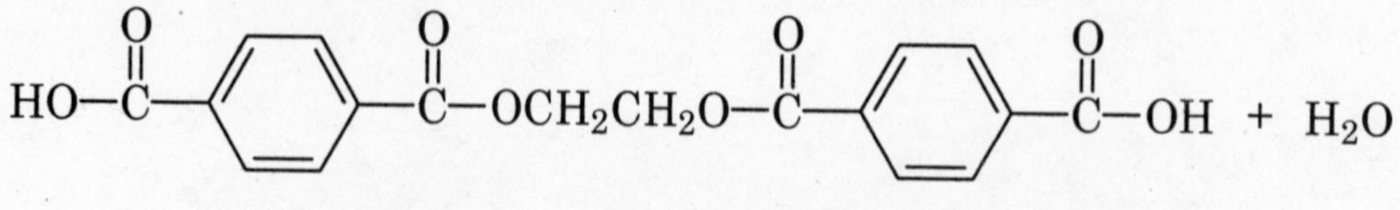

segment of a condensation polymer chain

The general formula for the polyester, Dacron, is:

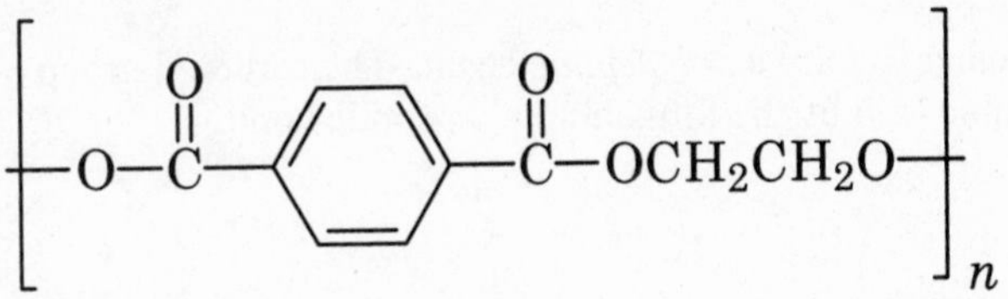

EXAMPLE 22.2
Draw structures for the monomers used to make the following polyester:

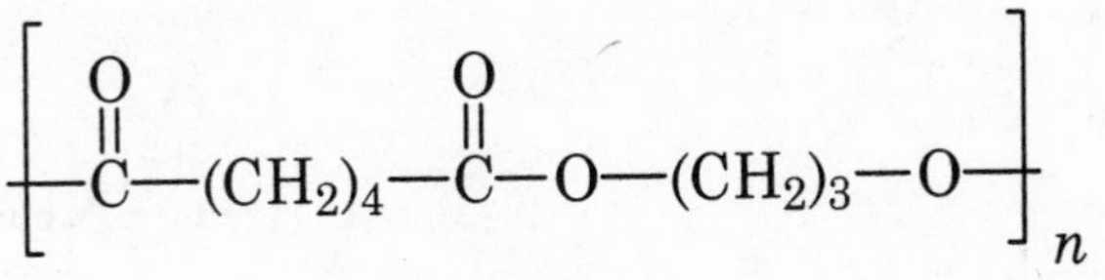

Answer

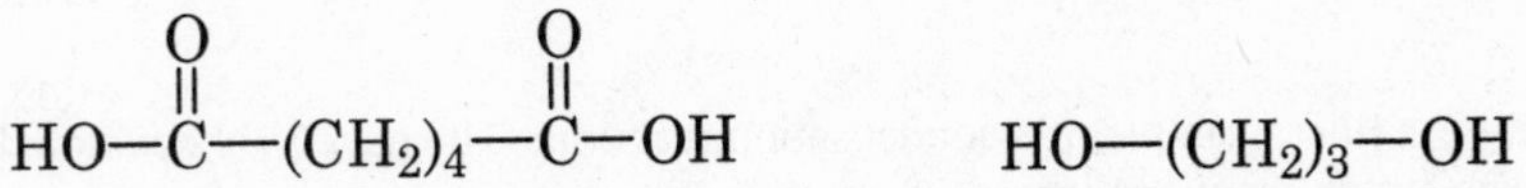

PRACTICE EXERCISE

2. List two examples of condensation polymers discussed in your textbook.

Text Problem: 12b

PROBLEM TYPE 2: PROTEINS

Proteins are *polymers of amino acids*. Proteins are truly giant molecules having molecular masses that range from about 10,000 to several million amu. Proteins play many roles in living organisms, where they function as catalysts (enzymes), transport molecules (hemoglobin), contractile fibers (muscle), protective agents (blood clots), hormones (chemical messengers), and structural members (feathers, horns, and nails). The word *protein* comes from the Greek work *proteios*, meaning "first". From the partial list of protein functions above, it is easy to see why proteins occupy "first place" among biomolecules in their importance to life.

Even though each protein is unique, all proteins are built from a set of only ***20* amino acids**. An amino acid consists of an amino group, a carboxylic acid group, a hydrogen atom, and a distinctive R group, all bonded to the same carbon atom.

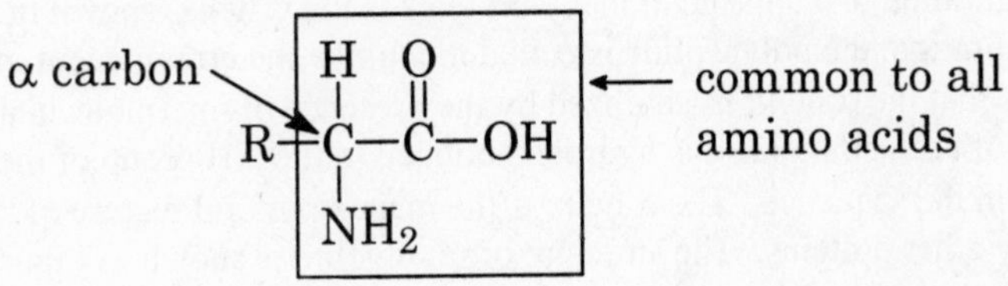

All amino acids in proteins have a common structural feature that is the attachment of the amino group ($-NH_2$) to the carbon atom adjacent to the carboxylic acid group. This carbon is called the α carbon. The difference in amino acids is due to different R groups. Twenty different R groups are found in proteins from natural sources. In fact, all proteins in all species, from bacteria to humans, are constructed from the same set of 20 amino acids.

In proteins, the amino acid units are joined together to form a *polypeptide* chain. The carboxyl group of one amino acid is joined to the amino group of another amino acid by the formation of a peptide bond.

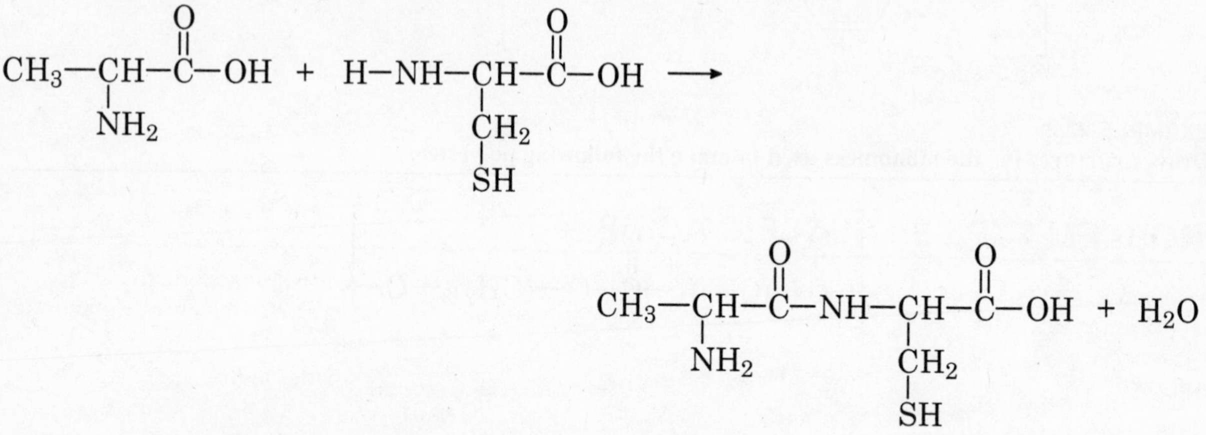

This type of reaction is another example of a condensation reaction. The new C–N covalent bond is called a *peptide* or an *amide* bond. The amide functional group present in all proteins is:

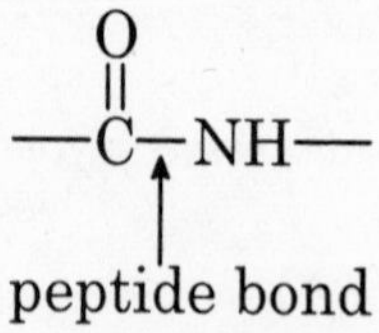

The molecule above, in which two amino acids are joined, is called a *dipeptide*. Peptides are structures intermediate in size between amino acids and proteins. The term *polypeptide* refers to long molecular chains containing many amino acid units. An amino acid unit in a polypeptide chain is called a *residue*.

Proteins are so complex that four levels of structural features have been identified. The structure of proteins is extremely important in determining how efficiently and effectively a protein will function. We will discuss the first two levels of protein structure.

1. **The Primary Structure**. Each protein has a unique amino acid sequence of its polypeptide chain. It is the amino acid sequence that distinguishes one protein from another. Proteins also differ in the numbers and types of amino acids, but the main difference is the sequence of amino acid residues.

2. **The Secondary Structure**. This refers to the spatial relationship of amino acid units that are close to one another in sequence. The configuration that appears in many proteins is the α helix, shown in Figure 22.16 of your textbook. In this configuration, the polypeptide is coiled much like the arrangement of stairs in a spiral staircase. Figure 22.16 also shows that the α helix is stabilized by the presence of intermolecular hydrogen bonds (dashed lines). The C=O group of each amino acid is hydrogen bonded to the NH group of the amino acid that is located four amino acids ahead in the sequence. The α helix is the main structural feature of the oxygen-storage protein, myoglobin, and of many other proteins. The structure of myoglobin is shown in Figure 22.17 of your text.

   The β-pleated sheet is another common secondary structure. In this structure, a polypeptide chain interacts strongly with adjacent chains by forming many hydrogen bonds. See Figure 22.18 of your text.

EXAMPLE 22.3
What are the five chemical elements found in proteins?

Proteins are made up of the same elements as amino acids. Therefore, proteins contain C, H, O, and N. Two amino acids, methionine and cysteine, also contain sulfur, S.

PRACTICE EXERCISE

3. Sketch a portion of the polypeptide chain consisting of the amino acids glycine, valine, and alanine, in that order.

Text Problem: 18

PROBLEM TYPE 3: NUCLEIC ACIDS

The chemical composition of the cell nucleus was first studied in the 1860s by Friedrich Miescher. He found the major components to be protein and a new material not previously isolated. This material was found to be acidic, and so it was referred to as **nucleic acid**.

Nucleic acids are polymers that store genetic information and control protein synthesis. There are two types of nucleic acids, deoxyribonucleic acid (DNA) and ribonucleic acid (RNA). ***DNA*** carries all the genetic information necessary to carry on reproduction and to sustain an organism throughout its lifetime. ***RNA*** transcribes these instructions and controls the synthesis of proteins needed to implement them.

DNA molecules are among the largest known with molar masses up to 10 billion g/mol. Hydrolysis of nucleic acids show that they are composed of only four types of building blocks: a phosphate group, a sugar, purines, and pyrimidines.

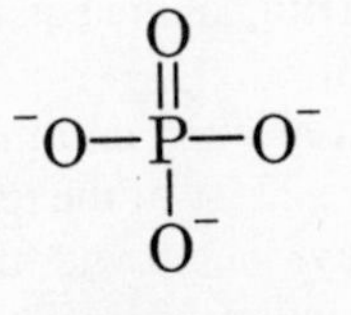

Phosphate

One of two sugars is present, ribose in RNA and deoxyribose in DNA.

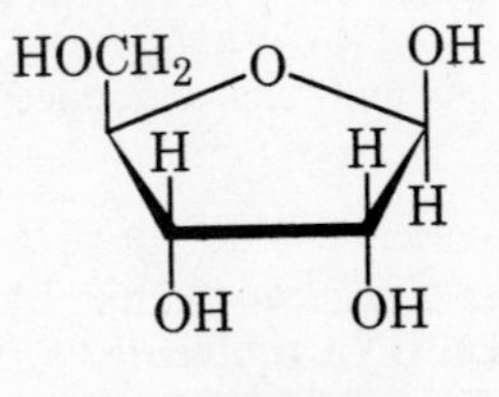

Ribose

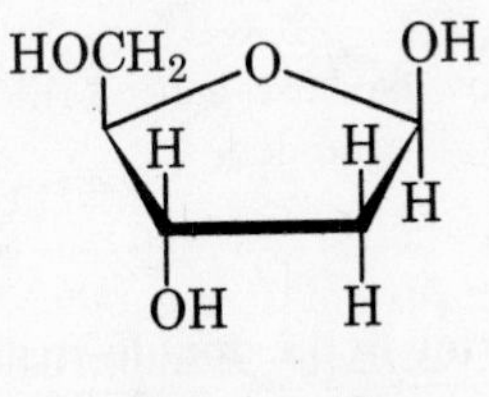

Deoxyribose

Two purines, adenine and guanine, are found in both DNA and RNA.

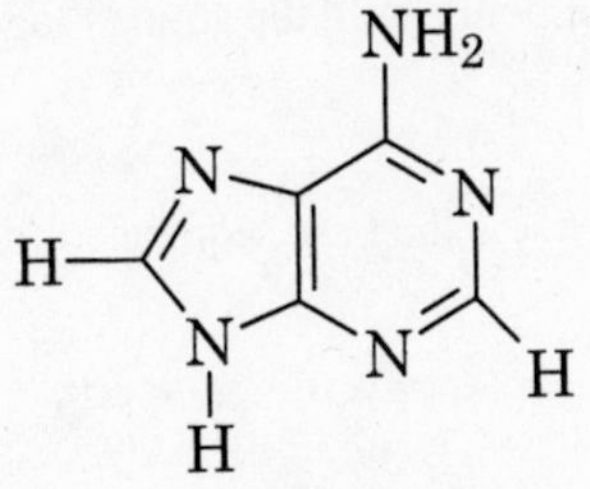

Adenine

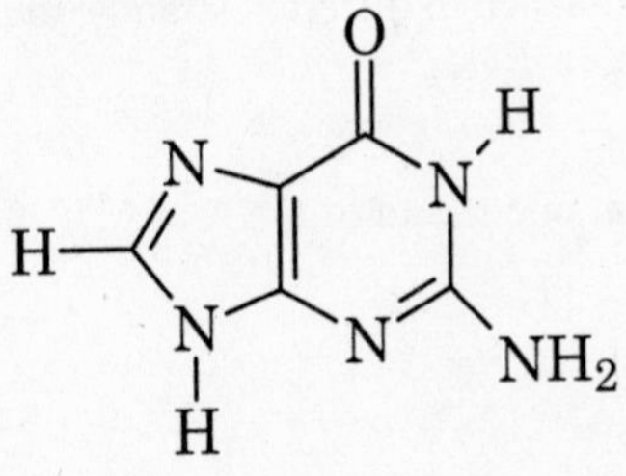

Guanine

DNA also contains the pyrimidines thymine and cytosine. RNA also contains thymine and another pyrimidine, uracil.

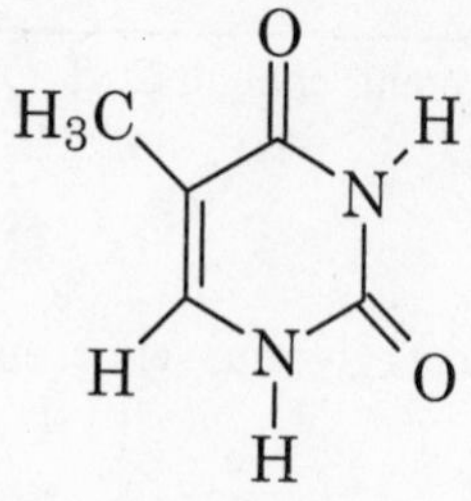

Thymine

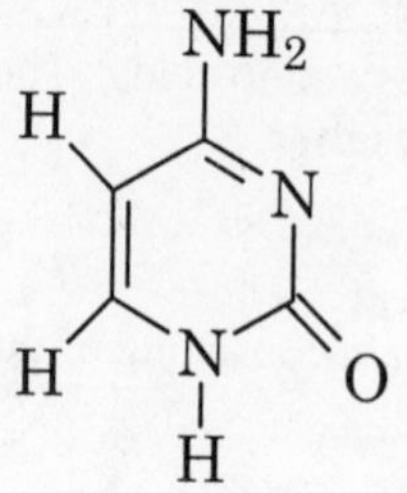

Cytosine

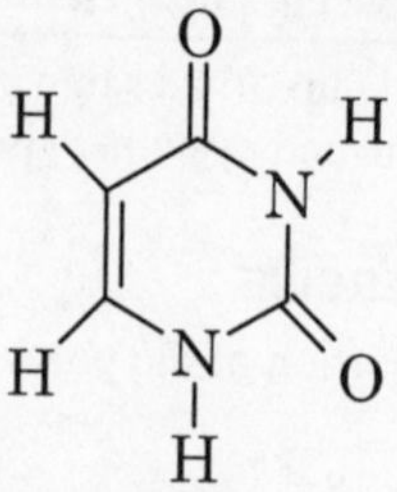

Uracil

The purines and pyrimidines are collectively called ***bases***. The table below summarizes the building blocks of DNA and RNA.

Building Blocks of DNA and RNA

	DNA	**RNA**
	Phosphate (P)	Phosphate (P)
Sugar	Deoxyribose (D)	Ribose (R)
Purines	Adenine (A)	Adenine (A)
	Guanine (G)	Guanine (G)
Pyrimidines	Thymine (T)	Thymine (T)
	Cytosine (C)	Uracil (U)

In the 1940s, Edwin Chargaff studied the composition of DNA. His analysis of the base composition of DNA showed that the amount of adenine always equaled that of thymine and that the amount of guanine equaled that of cytosine.

In 1953, Watson and Crick proposed a double-helical structure for DNA. The repeating unit in each strand is called a **nucleotide**, which consists of a base-deoxyribose-phosphate linkage (see Figure 22.21 of the text). The two strands are *not* identical, rather they are complementary. Adenine in one strand is always paired with thymine in the other strand. Guanine is always paired with cytosine. The base pairing is consistent with Chargaff's rules. Base pairing and the resulting association of the two strands are the result of *hydrogen bonding*. Hydrogen atoms in a base in one strand are attracted to unshared electron pairs on oxygen and nitrogen atoms of the base attached to the other strand (see Figure 22.22 of the text).

RNA, on the other hand, does not follow the base-pairing rules. X ray data and other evidence ruled out a double-helical structure for RNA. RNA is single-stranded.

EXAMPLE 22.4
What types of forces cause base pairing in the double-stranded helical DNA molecule?

The purine bases (adenine and guanine) in one strand of DNA form hydrogen bonds to the pyrimidine bases (thymine and cytosine) in the other DNA strand. Hydrogen atoms covalently bonded to nitrogen carry a partial positive charge. These H atoms are attracted to lone-electron pairs on oxygen and nitrogen atoms of another base. Since two complementary bases are attached to different strands, the hydrogen bonds hold the strands together.

PRACTICE EXERCISE

4. If the base sequence in one strand of DNA is A, T, G, C, T, what is the base sequence in the complementary strand?

Text Problem: 28

ANSWERS TO PRACTICE EXERCISES

1. (a)

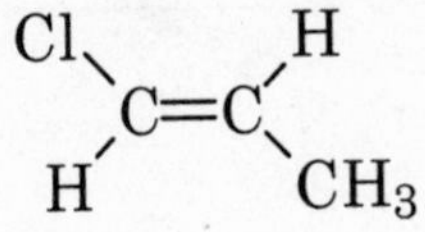

(b)

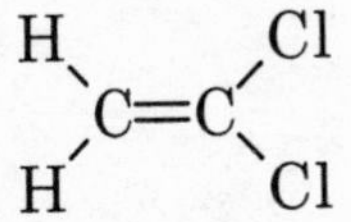

2. Nylon 66 (first prepared in 1931) and polyester.

3.

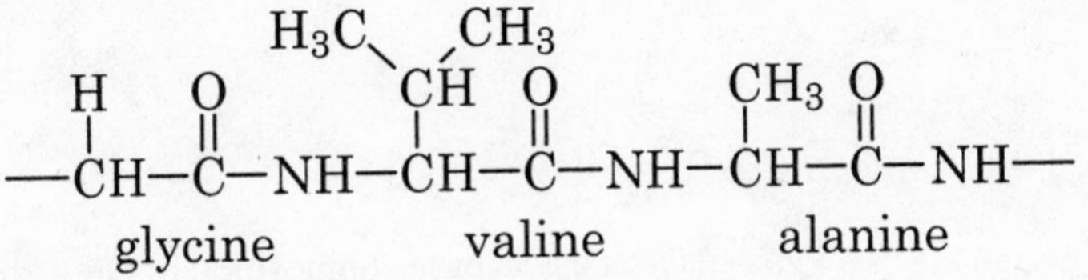

4. Recall that according to Chargaff's rules, adenine (A) is always paired with thymine (T) and guanine (G) is always paired with cytosine (C). The complementary base sequence is T, A, C, G, A.

SOLUTIONS TO SELECTED TEXT PROBLEMS

22.8 The repeating structural unit of the polymer is:

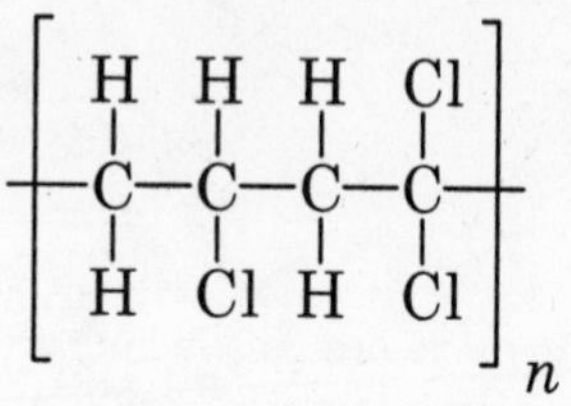

Does each carbon atom still obey the octet rule?

22.10 Polystyrene is formed by an addition polymerization reaction with the monomer, styrene, which is a phenyl–substituted ethylene. The structures of styrene and polystyrene are shown in Table 22.1 of your text.

22.12 The structures are shown.

(a) $H_2C{=}CH{-}CH{=}CH_2$

(b)

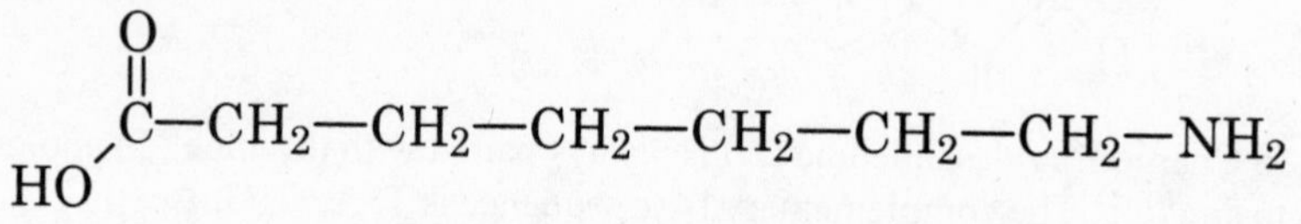

22.18 The main backbone of a polypeptide chain is made up of the α carbon atoms and the amide group repeating alternately along the chain.

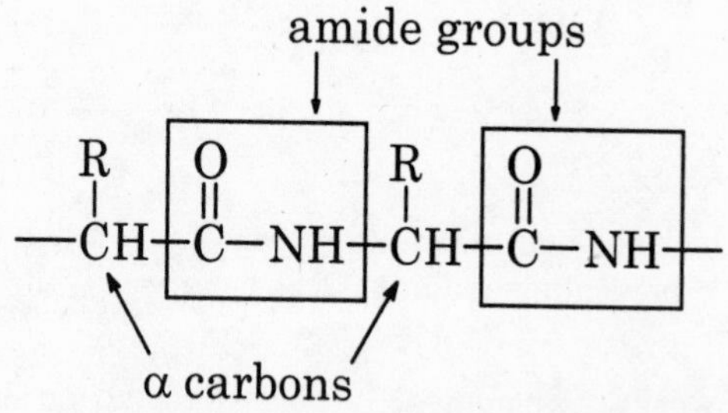

For each R group shown above, substitute the distinctive side groups of the two amino acids. Their are two possible dipeptides depending on how the two amino acids are connected, either glycine–valine or valine–glycine. The structures of the dipeptides are:

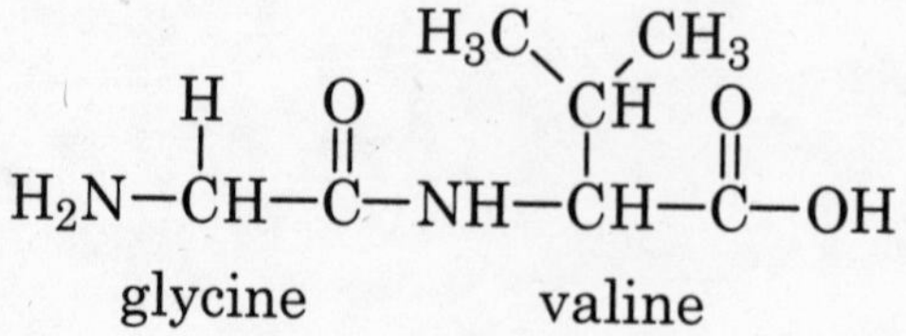

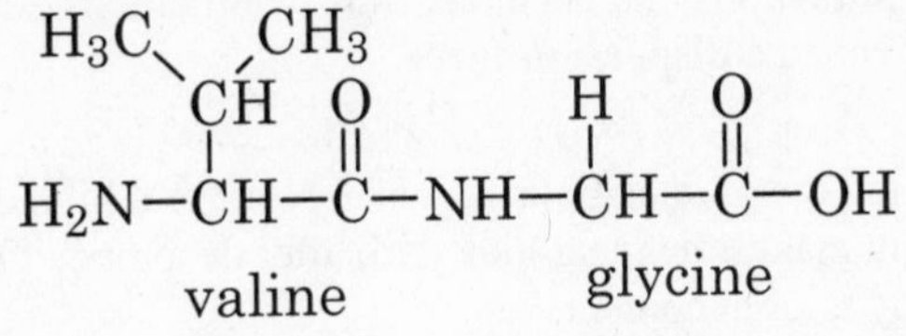

22.20 The rate increases in an expected manner from 10°C to 30°C and then drops rapidly. The probable reason for this is the loss of catalytic activity of the enzyme because of denaturation at high temperature.

22.26 Nucleic acids play an essential role in protein synthesis. Compared to proteins, which are made of up to 20 different amino acids, the composition of nucleic acids is considerably simpler. A DNA or RNA molecule contains only four types of building blocks: purines, pyrimidines, furanose sugars, and phosphate groups. Nucleic acids have simpler, uniform structures because they are primarily used for protein synthesis, whereas proteins have many uses.

22.28 The sample that has the higher percentage of C–G base pairs has a higher melting point because C–G base pairs are held together by three hydrogen bonds. The A–T base pair interaction is relatively weaker because it has only two hydrogen bonds. Hydrogen bonds are represented by dashed lines in the structures below.

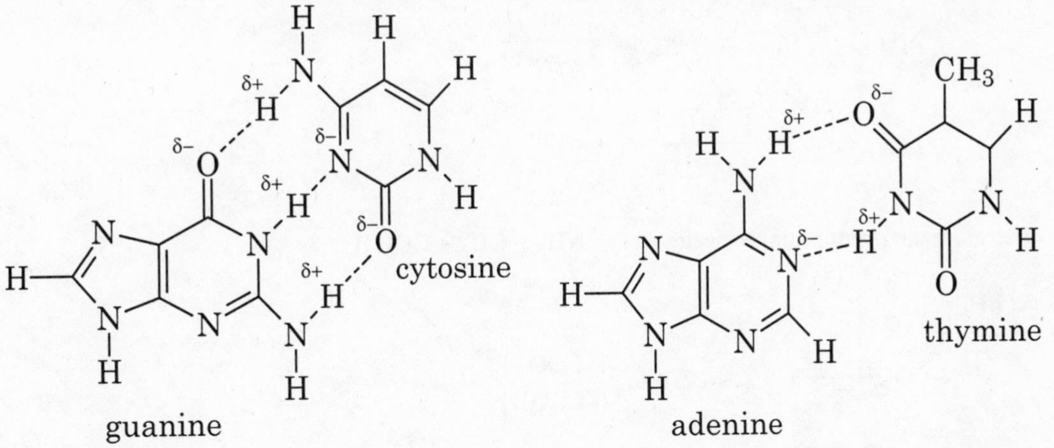

22.30 Both salt and vinegar (which contains acetic acid) denature the proteins resulting in gel formation. The gel effectively seals the crack, preventing additional leakage of egg white.

22.32 Insects have blood that contains no hemoglobin. Thus, they rely on simple diffusion to supply oxygen. It is unlikely that a human–sized insect could obtain sufficient oxygen by diffusion alone to sustain its metabolic requirements.

22.34 Calculating percent composition of a compound, Problem Type 4A, Chapter 3.

$$\%\ \text{Fe} = \frac{\text{mass of Fe}}{\text{mass of compound (hemoglobin)}} \times 100\%$$

$$0.34\% = \frac{55.85\ \text{g}}{\text{mass of hemoglobin}} \times 100\%$$

$$\textbf{minimum mass of hemoglobin} = \mathbf{1.6 \times 10^4\ g}$$

Hemoglobin must contain four Fe atoms per molecule for the actual molar mass to be four times the minimum value calculated.

22.36 The type of intermolecular attractions that occur are mostly attractions between nonpolar groups. This type of intermolecular attraction is called a **dispersion force**.

22.38 Leg muscles are active having a high metabolism, which requires a high concentration of myoglobin. The high iron content from myoglobin makes the meat look dark after decomposition due to heating. The breast meat is "white" because of a low myoglobin content.

22.40 No, the milk would *not* be fit to drink. Enzymes only act on one of two optical isomers of a compound.

22.42 $$\text{pH} = \text{p}K_a + \log\frac{[\text{conjugate base}]}{[\text{acid}]}$$

At pH = 1,

–COOH $$1 = 2.3 + \log\frac{[\text{-COO}^-]}{[\text{-COOH}]}$$

$$\frac{[\text{-COOH}]}{[\text{-COO}^-]} = 20$$

$-NH_3^+$ $$1 = 9.6 + \log\frac{[\text{-NH}_2]}{[\text{-NH}_3^+]}$$

$$\frac{[\text{-NH}_3^+]}{[\text{-NH}_2]} = 4 \times 10^8$$

Therefore the **predominant species** is: $\mathbf{^+NH_3 - CH_2 - COOH}$

At pH = 7,

–COOH $$7 = 2.3 + \log\frac{[\text{-COO}^-]}{[\text{-COOH}]}$$

$$\frac{[\text{-COO}^-]}{[\text{-COOH}]} = 5 \times 10^4$$

$-NH_3^+$ $$7 = 9.6 + \log\frac{[\text{-NH}_2]}{[\text{-NH}_3^+]}$$

$$\frac{[\text{-NH}_3^+]}{[\text{-NH}_2]} = 4 \times 10^2$$

Predominant species: $\mathbf{^+NH_3 - CH_2 - COO^-}$

At pH = 12,

$-COOH$ $$12 = 2.3 + \log\frac{[-COO^-]}{[-COOH]}$$

$$\frac{[-COO^-]}{[-COOH]} = 5 \times 10^9$$

$-NH_3^+$ $$12 = 9.6 + \log\frac{[-NH_2]}{[-NH_3^+]}$$

$$\frac{[-NH_2]}{[-NH_3^+]} = 2.5 \times 10^2$$

Predominant species: $\mathbf{NH_2 - CH_2 - COO^-}$